Cancer Sensitizing Agents for Chemotherapy

ANTIANGIOGENIC DRUGS AS CHEMOSENSITIZERS IN CANCER THERAPY

VOLUME 18

Cancer Sensitizing Agents for Chemotherapy Series

Series Editor: Benjamin Bonavida, PhD

Volume 1: Nitric Oxide (Donor/Induced) in Chemosensitization
Edited by Benjamin Bonavida, PhD

Volume 2: Role of Nutraceuticals in Cancer Chemosensitization
Edited by Alok Chandra Bharti and Bharat Bhushan Aggarwal

Volume 3: Targeting Cell Survival Pathways to Enhance Response to Chemotherapy
Edited by Daniel E. Johnson, PhD

Volume 4: Protein Kinase Inhibitors as Sensitizing Agents for Chemotherapy
Edited by Zhe-Sheng Chen and Dong-Hua Yang

Volume 5: Breaking Tolerance to Pancreatic Cancer Unresponsiveness to Chemotherapy
Edited by Ganji Purnachandra Nagaraju, PhD, DSc

Volume 6: Improving the Therapeutic Ratio in Head and Neck Cancer
Edited by Randall J. Kimple, MD, PhD

Volume 7: Drug Efflux Pumps in Cancer Resistance Pathways: From Molecular Recognition and Characterization to Possible Inhibition Strategies in Chemotherapy
Edited by Alejandro Sosnik, PhD and Reina Bendayan, PharmD

Volume 8: Drug Resistance in Colorectal Cancer: Molecular Mechanisms and Therapeutic Strategies
Edited by Chi Hin Cho and Tao Hu, PhD

Volume 9: Novel Therapies in Head and Neck Cancer: Beyond the Horizon
Edited by Maie A. St. John, MD, PhD, FACS and No-Hee Park, DMD, PhD

Volume 10: pH-Interfering Agents as Chemosensitizers in Cancer Therapy
Edited by Claudiu T. Supuran, PhD and Simone Carradori, PhD

Volume 11: Overcoming Ovarian Cancer Chemoresistance
Edited by Goli Samimi, PhD, MPH and Christina Annunziata, MD, PhD

Volume 12: Biological Mechanisms and the Advancing Approaches to Overcoming Cancer Drug Resistance
Edited by Andrew Freywald, PhD and Franco J. Vizeacoumar, PhD

Volume 13: Therapeutic Strategies to Overcome ALK Resistance in Cancer
Edited by Luc Friboulet, PhD

Volume 14: New Targeting in the Reversal of Resistant Glioblastomas
Edited by Ali Syed Arbab, MD, PhD

Volume 15: Glioblastoma Resistance to Chemotherapy: Molecular Mechanisms and Innovative Reversal Strategies
Edited by Ramasamy Paulmurugan, PhD and Tarik F. Massoud, MD, PhD

Volume 16: Epigenetic Regulation in Overcoming Chemoresistance
Edited by Chunfu Wu, PhD and Lihui Wang, PhD

Volume 17: Overcoming Drug Resistance in Gynecologic Cancers
Edited by Riyaz Basha, PhD, FAACC, FABAP and Sarfraz Ahmad, PhD, FAACC, FABAP

Volume 18: Antiangiogenic Drugs as Chemosensitizers in Cancer Therapy
Edited by Lucia Morbidelli, PhD

Upcoming Volumes:

Terpenoids as Chemosensitizers Against Drug-resistant Tumors
Edited by Thomas Efferth

Chemotherapy Sensitizing Agents for Prostate Cancer
Edited by K. Ketola

Chemosensitization by TRK Inhibitors
Edited by Dirk Hempel

Chemosensitization by BTK Inhibitors
Edited by Varinder Kaur

Chemosensitization by IDH1 Inhibitors
Edited by Robert Musiol

Chemosensitization by EGFR Inhibitors
Edited by M. Pancione

Chemosensitization by IDH2 Inhibitors
Edited by Sanjee Singh

Chemosensitization by Bcl2 Inhibitors
Edited by Yu Yu

Chemosensitization by BTK Inhibitors (II)
Edited by Jean Koff

Overcoming Resistance to EGFR Inhibitors in EGFR Mutant NSCLC
Edited by Anthony Faber

Cancer Sensitizing Agents for Chemotherapy

ANTIANGIOGENIC DRUGS AS CHEMOSENSITIZERS IN CANCER THERAPY

VOLUME 18

Edited by

LUCIA MORBIDELLI, PhD
Department of Life Sciences, University of Siena, Siena, Italy

ELSEVIER

ACADEMIC PRESS
An imprint of Elsevier

Academic Press is an imprint of Elsevier
125 London Wall, London EC2Y 5AS, United Kingdom
525 B Street, Suite 1650, San Diego, CA 92101, United States
50 Hampshire Street, 5th Floor, Cambridge, MA 02139, United States
The Boulevard, Langford Lane, Kidlington, Oxford OX5 1GB, United Kingdom

Library of Congress Cataloging-in-Publication Data
A catalog record for this book is available from the Library of Congress

British Library Cataloguing-in-Publication Data
A catalogue record for this book is available from the British Library

ISSN: 2468-3183

ISBN: 978-0-323-90190-1

For information on all Academic Press publications
visit our website at https://www.elsevier.com/books-and-journals

Publisher: Stacy Masucci
Senior Acquisitions Editor: Rafael Teixeira
Editorial Project Manager: Samantha Allard
Production Project Manager: Punithavathy Govindaradjane
Cover Designer: Christian Bilbow

Typeset by STRAIVE, India

Dedication

This book is dedicated to the memory of Prof. Antonio Giachetti.

Thank you for your valuable advice, for fueling our curiosity, and for your generous tenacity in teaching us how to do "science."

Cover Image Insert

Immunofluorescence analysis of tumor vessels in a human melanoma xenograft in immunodeficient mice. A375 melanoma cells were implanted in immunodeficient nude mice and left to grow for 20 days. Tumors were collected, embedded in Tissue-Tek O.C.T., cooled in isopentane, and frozen in liquid nitrogen for histology. Seven-micrometer thick cryostat sections were used for double-immunofluorescence staining with anti-CD31 and anti-NG2 antibodies to label endothelial cells (green) and pericytes (red), respectively. The green fluorescence documented the presence of a continuous endothelial layer, delimitating the newly formed and dilated vessels (black), while pericytes were scarce.

The book cover image was taken using confocal microscopy in Professor Lucia Morbidelli's Laboratory of Pharmacology of Angiogenesis and Microcirculation, Department of Life Sciences, University of Siena, Italy.

Aims and Scope for Series "Cancer Sensitizing Agents for Chemotherapy"

Current cancer management strategies fail to adequately treat malignancies with chemotherapy, with multivariable dose-restrictive factors such as systemic toxicity and multidrug resistance; hence limiting therapeutic benefits, quality of life, and complete long-term remission rates. The resistance of cancer cells to anticancer drugs is one of the major reasons for the failure of traditional cancer treatments. Cellular components and dysregulation of signaling pathways contribute to drug resistance. If modulated, such perturbations may restore the drug response and its efficacy. The recent understanding of the molecular mechanisms and targets that are implicated for cancer chemoresistance has paved the way to develop a large battery of small molecules (sensitizing agents) that can target resistant factors and reduce the threshold of resistance, thus allowing their combination with chemotherapeutic drugs to be effective and to reverse chemoresistance. A large variety of chemotherapy-sensitizing agents has been developed, and several have been shown to be effective in experimental models and in cancer patients.

The main objective of the proposed series "Cancer Sensitizing Agents for Chemotherapy" is to publish individualized and focused volumes, whereby each volume is edited by an invited expert Editor(s). Each volume will dwell on specific chemosensitizing agents with similar targeting activities. The combination treatment of the sensitizing agent with chemotherapy may result in a synergistic/additive activity and the reversal of tumor cells' resistance to drugs.

The Editor(s) will compile nonoverlapping review chapters on reported findings in various cancers, both experimentally and clinically, with particular emphases on underlying biochemical, genetic, and molecular mechanisms of the sensitizing agent and the combination treatment.

The scope of the series is to provide scientists and clinicians with updated basic and clinical information that will be valuable in their quest to investigate, develop, and apply novel combination therapies to reverse drug resistance and, thereby, prolonging survival and even resulting in a cure in cancer patients.

Benjamin Bonavida, PhD (Series Editor)

About the Series Editor

Dr. Benjamin Bonavida, PhD, (Series Editor), is currently Distinguished Research Professor at the University of California, Los Angeles (UCLA). He is affiliated with the Department of Microbiology, Immunology and Molecular Genetics, UCLA David Geffen School of Medicine. His research career, thus far, has focused on investigations in the fields of basic immunochemistry and cancer immunobiology. His research investigations have ranged from the biochemical, molecular, and genetic mechanisms of cell-mediated killing and tumor cell resistance to chemoimmuno cytotoxic drugs. The reversal of tumor cell resistance was investigated by the use of various selected sensitizing agents based on molecular mechanisms of resistance. In these investigations, there was the newly characterized dysregulated NF-κB/Snail/YY1/RKIP/PTEN loop in many cancers that was reported to regulate cell survival, proliferation, invasion, metastasis, and resistance. Emphasis was focused on the roles of the tumor suppressor Raf Kinase Inhibitor Protein (RKIP), the tumor promoter Yin Yang 1 (YY1), and the role of nitric oxide as a chemo-immuno-sensitizing factor. Many of the earlier mentioned studies are centered on the clinical challenging features of cancer patients' failure to respond to both conventional and targeted therapies.

The Editor has been active in the organization of regular sequential international miniconferences that are highly focused on the roles of YY1, RKIP, and nitric oxide in cancer and their potential therapeutic applications. Several books edited or coedited by the Editor have been published. In addition, the Editor has been the Series Editor of books (over 23) published by Springer on *Resistance to Anti-Cancer Targeted Therapeutics*. In addition, the Editor is presently the Series Editor of three series published by Elsevier/Academic Press on *Cancer Sensitizing Agents for Chemotherapy* and *Breaking Tolerance to Anti-Cancer Cell-Mediated Immunotherapy*. Lastly, the Editor is the Editor-in-Chief of the Journal Critical Reviews in Oncogenesis. The Editor has published over 500 research publications and reviews in various scientific journals of high impact.

Acknowledgments: The Series Editor acknowledges the Department of Microbiology, Immunology and Molecular Genetics and the UCLA David Geffen School of Medicine for their continuous support. The Series Editor also acknowledges the assistance of Mr. Rafael Teixeira, Acquisitions Editor for

Elsevier/Academic Press, and the excellent assistance and support of Ms. Samantha Allard, Editorial Project Manager for Elsevier/Academic Press, for their continuous cooperation throughout the development of this book.

Aims and Scope of the Volume

The aim of this volume is to update the rapidly evolving field of tumor angiogenesis and related drugs, used alone or in combination with conventional anticancer drugs. Several internationally recognized scientists/clinicians have contributed significantly to this book.

Angiogenesis, the formation of new vessels from preexisting ones, influences the progression of tumors since it guarantees the oxygen and nutrient intakes and an effective way for neoplastic cells to reach the bloodstream and metastasize. Antiangiogenic drugs have been designed and approved to treat tumors with some efficacy. Interestingly, antiangiogenic therapeutics (antibodies, small chemicals) have been designed either to be specific against selected angiogenic factors as VEGF or to inhibit key signaling kinases in endothelial and tumor cells. These therapeutics are clinically functional whether used alone or in combination with other antitumor strategies. The beneficial effects of the combination of antiangiogenic drugs and chemotherapies can be explained by the multi-targeting hits in the complex tumor microenvironment and, additionally, by the chemo-sensitization of drug-resistant cancer cells, often demonstrating a synergistic effect. This synergistic activity has the advantage to decrease the dose of the conventional chemotherapeutics, thus reducing adverse side effects and relapses. From a mechanistic point of view, the combination with antiangiogenic drugs works through multiple mechanisms, for example, by remodeling the tumor microenvironment, by allowing vessel normalization, by inducing shifts in the tumor metabolism, by enhancing antitumor immune cell infiltrates within the tumor, and by inhibiting the activity of p-gp/MDR or key enzymes involved in chemotherapy activation or metabolism. Finally, the recent fields that are under active research are the investigations of nutraceuticals and natural products, which can act on both angiogenesis and the tumor microenvironment, and the use of metronomic therapy combined with antiangiogenics. These established and innovative approaches are discussed together with the ongoing clinical trials. A critical issue is the finding that antiangiogenic drugs' use can result in resistance, thus necessitating therapy adjustments and novel targets in the evolving tumor cells.

From a clinical point of view, the validation of appropriate and predictive biomarkers for drug choice and clinical responses is also challenging. A molecular comprehensive signature of new multi-"omic" biomarkers predictive of drug outcome together with clinical, pathological, hemato-chemical, and standard molecular determinants would be of great use for personalizing drug treatment for cancer patients.

The book proposes examples of studies on the use of antiangiogenic strategies, molecules that inhibit the growth of vessels inside a tumor, together with antineoplastic drugs, in order to sensitize the resistant tumor to their cytotoxic effects, thus reducing adverse side reactions and resistance and improving the therapeutic response of cancer patients.

This book provides significant novel data in this rapidly evolving field that will be highly useful for both scientists and clinicians as well as pharmaceutical companies involved in the development of new anticancer protocols and unique strategies.

Lucia Morbidelli, PhD

About the Volume Editor

Professor Lucia Morbidelli is currently Associate Professor of Pharmacology at the Department of Life Sciences, University of Siena, Italy. Prof. Morbidelli's research experience is in the field of pharmacology of angiogenesis and microcirculation using a plethora of in vitro and in vivo models. Her research interests include the definition of the role of nitric oxide and prostanoid pathways in the process of physiological and tumor angiogenesis. She has contributed to characterizing the activity of synthetic molecules and natural bioactives as pro- or antiangiogenic strategies. In the last decade, her research has been focused on the gaseous transmitters nitric oxide and hydrogen sulfide by charactering their activity in the maintenance of endothelial survival and functionality or as antitumor agents.

She is the author of over 160 scientific publications comprising peer-reviewed journal articles, reviews, and chapters in books on the molecular, biochemical, and cellular aspects of angiogenesis and endothelial pharmacology. Prof. Morbidelli was Guest Editor of the Special Issue of *Pharmacological Research* "Drugs and Druggable Signaling in Angiogenesis" and of the Special Issue of *Molecules* on "Natural Products as Preventive or Therapeutic Tools for Angiogenesis Related Diseases." Together with Dr. Benjamin Bonavida, she was the editor of the book *Therapeutic Application of Nitric Oxide in Cancer and Inflammatory Disorders* published by Elsevier. Apart from scientific production, Prof. Morbidelli's international collaborations include participation at the COST Action BM1005 European Network on Gasotransmitters (2012–2015) and representation as member responsible for pharmacological countermeasures of the ESA topical team "Tissue Healing in Space: Techniques for Promoting and Monitoring Tissue Repair and Regeneration" (since 2017) and "Pharmacological Countermeasures" (since 2021).

Her current research topics are related to the (1) definition of new targets within the tumor microenvironment responsible for the control of tumor angiogenesis and (2) identification of drugs and medical devices (natural extracts, nutraceuticals) to recover endothelial and tissue function following injury, or disease conditions related to inflammation and oxidative stress, or unloading conditions.

Preface

This book presents recent advances in the field of angiogenesis and antiangiogenic therapies. Clinical studies clearly confirm that tumor growth, metastasis, and progression depend on angiogenesis. These findings document that therapies using antiangiogenic drugs are of particular importance because they can be used against various types of cancers, both solid and hematological. A number of approaches have been developed to inhibit tumor angiogenesis, and numerous compounds were discovered to inhibit angiogenesis, but only a few of them proved clinically effective and to induce tumor regression.

The first part of the book summarizes the knowledge acquired on the molecular entities that play a role in tumor angiogenesis and their mechanisms of action. Particular attention is given to the complexity of the tumor microenvironment composition and the communications routes and signals within it. The book then proposes examples of studies on the use of antiangiogenic strategies, molecules that inhibit the growth of vessels inside a tumor, together with antineoplastic drugs, in order to sensitize the resistant tumor to their cytotoxic effects, thus reducing adverse side reactions and resistance and improving the therapeutic response of cancer patients. The clinical applications of antiangiogenic treatments are documented for a wide variety of tumors. In particular, the modes of combination with other targeted and/or cytotoxic cancer therapies, including the metronomic approach, are presented. Furthermore, the identification of molecular markers as well as their validations as predictive and/or prognostic tools is reviewed.

The book aims to unravel these aspects by providing examples of clinical practices or ongoing clinical trials on a large panel of malignancies. Moreover, attention is also pointed to identifying novel circulating biomarkers for the proper drug choice and prediction of the patient response.

The book consists of 11 review chapters by experts in the field, which are summarized as follows:

In the Intoduction (Chapter 1), Morbidelli and Donnini introduce the topic of tumor neovascularization and a plethora of factors and communication signals within the tumor microenvironment involved in endothelial cell activation during the so-called angiogenic switch. The receptors and intracellular signaling pathways are described together with the endogenous angiogenesis inhibitors. The chapter discusses the hierarchy of the various factors against which specific therapies have been developed. From a clinical point of view, the focus of the introductory chapter is on the angiogenic switch that characterizes the occurrence of tumor angiogenesis and is also responsible for resistance to antiangiogenic therapy.

In Chapter 2, Donnini and colleagues review the selective and multi-targeting antiangiogenic drugs and their pharmacokinetics and pharmacodynamics features. Their approved clinical use in monotherapy or in combination with the other antitumor strategies (conventional chemotherapy, target therapy, immunotherapy, hormonal therapy) is reported, along with side effects and contraindications, if any. A general up-to-date scenario is proposed with the experimental findings and, most importantly, with the ongoing clinical trials.

In Chapter 3, Petrioli and colleagues discuss the antiangiogenic drugs as chemosensitizers in gastrointestinal cancers, reporting on the use of combination therapies against both upper GI and colorectal cancer. They summarize the clinical evidence supporting the use of antiangiogenic agents in patients with colorectal cancer and the promising activity of antiangiogenic drugs, alone or in combination with conventional therapy, exhibited in patients with advanced gastric cancer (GC) or gastroesophageal junction adenocarcinoma (GEJ). Recently, the combination of antiangiogenic agents plus immunotherapy exhibited promising results in patients with advanced GC or GEJ adenocarcinoma. The ongoing trials will help clarify the real efficacy of the combination of antiangiogenic agents with conventional anticancer treatments and with immune checkpoint inhibitors (ICIs) in advanced GC and GEJ adenocarcinoma and the role of predictive and prognostic biomarkers.

In Chapter 4, Roviello and colleagues review the antiangiogenic drugs as chemosensitizers in renal cell carcinoma (RCC). The understanding of the molecular pathogenesis of RCC allowed to identify targets for therapeutic intervention with the development of multiple agents, having an important role in the management of patients with advanced disease. Among them, antiangiogenic drugs have changed the management of these patients considering the central role of angiogenesis in the metastatic RCC tumorigenesis and progression. In the chapter, an overview on the use of antiangiogenic drugs in the treatment of RCC, with regard to their chemosensitizer activities, is reported.

In Chapter 5, Sessa and coworkers describe the role of angiogenesis in genitourinary cancers. New strategies have focused on the combination of antiangiogenic drugs with chemotherapy or immunotherapy. The state-of-the-art therapies and ongoing clinical trials are reported for urothelial cancer, testicular germ cell tumor, and prostate cancer.

In Chapter 6, Quaresmini and Guida review the biological bases of angiogenesis in melanoma and the currently available data on the efficacy of antiangiogenic drugs in the treatment of advanced melanoma. Melanoma can induce neoangiogenesis in multiple ways, including vascular mimicry, consisting in the capacity of melanoma cells to assume endothelial-like characteristics and functionally express pro-angiogenic mediators and receptors that participate in the interplay with the extracellular matrix components. Despite the unavoidable biological role of angiogenesis and its association with worse outcomes, in melanoma, the combinations of antiangiogenic and chemotherapeutic agents have been investigated in clinical trials with only poor improvement of the already unsatisfactory outcomes. In recent years, immunotherapy has induced previously unexpected results in the treatment of melanoma. Recent clinical trials have been investigating the efficacy of the association between antiangiogenic agents and immune checkpoint inhibitors in advanced melanoma, with interesting outcomes. Early data from a recent trial, in particular, show the possible role of antiangiogenic treatment in overcoming the resistance to immune checkpoint inhibitors.

In Chapter 7, Desantis and colleagues focus on the role of circulating EPCs, which mediate the cross talk between cancer angiogenesis and neoplastic clone, as potential novel targets for antiangiogenic drugs with particular relevance for hematological malignancies. The state of the art on the approved therapies is reported to be focusing on tyrosine kinase inhibitors, immunomodulatory drugs, and monoclonal antibodies. The promising activity of innovative drugs is also reviewed.

In Chapter 8, Bocci and colleagues discuss metronomic therapy, a new type of chemotherapy in which anticancer drugs are administered in a lower dose than the maximum tolerated dose repetitively over a long period to treat cancers with fewer side effects. Metronomic therapy is shown to affect both the tumor microenvironment and tumor cells to achieve its therapeutic effect. The results of the clinical trials with the combination of metronomic therapy and antiangiogenics are discussed, pointing to the necessity to clearly define the therapeutic efficacy of the combination.

In Chapter 9, Koltai and colleagues review the topic of resistance to antiangiogenic treatments. Antiangiogenic treatments can undergo failure, and a deeper knowledge of the mechanisms of resistance is necessary to delay or eliminate this risk. Two different mechanisms have been identified: (i) primary or evasive resistance and (ii) secondary or adaptive resistance. The authors describe in detail the canonical and noncanonical pathways responsible for nonresponsiveness to classical antiangiogenic treatments. They discuss this evidence as a central issue in antiangiogenic treatment failure, and some nonconventional or novel pharmacologic strategies against resistance are considered.

In Chapter 10, Mini and colleagues report recent preclinical research findings on circulating predictive biomarkers of drug response, taking as example the colorectal cancer. Colorectal cancer represents a model of a tumor with a high impact on health, being one of the most incident and lethal cancers worldwide. Elevated death rates depend on neoplastic progression guided by several dysregulated biomolecular events, including the evolutionary metastatic process occurring in about 50% of the patients. Despite the availability of active cytotoxic and targeted drugs, including antiangiogenic agents, the efficacy of pharmacological treatment of metastatic, and even early, colorectal cancer is hampered by the occurrence of intrinsic and acquired drug resistance. A molecular biomarker signature predictive of response linked to standard clinical, pathological, biochemical, and molecular markers would be useful for patient stratification and drug treatment customization to select personalized therapies with greater efficacy and tolerability. The chapter describes the state of the art and promising perspectives of known and potential biomarkers predictive of drug response and tolerability in colorectal cancer from various "omics" sciences.

Finally, in **General Conclusions and Future Perspectives** (Chapter 11), Morbidelli discusses the novel targets within the tumor microenvironment, which influence the neovascularization and try to identify future perspectives to optimize antiangiogenic strategies. Some examples of therapeutic strategies acting on novel targets are reported, documenting the growing interest of the global tumor ecosystem to get rid of the pathology.

The editor L. Morbidelli acknowledges the supervision and guidance of Prof. Benjamin Bonavida and the technical assistance of Samantha Allard (Editorial Project Manager) for the excellent editing and formatting of the various contents of the book.

Lucia Morbidelli, PhD

Contents

7. Antiangiogenic drugs as chemosensitizers in hematological tumors

Vanessa Desantis, Aurelia Lamanuzzi, Ilaria Saltarella, Maria Antonia Frassanito, Angelo Vacca, Monica Montagnani, and Domenico Ribatti

8. Metronomic chemotherapy and antiangiogenic drugs: Preclinical and clinical data

Guido Bocci, Marta Banchi, Diana Gonzalez Garcia, Serina Batson, and Giulio Francia

9. Resistance to antiangiogenic treatments: A review

Tomas Koltai, Stephan Joel Reshkin, and Rosa Angela Cardone

10. Predictive "omic" biomarkers of drug response: Colorectal cancer as a model

Enrico Mini, Ida Landini, Antonello Di Paolo, Gloria Ravegnini, Simona Saponara, Maria Frosini, Andrea Lapucci, and Stefania Nobili

11. General conclusions and future perspectives

Lucia Morbidelli

Contributors

Marta Banchi Dipartimento di Medicina Clinica e Sperimentale, Università di Pisa, Pisa, Italy

Serina Batson Department of Biological Sciences/BBRC, University of Texas at El Paso (UTEP), El Paso, TX, United States

Claudio Bisegna Unit of Urological Robotic Surgery and Renal Transplantation, Careggi Hospital, Florence, Italy

Guido Bocci Dipartimento di Medicina Clinica e Sperimentale, Università di Pisa, Pisa, Italy

Rosa Angela Cardone Department of Biosciences, Biotechnology and Biopharmaceutics, University of Bari, Bari, Italy

Martina Catalano Department of Medical Oncology, Careggi Hospital, Florence, Italy

Valerio Ciccone Department of Life Sciences, University of Siena, Siena, Italy

Vanessa Desantis Department of Biomedical Sciences and Human Oncology, Unit of Internal Medicine and Clinical Oncology, University of Bari Medical School; Department of Biomedical Sciences and Human Oncology, Pharmacology Section, Medical School, University of Bari "Aldo Moro", Bari, Italy

Beatrice Detti Radiation Oncology Unit - Oncology Department, Careggi Hospital, Florence, Italy

Antonello Di Paolo Department of Clinical and Experimental Medicine, University of Pisa, Pisa, Italy

Sandra Donnini Department of Life Sciences, University of Siena, Siena, Italy

Arianna Filippelli Department of Life Sciences, University of Siena, Siena, Italy

Giulio Francia Department of Biological Sciences/BBRC, University of Texas at El Paso (UTEP), El Paso, TX, United States

Maria Antonia Frassanito Department of Biomedical Sciences and Human Oncology, General Pathology Unit, University of Bari Medical School, Bari, Italy

Maria Frosini Department of Life Sciences, University of Siena, Siena, Italy

Ilaria Camilla Galli Division of Pathological Anatomy, University of Florence, Florence, Italy

Diana Gonzalez Garcia Department of Biological Sciences/BBRC, University of Texas at El Paso (UTEP), El Paso, TX, United States

Michele Guida Rare Tumors and Melanoma Unit, IRCCS Istituto Oncologico "Giovanni Paolo II", Bari, Italy

Ester Illiano Department of Surgical and Biomedical Science, Andrological and Urogynecological Clinic, Santa Maria Terni Hospital, University of Perugia, Terni, Italy

Tomas Koltai Hospital del Centro Gallego de Buenos Aires, Buenos Aires, Argentina

Aurelia Lamanuzzi Department of Biomedical Sciences and Human Oncology, Unit of Internal Medicine and Clinical Oncology, University of Bari Medical School, Bari, Italy

Ida Landini Department of Health Sciences; DENOTHE Excellence Center, University of Florence, Florence, Italy

Andrea Lapucci Department of Health Sciences; DENOTHE Excellence Center, University of Florence, Florence, Italy

Andrea Liaci Unit of Urological Robotic Surgery and Renal Transplantation, Careggi Hospital, Florence, Italy

Ignazio Martellucci Medical Oncology Unit, Medical Hospital and University of Siena, Siena, Italy

Salvatora Tindara Miano Medical Oncology Unit, Medical Hospital and University of Siena, Siena, Italy

Enrico Mini Department of Health Sciences; DENOTHE Excellence Center, University of Florence, Florence, Italy

Monica Montagnani Department of Biomedical Sciences and Human Oncology, Pharmacology Section, Medical School, University of Bari "Aldo Moro", Bari, Italy

Lucia Morbidelli Department of Life Sciences, University of Siena, Siena, Italy

Gabriella Nesi Division of Pathological Anatomy, University of Florence, Florence, Italy

Stefania Nobili Department of Neurosciences, Imaging and Clinical Sciences, "G. D'Annunzio" University of Chieti-Pescara, Chieti, Italy

Roberto Petrioli Medical Oncology Unit, Medical Hospital and University of Siena, Siena, Italy

Davide Quaresmini Rare Tumors and Melanoma Unit, IRCCS Istituto Oncologico "Giovanni Paolo II", Bari, Italy

Gloria Ravegnini Department of Pharmacy and Biotechnology, University of Bologna, Bologna, Italy

Stephan Joel Reshkin Department of Biosciences, Biotechnology and Biopharmaceutics, University of Bari, Bari, Italy

Domenico Ribatti Department of Basic Medical Sciences, Neurosciences and Sensory Organs, University of Bari Medical School, Bari, Italy

Emma Ristori Department of Internal Medicine, Section of Cardiology, Yale University School of Medicine, Yale Cardiovascular Research Center, New Haven, CT, United States

Giandomenico Roviello Department of Medical Oncology, Careggi Hospital, Florence, Italy

Ilaria Saltarella Department of Biomedical Sciences and Human Oncology, Unit of Internal Medicine and Clinical Oncology, University of Bari Medical School, Bari, Italy

Simona Saponara Department of Life Sciences, University of Siena, Siena, Italy

Francesco Sessa Unit of Urological Robotic Surgery and Renal Transplantation, Careggi Hospital; Department of Experimental and Clinical Medicine, University of Florence, Florence, Italy

Pietro Spatafora Unit of Urological Robotic Surgery and Renal Transplantation, Careggi Hospital, Florence, Italy

Andrea Spini Bordeaux Population Health Center, University of Bordeaux, Bordeaux, France; Department of Medicine, Surgery and Neurosciences, University of Siena, Siena, Italy

Angelo Vacca Department of Biomedical Sciences and Human Oncology, Unit of Internal Medicine and Clinical Oncology, University of Bari Medical School, Bari, Italy

Lorenzo Verdelli Unit of Urological Robotic Surgery and Renal Transplantation, Careggi Hospital, Florence, Italy

Graziano Vignolini Unit of Urological Robotic Surgery and Renal Transplantation, Careggi Hospital, Florence, Italy

Donata Villari Unit of Urological Robotic Surgery and Renal Transplantation, Careggi Hospital; Department of Experimental and Clinical Medicine, University of Florence, Florence, Italy

Marina Ziche Department of Medicine, Surgery and Neurosciences, University of Siena, Siena, Italy

CHAPTER

1

Introduction

How tumor-endothelial cell communication within the tumor microenvironment affects angiogenesis

Lucia Morbidelli and Sandra Donnini

Department of Life Sciences, University of Siena, Siena, Italy

Abstract

Angiogenesis, the formation of new blood vessels from preexisting ones, occurs when the high proliferation rate of cancer cells creates hypoxic zones in the tumor mass and influences the progression of tumors since it guarantees the oxygen and nutrient delivery and an efficient way for neoplastic cells to reach the bloodstream and metastasize. Tumor neovascularization is the result of an angiogenic switch, meaning the unbalance between pro- and antiangiogenic factors, in favor of the first ones. The majority of the angiogenesis modulators acting on endothelial cells derive from neoplastic cells, but a growing contribution is coming from other cells of the tumor microenvironment as cancer-associated fibroblasts, pericytes, macrophages, and immune cells. On their site, tumor endothelial cells communicate with the other cellular components in favor of tumor progression. In this chapter, the description of the major proangiogenic factors is provided together with the one on antiangiogenic molecules. The transcriptional and not transcriptional mechanisms responsible for their accumulation or downregulation are reported. The signaling pathways depending on the receptor (and coreceptor) activation are described. All these findings have allowed the development of antiangiogenic strategies, but at the same time may explain the mechanisms of antiangiogenic drug failure and resistance.

Abbreviations

ALDH	aldehyde dehydrogenase
Ang	angiopoietin
CAF	cancer-associated fibroblast
COX	cyclooxygenase
CXCR4	CXC chemokine receptor type 4
DC	dendritic cell
EC	endothelial cell
ECM	extracellular matrix
EGF	epidermal growth factor
eNOS	endothelial nitric oxide synthase

Antiangiogenic Drugs as Chemosensitizers in Cancer Therapy
https://doi.org/10.1016/B978-0-323-90190-1.00018-4

EPC	endothelial progenitor cell
ERK	extracellular signal-related kinase
ET	endothelin
FAK	focal adhesion kinase
FGF	fibroblast growth factor
HGF	hepatocyte growth factor
HIF-1α	hypoxia-inducible factor-α
K5	Kringle 5
IL	interleukin
iNOS	inducible nitric oxide synthase
MAPK	mitogen-activated protein kinase
MDR	multidrug resistance
MMP	matrix metalloproteases
MVD	microvessel density
NEC	normal endothelial cell
PDGF	platelet-derived growth factor
PDGFR	platelet-derived growth factor receptor
PGE2	prostaglandin E2
PHD	prolyl hydroxylase
PI-3K	phosphatidyl inositol-3 kinase
PKB	protein kinase B
PKC	protein kinase C
PLC	phospholipase C
PlGF	placental growth factor
ROS	reactive oxygen species
SDF-1	stromal cell-derived factor-1
SH2	Src homology 2 domain
sVEGFR-1	soluble VEGFR-1
TAM	tumor-associated macrophage
TEC	tumor endothelial cell
TK	tyrosine kinase
TME	tumor microenvironment
TNF	tumor necrosis factor
TSP1	thrombospondin 1
VEGF	vascular endothelial growth factor
VEGFR	vascular endothelial growth factor receptor
vHL	von-Hippel-Lindau

Conflict of interest

No potential conflicts of interest were disclosed.

Introduction

It is well known how microvessel density (MVD), the concentration of small blood vessels in tumor sections, reflects the amount of angiogenesis within the tumor and correlates with the ability of tumors to grow and metastasize [1]. An inverse association between MVD and patient survival has been reported for several malignancies, including breast [2], head and neck [3] and melanoma [4], as well as prostate [5] and bladder [6] cancer.

Different mechanisms can contribute to the formation of a new vessel network in tumors, as sprouting angiogenesis, vasculogenesis [homing of endothelial precursor cells (EPCs) in the nascent vessels], intussusceptive angiogenesis (the splitting of preexisting vessel into new vascular structures), vascular co-option (cancer cells hijack the vessels of the surrounding normal tissue), and vasculogenic mimicry [differentiation of cancer stem cells into endothelial cells (ECs)] [7–9].

Sprouting angiogenesis is the best known mechanism of tumor neovascularization induced by a growing tumor and finalized to nourish itself. The signal exchange between tumor and endothelial cells is critical to the development of tumor angiogenesis. Nowadays, a tumor is considered a real ecosystem composed by the combination of tumor and stromal cells, extracellular matrix (ECM), and secreted factors [10]. It is now emerging that all these components participate in and control tumor angiogenesis. In this chapter, the description of the basic mechanisms of the so-called angiogenic switch occurring during sprouting angiogenesis and the plethora of pro- and antiangiogenic endogenous molecules and related signaling is reported, pointing to the communications among the various components of the tumor microenvironment (TME).

Tumor microenvironment components

Cells composing the tumor stroma are cancer-associated fibroblasts (CAFs), ECs, pericytes, adipocytes, and immune cells including macrophages, monocytes, and lymphocytes, among others (Fig. 1). These cells are cemented within the ECM and their function is controlled by soluble mediators and biophysical parameters including oxygenation and pH [11–14]. While tumor cells are the main producers of the proangiogenic molecules, the contribution of other cellular components is starting to become clear throughout an integrative approach.

Tumor endothelial cells: The existence of two different endothelial lineages within the tumor microenvironment is a consolidated concept: normal endothelial cells (NECs), representing the physiological form of vascular endothelium, and tumor endothelial cells (TECs), which are strongly promoted by the tumor microenvironment and are biologically different from NECs [15]. NECs derive from the effects of the angiogenic switch players as angiogenic factors, cytokines, and growth factors secreted into the TME on resident ECs or

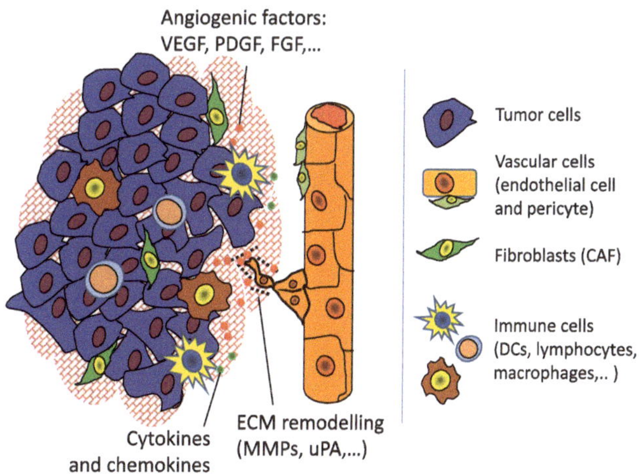

FIG. 1 Schematic representation of the cellular, soluble, and matrix components of the tumor microenvironment involved in the induction of tumor angiogenesis.

EPCs or cancer stem cells. TEC-derived vessels are fragile and leaky as compared to NEC-derived vessels, giving rise to an altered blood flow. TECs are irregular in morphology and size with long cytoplasmic projections extending across the vessel lumen [15]. TEC-derived enlarged vessels with gaps and fenestrae lead to hemorrhage, low tissue perfusion, and plasma leakage within the tumor, thus increasing the interstitial pressure [16]. From a molecular point of view, TECs exhibit high responsiveness to growth factors such as vascular endothelial growth factor (VEGF), epidermal growth factor (EGF), and fibroblast growth factor (FGF), with a clear proangiogenic and antiapoptotic phenotype [17], not observed in NECs, characterized by slow proliferation rate and long life span. Additionally, aldehyde dehydrogenase (ALDH) overexpression has been detected in TECs in comparison to NECs, which further confers stemness and proangiogenic properties [18].

Hypoxia, growth factors, reactive oxygen species (ROS), and cytokines present in TME can cause genetic instability in TECs. Aneuploidy and abnormal centrosomes are described in TECs which may explain their resistance to conventional chemotherapeutics [19] and the further variability inside the tumor vasculature. Higher genetic alterations in TECs are usually detected in aggressive human malignant tumors [20]. TECs take part in drug resistance also by expressing the multidrug resistance protein 1 gene (*MDR1*) and its product p-glycoprotein [21]. This upregulation seems to be under the control of Akt activation induced by tumor cell-secreted VEGF [21]. On their turn, TECs may also accelerate the conversion of tumor cells to chemoresistant stemlike tumor cells, with upregulation for example of insulin-like growth factor-1 [22].

From a metabolic point of view, TECs are described as a hyperglycolytic lineage, undergoing a strong metabolic reprogramming [9,23,24]. Lactic acid produced by TECs, or released by tumor cells, stimulates TEC proliferation, as demonstrated by increased carbonic anhydrase II expression [25].

TECs actively participate in cancer progression. Through the release of angiocrine factors (interleukins and other factors) in the TME, they stimulate tumor cells to grow and metastasize [26–28].

TECs, instructed by tumor cells, are able to direct the immune cell pattern inside tumors with a clear tumor immune tolerance feature [29]. TECs preferentially promote the extravasation and recruitment of regulatory Treg and myeloid-derived suppressor cells in TME, by the upregulation of selective complex on their cell membrane, as well as the secretion of soluble chemotactic factors. A negative modulation of the immune response against the tumor is now demonstrated for TEC functions, and downregulation of adhesion molecules for immune cells has been demonstrated in response to VEGF-A [30–34]. For further details see [13,14] and Chapter 2 of the book.

Cancer-associated fibroblasts normally originate from resident stroma cells, but they can also be differentiated from bone marrow mesenchymal stem cells or from cancer cells after epithelial-mesenchymal transition. Cancer cells can activate resident fibroblasts by targeting them with cancer-secreted factors, such as transforming growth factor-β (TGF-β) and CXC motif chemokine 12 (CXCL12)/stromal cell-derived factor-1 (SDF-1) [35]. CAFs secrete proteases, contributing to ECM degradation, while they also produce and deposit ECM, documenting the dual role of these cells in ECM remodeling. Besides, CAFs also secrete various angiogenic mediators participating in tumor angiogenesis [36]. Due to their localization at the leading edge of the tumor, where expanded vessel supply is demanded, the CAF contribution to angiogenesis is pivotal [37,38]. One of the most important molecules secreted by stromal CAFs is VEGF-A [39], but others have also been reported. Tang and collaborators [40] showed that CAFs in gastric cancer highly express galectin-1, a proangiogenic protein. In colorectal cancer, CAFs have been shown to promote tumor angiogenesis by overexpressing Wnt family member 2 (WNT2) [41–43]. In ovarian carcinomas, most angiogenic growth factors are provided by CAFs rather than by malignant cells [44]. CAFs also supply other factors such as angiopoietin-1 and -2 (Ang-1, Ang-2), which control neovascular stabilization [45].

Immune cells, within the TME, are a heterogenous population, resulting in complicated interactions between tumor cells and stroma. Immune cells have a remarkable role during the regulation of different aspects of tumor growth, such as modulation of angiogenesis and tumor cell-immune system evasion [46]. Tumor-associated macrophages (TAMs) represent one of the most abundant leukocyte populations in the tumor, and their presence correlates with a reduction in survival in most tumor types [47]. Macrophages can be classified into two alternative phenotypes, M1 and M2. Whereas M2 macrophages show a proangiogenic phenotype, M1 macrophages have been described with antitumor properties [48]. TAMs are thus important suppliers of angiogenic factors and cytokines and ECM remodeling molecules [49–51]. Indeed, macrophage infiltration has been correlated with high microvessel density in different tumors [52,53]. Apart from the canonical signaling pathways, alternative proangiogenic molecules such as semaphorins and plexins have also been described as mediators of the macrophage-EC cross talk [54].

Dendritic cells (DCs), due to their potent antigen-presenting ability, are considered a critical factor in antitumor immunity [55]. Nevertheless, defective myelopoiesis inside the tumor ecosystem renders DCs incompetent [56]. A role of DCs in tumor angiogenesis has been described after the finding that immature DCs increased neovascularization in implanted tumor models, while depletion of DCs revoked angiogenesis [57].

Mast cells were found more than 30 years ago to be accumulated in tumors before the onset of angiogenesis, residing in close proximity of blood vessels [58]. Those granulocytes participate in tumor rejection by production of interleukin-1 (IL-1), IL-4, IL-6, and tumor necrosis factor-α (TNF-α). However, mast cells also promote tumor growth by promoting angiogenesis, degradation of the ECM, and immunosuppression [59]. Indeed, mast cells release angiogenic cytokines, such as VEGF, FGF-2, and TGF-β, among others [60].

Depending on the tumor stage and pharmacological treatment, other immune cells are variably present in the TME as lymphocytes, Treg, natural killer, and T cells. The infiltration is highly dependent on the expression of adhesion molecules on the EC of tumor vessels, on its turn controlled by the TME components, the proportion of TECs, and the responsiveness to therapies [14].

Other cells: Even though ECs are the main players of tumor angiogenesis, other components of the vascular system, such as platelets and pericytes, are also necessary for the proangiogenic switch. For instance, platelets have been described as proangiogenic cells. Upon interaction with tumor cells, platelets are able to release VEGF and platelet-derived growth factor (PDGF) from the α granules [61,62].

The contractile cells that surround the basement membrane of vessels are known as pericytes. In quiescent vessels, pericytes are present and express proteins such as platelet-derived growth factor receptor-β (PDGFRβ), NG2, and desmin, but not α-smooth muscle actin (α-SMA). Upon the activation of angiogenic signaling via PDGF, TGF-β, angiopoietin, and

Notch, tumor pericytes loosen their attachment to the vessel, leading to a higher permeability of blood vessels [63,64]. Particularly, the recruitment of pericytes to the tumors highly depends on PDGF-B ligand production by ECs, being responsible for vessel maturation and normalization [65,66].

As mentioned earlier, the communication players among the different components of the TME can be in the form of soluble factors, enzymes, matrix-bound molecules, and cell-derived vesicles. They are under the control of genetic and epigenetic determinants, as well as of biophysical phenomena.

Biophysical determinants of TME for the control of angiogenesis

The accelerated proliferation and growth of tumor cells in a tissue make the preexistent circulatory system inefficient to support the oxygenation for volumes of more than 1–2 mm^3. Consequently, hypoxia (oxygen tension level lower than 5–10 mmHg) develops, and a complete program of events coordinated by the transcription factor hypoxia-inducible factor 1 alpha (HIF-1α) starts. Under normoxic condition, HIF-1α is hydroxylated by prolyl hydroxylase (PHD). The hydroxylated HIF-1α, in turn, is degraded by von-Hippel-Lindau (vHL)-E3 ubiquitin ligase [67,68]. In hypoxic cells, HIF-1α is not degraded, is released from the vHL protein, and migrates into the nucleus where it dimerizes with HIF-1β, acting as a transcription factor [69,70]. Among the genes harboring the hypoxia-responsive element (HRE) sequence in their promoter region, there are VEGF, PDGF, FGF-2, angiopoietins, carbonic anhydrase IX, and endothelial and inducible nitric oxide synthase (eNOS and iNOS), which result upregulated [71–73].

Acidity of the TME, due to the high metabolic rate of dividing cancer cells, is another stimulus for angiogenic factor upregulation. As a consequence of the Warburg effect, tumor cell metabolism becomes highly acidic [74]. The increase in glucose consumption induces lactic acid release, thus reducing extracellular pH [75]. Acidic pH in the TME is correlated with epithelial-to-mesenchymal transition, ECM degradation, and metastasis of several types of tumors [76–88]. As for hypoxia, acid microenvironment is a stimulus for increased VEGF expression [79]. Indeed, the final aim is to adapt the cells to the unfavorable environmental conditions, leading to metabolic changes that decrease oxygen consumption and stimulate the formation of new vessels to bring oxygen [69,70,80].

Tumor angiogenesis, defined as the formation of new vessels from preexisting capillaries grown into the tumor ecosystem [7], requires a series of specific and unspecific regulatory pathways within ECs, the major cellular components involved in vessel growth. The best known and principal angiogenic player is VEGF, interacting with its receptors VEGFRs [81]. Although imperfect and inefficient, these new vessels are essential for tumor initiation and progression. The newly formed blood vessels satisfy the requirement to deliver essential nutrients and oxygen to the highly proliferating tumor cells and to remove waste products. Together with angiogenesis, tumors stimulate the growth of new lymphatic vessels, a phenomenon called lymphangiogenesis, with some overlapping features [4,82]. Additionally, due to the high permeability of neovasculature, tumor cells gain access to the circulation and spread in the body, giving rise to metastasis [7]. Indeed, to induce neovessel stability, the involvement of other stromal cells as pericytes is needed.

The concept of angiogenesis was introduced by the English physician and anatomist John Hunter in 1787 [83]. Now we know that it is a complex multistep process, characterized by EC migration, proliferation, and organization in a functional network [84,85]. In physiological conditions, a balance between pro- and antiangiogenic factors thinly regulates the angiogenesis process. The prevalence of proangiogenic factors allows quiescent ECs to activate toward an angiogenic phenotype, as during development, menstrual cycle, or wound healing. Instead, during tumor development, the balance of pro- and antiangiogenic factors heavily shifts toward proangiogenic molecules, through the "angiogenic switch," a term coined by Folkman in the 1970s [86,87], to define the turning point from the avascular phase of a dormant tumor to a vascular phase characterized by exponential tumor growth. This pathological transition is caused by an increased and persistent release of proangiogenic factors in the TME mainly under the control of hypoxia, producing an inefficient organization of blood vessels, which became aberrant. Indeed, the newly formed vascular networks fail to mature and normalize, the vessel caliber size is markedly heterogeneous, and the blood flow is chaotic [88]. Increased permeability and interstitial fluid pressure are due to disrupted endothelial junctions and erratic presence of pericytes partially covering the endothelial monolayer [89]. The intratumor vessels are fragile with an enhanced risk of hemorrhage [90–92]. The defective microcirculation profoundly alters the tumor microenvironment with irregular blood flow within the tumor parenchyma, thus resulting in areas of persistent or intermittent hypoxia, further sustaining an aberrant neovascularization.

Furthermore, mobilization of bone marrow-derived precursor cells and recruitment of immune cells to the tumor ecosystem are also positively controlled by hypoxia [93]. By changing the cytokine milieu, hypoxia can also induce an immunosuppressive microenvironment, allowing immune system evasion by cancer cells [94]. Hypoxia also produces a metabolic switch in cancer cells to anaerobic metabolism, apoptosis inhibition, increased invasiveness, EMT, and metastasis [95]. A stemlike phenotype is induced in cancer cells with the release of cytokines like IL-6 by stromal cells. Consistently, hypoxia-driven expression of VEGF, matrix metalloproteases (MMPs), and angiopoietin-like 4 (ANGPTL4) in stromal cells is crucial for tumor cell intravasation, accompanied by downregulation of the tissue inhibitors of metalloproteases (TIMP)-2 and -3, again under the control of HIF-1α. In detail, ANGPTL4 expression disrupts vascular endothelial tight junctions and augments permeability, thereby altering the trans-endothelial barrier, together with ECM rearrangements [96].

Besides the crucial control by hypoxia and acidosis, other factors are responsible for altered gene transcription or genetic instability in cells of tumor ecosystem. Epigenetic enzymes have been identified to modulate gene transcription by acting on DNA methylation and histone posttranslational modification of a pattern of genes involved in angiogenesis and endothelial responsiveness [97]. Different transcription factors involved in tumorigenesis such as p53, c-Myc, nuclear receptors, and RAS control the expression of angiogenic modulators and miRNAs. Among them are ROS, change in pH, cytokines, chemokines, inflammatory mediators, and growth factors released in the TME [97,98]. In particular, miRNAs derived from neoplastic cells, ECs, and cells of the TME regulate tumor angiogenesis, acting as pro-angiomiR (by targeting and inhibiting negative regulators) or anti-angiomiR (by targeting positive regulators) [97,99].

miRNA, together with the other signaling molecules, can be secreted alone or inside of vesicles as exosomes, 30–100 nm dimeter vesicles diffused by all the cellular components of TME in the extracellular fluid. Exosomes are nowadays considered important communication cargo able to influence also tissues at distance. Findings report the ability of exosomes to induce angiogenesis and to predispose pre-metastatic niche [100–102].

Nowadays, angiogenesis is included within the hallmarks of cancer [11] and the proangiogenic shift definitely represents an early event in tumor progression and metastasis.

Soluble proangiogenic factors

Early in their development, tumors undergo an angiogenic switch, which is the result of a change in the balance between pro- and antiangiogenic factors in favor of the former [103,104]. Initially, the pro- and antiangiogenic systems seemed to be simple, meaning that VEGF was the main proangiogenic protein [81], and thrombospondin and angiostatin were the endogenous antiangiogenic counterparts [105,106]. Many proangiogenic and antiangiogenic factors have been identified since then. By acting on specific receptors present on ECs, the angiogenesis modulators modify the various EC functions involved in vessel growth. The events occurring during sprouting angiogenesis include (i) disintegration of the existing blood vessels by digestion of basement membrane, (ii) migration and proliferation of ECs in response to a gradient of the angiogenic stimuli, and (iii) differentiation of the newly formed solid buds into functional blood vessels [84]. The second phase corresponds to endothelial-mesenchymal transition with tip ECs leading the spouting process and stalk ECs providing the highly proliferative counterpart. In the third phase, the opposite mesenchymal-endothelial transition program occurs. In physiological angiogenesis, after EC migration, proliferation, and sprouting, the new blood vessels fuse with blood circulation, while some vessels can undergo pruning. Finally, the newly formed vessels are stabilized through mural cell (pericytes and vascular smooth muscle cells) covering and synthesis of a new basement membrane also with the collaboration of stalk cells [107]. Instead, in pathological angiogenesis, the angiogenic switch remains turned on, NECs transform in TECs, and the newly formed vessels are disorganized, immature, leaky, and chaotic.

Among the different proangiogenic factors underlying the development of malignant tumors, accomplished with vascularization ingrowth, the major stimulus for EC activation toward an angiogenic phenotype is VEGF, expressed under the control of tissue hypoxia and low pH. Indeed, low oxygen tension, a condition that promotes activation and stabilization of HIF-1α, controls the transcription of VEGF, promoting tumor angiogenesis, tumor cell growth, and, finally, cancer metastasis [108]. Even before the discovery of VEGF, 40 years ago Folkman proposed tumor angiogenesis as a valid target to arrest tumor growth and metastasis [86] and, despite the differences in angiogenesis dependency among different types of tumors, inhibition of VEGF/VEGFR signaling pathway rapidly became the focus of antiangiogenic therapy. However, it took 30 years for the first antiangiogenic drugs to become available for clinical use. Importantly, they were successful, prolonging overall survival in many patients, but suffered of some drawbacks related to resistance and failure [109].

Nevertheless, if VEGF is the most important stimulus for proliferation and migration of ECs, other growth factors, such as the PDGF and Ang-2 and the related receptors Tie-1 and -2, facilitate recruitment of pericytes and smooth muscle cells, important for maturation and stability of the neovasculature.

The complex modulation of the angiogenesis process also includes cytokines, inflammatory and vasoactive mediators, and other growth factors, such as FGF-2, EGF, hepatocyte growth factor (HGF) released by tumor or stromal cells in the TME, and negative modulators such as thrombospondin, angiostatin, endostatin, and Ang-1 [88]. It is now clear that the complexity of angiogenesis regulation is also related to epigenetic modulation of gene transcription for angiogenesis modulators, and the fact that various factors are involved in angiogenesis regulation either directly, through specific receptors expressed on EC surface, or indirectly, by modulating angiogenic factor expression and receptor availability and functioning.

A schematic description of the pro- and antiangiogenic factors is here reported together with not soluble modulators and related intracellular signaling pathways.

VEGF

The glycoprotein VEGF, a mitogen specific for ECs, is secreted as either soluble or ECM-bound form [81,110]. The VEGF sequestered by ECM is activated by the action of MMP-9, a proteolytic enzyme involved in cell migration and activated during angiogenesis [71]. Besides hypoxia, VEGF expression by tumor cells is controlled by inflammatory mediators (cytokines, PGE2, ROS), growth factors as EGF, and oncogenes as RAS [111,112] or after loss of tumor suppressor phosphatase and tensin homolog (PTEN) and vHL [113,114]. Recent literature documents a crucial role in VEGF regulation by miRNAs (miR-15a/16, miR-17/92, miR-29b, miR-107, miR-126, miR-134, miR-145, miR-195, miR-204, and miR-497 have antiangiogenic effects; miR-21, miR-24, miR-146a, and miR-378 have proangiogenic effects, depending on the cancer type) [97,99]. To make the scenario more complex, the same or other miRNAs control HIF-1α expression: miR-17/92, miR-204 (antiangiogenic), miR-21, and miR-210 (proangiogenic) [97,99]. These findings document the complex cross talk among the various components of the TME: cells, physical condition, and soluble and matrix-bound mediators.

The VEGF family currently consists of six growth factors, VEGF-A, -B, -C, -D, -E, and placental growth factor (PlGF), and three tyrosine kinase (TK) receptors: VEGF receptor 1 (VEGFR-1) (Flt-1), VEGFR-2 (KDR/Flk-1), and VEGFR-3 (Flt-3). The most relevant isoform for the angiogenic process is VEGF-A, with the splicing variant VEGF-A165 being the mostly expressed one. It binds VEGFR-1 and VEGFR-2, localized on the EC surface, with VEGFR-2 being the most relevant for angiogenesis occurrence and binding all VEGF-A isoforms, VEGF-C and VEGF-D [115]. VEGFR-1, mainly expressed on stalk cells, binds VEGF-A with stronger affinity than VEGFR-2, highly expressed on tip cells, and is the only receptor for PlGF-1 [116]. VEGFR-3 is mainly expressed on lymphatic ECs and is involved in lymphangiogenesis in response to VEGF-C and -D [117]. VEGFR-1 is not only implicated in EC proliferation, but it controls the chemotaxis of monocytes/macrophages and the mobilization of ECs and recruitment of hematopoietic stem cells and EPCs from the bone marrow [118–120]. VEGFR-1 exists also as a soluble form (sVEGFR-1) and functions as a decoy receptor sequestering the ligand and preventing its interaction with its receptors, thus impairing VEGF-A angiogenic functions, for example in the cornea and cartilages [121,122].

Moreover, two nonenzymatic co-receptors for the VEGF family, neuropilin-1 and -2, cooperate in blood vessel development and potentiate the signaling due to VEGFR-2 [123]. VEGFR-2 interacts also with other proteins as integrins and cadherins, and heparan sulfate proteoglycans. In the complex, VEGFR-2 and its ligand VEGF-A represent the main angiogenic system, triggering this process during tumor development. In common with many TK receptors, the kinase activity of VEGFR-2 determines the activation of a broad spectrum of biological signaling [124]. Indeed, in TK receptors, the phosphorylation of intracellular specific tyrosine residues forms an anchorage site for proteins containing the Src homology 2 domain (SH2) and involved in many post-receptor signaling [125]. For example, c-Src is a typical protein containing the SH2 domain. The activation of c-Src plays a key role in VEGF-mediated angiogenesis and vessel permeability, given that it coordinates the VEGF-induced MAPK cascade, the focal adhesion kinase (FAK)-$\alpha v\beta 5$ integrin cascade, and eNOS activation [126,127]. Similar to c-Src, growth factor receptor-bound protein 2 (GRB2), which contains SH2 and SH3 domains, and phospholipase C-γ (PLCγ), which contains SH2, are adapter proteins involved in the activation of VEGFR-downstream Ras/Raf/MEK/MAPK and PI-3K/Akt signaling [128–133]. In particular, phosphorylation of eNOS by Akt/protein kinase B (PKB) is leading to angiogenesis, vasodilation, and vascular remodeling [134]. The activation of protein kinase C (PKC) pathway has also been reported [132]. After stimulation with VEGF, VE-cadherin is internalized and the tight junctions between cells loosen, inducing vascular permeability [135]. In TECs, VEGF-A seems to mediate a clustering defect of the adhesion molecules, like ICAM-1 and VCAM-1, which hampers immune cell extravasation [136,137].

Other proteins interacting with VEGFR-2, such as the T-cell-specific adaptor protein, p38 MAPK, and FAK, not only maintain survival signals, but also regulate cell migration and vascular permeability, through the upregulation and activation of MMPs which degrade the ECM, modulation of cytoskeleton, and focal adhesion sites and other actions.

VEGFR2 levels can be upregulated by the same post-receptor signaling. FAK, for example, translocates to the nucleus and takes part in RNA polymerase II complex on the VEGFR2 promoter [138]. The activities regulated by VEGFR2 are influenced by posttranslational modifications as acetylation, methylation, and ubiquitination [139–141]. Moreover, VEGF-induced intracellular VEGFR2 trafficking from and to the plasma membrane is key to proper signaling. Although considered mainly a clathrin-mediated event, VEGF-induced VEGFR2 internalization has now been shown to preferentially occur through cell division cycle 42-mediated micropinocytosis [142].

PDGF

PDGFs are a group of peptides, composed of PDGF-A, -B, -C, and -D, which act as homo- or heterodimers and interact with the PDGF transmembrane receptors α and β. PDGF is synthetized and stored in the alpha granules of platelets from which it is released upon activation [61,62]. It is also produced by stromal cells of the TME as activated macrophages, smooth muscle cells/pericytes, and ECs. The PDGF/PDGFR downstream signaling elicits the activation of Ras/Raf/MAPK, PKC, PI-3K, and PLCγ, signaling pathways very similar to the VEGFR-mediated ones [143]. PDGF is the major mitogen for mesenchymal cells, including fibroblasts, glial cells, and smooth muscle cells. The binding of PDGF-B on PDGFR-β is the first event implied in pericyte recruitment to support the growth of neovessels and the release of several proangiogenic factors [144,145]. It is also capable of upregulating its own receptors and stimulating the proliferation of pericytes and smooth muscle cells [103,146].

miRNAs reported to control PDGF production are miR204 and mir-29b (antiangiogenic), while miRNA controlling PDGFR expression is miR-146a (proangiogenic) [97].

The overactivation of PDGF signaling in the TME promotes tumor growth through angiogenesis [144,145] and, in particular, a constitutively active PDGFR mutation supports the progression of gastrointestinal stroma tumor [147]. Moreover, PDGF-C isoform produced by CAFs has been shown to confer resistance to anti-VEGF therapies [148].

FGF

FGFs are a family of 22 proteins, secreted by stem cells and damaged cardiac myocytes and vascular ECs. FGFs are mitogens which are known to play an essential role in angiogenesis, stimulating the proliferation of ECs and fibroblasts [149]. The major FGF isoforms involved in angiogenesis are FGF-1 (or acidic FGF) and FGF-2 (or basic FGF), which transcription is under the control of hypoxia. FGF-2 on its turn regulates HIF-1α, thus creating a feedback loop. FGF-2/FGFR1 axis is involved in EC proliferation, migration, and differentiation during tumor angiogenesis [150]. FGF-2, a heparin-binding growth factor, is secreted mainly by fibroblasts and interacts with TK receptors on ECs. Heparin sulfate proteoglycans and syndecans also play a key role in regulating the angiogenic activity of the factors VEGF and FGF-2 [151]. The FGF ligand binding to its receptors (FGFR1–4) initiates the autophosphorylation of the receptors and leads to the activation of downstream signaling, such as the PI-3K/Akt and Ras/Raf/MEK/MAPK, which are involved in the proliferation, migration, and apoptosis of ECs [150].

FGF overexpression seems one of the factors responsible for resistance to anti-VEGF antiangiogenic therapies [152].

HGF/C-Met

Originally discovered as a potent mitogen for adult hepatocytes, HGF/scatter factor is a pleiotropic cytokine that shows a variety of cellular effects upon its receptor activation. c-Met (sometimes termed Met or MET) is a TK receptor that, when bound to its ligand HGF, activates a wide range of different cellular signaling pathways leading to angiogenesis [153]. Indeed c-Met is expressed in ECs, smooth muscle cells, and bone marrow-derived endothelial progenitor cells and its transcription is under the control of hypoxia. Zhang et al. demonstrated that HGF was able to modulate neovascularization through increased VEGF and reduced thrombospondin-1, a negative regulator of angiogenesis [154]. High tumor grade and clinical stage in renal cell carcinoma correlate with increased c-Met levels [155]. Expression of c-Met was higher in ECs than in tumor cells in a human tumor experimental model [156]. Again, overexpression of HGF and/or c-Met has been reported to account for resistance to anti-VEGF drugs [157].

Inflammatory mediators

TME is particular rich of inflammatory mediators which affect all the cellular components herein. Indeed, chronic inflammation is a known trigger of tumor progression and angiogenesis. On ECs the cross talk among TK receptors and G protein-coupled receptors for both vasoactive and inflammatory mediators has also been described. Following is a partial list of the main inflammatory mediators involved in tumor angiogenesis occurrence.

Prostanoids: Many solid tumors show an overexpression of the main mediators of the inflammatory process, such as cyclooxygenase 1 and 2 (COX1 and COX2) and prostanoids, especially prostaglandin E2 (PGE2), the major final product of COX activity. The overexpression of COX2 in tumors is accompanied by enhanced expression and production of angiogenic factors, such as VEGF, FGF-2, MMPs, and vascular modulators, i.e., angiopoietins [158]. The proangiogenic role of PGE2 results in the direct induction of EC proliferation, migration, and tube formation, mediated by both EP receptors and transactivation of FGFR in ECs through underneath signaling [159–161]. Moreover, PGE2 increases the barrier function of TECs by inhibiting trans-endothelial migration of cytotoxic tumor-infiltrating lymphocytes [18].

Interleukin 8 (IL-8): Tumor-associated macrophages and myeloid angiogenic cells promote angiogenesis through IL-8 production [162–164]. Extracellular IL-8 has been found to transactivate VEGFR2 [164]. Transfection of IL-8 genes in gastric cancer cells promoted angiogenesis [165]. Nowadays, IL-8 production by cancer and stromal cells is universally recognized for its angiogenic potential and correlation with tumor progression and also for its ability to recruit immunosuppressive cells to the tumor. IL-8 expression is under the control of hypoxia, acidosis, and NF-kB. Importantly, its overexpression mediates resistance to antiangiogenic treatments and immune checkpoint inhibitors [166].

SDF-1: Some chemokines like stromal cell-derived factor-1 (SDF-1)/CXCL12, interacting with chemokine receptor 4 (CXCR4), are also angiogenesis inducers [167,168]. Hypoxia upregulates the level of SDF-1, a stem factor for pericytes and smooth muscle cells, and CXCR4 in bone marrow-derived cells, EPCs and TECs, with the chemokine receptor 2 (CCR2) being a marker of TECs [169]. The binding of SDF-1 secreted by EPCs to CXCR4 in tumor cells allows the upregulation of proangiogenic factors and enhances tumor cell invasion with the development of pre-metastatic niche [170,171]. Moreover, CXCR7 regulates CXCL12–CXCR4-mediated tumor trans-endothelial migration, and high expression of CXCR7 in tumor vasculature is definitely linked with tumor angiogenesis and metastasis in various tumor types [172–174].

Other regulatory systems

Angiopoietins/Tie receptors

Within the tumor ecosystem, CAFs secrete Ang-1 and -2, involved in vessel maturation [45]. Angiopoietin-Tie-2 signaling is a vascular specific TK receptor system essential for blood vessel stabilization and remodeling, and modulation of vascular permeability. The system is based on the dual regulation of Tie-2 receptor. Interaction of Ang-1 with Tie-2 receptor is necessary for the vessel stabilization and maturation [175]. Ang-1-mediated Tie-2 signaling also interacts with NF-κB activity, leading to an anti-inflammatory effect [176]. Ang-2 expression is increased under hypoxic condition and in TEC, particularly in the highly specialized tip cells of sprouting vessels [177,178]. Ang-2 antagonizes the Ang-1-induced Tie-2 phosphorylation [179]. Under VEGF exposure, Ang-2 interaction with Tie receptor stimulates pericyte detachment followed by sprouting of blood vessels, whereas in its absence, Ang-2 leads to cell apoptosis [177,180].

Tie-2 receptor is expressed on ECs of both blood and lymphatic vessels. Tie-2 receptor activation regulates downstream signaling pathways such as the PI-3K/Akt and/or Ras/Raf/MEK/ERK cascade [181]. The Ang/Tie system plays a crucial role in the pathophysiology of the tumor vasculature, as well as normal vasculature, and Ang-2 expression is found to be upregulated in many types of cancers [182].

Moreover, it has been reported that the endothelial secreted Ang-2 recruits innate immune cells in solid tumors [183]. In tumors, recruited macrophages are reprogrammed to TAM which secrete proangiogenic factors like VEGF, FGF, and MMP-9 [184].

Endothelin (ET) and endothelin receptors (ETR)

Endothelins (ETs) are 21-amino acid peptides, mainly produced by ECs, that have powerful vasoconstrictive abilities and participate in blood pressure homeostasis via endothelin binding to its receptors, namely ET-A and ET-B [185]. In addition to its physiological effects, the ET-ETR system has been found to participate actively in tumor progression and angiogenesis. ET is produced by tumor, endothelial, and stromal CAFs and TAMs. ET binds ETRs, which initiate signaling cascades that produce the abovementioned effects [185]. There are different ETs and ETRs. The most frequently involved in cancer are ET-1 and ET-2, and the receptor usually associated with angiogenesis is the ET-B [186]. Indeed, endothelin receptors, particularly ET-B, play a pivotal role in tumor progression, as well as in angiogenesis and immunological escape and probably in resistance to angiogenic treatments [187]. These receptors seem to be the nexus between angiogenesis and immune escape in cancer. ETs through their receptors definitely control vascular tone, mitosis, cell survival, angiogenesis, and immune cell regulation in a large variety of cancers [188].

Surface-associated modulators of angiogenesis

Integrins: Integrins are heterodimeric transmembrane glycoproteins that act as adhesion molecules cementing the new vessels with the components of the basement membrane and allow adhesion of pericytes around the newly formed vessel, decreasing leakage and stabilizing the new structure. Cell adhesion receptors are essential proteins that contribute not only to the adhesive process between cells but also to the activation of intracellular signaling pathways. These receptors can activate enzymatic proteins located in the intracellular domain of the receptor and can also bind to adaptor proteins and to the intracellular cytoskeleton (outside-in signaling). On the other hand, adhesion receptors are regulated by intracellular signals as they exist in protein complexes that are dynamically adjusted in response to a stimulus, being translocated into the membrane sites of the adhesion (inside-out signaling).

Adhesion receptors are critical in the regulation of migration. Therefore, the dynamic regulation of these adhesion complexes is a crucial feature of angiogenesis, and the expression of different integrins on the spouting ECs is essential to produce new vessels [189]. Various integrins have been identified and specific integrin antagonists have been designed to seriously impair angiogenesis and regulate monocyte trafficking [190–192]. Their development in combination with chemotherapy is still experimental but promising [193].

Neuropilin-ABL1 pathway: Neuropilins are transmembrane glycoprotein receptors that drive neurogenesis and angiogenesis. Their function is pleiotropic, and they can interact with almost all the growth factors [194]. Neuropilins have been found to be over-expressed in many tumors [195–197]. Semaphorin 3 is the main ligand of neuropilins, but various growth factors can also bind and activate them [198]. Neuropilins behave as co-receptors since their short intracellular portion requires an associated receptor in order to transmit signals. As an example, neuropilin-1 potentiates the phosphorylation activity of VEGFR, thus leading to EC migration. While this is a VEGF-dependent pathway, neuropilins take part in VEGF-independent pathways, ultimately activating the cytosolic tyrosine kinase ABL1 protein to exert angiogenic functions [199].

Delta-Notch Pathway: This is an evolutionary ancient pathway mainly associated with cell differentiation, proliferation, and signal integration during embryonic life [200]. While the Notch signaling acts as a coordinator of the multiple steps of vascular development, the main function in angiogenesis is to suppress nonfunctional sprouts and stimulate blood vessel formation. The mechanism is based on Notch, a membrane receptor, and Delta, a one-pass transmembrane protein acting as a ligand. Ligand binding releases the intracellular portion of Notch, which then acts as a transcription factor for genes related to angiogenesis and other functions. Blocking the Delta-Notch pathway does not impair angiogenesis, but induces nonproductive vessels, acting paradoxically as an antiangiogenic alternative [201,202]. Specifically, the Delta-like 4 ligand (Dll4) is the responsible agent for the differentiation of the two types of ECs that participate in sprouting angiogenesis, namely tip and stalk cells [203,204]. VEGF and hypoxia promote the expression of Dll4 in EC membrane of tip cells by binding to Notch-1 expressed in stalk cells. On its turn, Notch regulates VEGFR-2 and -3 and neuropilin 1 expression [203,204]. Hainoud et al. [205] have shown that the Dll4-Notch 4-ephrin B2 pathway is essential for the building of functional vessels, since Notch decreases cell migration and enhances pericyte recruitment There is evidence showing that Notch is part of a VEGF-VEGFR downstream regulating signaling pathway [206,207], and there are other factors that can activate Notch signaling, such as FGF and Wnt pathway [208].

The Notch signaling is part of the direct tumor cell-TEC communication. It was demonstrated in vitro that JAGGED1 transmembrane protein expressed in breast cancer tumor cells can activate the Notch receptor in ECs, triggering an angiogenic cascade [209]. TECs also regulate tumor cell intravasation and metastasis. Notch-1 in TECs was shown to promote lung metastasis. Moreover, the elevated expression of Notch-1 in TECs is correlated with poor prognosis for melanoma, lung adenocarcinoma, and breast carcinoma, as Notch-1 activity facilitates metastasis [210].

Ephrins and ephrin receptors: Ephrins and their receptors have been reported to be promoters of EC migration and proliferation [211,212] and to mediate VEGF actions on angiogenesis and lymphangiogenesis [213,214].

Endogenous negative regulators of angiogenesis

A series of angiogenesis inhibitors are produced by the specific cleavage of soluble or matrix-bound large glycoproteins. Thus, the expression and activation of specific proteases in the TME are crucial for their availability. Examples of such molecules that facilitate the termination of neovascular growth through anti-apoptosis gene suppression are thrombospondins, angiostatin, endostatin, platelet factor 4, interleukin-12, and tumstatin among others.

Thrombospondin: 40 years ago, the glycoprotein thrombospondin 1 (TSP1) was found on platelet membranes and further characterized as a powerful antiangiogenic mediator [215]. It was identified as an endogenous natural antagonist of angiogenesis [105], through inhibition of the synthesis and activity of VEGF [216–218] and FGF-2 [219–221] on their receptors or heparan sulfate proteoglycans. CD36, CD47, and integrins are some of its membrane receptors. Its endogenous expression is under the control of miRNAs: let-7f and mir17/92 (proangiogenic) [97,99]. However, its large size and difficult handling make recombinant TSP or TSP fragments inappropriate for clinical use.

Angiostatin: Endogenous angiostatin is a 38 kDa amino-terminal fragment of plasminogen. It was originally isolated from tumor-bearing mice [106] and has both potent antiangiogenic and antiproliferative activity toward cancer cells [222,223]. Angiostatin has proved to inhibit EC proliferation and migration, tube formation, MMP expression, neutrophil activation and migration, monocyte and macrophage trafficking, and leukocyte recruitment. Additionally, angiostatin induces EC apoptosis and production of antiangiogenic factors such as TSP1 and reduces VEGF expression by binding to ATP synthase and integrins and by preventing cell cycle progression. Plasminogen contains five kringle (K) domains of ~80 residues each. Studies using recombinant angiostatin demonstrated that the tumor inhibitory activity resides in a fragment of K1-3 [224,225].

Kringle 5: Like angiostatin, it is a by-product of the proteolytic cleavage of plasminogen and was shown to exert antiangiogenic and antitumor activities [226]. Ansell et al. demonstrated that Kringle 5 (K5) functions as a competitive antagonist of HGF [227]. In addition to its potent angiostatic role, K5 can direct a potent antitumor response with its ability to recruit tumor-associated neutrophils and natural killer T cells [228].

Endostatin: Endostatin is a 20 kDa carboxyl-terminal fragment of type XVIII collagen located in the basement membrane of blood vessels and plays an important role in EC adhesion and cytoskeletal organization [229,230]. Endogenous endostatin inhibits migration, MMPs, and MAPK signaling; induces apoptosis of proliferating ECs; inhibits tumor growth; and impairs blood vessel maturation in wound healing [231,232]. It is thought to directly interfere with the proangiogenic action of growth factors such as FGF-2 and VEGF (through binding to VEGFR-2) and is known to inhibit >60 different tumor types [233]. Additionally, a study measuring changes in gene expression in human dermal microvascular cells following treatment with endostatin demonstrated downregulation of several proangiogenic pathways as well as upregulation of many antiangiogenic genes [233]. We have contributed to find the domain responsible for the antiangiogenic activity of the whole endostatin, documenting the antiangiogenic activity of the peptide fragment covering the COOH-terminal endostatin region [234].

Besides the promising results of the many in vitro and experimental animal studies, these molecules and their recombinant formulations (i.e., Endostar) did not arrive to the clinic use in oncology due to the failure of most of the clinical trials.

Heterogeneity of endothelial component as a modifier of antiangiogenic therapy

Due to the complexity of factors and communications affecting EC function and the existence of various types of tumor angiogenesis, it is now comprehensible that antiangiogenic agents are not uniformly active across all tumor types, with lack of efficacy and failure of antiangiogenic treatments specifically designed against VEGF/VEGFR. We have to consider that naïve NECs are basically heterogenous depending on the district (capillary, venous, arteriosus) and, especially for capillaries, on the organs and tissues. The term organotypic heterogeneity has been introduced meaning that ECs exhibit organ-specific molecular and functional properties due to transcriptional programs that provide heterogeneity in relation to the vascular tree [9]. We must also consider that inside tumors there are different proportions and stages of NECs and TECs, and on their side TECs are characterized by extreme variability and heterogeneity due to their responsiveness to the TME particularly rich in stimuli and damaging noxa. We have to consider that inside the tumor mass the population of EC phenotypes is mixed. There are naïve EC, early TECs with high proliferation capacity, and TECs with stemness markers (ALDH+) [18]. Additionally, there is a great inter-tumor variability depending on the organ of origin, the stage of the tumor, and the patient medical history and comorbidities (i.e., diabetes, obesity). If the tumor has been treated with conventional or target therapy, a proportion (not all) of TECs have been inhibited and/or induced to senescence. Thus, it is difficult to consider that the further addition of an antiangiogenic drug as second/third-line treatment will find all the ECs responsive in the same manner.

In this complex scenario, more studies are needed to understand the TEC characteristics and behavior by an integrated approach using single-cell RNA sequencing, proteomics/transcriptomics, and functional assays in 2D and 3D tumor models [235]. This will probably allow the design of antiangiogenic drugs able to target TECs or pericytes, hopefully with a personalized approach.

Conclusions

In summary, EC migration and proliferation are essential events during angiogenesis, associated with activation of complex intracellular pathways, mostly resulting from VEGF binding to its receptor on EC membrane and integrin inside-out signaling. These pathways involve activation of small GTPases of the Rho family, PI3K/Akt/eNOS, p38, FAK, and MAPK/ERK1/2 signaling cascades, among others. But VEGF is not working alone due to the redundancy of proangiogenic factors. Based on the complexity of factors involved in the control of angiogenesis, it has been possible to design and develop antiangiogenic strategies acting on each specific factor or, otherwise, on multiple targets (see Chapter 2 of the book). Specific examples of successful drugs or drug combinations are reported in this book for breast and lung cancers and various other neoplasms in Chapter 2, gastrointestinal tumors and colorectal cancer (CRC) in Chapter 3, renal and urogenital cancer in Chapters 4 and 5, melanoma in Chapter 6, and hematological tumors in Chapter 7. Anti-VEGF drugs are consolidated pharmacological strategies for targeting the TME but failed to produce universal durable responses due to the complex variability of the same TME. Indeed, the redundancy of positive modulators of tumor angiogenesis and the occurrence of different vasculogenic mechanisms (i.e., vascular mimicry) explain the failure of antiangiogenic treatments due to resistance development and escape mechanisms [109] (Chapter 9 of this book).

Tumors are now regarded as a heterogeneous association of both tumor and stromal cells that contribute in an interconnected fashion to malignant progression. Undoubtedly, the normalization of the tumor ecosystem is an important new aspect for integrated cancer treatment (see Chapter 11). Tumor blood vessel heterogeneity as well as TEC heterogeneity (including nonresponsive and senescence TECs) may be responsible for mediating the variable response to anti-VEGF treatments and anti-VEGF drugs in combination with other therapeutics.

However, the combination of antiangiogenic strategies with conventional chemotherapy (or target therapy or immune checkpoint inhibitors) or metronomic chemotherapy (see Chapter 8) is showing promising synergistic antitumor effects with the potential of antiangiogenic drugs to sensitize tumor and stromal/immunological cells to the other therapies. The active clinical research on the definition of prognostic markers of drug response (see Chapter 10) will greatly help in this direction.

References

[1] Hlatky L, Hahnfeldt P, Folkman J. Clinical application of antiangiogenic therapy: microvessel density, what it does and doesn't tell us. J Natl Cancer Inst 2002;94:883–93.

[2] Zhou D, Cheng SQ, Ji HF, Wang JS, Xu HT, Zhang GQ, Pang D. Evaluation of protein pigment epithelium-derived factor (PEDF) and microvessel density (MVD) as prognostic indicators in breast cancer. J Cancer Res Clin Oncol 2010;136:1719–27.

[3] Tae K, El-Naggar AK, Yoo E, Feng L, Lee JJ, Hong WK, Hittelman WN, Shin DM. Expression of vascular endothelial growth factor and microvessel density in head and neck tumorigenesis. Clin Cancer Res 2000;6:2821–8.

[4] Pastushenko I, Vermeulen PB, Carapeto FJ, Van den Eynden G, Rutten A, Ara M, Dirix LY, Van Laere S. Blood microvessel density, lymphatic microvessel density and lymphatic invasion in predicting melanoma metastases: systematic review and meta-analysis. Br J Dermatol 2014;170:66–77.

[5] Miyata Y, Sakai H. Reconsideration of the clinical and histopathological significance of angiogenesis in prostate cancer: usefulness and limitations of microvessel density measurement. Int J Urol 2015;22:806–15.

[6] Huang J, Ma X, Chen X, Liu X, Zhang B, Minmin L, Nie W, Zhang L, Liu L. Microvessel density as a prognostic factor in bladder cancer: a systematic review of literature and meta-analysis. Cancer Biomark 2014;14:505–14.

[7] Carmeliet P, Jain RK. Angiogenesis in cancer and other diseases. Nature 2000;407:249–57. https://doi.org/10.1038/35025220.

[8] Ribatti D, Crivellato E. "Sprouting angiogenesis", a reappraisal. Dev Biol 2012;372:157–65.

[9] Eelen G, Treps L, Li X, Carmeliet P. Basic and therapeutic aspects of angiogenesis updated. Circ Res 2020; 127(2):310–29. https://doi.org/10.1161/CIRCRESAHA.120.316851.

[10] Zuazo-Gaztelu I, Casanovas O. Unraveling the role of angiogenesis in cancer ecosystems. Front Oncol 2018;8:248. https://doi.org/10.3389/fonc.2018.00248.

[11] Hanahan D, Weinberg RA. Hallmarks of cancer: the next generation. Cell 2011;144(5):646–74.

[12] Quail DF, Joyce JA. Microenvironmental regulation of tumor progression and metastasis. Nat Med 2013;19:1423–37. https://doi.org/10.1038/nm.3394.

[13] Lambrechts D, Wauters E, Boeckx B, Aibar S, Nittner D, Burton O, Bassez A, Decaluwé H, Pircher A, Van den Eynde K, Weynand B, Verbeken E, De Leyn P, Liston A, Vansteenkiste J, Carmeliet P, Aerts S, Thienpont B. Phenotype molding of stromal cells in the lung tumor microenvironment. Nat Med 2018;24:1277–89. https://doi.org/10.1038/s41591-018-0096-5.

[14] Nagl L, Horvath L, Pircher A, Wolf D. Tumor endothelial cells (TECs) as potential immune directors of the tumor microenvironment – new findings and future perspectives. Front Cell Dev Biol 2020;8:766. https://doi.org/10.3389/fcell.2020.00766.
[15] Hida K, Maishi N, Torii C, Hida Y. Tumor angiogenesis-characteristics of tumor endothelial cells. Int J Clin Oncol 2016;21:206–12.
[16] De Sanctis F, Ugel S, Facciponte J, Facciabene A. The dark side of tumor-associated endothelial cells. Semin Immunol 2018;35:35–47.
[17] Matsuda K, Ohga N, Hida Y, Muraki C, Tsuchiya K, Kurosu T, Akino T, Shih SC, Totsuka Y, Klagsbrun M, Shindoh M, Hida K. Isolated tumor endothelial cells maintain specific character during long-term culture. Biochem Biophys Res Commun 2010;394(4):947–54. https://doi.org/10.1016/j.bbrc.2010.03.089.
[18] Ohmura-Kakutani H, Akiyama K, Maishi N, Ohga N, Hida Y, Kawamoto T, Iida J, Shindoh M, Tsuchiya K, Shinohara N, Hida K. Identification of tumor endothelial cells with high aldehyde dehydrogenase activity and a highly angiogenic phenotype. PLoS ONE 2014;9, e113910.
[19] Hida K, Maishi N, Annan D, Hida Y. Contribution of tumor endothelial cells in cancer progression. Int J Mol Sci 2018;19:1272.
[20] Akino T, Hida K, Hida Y, Tsuchiya K, Freedman D, Muraki C, Ohga N, Matsuda K, Akiyama K, Harabayashi T, Shinohara N, Nonomura K, Klagsbrun M, Shindoh M. Cytogenetic abnormalities of tumor-associated endothelial cells in human malignant tumors. Am J Pathol 2009;175(6):2657–67. https://doi.org/10.2353/ajpath.2009.090202.
[21] Hida K, Akiyama K, Ohga N, Maishi N, Hida Y. Tumour endothelial cells acquire drug resistance in a tumour microenvironment. J Biochem 2013;153:243–9.
[22] Cao Z, Scandura JM, Inghirami GG, Shido K, Ding B-S, Rafii S. Molecular checkpoint decisions made by subverted vascular niche transform indolent tumor cells into chemoresistant cancer stem cells. Cancer Cell 2017;31:110–26.
[23] Zecchin A, Kalucka J, Dubois C, Carmeliet P. How endothelial cells adapt their metabolism to form vessels in tumors. Front Immunol 2017;8:1750.
[24] Cantelmo AR, Conradi LC, Brajic A, Goveia J, Kalucka J, Pircher A, Chaturvedi P, Hol J, Thienpont B, Teuwen LA, Schoors S, Boeckx B, Vriens J, Kuchnio A, Veys K, Cruys B, Finotto L, Treps L, Stav-Noraas TE, Bifari F, Stapor P, Decimo I, Kampen K, De Bock K, Haraldsen G, Schoonjans L, Rabelink T, Eelen G, Ghesquière B, Rehman J, Lambrechts D, Malik AB, Dewerchin M, Carmeliet P. Inhibition of the glycolytic activator PFKFB3 in endothelium induces tumor vessel normalization, impairs metastasis, and improves chemotherapy. Cancer Cell 2016;30:968–85. https://doi.org/10.1016/j.ccell.2016.10.006.
[25] Annan DA, Maishi N, Soga T, Dawood R, Li C, Kikuchi H, Hojo T, Morimoto M, Kitamura T, Alam MT, Minowa K, Shinohara N, Nam JM, Hida Y, Hida K. Carbonic anhydrase 2 (CAII) supports tumor blood endothelial cell survival under lactic acidosis in the tumor microenvironment. Cell Commun Signal 2019;17:169. https://doi.org/10.1186/s12964-019-0478-4.
[26] Butler JM, Kobayashi H, Rafii S. Instructive role of the vascular niche in promoting tumour growth and tissue repair by angiocrine factors. Nat Rev Cancer 2010;10:138–46.
[27] Pirtskhalaishvili G, Nelson JB. Endothelium-derived factors as paracrine mediators of prostate cancer progression. Prostate 2000;44:77–87.
[28] Maishi N, Ohba Y, Akiyama K, Ohga N, Hamada J, Nagao-Kitamoto H, Alam MT, Yamamoto K, Kawamoto T, Inoue N, Taketomi A, Shindoh M, Hida Y, Hida K. Tumour endothelial cells in high metastatic tumours promote metastasis via epigenetic dysregulation of biglycan. Sci Rep 2016;6:28039. https://doi.org/10.1038/srep28039.
[29] Motz GT, Santoro SP, Wang LP, Garrabrant T, Lastra RR, Hagemann IS, Lal P, Feldman MD, Benencia F, Coukos G. Tumor endothelium FasL establishes a selective immune barrier promoting tolerance in tumors. Nat Med 2014;20(6):607–15. https://doi.org/10.1038/nm.3541.
[30] Dirkx AE, Oude Egbrink MG, Kuijpers MJ, van der Niet ST, Heijnen VV, Bouma-ter Steege JC, Wagstaff J, Griffioen AW. Tumor angiogenesis modulates leukocyte-vessel wall interactions in vivo by reducing endothelial adhesion molecule expression. Cancer Res 2003;63(9):2322–9.
[31] Shrimali RK, Yu Z, Theoret MR, Chinnasamy D, Restifo NP, Rosenberg SA. Antiangiogenic agents can increase lymphocyte infiltration into tumor and enhance the effectiveness of adoptive immunotherapy of cancer. Cancer Res 2010;70(15):6171–80. https://doi.org/10.1158/0008-5472.CAN-10-0153.

[32] Millán J, Hewlett L, Glyn M, Toomre D, Clark P, Ridley AJ. Lymphocyte transcellular migration occurs through recruitment of endothelial ICAM-1 to caveola- and F-actin-rich domains. Nat Cell Biol 2006;8(2):113–23. https://doi.org/10.1038/ncb1356.

[33] Chouaib S, Kieda C, Benlalam H, Noman MZ, Mami-Chouaib F, Rüegg C. Endothelial cells as key determinants of the tumor microenvironment: interaction with tumor cells, extracellular matrix and immune killer cells. Crit Rev Immunol 2010;30(6):529–45. https://doi.org/10.1615/critrevimmunol.v30.i6.30.

[34] Fukumura D, Kloeppe J, Amoozgar Z, Duda DG, Jain RK. Enhancing cancer immunotherapy using antiangiogenics: opportunities and challenges. Nat Rev Clin Oncol 2018;15(5):325–40.

[35] Kojima Y, Acar A, Eaton EN, Mellody KT, Scheel C, Ben-Porath I, Onder TT, Wang ZC, Richardson AL, Weinberg RA, Orimo A. Autocrine TGFβ and stromal cell-derived factor-1 (SDF-1) signaling drives the evolution of tumor-promoting mammary stromal myofibroblasts. Proc Natl Acad Sci U S A 2010;107:20009–14.

[36] Watnick RS. The role of the tumor microenvironment in regulating angiogenesis. Cold Spring Harb Perspect Med 2012;2. https://doi.org/10.1101/cshperspect.a006676, a006676.

[37] Sahai E, Astsaturov I, Cukierman E, DeNardo DG, Egeblad M, Evans RM, Fearon D, Greten FR, Hingorani SR, Hunter T, Hynes RO, Jain RK, Janowitz T, Jorgensen C, Kimmelman AC, Kolonin MG, Maki RG, Powers RS, Puré E, Ramirez DC, Scherz-Shouval R, Sherman MH, Stewart S, Tlsty TD, Tuveson DA, Watt FM, Weaver V, Weeraratna AT, Werb Z. A framework for advancing our understanding of cancer-associated fibroblasts. Nat Rev Cancer 2020;20:174–86. https://doi.org/10.1038/s41568-019-0238-1.

[38] Chen X, Song E. Turning foes to friends: targeting cancer-associated fibroblasts. Nat Rev Drug Discov 2019;18:9–115. https://doi.org/10.1038/s41573-018-0004-1.

[39] Fukumura D, Xavier R, Sugiura T, Chen Y, Park EC, Lu N, Selig M, Nielsen G, Taksir T, Jain RK, Seed B. Tumor induction of VEGF promoter activity in stromal cells. Cell 1998;94:715–25. https://doi.org/10.1016/S0092-8674(00)81731-6.

[40] Tang D, Gao J, Wang S, Ye N, Chong Y, Huang Y, Wang J, Li B, Yin W, Wang D. Cancer-associated fibroblasts promote angiogenesis in gastric cancer through galectin-1 expression. Tumour Biol 2016;37(2):1889–99. https://doi.org/10.1007/s13277-015-3942-9.

[41] Unterleuthner D, Neuhold P, Schwarz K, Janker L, Neuditschko B, Nivarthi H, Crncec I, Kramer N, Unger C, Hengstschläger M, Eferl R, Moriggl R, Sommergruber W, Gerner C, Dolznig H. Cancer-associated fibroblast-derived WNT2 increases tumor angiogenesis in colon cancer. Angiogenesis 2020;23(2):159–77. https://doi.org/10.1007/s10456-019-09688-8.

[42] Aizawa T, Karasawa H, Funayama R, Shirota M, Suzuki T, Maeda S, Suzuki H, Yamamura A, Naitoh T, Nakayama K, Unno M. Cancer-associated fibroblasts secrete Wnt2 to promote cancer progression in colorectal cancer. Cancer Med 2019;8(14):6370–82. https://doi.org/10.1002/cam4.2523.

[43] Kramer N, Schmöllerl J, Unger C, Nivarthi H, Rudisch A, Unterleuthner D, Scherzer M, Riedl A, Artaker M, Crncec I, Lenhardt D, Schwarz T, Prieler B, Han X, Hengstschläger M, Schüler J, Eferl R, Moriggl R, Sommergruber W, Dolznig H. Autocrine WNT2 signaling in fibroblasts promotes colorectal cancer progression. Oncogene 2017;36(39):5460–72. https://doi.org/10.1038/onc.2017.144.

[44] Thijssen VLJL, Brandwijk RJMGE, Dings RPM, Griffioen AW. Angiogenesis gene expression profiling in xenograft models to study cellular interactions. Exp Cell Res 2004;299:286–93. https://doi.org/10.1016/j.yexcr.2004.06.014.

[45] Gilad AA, Israely T, Dafni H, Meir G, Cohen B, Neeman M. Functional and molecular mapping of uncoupling between vascular permeability and loss of vascular maturation in ovarian carcinoma xenografts: the role of stroma cells in tumor angiogenesis. Int J Cancer 2005;117:202–11. https://doi.org/10.1002/ijc.21179.

[46] Albini A, Bruno A, Noonan DM, Mortara L. Contribution to tumor angiogenesis from innate immune cells within the tumor microenvironment: implications for immunotherapy. Front Immunol 2018;9:527. https://doi.org/10.3389/fimmu.2018.00527.

[47] Petty AJ, Yang Y. Tumor-associated macrophages: implications in cancer immunotherapy. Immunotherapy 2017;9:289–302. https://doi.org/10.2217/imt-2016-0135.

[48] Schmid MC, Varner JA. Myeloid cells in the tumor microenvironment: modulation of tumor angiogenesis and tumor inflammation. J Oncol 2010;2010:1–10. https://doi.org/10.1155/2010/201026.

[49] Giraudo E, Inoue M, Hanahan D. An amino-bisphosphonate targets MMP-9-expressing macrophages and angiogenesis to impair cervical carcinogenesis. J Clin Invest 2004;114:623–33. https://doi.org/10.1172/JCI200422087.

[50] Hildenbrand R, Dilger I, Hörlin A, Stutte HJ. Urokinase and macrophages in tumour angiogenesis. Br J Cancer 1995;72:818–23. https://doi.org/10.1038/bjc.1995.419.
[51] Sunderkötter C, Steinbrink K, Goebeler M, Bhardwaj R, Sorg C. Macrophages and angiogenesis. J Leukoc Biol 1994;55:410–22. https://doi.org/10.1002/jlb.55.3.410.
[52] Leek RD, Harris AL. Tumor-associated macrophages in breast cancer. J Mammary Gland Biol Neoplasia 2002;7:177–89. https://doi.org/10.1023/A:1020304003704.
[53] Nishie A, Ono M, Shono T, Fukushi J, Otsubo M, Onoue H, Ito Y, Inamura T, Ikezaki K, Fukui M, Iwaki T, Kuwano M. Macrophage infiltration and heme oxygenase-1 expression correlate with angiogenesis in human gliomas. Clin Cancer Res 1999;5:1107–13.
[54] Sierra JR, Corso S, Caione L, Cepero V, Conrotto P, Cignetti A, Piacibello W, Kumanogoh A, Kikutani H, Comoglio PM, Tamagnone L, Giordano S. Tumor angiogenesis and progression are enhanced by Sema4D produced by tumor-associated macrophages. J Exp Med 2008;205:1673–85. https://doi.org/10.1084/jem.20072602.
[55] Veglia F, Gabrilovich DI. Dendritic cells in cancer: the role revisited. Curr Opin Immunol 2017;45:43–51. https://doi.org/10.1016/j.coi.2017.01.002.
[56] Gabrilovich D. Mechanisms and functional significance of tumour-induced dendritic-cell defects. Nat Rev Immunol 2004;4:941–52. https://doi.org/10.1038/nri1498.
[57] Fainaru O, Almog N, Yung CW, Nakai K, Montoya-Zavala M, Abdollahi A, D'Amato R, Ingber DE. Tumor growth and angiogenesis are dependent on the presence of immature dendritic cells. FASEB J 2010;24: 1411–8. https://doi.org/10.1096/fj.09-147025.
[58] Kessler DA, Langer RS, Pless NA, Folkman J. Mast cells and tumor angiogenesis. Int J Cancer 1976;18: 703–9. https://doi.org/10.1002/ijc.2910180520.
[59] Wroblewski M, Bauer R, Cubas Córdova M, Udonta F, Ben-Batalla I, Legler K, Hauser C, Egberts J, Janning M, Velthaus J, Schulze C, Pantel K, Bokemeyer C, Loges S. Mast cells decrease efficacy of anti-angiogenic therapy by secreting matrix-degrading granzyme B. Nat Commun 2017;8:269. https://doi.org/10.1038/s41467-017-00327-8.
[60] Cimpean AM, Tamma R, Ruggieri S, Nico B, Toma A, Ribatti D. Mast cells in breast cancer angiogenesis. Crit Rev Oncol Hematol 2017;115:23–6. https://doi.org/10.1016/j.critrevonc.2017.04.009.
[61] Battinelli EM, Markens BA, Italiano JE. Release of angiogenesis regulatory proteins from platelet alpha granules: modulation of physiologic and pathologic angiogenesis. Blood 2011;118:1359–69. https://doi.org/10.1182/blood-2011-02-334524.
[62] Italiano JE, Richardson JL, Patel-Hett S, Battinelli E, Zaslavsky A, Short S, Ryeom S, Folkman J, Klement GL. Angiogenesis is regulated by a novel mechanism: pro- and antiangiogenic proteins are organized into separate platelet alpha granules and differentially released. Blood 2008;111:1227–33. https://doi.org/10.1182/blood-2007-09-113837.
[63] Gaengel K, Genové G, Armulik A, Betsholtz C. Endothelial-mural cell signaling in vascular development and angiogenesis. Arterioscler Thromb Vasc Biol 2009;29:630–8. https://doi.org/10.1161/ATVBAHA.107.161521.
[64] Morikawa S, Baluk P, Kaidoh T, Haskell A, Jain RK, McDonald DM. Abnormalities in pericytes on blood vessels and endothelial sprouts in tumors. Am J Pathol 2002;160:985–1000. https://doi.org/10.1016/S0002-9440(10)64920-6.
[65] Abramsson A, Lindblom P, Betsholtz C. Endothelial and nonendothelial sources of PDGF-B regulate pericyte recruitment and influence vascular pattern formation in tumors. J Clin Invest 2003;112:1142–51. https://doi.org/10.1172/JCI200318549.
[66] Song S, Ewald AJ, Stallcup W, Werb Z, Bergers G. PDGFRbeta+ perivascular progenitor cells in tumours regulate pericyte differentiation and vascular survival. Nat Cell Biol 2005;7:870–9. https://doi.org/10.1038/ncb1288.
[67] Iliopoulos O, Levy AP, Jiang C, Kaelin WG, Goldberg MA. Negative regulation of hypoxia- inducible genes by the von Hippel-Lindau protein. Proc Natl Acad Sci U S A 1996;93:10595–9. https://doi.org/10.1073/pnas.93.20.10595.
[68] Ivan M, Kondo K, Yang H, Kim W, Valiando J, Ohh M, Salic A, Asara JM, Lane WS, Kaelin Jr WG. HIFα targeted for VHL-mediated destruction by proline hydroxylation: implications for O_2 sensing. Science 2001;292: 464–8. https://doi.org/10.1126/science.1059817.
[69] Wenger RH. Cellular adaptation to hypoxia: O_2-sensing protein hydroxylases, hypoxia-inducible transcription factors, and O_2-regulated gene expression. FASEB J 2002;16(10):1151–62.

[70] Eales KL, Hollinshead KER, Tennant DA. Hypoxia and metabolic adaptation of cancer cells. Oncogenesis 2016;5(1):e190.
[71] Bergers G, Brekken R, McMahon G, Vu TH, Itoh T, Tamaki K, Tanzawa K, Thorpe P, Itohara S, Werb Z, Hanahan D. Matrix metalloproteinase-9 triggers the angiogenic switch during carcinogenesis. Nat Cell Biol 2000;2:737. https://doi.org/10.1038/35036374.
[72] Hu CJ, Wang LY, Chodosh LA, Keith B, Simon MC. Differential roles of hypoxia-inducible factor 1α (HIF-1α) and HIF-2α in hypoxic gene regulation. Mol Cell Biol 2003;23:9361–74. https://doi.org/10.1128/MCB.23.24.9361-9374.2003.
[73] Liao D, Johnson RS. Hypoxia: a key regulator of angiogenesis in cancer. Cancer Metastasis Rev 2007;26:281–90. https://doi.org/10.1007/s10555-007-9066-y.
[74] Warburg O, Wind F, Negelein E. The metabolism of tumors in the body. J Gen Physiol 1927;8:519–30. https://doi.org/10.1085/jgp.8.6.519.
[75] Payen VL, Porporato PE, Baselet B, Sonveaux P. Metabolic changes associated with tumor metastasis, part 1: tumor pH, glycolysis and the pentose phosphate pathway. Cell Mol Life Sci 2016;73:1333–48. https://doi.org/10.1007/s00018-015-2098-5.
[76] Peppicelli S, Bianchini F, Calorini L. Extracellular acidity, a "reappreciated" trait of tumor environment driving malignancy: perspectives in diagnosis and therapy. Cancer Metastasis Rev 2014;33:823–32. https://doi.org/10.1007/s10555- 014-9506-4.
[77] Suzuki A, Maeda T, Baba Y, Shimamura K, Kato Y. Acidic extracellular pH promotes epithelial mesenchymal transition in Lewis lung carcinoma model. Cancer Cell Int 2014;14:129. https://doi.org/10.1186/s12935-014-0129-1.
[78] Walenta S, Mueller-Klieser WF. Lactate: mirror and motor of tumor malignancy. Semin Radiat Oncol 2004;14:267–74. https://doi.org/10.1016/j.semradonc. 2004.04.004.
[79] Shi Q, Le X, Wang B, Abbruzzese JL, Xiong Q, He Y, Xie K. Regulation of vascular endothelial growth factor expression by acidosis in human cancer cells. Oncogene 2001;20:3751–6. https://doi.org/10.1038/sj.onc.1204500.
[80] Pouyssegur J, López-Barneo J. Hypoxia in health and disease. Mol Aspects Med 2016;47:1–2.
[81] Ferrara N. VEGF and the quest for tumour angiogenesis factors. Nat Rev Cancer 2002;2:795. https://doi.org/10.1038/nrc909.
[82] Gutierrez-Miranda L, Yaniv K. Cellular origins of the lymphatic endothelium: implications for cancer lymphangiogenesis. Front Physiol 2020;11:577584. https://doi.org/10.3389/fphys.2020.577584.
[83] Folkman J. History of angiogenesis. In: Figg WD, Folkman J, editors. Angiogenesis. Boston: Springer; 2008. p. 1–14.
[84] Risau W. Mechanisms of angiogenesis. Nature 1997;386:671–4.
[85] Hanahan D, Folkman J. Patterns and emerging mechanisms of the angiogenic switch during tumorigenesis. Cell 1996;86(3):353–64.
[86] Folkman J. Tumor angiogenesis: therapeutic implications. N Engl J Med 1971;285(21):1182–6.
[87] Folkman J, Merler E, Abernathy C, Williams G. Isolation of a tumor factor responsible for angiogenesis. J Exp Med 1971;133(2):275–88.
[88] Al-Abd AM, Alamoudi AJ, Abdel-Naim AB, Neamatallah TA, Ashour OM. Anti-angiogenic agents for the treatment of solid tumors: potential pathways, therapy and current strategies – a review. J Adv Res 2017; 8(6):591–605.
[89] Hashizume H, Baluk P, Morikawa S, McLean JW, Thurston G, Roberge S, Jain RK, McDonald DM. Openings between defective endothelial cells explain tumor vessel leakiness. Am J Pathol 2000;156(4):1363–80.
[90] Abramsson A, Berlin O, Papayan H, Paulin D, Shani M, Betsholtz C. Analysis of mural cell recruitment to tumor vessels. Circulation 2002;105(1):112–7.
[91] Morikawa S, Baluk P, Kaidoh T, Haskell A, Jain RK, McDonald DM. Abnormalities in pericytes on blood vessels and endothelial sprouts in tumors. Am J Pathol 2002;160(3):985–1000.
[92] Baluk P, Morikawa S, Haskell A, Mancuso M, McDonald DM. Abnormalities of basement membrane on blood vessels and endothelial sprouts in tumors. Am J Pathol 2003;163(5):1801–15.
[93] Blouw B, Song H, Tihan T, Bosze J, Ferrara N, Gerber HP, Johnson RS, Bergers G. The hypoxic response of tumors is dependent on their microenvironment. Cancer Cell 2003;4:133–46. https://doi.org/10.1016/S1535-6108(03)00194-6.

[94] Mohme M, Riethdorf S, Pantel K. Circulating and disseminated tumour cells – mechanisms of immune surveillance and escape. Nat Rev Clin Oncol 2017;14:155–67. https://doi.org/10.1038/nrclinonc.2016.144.
[95] Mimeault M, Batra SK. Hypoxia-inducing factors as master regulators of stemness properties and altered metabolism of cancer- and metastasis-initiating cells. J Cell Mol Med 2013;17:30–54. https://doi.org/10.1111/jcmm.12004.
[96] Padua D, Zhang XH-F, Wang Q, Nadal C, Gerald WL, Gomis RR, Massagué J. TGFbeta primes breast tumors for lung metastasis seeding through angiopoietin-like 4. Cell 2008;133:66–77. https://doi.org/10.1016/j.cell.2008.01.046.
[97] Ciesielski O, Biesiekierska M, Panthu B, Vialichka V, Pirola L, Balcerczyk A. The epigenetic profile of tumor endothelial cells. Effects of combined therapy with antiangiogenic and epigenetic drugs on cancer progression. Int J Mol Sci 2020;21(7):2606. https://doi.org/10.3390/ijms21072606.
[98] Lopes-Bastos BM, Jiang WG, Cai J. Tumour-endothelial cell communications: important and indispensable mediators of tumour angiogenesis. Anticancer Res 2016;36(3):1119–26.
[99] Annese T, Tamma R, De Giorgis M, Ribatti D. microRNAs biogenesis, functions and role in tumor angiogenesis. Front Oncol 2021;10:581007. https://doi.org/10.3389/fonc.2020.581007.
[100] Bao L, You B, Shi S, Shan Y, Zhang Q, Yue H, Zhang J, Zhang W, Shi Y, Liu Y, Wang X, Liu D, You Y. Metastasis-associated miR-23a from nasopharyngeal carcinoma-derived exosomes mediates angiogenesis by repressing a novel target gene TSGA10. Oncogene 2018;37(21):2873–89. https://doi.org/10.1038/s41388-018-0183-6.
[101] Mao Y, Wang Y, Dong L, Zhang Y, Zhang Y, Wang C, Zhang Q, Yang S, Cao L, Zhang X, Li X, Fu Z. Hypoxic exosomes facilitate angiogenesis and metastasis in esophageal squamous cell carcinoma through altering the phenotype and transcriptome of endothelial cells. J Exp Clin Cancer Res 2019;38(1):389. https://doi.org/10.1186/s13046-019-1384-8.
[102] Cao Q, Liu Y, Wu Y, Hu C, Sun L, Wang J, Li C, Guo M, Liu X, Lv J, Huo X, Yue J, Du X, Chen Z. Profilin 2 promotes growth, metastasis, and angiogenesis of small cell lung cancer through cancer-derived exosomes. Aging (Albany NY) 2020;12(24):25981–99. https://doi.org/10.18632/aging.202213.
[103] Folkman J. Angiogenesis: an organizing principle for drug discovery? Nat Rev Drug Discov 2007;6:273–86. https://doi.org/10.1038/nrd2115.
[104] Naumov GN, Akslen LA, Folkman J. Role of angiogenesis in human tumor dormancy: animal models of the angiogenic switch. Cell Cycle 2006;5(16):1779–87.
[105] Good DJ, Polverini PJ, Rastinejad F, Le Beau MM, Lemons RS, Frazier WA, Bouck NP. A tumor suppressor-dependent inhibitor of angiogenesis is immunologically and functionally indistinguishable from a fragment of thrombospondin. Proc Natl Acad Sci U S A 1990;87(17):6624–8.
[106] O'Reilly MS, Holmgren L, Shing Y, Chen C, Rosenthal RA, Moses M, Lane WS, Cao Y, Sage EH, Folkman J. Angiostatin: a novel angiogenesis inhibitor that mediates the suppression of metastases by a Lewis lung carcinoma. Cell 1994;79(2):315–28.
[107] Jain RK. Molecular regulation of vessel maturation. Nat Med 2003;9:685–93. https://doi.org/10.1038/nm0603-685.
[108] Shinojima T, Oya M, Takayanagi A, Mizuno R, Shimizu N, Murai M. Renal cancer cells lacking hypoxia inducible factor (HIF)–1alpha expression maintain vascular endothelial growth factor expression through HIF–2alpha. Carcinogenesis 2007;28:529–36.
[109] Filippelli A, Ciccone V, Donnini S, Ziche M, Morbidelli L. Molecular mechanisms of resistance to anti-angiogenic drugs. Crit Rev Oncog 2021;26(2):39–66. https://doi.org/10.1615/CritRevOncog.2020035422.
[110] Keyt BA, Berleau LT, Nguyen HV, Chen H, Heinsohn H, Vandlen R, Ferrara N. The carboxyl-terminal domain (111165) of vascular endothelial growth factor is critical for its mitogenic potency. J Biol Chem 1996;271: 7788–95. https://doi.org/10.1074/jbc.271.13.7788.
[111] Milanini J, Vinals F, Pouyssegur J, Pages G. p42/p44 MAP kinase module plays a key role in the transcriptional regulation of the vascular endothelial growth factor gene in fibroblasts. J Biol Chem 1998;273:18165–72.
[112] Rak J, Mitsuhashi Y, Sheehan C, Tamir A, Viloria-Petit A, Filmus J, Mansour SJ, Ahn NG, Kerbel RS. Oncogenes and tumor angiogenesis: differential modes of vascular endothelial growth factor up-regulation in ras-transformed epithelial cells and fibroblasts. Cancer Res 2000;60(2):490–8.
[113] Karar J, Maity A. PI3K/AKT/mTOR pathway in angiogenesis. Front Mol Neurosci 2011;4:51.
[114] Kaelin Jr WG. The von Hippel-Lindau tumor suppressor protein and clear cell renal carcinoma. Clin Cancer Res 2007;13:680s–4s.
[115] Sullivan LA, Brekken RA. The VEGF family in cancer and antibody-based strategies for their inhibition. MAbs 2010;2(2):165–75.

[116] Ziche M, Maglione D, Ribatti D, Morbidelli L, Lago CT, Battisti M, Paoletti I, Barra A, Tucci M, Parise G, Vincenti V, Granger HJ, Viglietto G, Persico MG. Placenta growth factor-1 is chemotactic, mitogenic, and angiogenic. Lab Invest 1997;76(4):517–31.

[117] Wang J, Huang Y, Zhang J, Wei Y, Mahoud S, Bakheet AM, Wang L, Zhou S, Tang J. Pathway-related molecules of VEGFC/D-VEGFR3/NRP2 axis in tumor lymphangiogenesis and lymphatic metastasis. Clin Chim Acta 2016;461:165–71. https://doi.org/10.1016/j.cca.2016.08.008.

[118] Ferrara N, Houck K, Jakeman LYN, Leung DW. Molecular and biological properties of the vascular endothelial growth factor family of proteins. Endocr Rev 1992;13:18–32. https://doi.org/10.1210/edrv-13-1-18.

[119] Sawano A, Iwai S, Sakurai Y, Ito M, Shitara K, Nakahata T, Shibuya M. Flt–1, vascular endothelial receptor 1, is a novel cell surface marker for the lineage of monocyte-macrophages in humans. Blood 2001;97:785–91.

[120] Lyden D, Hattori K, Dias S, Costa C, Blaikie P, Butros L, Chadburn A, Heissig B, Marks W, Witte L, Wu Y, Hicklin D, Zhu Z, Hackett NR, Crystal RG, Moore MA, Hajjar KA, Manova K, Benezra R, Rafii S. Impaired recruitment of bone–marrow–derived endothelial and hematopoietic precursor cells blocks tumor angiogenesis and growth. Nat Med 2001;7:1194–201.

[121] Takahashi H, Shibuya M. The vascular endothelial growth factor (VEGF)/VEGF receptor system and its role under physiological and pathological conditions. Clin Sci 2005;109(3):227–41.

[122] Stevens M, Oltean S. Modulation of receptor tyrosine kinase activity through alternative splicing of ligands and receptors in the VEGF–A/VEGFR axis. Cell 2019;8:288.

[123] Niland S, Eble JA. Neuropilins in the context of tumor vasculature. Int J Mol Sci 2019;20(3):639. https://doi.org/10.3390/ijms20030639.

[124] Schlessinger J. Cell signaling by receptor tyrosine kinases. Cell 2000;103(2):211–25.

[125] Takahashi T, Yamaguchi S, Chida K, Shibuya M. A single autophosphorylation site on KDR/Flk–1 is essential for VEGF–A–dependent activation of PLC–gamma and DNA synthesis in vascular endothelial cells. EMBO J 2001;20:2768–78.

[126] Ziche M, Morbidelli L, Choudhuri R, Zhang HT, Donnini S, Granger HJ, Bicknell R. Nitric oxide synthase lies downstream from vascular endothelial growth factor–induced but not basic fibroblast growth factor induced angiogenesis. J Clin Invest 1997;99:2625–34.

[127] Morbidelli L, Donnini S, Ziche M. Role of nitric oxide in the modulation of angiogenesis. Curr Pharm Des 2003;9 (7):521–30.

[128] Ravichandran KS, Lorenz U, Shoelson SE, Burakoff SJ. Interaction of Shc with Grb2 regulates association of Grb2 with mSOS. Mol Cell Biol 1995;15:593–600.

[129] Gerber HP, McMurtrey A, Kowalski J, Yan M, Keyt BA, Dixit V, Ferrara N. Vascular endothelial growth factor regulates endothelial cell survival through the phosphatidylinositol 3′- kinase/Akt signal transduction pathway requirement for Flk-1/KDR activation. J Biol Chem 1998;273:30336–43. https://doi.org/10.1074/jbc.273.46.30336.

[130] Pedram A, Razandi M, Levin ER. Extracellular signal-regulated protein kinase/Jun kinase cross-talk underlies vascular endothelial cell growth factor-induced endothelial cell proliferation. J Biol Chem 1998;273:26722–8. https://doi.org/10.1074/jbc.273.41.26722.

[131] Wu LW, Mayo LD, Dunbar JD, Kessler KM, Baerwald MR, Jaffe EA, Wang D, Warren RS, Donner DB. Utilization of distinct signaling pathways by receptors for vascular endothelial cell growth factor and other mitogens in the induction of endothelial cell proliferation. J Biol Chem 2000;275:5096–103. https://doi.org/10.1074/jbc.275.7.5096.

[132] Zachary I, Gliki G. Signaling transduction mechanisms mediating biological actions of the vascular endothelial growth factor family. Cardiovasc Res 2001;49:568–81. https://doi.org/10.1016/S0008-6363(00)00268-6.

[133] Shibuya M. Vascular endothelial growth factor-dependent and -independent regulation of angiogenesis. BMB Rep 2008;41:278–86.

[134] Dimmeler S, Dernbach E, Zeiher AM. Phosphorylation of the endothelial nitric oxide synthase at ser-1177 is required for VEGF-induced endothelial cell migration. FEBS Lett 2000;477(3):258–62.

[135] Gavard J, Gutkind JS. VEGF controls endothelial-cell permeability by promoting the β-arrestin-dependent endocytosis of VE-cadherin. Nat Cell Biol 2006;8:1223–34.

[136] Chouaib S, Kieda C, Benlalam H, Noman MZ, Mami-Chouaib F, Rüegg C. Endothelial cells as key determinants of the tumor microenvironment: interaction with tumor cells, extracellular matrix and immune killer cells. Crit Rev Immunol 2010;30(6):529–45. https://doi.org/10.1615/critrevimmunol.v30.i6.30.

[137] Fukumura D, Kloeppe J, Amoozgar Z, Duda DG, Jain RK. Enhancing cancer immunotherapy using antiangiogenics: opportunities and challenges. Nat Rev Clin Oncol 2018;15(5):325–40.

[138] Sun S, Wu HJ, Guan JL. Nuclear FAK and its kinase activity regulate VEGFR2 transcription in angiogenesis of adult mice. Sci Rep 2018;8:2550. https://doi.org/10.1038/s41598-018-20930-z.
[139] Simons M, Gordon E, Claesson-Welsh L. Mechanisms and regulation of endothelial VEGF receptor signalling. Nat Rev Mol Cell Biol 2016;17:611–25. https://doi.org/10.1038/nrm.2016.87.
[140] Apte RS, Chen DS, Ferrara N. VEGF in signaling and disease: beyond discovery and development. Cell 2019;176:1248–64. https://doi.org/10.1016/j.cell.2019.01.021.
[141] Hartsough EJ, Meyer RD, Chitalia V, Jiang Y, Marquez VE, Zhdanova IV, Weinberg J, Costello CE, Rahimi N. Lysine methylation promotes VEGFR-2 activation and angiogenesis. Sci Signal 2013;6. https://doi.org/10.1126/scisignal.2004289, ra104.
[142] Basagiannis D, Zografou S, Murphy C, Fotsis T, Morbidelli L, Ziche M, Bleck C, Mercer J, Christoforidis S. VEGF induces signalling and angiogenesis by directing VEGFR2 internalisation through macropinocytosis. J Cell Sci 2016;129:4091–104. https://doi.org/10.1242/jcs.188219.
[143] Yu J, Ustach C, Kim HRC. Platelet-derived growth factor signaling and human cancer. J Biochem Mol Biol 2003;36:49–59.
[144] Bran B, Bran G, Hörmann K, Riedel F. The platelet–derived growth factor receptor as a target for vascular endothelial growth factor–mediated anti–angiogenetic therapy in head and neck cancer. Int J Oncol 2009;34:255–61.
[145] Zhao Y, Adjei AA. Targeting angiogenesis in cancer therapy: moving beyond vascular endothelial growth factor. Oncologist 2015;20:660–73.
[146] Edelberg JM, Aird WC, Wu W, Rayburn H, Mamuya WS, Mercola M, Rosenberg RD. PDGF mediates cardiac microvascular communication. J Clin Invest 1998;102:837–43. https://doi.org/10.1172/JCI3058.
[147] Liegl B, Kepten I, Le C, Zhu M, Demetri GD, Heinrich MC, Fletcher CD, Corless CL, Fletcher JA. Heterogeneity of kinase inhibitor resistance mechanisms in GIST. J Pathol 2008;216(1):64–74.
[148] Crawford Y, Kasman I, Yu L, Zhong C, Wu X, Modrusan Z, Kaminker J, Ferrara N. PDGF-C mediates the angiogenic and tumorigenic properties of fibroblasts associated with tumors refractory to anti-VEGF treatment. Cancer Cell 2009;15:21–34.
[149] Gospodarowicz D, Bialecki H, Thakral TK. The angiogenic activity of the fibroblast and epidermal growth factor. Exp Eye Res 1979;28(5):501–14.
[150] Ronca R, Giacomini A, Rusnati M, Presta M. The potential of fibroblast growth factor/fibroblast growth factor receptor signaling as a therapeutic target in tumor angiogenesis. Expert Opin Ther Targets 2015;19(10):1361–77. https://doi.org/10.1517/14728222.2015.106247.
[151] Corti F, Wang Y, Rhodes JM, Atri D, Archer-Hartmann S, Zhang J, Zhuang ZW, Chen D, Wang T, Wang Z, Azadi P, Simons M. N-terminal syndecan-2 domain selectively enhances 6-O heparan sulfate chains sulfation and promotes VEGFA(165)-dependent neovascularization. Nat Commun 2019;10(1):1562. https://doi.org/10.1038/s41467-019-09605-z.
[152] Gyanchandani R, Ortega Alves MV, Myers JN, Kim S. A proangiogenic signature is revealed in FGF-mediated bevacizumab-resistant head and neck squamous cell carcinoma. Mol Cancer Res 2013;11(12):1585–96. https://doi.org/10.1158/1541-7786.MCR-13-0358.
[153] Bussolino F, Di Renzo MF, Ziche M, Bocchietto E, Olivero M, Naldini L, Gaudino G, Tamagnone L, Coffer A, Comoglio PM. Hepatocyte growth factor is a potent angiogenic factor which stimulates endothelial cell motility and growth. J Cell Biol 1992;119(3):629–41.
[154] Zhang YW, Su Y, Volpert OV, Woude GFV. Hepatocyte growth factor/scatter factor mediates angiogenesis through positive VEGF and negative thrombospondin 1 regulation. Proc Natl Acad Sci U S A 2003;100(22):12718–23.
[155] Gibney GT, Aziz SA, Camp RL, Conrad P, Schwartz BE, Chen CR, Kelly WK, Kluger HM. c–Met is a prognostic marker and potential therapeutic target in clear cell renal cell carcinoma. Ann Oncol 2013;24(2):343–9.
[156] Singla M, Bhattacharyya S. Autophagy as a potential therapeutic target during epithelial to mesenchymal transition in renal cell carcinoma: an in vitro study. Biomed Pharmacother 2017;94:332–40.
[157] Carvalho B, Lopes JM, Silva R, Peixoto J, Leitão D, Soares P, Fernandes AC, Linhares P, Vaz R, Lima J. The role of c-Met and VEGFR2 in glioblastoma resistance to bevacizumab. Sci Rep 2021;11(1):6067. https://doi.org/10.1038/s41598-021-85385-1.
[158] Frejborg E, Salo T, Salem A. Role of Cyclooxygenase-2 in head and neck tumorigenesis. Int J Mol Sci 2020;21(23):9246. https://doi.org/10.3390/ijms21239246.

[159] Finetti F, Donnini S, Giachetti A, Morbidelli L, Ziche M. Prostaglandin E(2) primes the angiogenic switch via a synergic interaction with the fibroblast growth factor–2 pathway. Circ Res 2009;105(7):657–66.
[160] Finetti F, Solito R, Morbidelli L, Giachetti A, Ziche M, Donnini S. Prostaglandin E2 regulates angiogenesis via activation of fibroblast growth factor receptor–1. J Biol Chem 2008;283:2139–46.
[161] Finetti F, Travelli C, Ercoli J, Colombo G, Buoso E, Trabalzini L. Prostaglandin E2 and Cancer: insight into tumor progression and immunity. Biology (Basel) 2020;9(12):434. https://doi.org/10.3390/biology9120434.
[162] Koch AE, Polverini PJ, Kunkel SL, Harlow LA, DiPietro LA, Elner VM, Elner SG, Strieter RM. Interleukin-8 as a macrophage-derived mediator of angiogenesis. Science 1992;258(5089):1798–801.
[163] Chen JJ, Yao PL, Yuan A, Hong TM, Shun CT, Kuo ML, Lee YC, Yang PC. Up-regulation of tumor interleukin-8 expression by infiltrating macrophages: its correlation with tumor angiogenesis and patient survival in non-small cell lung cancer. Clin Cancer Res 2003;9(2):729–37.
[164] Medina RJ, O'Neill CL, O'Doherty TM, Knott H, Guduric-Fuchs J, Gardiner TA, Stitt AW. Myeloid angiogenic cells act as alternative M2 macrophages and modulate angiogenesis through interleukin-8. Mol Med 2011;17 (9):1045–55.
[165] Kitadai Y, Haruma K, Sumii K, Yamamoto S, Ue T, Yokozaki H, Yasui W, Ohmoto I, Kajiyama G, Fidler IJ, Tahara E. Expression of interleukin-8 correlates with vascularity in human gastric carcinomas. Am J Pathol 1998;152(1):93.
[166] Fousek K, Horn LA, Palena C. Interleukin-8: a chemokine at the intersection of cancer plasticity, angiogenesis, and immune suppression. Pharmacol Ther 2021;219:107692. https://doi.org/10.1016/j.pharmthera.2020.107692.
[167] Deshane J, Chen S, Caballero S, Grochot-Przeczek A, Was H, Li Calzi S, Lach R, Hock TD, Chen B, Hill-Kapturczak N, Siegal GP, Dulak J, Jozkowicz A, Grant MB, Agarwal A. Stromal cell-derived factor 1 promotes angiogenesis via a heme oxygenase 1-dependent mechanism. J Exp Med 2007;204(3):605–18. https://doi.org/10.1084/jem.20061609.
[168] Laiva AL, Raftery RM, Keogh MB, O'brien FJ. Pro-angiogenic impact of SDF-1α gene-activated collagen-based scaffolds in stem cell driven angiogenesis. Int J Pharm 2018;544(2):372–9.
[169] Salazar N, Zabel BA. Support of tumor endothelial cells by chemokine receptors. Front Immunol 2019;10:147. https://doi.org/10.3389/fimmu.2019.00147.
[170] Petit I, Jin D, Rafii S. The SDF–1–CXCR4 signaling pathway: a molecular hub modulating neo–angiogenesis. Trends Immunol 2007;28(7):299–307.
[171] Jin F, Brockmeier U, Otterbach F, Metzen E. New insight into the SDF-1/ CXCR4 axis in a breast carcinoma model: hypoxia-induced endothelial SDF-1 and tumor cell CXCR4 are required for tumor cell intravasation. Mol Cancer Res 2012;10:1021–31. https://doi.org/10.1158/1541-7786.MCR-11-0498.
[172] Maishi N, Ohga N, Hida Y, Akiyama K, Kitayama K, Osawa T, Onodera Y, Shinohara N, Nonomura K, Shindoh M, Hida K. CXCR7: a novel tumor endothelial marker in renal cell carcinoma. Pathol Int 2012;62(5):309–17. https://doi.org/10.1111/j.1440-1827.2012.02792.x.
[173] Berahovich RD, Zabel BA, Lewén S, Walters MJ, Ebsworth K, Wang Y, Jaen JC, Schall TJ. Endothelial expression of CXCR7 and the regulation of systemic CXCL12 levels. Immunology 2013;141:111–22.
[174] Stacer AC, Fenner J, Cavnar SP, Xiao A, Zhao S, Chang SL, Salomonnson A, Luker KE, Luker GD. Endothelial CXCR7 regulates breast cancer metastasis. Oncogene 2015;35:1716–24.
[175] Lobov IB, Brooks PC, Lang RA. Angiopoietin-2 displays VEGF-dependent modulation of capillary structure and endothelial cell survival in vivo. Proc Natl Acad Sci U S A 2002;99:11205–10. https://doi.org/10.1073/pnas.172161899.
[176] Hughes DP, Marron MB, Brindle NP. The antiinflammatory endothelial tyrosine kinase Tie2 interacts with a novel nuclear factor-κB inhibitor ABIN-2. Circ Res 2003;92:630–6. https://doi.org/10.1161/01.RES.0000063422.38690.DC.
[177] Holash J, Maisonpierre PC, Compton D, Boland P, Alexander CR, Zagzag D, Yancopoulos GD, Wiegand SJ. Vessel cooption, regression, and growth in tumors mediated by angiopoietins and VEGF. Science 1999;284:1994–8. https://doi.org/10.1126/science.284.5422.1994.
[178] del Toro R, Prahst C, Mathivet T, Siegfried G, Kaminker JS, Larrivee B, Breant C, Duarte A, Takakura N, Fukamizu A, Penninger J, Eichmann A. Identification and functional analysis of endothelial tip cell-enriched genes. Blood 2010;116:4025–33. https://doi.org/10.1182/blood-2010-02-270819.
[179] Reiss Y, Droste J, Heil M, Tribulova S, Schmidt MH, Schaper W, Dumont DJ, Plate KH. Angiopoietin-2 impairs revascularization after limb ischemia. Circ Res 2007;101:88–96. https://doi.org/10.1161/CIRCRESAHA.106.143594.

[180] Hammes HP, Lin J, Wagner P, Feng Y, Vom Hagen F, Krzizok T, Renner O, Breier G, Brownlee M, Deutsch U. Angiopoietin-2 causes pericyte dropout in the normal retina: evidence for involvement in diabetic retinopathy. Diabetes 2004;53:1104–10. https://doi.org/10.2337/diabetes.53.4.1104.

[181] Eklund L, Kangas J, Saharinen P. Angiopoietin–tie signalling in the cardiovascular and lymphatic systems. Clin Sci 2017;131(1):87–103.

[182] Itatani Y, Kawada K, Yamamoto T, Sakai Y. Resistance to anti–angiogenic therapy in cancer-alterations to anti-VEGF pathway. Int J Mol Sci 2018;19:1232.

[183] Coffelt SB, Tal AO, Scholz A, De Palma M, Patel S, Urbich C, Biswas SK, Murdoch C, Plate KH, Reiss Y, Lewis CE. Angiopoietin-2 regulates gene expression in TIE2-expressing monocytes and augments their inherent proangiogenic functions. Cancer Res 2010;70:5270–80. https://doi.org/10.1158/0008-5472.CAN-10-0012.

[184] Kzhyshkowska J, Riabov V, Gudima A, Wang N, Orekhov A, Mickley A. Role of tumor associated macrophages in tumor angiogenesis and lymphangiogenesis. Front Physiol 2014;5:75. https://doi.org/10.3389/fphys.2014.00075.

[185] Horinouchi T, Terada K, Higashi T, Miwa S. Endothelin receptor signaling: new insight into its regulatory mechanisms. J Pharmacol Sci 2013;123(2):85–101. https://doi.org/10.1254/jphs.13r02cr.

[186] Morbidelli L, Orlando C, Maggi CA, Ledda F, Ziche M. Proliferation and migration of endothelial cells is promoted by endothelins via activation of ETB receptors. Am J Physiol 1995;269(2 Pt 2):H686–95. https://doi.org/10.1152/ajpheart.1995.269.2.H686.

[187] Kandalaft LE, Facciabene A, Buckanovich RJ, Coukos G. Endothelin B receptor, a new target in cancer immune therapy. Clin Cancer Res 2009;15(14):4521–8.

[188] Kandalaft LE, Motz GT, Duraiswamy J, Coukos G. Tumor immune surveillance and ovarian cancer: lessons on immune mediated tumor rejection or tolerance. Cancer Met Rev 2011;30(1):141–51. https://doi.org/10.1007/s10555-011-9289-9.

[189] Weis SM, Cheresh DA. αV integrins in angiogenesis and cancer. Cold Spring Harb Perspect Med 2011;1(1):a006478. https://doi.org/10.1101/cshperspect.a006478.

[190] Carlson TR, Feng Y, Maisonpierre PC, Mrksich M, Morla AO. Direct cell adhesion to the angiopoietins mediated by integrins. J Biol Chem 2001;276(28):26516–25.

[191] Senger DR, Perruzzi CA, Streit M, Koteliansky VE, de Fougerolles AR, Detmar M. The α1β1 and α2β1 integrins provide critical support for vascular endothelial growth factor signaling, endothelial cell migration, and tumor angiogenesis. Am J Pathol 2002;160(1):195–204.

[192] Jin H, Su J, Garmy-Susini B, Kleeman J, Varner J. Integrin α4β1 promotes monocyte trafficking and angiogenesis in tumors. Cancer Res 2006;66(4):2146–52.

[193] Debreli Coskun M, Sudha T, Bharali DJ, Celikler S, Davis PJ, Mousa SA. αvβ3 integrin antagonists enhance chemotherapy response in an orthotopic pancreatic cancer model. Front Pharmacol 2020;11:95. https://doi.org/10.3389/fphar.2020.00095.

[194] Niland S, Eble JA. Neuropilins in the context of tumor vasculature. Int J Mol Sci 2019;20(3):639. https://doi.org/10.3390/ijms20030639.

[195] Hansel DE, Wilentz RE, Yeo CJ, Schulick RD, Montgomery E, Maitra A. Expression of neuropilin-1 in high-grade dysplasia, invasive cancer, and metastases of the human gastrointestinal tract. Am J Surg Pathol 2004;28(3):347–56.

[196] Hong TM, Chen YL, Wu YY, Yuan A, Chao YC, Chung YC, Wu MH, Yang SC, Pan SH, Shih JY, Chan WK, Yang PC. Targeting neuropilin 1 as an antitumor strategy in lung cancer. Clin Cancer Res 2007;13(16):4759–68. https://doi.org/10.1158/1078-0432.CCR-07-0001.

[197] Prud'homme GJ, Glinka Y. Neuropilins are multifunctional coreceptors involved in tumor initiation, growth, metastasis and immunity. Oncotarget 2012;3:921–39.

[198] Geretti E, Shimizu A, Klagsbrun M. Neuropilin structure governs VEGF and semaphorin binding and regulates angiogenesis. Angiogenesis 2008;11:31–9. https://doi.org/10.1007/s10456-008-9097-1.

[199] Raimondi C, Fantin A, Lampropoulou A, Denti L, Chikh A, Ruhrberg C. Imatinib inhibits VEGF-independent angiogenesis by targeting neuropilin 1–dependent ABL1 activation in endothelial cells. J Exp Med 2014;211(6):1167–83.

[200] Artavanis-Tsakonas S, Rand MD, Lake RJ. Notch signaling: cell fate control and signal integration in development. Science 1999;284(5415):770–6.

[201] Noguera-Troise I, Daly C, Papadopoulos NJ, Coetzee S, Boland P, Gale NW, Lyn HC, Yancopoulos GD, Thurston G. Blockade of Dll4 inhibits tumour growth by promoting non-productive angiogenesis. Nature 2006;444(7122):1032–7.
[202] Thurston G, Kitajewski J. VEGF and Delta-Notch: interacting signalling pathways in tumour angiogenesis. Br J Cancer 2008;99(8):1204–9.
[203] Sainson RC, Aoto J, Nakatsu MN, Holderfield M, Conn E, Koller E, Hughes CC. Cell-autonomous notch signaling regulates endothelial cell branching and proliferation during vascular tubulogenesis. FASEB J 2005;19 (8):1027–9.
[204] Hellström M, Phng LK, Hofmann JJ, Wallgard E, Coultas L, Lindblom P, Alva J, Nilsson AK, Karlsson L, Gaiano N, Yoon K, Rossant J, Iruela-Arispe ML, Kalén M, Gerhardt H, Betsholtz C. Dll4 signalling through Notch1 regulates formation of tip cells during angiogenesis. Nature 2007;445(7129):776–80. https://doi.org/10.1038/nature05571.
[205] Hainaud P, Contrerès JO, Villemain A, Liu LX, Plouët J, Tobelem G, Dupuy E. The role of the vascular endothelial growth factor–Delta-like 4 ligand/Notch4-ephrin B2 cascade in tumor vessel remodeling and endothelial cell functions. Cancer Res 2006;66(17):8501–10.
[206] Lobov IB, Renard RA, Papadopoulos N, Gale NW, Thurston G, Yancopoulos GD, Wiegand SJ. Delta-like ligand 4 (Dll4) is induced by VEGF as a negative regulator of angiogenic sprouting. Proc Natl Acad Sci U S A 2007;104 (9):3219–24.
[207] Bridges E, Oon CE, Harris A. Notch regulation of tumor angiogenesis. Future Oncol 2011;7(4):569–88.
[208] Thurston G, Noguera-Troise I, Yancopoulos GD. The Delta paradox: DLL4 blockade leads to more tumour vessels but less tumour growth. Nat Rev Cancer 2007;7(5):327–31.
[209] Chen H, Liu D, Yang Z, Sun L, Deng Q, Yang S, Qian L, Guo L, Yu M, Hu M, Shi M, Guo N. Adrenergic signaling promotes angiogenesis through endothelial cell–tumor cell crosstalk. Endocr Relat Cancer 2014;21:783–95.
[210] Wieland E, Rodriguez-Vita J, Liebler SS, Mogler C, Moll I, Herberich SE, Espinet E, Herpel E, Menuchin A, Chang-Claude J, Hoffmeister M, Gebhardt C, Brenner H, Trumpp A, Siebel CW, Hecker M, Utikal J, Sprinzak D, Fischer A. Endothelial Notch1 activity facilitates metastasis. Cancer Cell 2017;31(3):355–67. https://doi.org/10.1016/j.ccell.2017.01.007.
[211] Steinle JJ, Meininger CJ, Forough R, Wu G, Wu MH, Granger HJ. Eph B4 receptor signaling mediates endothelial cell migration and proliferation via the phosphatidylinositol 3-kinase pathway. J Biol Chem 2002;277 (46):43830–5.
[212] Steinle JJ, Meininger CJ, Chowdhury U, Wu G, Granger HJ. Role of ephrin B2 in human retinal endothelial cell proliferation and migration. Cell Signal 2003;15(11):1011–7.
[213] Sawamiphak S, Seidel S, Essmann CL, Wilkinson GA, Pitulescu ME, Acker T, Acker-Palmer A. Ephrin-B2 regulates VEGFR2 function in developmental and tumour angiogenesis. Nature 2010;465(7297):487–91.
[214] Wang Y, Nakayama M, Pitulescu ME, Schmidt TS, Bochenek ML, Sakakibara A, Adams S, Davy A, Deutsch U, Lüthi U, Barberis A, Benjamin LE, Mäkinen T, Nobes CD, Adams RH. Ephrin-B2 controls VEGF-induced angiogenesis and lymphangiogenesis. Nature 2010;465(7297):483–6. https://doi.org/10.1038/nature09002.
[215] Baenziger NL, Brodie GN, Majerus PW. A thrombin-sensitive protein of human platelet membranes. Proc Natl Acad Sci U S A 1971;68(1):240–3.
[216] Kaur S, Martin-Manso G, Pendrak ML, Garfield SH, Isenberg JS, Roberts DD. Thrombospondin-1 inhibits VEGF receptor-2 signaling by disrupting its association with CD47. J Biol Chem 2010;285(50):38923–32.
[217] Greenaway J, Lawler J, Moorehead R, Bornstein P, Lamarre J, Petrik J. Thrombospondin-1 inhibits VEGF levels in the ovary directly by binding and internalization via the low density lipoprotein receptor-related protein-1 (LRP-1). J Cell Physiol 2007;210(3):807–18.
[218] Chu LY, Ramakrishnan DP, Silverstein RL. Thrombospondin-1 modulates VEGF signaling via CD36 by recruiting SHP-1 to VEGFR2 complex in microvascular endothelial cells. Blood 2013;122(10):1822–32.
[219] Taraboletti G, Morbidelli L, Donnini S, Parenti A, Granger HJ, Giavazzi R, Ziche M. The heparin binding 25 kDa fragment of thrombospondin-1 promotes angiogenesis and modulates gelatinase and TIMP-2 production in endothelial cells. FASEB J 2000;14(12):1674–6. https://doi.org/10.1096/fj.99-0931fje.
[220] Donnini S, Morbidelli L, Taraboletti G, Ziche M. ERK1-2 and p38 MAPK regulate MMP/TIMP balance and function in response to thrombospondin-1 fragments in the microvascular endothelium. Life Sci 2004;74 (24):2975–85. https://doi.org/10.1016/j.lfs.2003.09.075.

[221] Colombo G, Margosio B, Ragona L, Neves M, Bonifacio S, Annis DS, Stravalaci M, Tomaselli S, Giavazzi R, Rusnati M, Presta M, Zetta L, Mosher DF, Ribatti D, Gobbi M, Taraboletti G. Non-peptidic Thrombospondin-1 mimics as fibroblast growth factor-2 inhibitors an integrated strategy for the development of new antiangiogenic compounds. J Biol Chem 2010;285(12):8733–42. https://doi.org/10.1074/jbc.M109.085605.
[222] Xu Z, Shi H, Li Q, Mei Q, Bao J, Shen Y, Xu J. Mouse macrophage metalloelastase generates angiostatin from plasminogen and suppresses tumor angiogenesis in murine colon cancer. Oncol Rep 2008;20(1):81–8.
[223] O'Reilly MS. Angiostatin: an endogenous inhibitor of angiogenesis and of tumor growth. EXS 1997;79:273–94.
[224] Gately S, Twardowski P, Stack MS, Cundiff DL, Grella D, Castellino FJ, Enghild J, Kwaan HC, Le F, Kramer RA, Volpert O, Bouck N, Soff GA. The mechanism of cancer-mediated conversion of plasminogen to the angiogenesis inhibitor angiostatin. Proc Natl Acad Sci U S A 1997;94(20):10868–72. https://doi.org/10.1073/pnas.94.20.10868.
[225] MacDonald NJ, Murad AC, Fogler WE, Lu Y, Sim B. The tumor-suppressing activity of angiostatin protein resides within kringles 1 to 3. Biochem Biophys Res Commun 1999;264(2):469–77. https://doi.org/10.1006/bbrc.1999.1486.
[226] Schmitz V, Raskopf E, Gonzalez-Carmona M, Vogt A, Kornek M, Sauerbruch T, Caselmann WH. Plasminogen derivatives encoding kringles 1-4 and Kringles 1-5 exert indirect antiangiogenic and direct antitumoral effects in experimental lung cancer. Cancer Invest 2008;26(5):464–70. https://doi.org/10.1080/07357900801970927.
[227] Ansell PJ, Zhang H, Davidson DJ, Harlan JE, Xue J, Brodjian S, Lesniewski R, McKeegan E. Recombinant kringle 5 from plasminogen antagonises hepatocyte growth factor-mediated signalling. Eur J Cancer 2010;46(5):966–73. https://doi.org/10.1016/j.ejca.2009.12.026.
[228] Perri SR, Martineau D, Francois M, Lejeune L, Bisson L, Durocher Y, Galipeau J. Plasminogen Kringle 5 blocks tumor progression by antiangiogenic and proinflammatory pathways. Molec Cancer Therap 2007;6(2):441–9. https://doi.org/10.1158/1535-7163.mct-06-0434.
[229] Miosge N, Sasaki T, Timpl R. Angiogenesis inhibitor endostatin is a distinct component of elastic fibers in vessel walls. FASEB J 1999;13(13):1743–50. https://doi.org/10.1096/fasebj.13.13.1743.
[230] Dixelius J, Cross M, Matsumoto T, Sasaki T, Timpl R, Claesson-Welsh L. Endostatin regulates endothelial cell adhesion and cytoskeletal organization. Cancer Res 2002;62:1944–7.
[231] O'Reilly MS, Boehm T, Shing Y, Fukai N, Vasios G, Lane WS, Flynn E, Birkhead JR, Olsen BR, Folkman J. Endostatin: an endogenous inhibitor of angiogenesis and tumor growth. Cell 1997;88(2):277–85. https://doi.org/10.1016/s0092-8674(00)81848-6.
[232] Bloch W, Huggel K, Sasaki T, Grose R, Bugnon P, Addicks K, Timpl R, Werner S. The angiogenesis inhibitor endostatin impairs blood vessel maturation during wound healing. FASEB J 2000;14(15):2373–6. https://doi.org/10.1096/fj.00-0490fje.
[233] Folkman J. Antiangiogenesis in cancer therapy? Endostatin and its mechanisms of action. Exp Cell Res 2006;312(5):594–607. https://doi.org/10.1016/j.yexcr.2005.11.015.
[234] Morbidelli L, Donnini S, Chillemi F, Giachetti A, Ziche M. Angiosuppressive and angiostimulatory effects exerted by synthetic partial sequences of endostatin. Clin Cancer Res 2003;9(14):5358–69.
[235] Goveia J, Rohlenova K, Taverna F, Treps L, Conradi LC, Pircher A, Geldhof V, de Rooij LPMH, Kalucka J, Sokol L, García-Caballero M, Zheng Y, Qian J, Teuwen LA, Khan S, Boeckx B, Wauters E, Decaluwé H, De Leyn P, Vansteenkiste J, Weynand B, Sagaert X, Verbeken E, Wolthuis A, Topal B, Everaerts W, Bohnenberger H, Emmert A, Panovska D, De Smet F, FJT S, Mclaughlin RJ, Impens F, Lagani V, Vinckier S, Mazzone M, Schoonjans L, Dewerchin M, Eelen G, Karakach TK, Yang H, Wang J, Bolund L, Lin L, Thienpont B, Li X, Lambrechts D, Luo Y, Carmeliet P. An integrated gene expression landscape profiling approach to identify lung tumor endothelial cell heterogeneity and angiogenic candidates. Cancer Cell 2020;37:21–36. https://doi.org/10.1016/j.ccell.2019.12.001.

CHAPTER

2

Antiangiogenic drugs: Chemosensitizers for combination cancer therapy

Sandra Donnini[a], *Arianna Filippelli*[a], *Valerio Ciccone*[a], *Andrea Spini*[b,c], *Emma Ristori*[d], *Marina Ziche*[c], *and Lucia Morbidelli*[a]

[a]Department of Life Sciences, University of Siena, Siena, Italy [b]Bordeaux Population Health Center, University of Bordeaux, Bordeaux, France [c]Department of Medicine, Surgery and Neurosciences, University of Siena, Siena, Italy [d]Department of Internal Medicine, Section of Cardiology, Yale University School of Medicine, Yale Cardiovascular Research Center, New Haven, CT, United States

Abstract

Angiogenesis is a key step in tumor progression and recurrence, and is known as a hallmark of solid tumors. The architecture of the tumor vasculature is deregulated, vessels are immature, and their functionality is defective, causing increase of interstitial pressure, hypoxia, and acidity in the tumor microenvironment, and favoring an enhanced tumor dissemination and metastasis. The defective tumor vasculature as well as the modifications of tumor microenvironment are a barrier for the effective delivery and accumulation of either the anticancer drugs, including antiangiogenic drugs, or immune-competent T cells in tumor area, thus promoting tumor resistance to therapy. Among the various approaches for targeting tumor angiogenesis, vascular normalization and vessel maturation are considered the most promising approaches, which favor both an effective delivery of anticancer drugs and accumulation of immune-competent T cells into the tumor area. Antiangiogenic therapy can downregulate continuous angiogenic signaling and result in vascular normalization, such as pruning, vascular maturation, and increased perfusion. Thus, antiangiogenic drugs are now considered promising chemosensitizers of anticancer strategies such as chemotherapy, target therapies, and immune checkpoint inhibitors in several advanced tumors. This review summarizes the current understanding and clinical development of combination therapy with antiangiogenic drugs and anticancer chemo-, targeted, and immune therapies.

https://doi.org/10.1016/B978-0-323-90190-1.00008-1

Abbreviations

Ang-2	angiopoietin-2
BC	breast cancer
BRAF	v-raf murine sarcoma viral oncogene homolog B1
CAFs	cancer-associated fibroblasts
CFDA	Chinese Food and Drug Administration
c-MET	hepatocyte growth factor receptor
CML	chronic myelogenous leukemia
CRC	colorectal cancer
CSF-1R	colony-stimulating factor receptor 1
CTLA-4	cytotoxic T lymphocyte antigen 4
DCs	dendritic cells
DOR	duration of response
ECs	endothelial cells
EGFR	epidermal growth factor receptor
EMA	European Medicines Agency
FDA	Food and Drug Administration
FGF-2	fibroblast growth factor-2
FGFR	fibroblast growth factor receptor
FLT3	FMS-like tyrosine kinase
GC	gastric cancer
GEJ	gastroesophageal junction adenocarcinoma
GIST	gastrointestinal stromal tumor
HCC	hepatocellular carcinoma
HIF-1α	hypoxia-inducible factor-1α
ICI	immune checkpoint inhibitor
ILs	interleukins
INF	interferon
KIT	CD117-cluster of differentiation 117
MTC	medullary thyroid cancer
NSCLC	non-small cell lung carcinoma
ORR	Objective response rate
OS	overall survival
PDGFR	platelet-derived growth factor receptor
PD-L1/L2	program death ligand 1 and 2
PFS	progression-free survival
PGE2	prostaglandin E2
pNET	pancreatic neuroendocrine tumor
RAF-1	v-raf-1 murine leukemia viral oncogene homolog 1
RAI-DTC	radioactive iodine-refractory differentiated thyroid cancer
RCC	renal cell carcinoma
RET	rearranged during transfection
ROS1	ROS proto-oncogene 1
RTKs	receptor tyrosine kinases
SCLC	small-cell lung carcinoma
STS	soft-tissue sarcoma
TAM	tumor-associated macrophage
TEC	tumor endothelial cell
Tie-2	angiopoietin-1 and -2 receptor
TILs	tumor-infiltrating lymphocytes
TIM-3	T-cell immunoglobulin mucin receptor 3
TKI	tyrosine kinase inhibitor

TME	tumor microenvironment
Treg	regulatory T cells
TRKB	neurotrophic tyrosine kinase receptor B
VEGF	vascular endothelial growth factor
VEGFRs	vascular endothelial growth factor receptors

Conflict of interest

No potential conflicts of interest were disclosed.

Introduction

It is now well established that cancer cells coexist with stromal cells in a complex environment. Their growth and dissemination depend on both tight and plastic interactions with components of the tumor microenvironment (TME), including fibroblasts, cells of the immune system, platelets, pericytes, and endothelial cells (ECs) of the blood and lymphatic vasculature. Within the TME, cancer cells maintain a bidirectional communication with all TME components through the release of soluble factors or through cancer-stromal cell contacts. TME components, in turn, drive the formation of a chronic inflammatory, immunosuppressive, and hypoxic pro-angiogenic niche that sustains cancer cell stemness, promotes dissemination of cancer cells, and limits the effect of various therapeutic interventions. Furthermore, the hypoxic TME promotes the expression of hypoxia-inducible factor-1α (HIF-1α), which, in turn, induces a "metabolic switch" of glucose to aerobic glycolysis in cancer and stromal cells, generally known as the Warburg effect, along with an altered pH and cancer stemness, through the expression of pluripotent factors and epigenetic modifiers, and tumor angiogenesis [1]. The aberrant tumor vessels, that are immature, thin walled, leaky, chaotically organized, and ill perfused, contribute to maintain the pro-tumorigenic niche and profoundly influence how cancer cells escape the anticancer immunosurveillance, metastasize, and respond to therapy [1].

Tumor angiogenesis is mainly driven by an imbalance between pro-angiogenic and antiangiogenic signaling in the TME. Key pro-angiogenic mediators include vascular endothelial growth factor-A (VEGF-A), fibroblast growth factor-2 (FGF-2), interleukins (ILs) such as IL-8 and IL-10, and inflammatory mediators such as prostaglandin E2 (PGE2) [2,3]. These molecules become abundant in the TME and overwhelm angiostatic signals, thereby inducing a pro-angiogenic switch in endothelial cells (ECs) of preexisting vessels (for details see Introduction of this book). Briefly, in hypoxic TME, the levels of VEGF-A, IL-10, and PGE2 are high, and collectively induce FasL expression on tumor ECs (TECs), a death ligand for activated T cells, which upon binding to Fas, expressed on T cells, triggers their apoptosis [1,4]. In tumor vessels, FasL selectively kills cytotoxic T cells but not regulatory T cells (Tregs) because of their high expression of c-FLIP, resulting in predominant Treg infiltration in the TME. It has also been reported that TECs preferentially promote the extravasation and recruitment of regulatory Tregs and myeloid-derived suppressor cells in TME, by upregulating a selective complex on their cell membrane, as well as the secretion of soluble chemotactic factors. The release of VEGF-A seems to mediate a clustering defect of adhesion molecules, like ICAM-1 and VCAM-1 in TECs, which hampers immune cell extravasation [5–9]. Similarly, PGE2 causes the formation of neovessels, in part synergizing with pro-angiogenic signaling pathways, such as the system FGF-2 and its receptor fibroblast growth factor receptor 1 (FGFR-1) [2], in part promoting the barrier function of TECs by inhibiting trans-endothelial migration of cytotoxic tumor-infiltrating lymphocytes (TILs) [10].

Additionally, hypoxia and the associated tumor angiogenesis are also involved in promoting an immune-suppressive TME, through induction of program death ligand 1 and 2 (PD-L1/PD-L2), T-cell immunoglobulin mucin receptor 3 (TIM-3), and cytotoxic T lymphocyte antigen 4 (CTLA-4), and leading to cytotoxic T-cell exhaustion and activation of signaling pathways that enhance Treg differentiation and attenuate cytotoxic activity of $CD8^+$ T cells and other protective responses of the immune system in both cancer and stromal cells, including TECs [11–13]. TEC expression of PD-L1 inactivates T cells, which become functionally anergic before migrating across the vessel wall and entering the TME [10].

Because of hypoxia and HIF-1α/VEGF expression, tumor-associated macrophages (TAMs) switch from M1-polarized phenotype to M2-polarized macrophages, which are associated with tumor angiogenesis and promotion of cancer progression, including vascular mimicry, fibroblast differentiation to cancer-associated fibroblasts (CAFs), and cancer cell infiltration and recovery from cancer therapy [14]. Concurrently, tumor-associated antigen presentation by dendritic cells (DCs) is impaired [15]. Moreover, increased plasma VEGF levels correlate with an increased number of immature DC precursors and fewer DCs in the peripheral blood of patients with cancer [15,16]. Considering this, the hypothesis of normalizing the microenvironment to fight the tumor by combinatorial approaches came to light, including the antiangiogenic strategy to normalize tumor vessels (see Conclusions of this book), which are crucial in perpetuating the vicious cycle established by increasing hypoxia and pro-angiogenic mediators in the TME.

The following is a non-exhaustive description of the major antiangiogenic drugs in use in the oncological setting in monotherapy and in combination with chemotherapy, target therapy, and immunotherapy.

Antiangiogenic drugs

Angiogenesis is a process regulated by pro- and antiangiogenic mediators. Among these, VEGF plays a central role not only in physiological but also in pathological angiogenesis. VEGF is crucial for tumor neoangiogenesis, and most antiangiogenic drugs are directed against this factor or its receptors, VEGFRs. VEGF activity is mediated by three cell surface tyrosine kinase receptors: VEGFR-1, VEGFR-2, and VEGFR-3. VEGFR-2 plays a major role in VEGF-proangiogenic activity, while VEGFR-1 causes VEGF sequestration or stimulation of hematopoietic stem-cell migration. VEGFR-3 mediates lymphangiogenesis, facilitating lymph node metastasis [17,18]. A total of 17 antiangiogenic agents, including anti-VEGF antibodies, anti-VEGFR antibodies, as well as VEGFR tyrosine kinase inhibitors (TKIs), have been approved by regulatory agencies for certain types of cancer (Table 1).

Bevacizumab, a humanized antibody against VEGF, has been the first antiangiogenic drug approved by the Food and Drug Administration (FDA) for the treatment of metastatic colon cancer in combination with chemotherapy [20]. The antibody binds VEGF and prevents its binding to VEGFR-1 and -2 on the surface of ECs, blocking the biological activity of VEGF. Bevacizumab treatment regresses tumor vascularization, normalizes residual tumor vessels,

TABLE 1 Antiangiogenic drugs approved by the EMA and their therapeutic indications.

Drug	Class	Target	Approved clinical use	References
Axitinib (Inlyta®)	Multi-TKI	VEGFR-1/2/3	RCC, plus avelumab or pembrolizumab as the first treatment in advanced RCC	[19]
Bevacizumab (Avastin®)	Humanized monoclonal antibody	VEGF-A	Metastatic CC, recurrent glioblastoma, HCC, advanced non-squamous NSCLC, OEC, fallopian tube/primary peritoneal cancer, RCC	[20–23]
Cabozantinib (Cometriq®)	Multi-TKI	VEGFR-2, c-Met, ROS1, TYRO3, MER, Ret, Kit, TRKB Flt-3, AXL, Tie-2	Advanced RCC, HCC	[30–32]
Dovitinib (DRP®)	Pan-TKI	VEGFR, FGFR, TKRs	RCC, HCC, GIST, BC, endometrial cancer	[24]
Lenvatinib (Lenvima®)	Multi-TKI	VEGFR1/2/3, FGFRs, PDGFR-α, c-Kit receptor, RET	Metastatic RCC, RAI-DTC, HCC.	[25–28]
Nintedanib (BIBF 1120, Ofev®)	Multi-TKI	VEGFR-1/2/3, PDGFR-α/β, FGFR-1/2	NSCLC (with docetaxel therapy)	[29,30]
Ponatinib (Iclusig™)	Multi-TKI	VEGF-R, SRC, ABL, FGF-R, PDGF-R. ABL-T135I mutation	CML, Ph^+ ALL	[31–33]
Pazopanib (Votrient®)	Multi-TKI	VEGF-R-1/2/3, PDGF-R-α/β, c-Kit	Advanced RCC, advanced STS	[34–37]
Ramucirumab (Cyramza®)	Humanized monoclonal antibody	VEGFR-2	CRC, NSCLC, HCC, GC or GEJ adenocarcinoma	[23,38,39]
Regorafenib (Stivarga®)	Multi-TKI	VEGFR-1/2/3, PDGFR-α/β, FGFR-1/2, Tie2, RAF-1, BRAF, $BRAF^{V600E}$, c-Kit receptor	mCRC, advanced GIST, HCC	[40–42]
Sorafenib (Nexavar®)	Multi-TKI	VEGFR-1/2/3, PDGFR-β, Raf serine/threonine kinases, c-Kit receptor	HCC, RCC, RAI-DTC	[43–47]
Sunitinib (Sutent®)	Multi-TKI	VEGFR-1/2/3, PDGFR-α/β, c-Kit receptor, RET, FLT3, CSF-1R	Imatinib-resistant GIST, RCC, pNET	[48–50]
Tivozanib (Fotivda®)	TKI	VEGFR-1/2/ 3	Relapsed or refractory RCC	[51–54]
Vandetanib (Caprelsa®)	Multi-TKI	VEGFR-2, EGFR, RET signaling	MTC	[55]

Continued

TABLE 1 Antiangiogenic drugs approved by the EMA and their therapeutic indications—cont'd

Drug	Class	Target	Approved clinical use	References
Ziv-aflibercept (Zaltrap®)	Soluble recombinant fusion protein	VEGF A-D, PlGF	Metastatic CRC (with FOLFIRI)	[56]

GC, *gastric cancer;* RCC, *renal cell carcinoma;* CC, *cervical cancer;* GC, *gastric cancer;* NSCLC, *non-small cell lung cancer;* HCC, *hepatocellular carcinoma;* MTC, *medullary thyroid cancer;* DTC, *refractory thyroid cancer;* STS, *soft-tissue sarcoma;* mCRC, *metastatic colorectal cancer;* GIST, *gastrointestinal stromal tumors;* CML, *chronic myeloid leukemia;* Ph+ ALL, *Philadelphia chromosome-positive acute lymphoblastic leukemia;* pNET, *pancreatic neuroendocrine tumor;* EOC, *epithelial ovarian cancer;* BC, *breast cancer;* TKI, *tyrosine kinase inhibitor.*

and inhibits the formation of new vasculature, thereby preventing tumor growth [20]. Administration of bevacizumab has demonstrated broad antitumor activity in human tumors [20]. However, in clinical practice, bevacizumab is not approved as a single agent for the treatment of any tumors [20–22].

Ramucirumab is a human receptor-targeted antibody that binds VEGFR-2 preventing VEGF binding. As a result, the antibody inhibits ligand-stimulated activation of VEGFR-2 signaling pathways and functions, including proliferation and migration of ECs [23]. Ramucirumab was approved in 2014 by the FDA as a single agent or in combination with paclitaxel for the treatment of advanced or metastatic gastric cancer (GC) or gastroesophageal junction (GEJ) adenocarcinoma with disease progression after prior fluoropyrimidine- or platinum-containing chemotherapy, and in 2019 as a single agent, for the treatment of hepatocellular carcinoma (HCC) in patients who have alpha fetoprotein level ≥400 ng/mL after prior treatment with sorafenib. In the REGARD multicentric phase III trial, ramucirumab was the first biological treatment given as a single drug that had survival benefits in patients with advanced GC or GEJ adenocarcinoma progressing after first-line chemotherapy (median OS: 5.2 vs. 3.8 months in patients in the ramucirumab arm and in those in the placebo arm, respectively) [38]. As a side effect, hypertension was higher in the ramucirumab arm than in the placebo group (16% vs. 8%), whereas rates of other adverse events were mostly similar between groups [38]. In the Reach-2 phase III trial, ramucirumab compared with placebo in patients with HCC and α-fetoprotein concentrations of at least 400 ng/mL who had previously received sorafenib improved both OS (8.5 months vs. 7.3 months) and PFS (2.8 months vs. 1.6 months) [39].

Anlotinib, a multitarget tyrosine kinase inhibitor (TKI), inhibits the tyrosine kinase activity of a number of RTKs, including the VEGF receptors (VEGFR-1, VEGFR-2, VEGFR-3), c-Kit, PDGFR-α, and fibroblast growth factor receptors (FGFR-1, FGFR-2, and FGFR-3). It was approved in 2018 by the China Food and Drug Administration (CFDA) for the treatment of patients with advanced non-small cell lung carcinoma (NSCLC) who have progressed after treatment with two or more lines of prior systemic chemotherapy (ALTER0303 phase III trial). The approval was based on a significant improvement in median OS with anlotinib (9.46 months) compared with placebo (6.37 months) [57]. The confirmed objective response rate (ORR) was 9.2% in the anlotinib arm and 0.7% in the placebo arm. The toxicity profile was consistent with that of known antiangiogenesis inhibitors. In the above trial, the common side effects were hypertension, hand-foot syndrome, hemoptysis, thyroid-stimulating hormone elevation, and corrected QT interval prolongation [57]. Recently, because of the

ALTER1202 phase II trial, anlotinib has been approved by CFDA as third- and further-line treatment option also for small-cell lung cancer (SCLC) patients in China, and the results of ALTER0703 phase III trial on the efficacy and safety of anlotinib on refractory metastatic colorectal cancer (CRC) have been published. Anlotinib as a third-line or subsequent treatment for Chinese patients with SCLC showed improved PFS (4.1 vs. 0.7 months) and OS (7.3 vs. 4.9 months) than placebo with favorable safety profile, while in Chinese patients with refractory metastatic CRC, anlotinib provided clinical benefits by prolonging PFS (4.1 months) over placebo group (1.5 months) and was well tolerated; however, OS was similar between the two groups [58,59].

Apatinib is a selective inhibitor of VEGFR-2. Like anlotinib, it is approved in China as monotherapy in heavily pretreated patients with GC or GEJ adenocarcinoma. A phase III trial showed that the antiangiogenic TKI for VEGFR prolonged the median OS (6.5 vs. 4.7 months) and median PFS (2.6 vs. 1.8 months) compared with placebo with moderate, reversible, and easily managed adverse events (hand-foot syndrome, proteinuria, and hypertension) [60].

Axitinib is a potent and highly selective inhibitor of VEGFR-1, -2, and -3. According to the results of AXIS 1032 phase III trial, this TKI is approved in patients with advanced or metastatic renal cell carcinoma (RCC), after a previous regimen for FDA or a previous treatment with sunitinib or cytokines for European Medicines Agency (EMA), while a combination with avelumab or pembrolizumab is indicated for the first-line treatment of patients with advanced RCC. AIS phase III trial showed that the patients who were treated with axitinib as the second-line treatment achieved significantly extended PFS, with a median PFS of 6.8 months, when compared to 4.7 months achieved by the patients treated with sorafenib. Furthermore, the superiority of axitinib was demonstrated in subgroups of patients treated by cytokines (median PSF: 12.1 versus 6.5 months) and in patients previously treated with sunitinib (median PSF: 4.8 compared to 3.4 months). The ORR was 19% for the axitinib arm and 9% for sorafenib [19]. In terms of tolerability, the toxicity profile of axitinib was consistent with that reported for VEGFR-TKI family, with grade 3–4 adverse effects such as hand-foot syndrome, fatigue, dyspnea, diarrhea, and hypertension.

Cabozantinib is a multi-TKI for c-MET (hepatocyte growth factor receptor), VEGFR-2, GAS6 receptor (AXL), rearranged during transfection (RET), ROS proto-oncogene 1 (ROS1), TYRO3, MER, KIT, neurotrophic tyrosine kinase receptor B (TRKB), FMS-like tyrosine kinase (FLT3), and Tie-2. Cabozantinib is indicated as monotherapy for the treatment of advanced RCC as first-line treatment only for adult patients with intermediate or poor risk, or as second-line treatment for all other patients, and for the treatment of HCC in adults who have previously been treated with sorafenib. The randomized phase III METEOR study demonstrated that cabozantinib has superior PFS (7.4 vs. 3.8 months), OS (21.4 vs. 16.5 months), and ORR (per independent radiology review in 17% vs. 3% of patients) compared to everolimus in patients with advanced RCC who progressed after previous VEGFR-targeted therapy, as sunitinib or pazopanib [61]. The most common grade 3 or 4 adverse events were hypertension, diarrhea, fatigue, palmar-plantar erythrodysesthesia syndrome, anemia, hyperglycemia, and hypomagnesemia. Cabozantinib was also approved for use in the first-line setting for patients with advanced RCC with intermediate/poor-risk disease. This was based on the randomized phase II CABOSUN trial, which demonstrated a significant improvement in median PFS (8.6 vs. 5.3 months) and ORR (per independent radiology committee in 20% vs. 9%) over the standard of care sunitinib for the International Metastatic RCC Database

Consortium (IMDC) intermediate/poor-risk patients [62]. The randomized phase III CELESTIAL trial included patients with HCC in the third-line setting and cabozantinib efficacy was confirmed in several post hoc analyses (median OS: 16.1 versus 9.1 months), irrespective of alpha-fetoprotein levels, albumin-bilirubin score, age, and duration of previous sorafenib treatment. The safety profile of cabozantinib was comparable with other multi-TKI used for HCC and the most frequent grade $\geq$3 adverse events were diarrhea, palmar-plantar erythrodysesthesia, fatigue, hypertension, and aspartate aminotransferase increase [63].

Dovitinib is a pan-TKI targeting FGFR-1, -2, and -3; VEGFR-1, -2, and -3; platelet-derived growth factor receptor (PDGFR)-β, FLT-3, c-KIT, and colony-stimulating factor receptor 1 (CSF-1R). Dovitinib is under evaluation in 5 different cancers: RCC, gastrointestinal stromal tumor (GIST), HCC, metastatic breast cancer (BC), and endometrial cancer. The TKI completed a phase III study in RCC (clinicaltrials.gov identifier NCT01223027) where it showed a non-inferiority activity of sorafenib in patients who had progressed on previous VEGF-targeted therapies and mTOR inhibitors. The drug has been recently approved by FDA for RCC treatment in the third-line setting, as a therapeutic alternative to sorafenib [24].

Fruquintinib is a potent, highly selective small-molecule inhibitor of VEGFR-1, -2, and -3 approved by the CFDA in 2018 [64]. In the FRESCO phase III trial, fruquintinib compared with placebo resulted in a statistically significant increase in median OS (9.3 vs. 6.6 months) and PFS (3.7 vs. 1.8 months) among Chinese patients with metastatic CRC who had tumor progression following at least 2 prior chemotherapy regimens without VEGFR inhibitors [25,64]. Fruquintinib also showed an acceptable safety and tolerability profile.

Lenvatinib is a multitargeted TKI of VEGFR-1, -2, and -3, FGFR-1 to -4, PDGFRα, and RET and c-KIT proto-oncogenes [26,27]. The TKI is indicated as a monotherapy for the treatment of patients with progressive, locally advanced, or metastatic, differentiated (papillary/follicular/Hürthle cell) thyroid carcinoma (DTC), refractory to radioactive iodine (RAI) and patients with advanced or unresectable HCC who have received no prior systemic therapy. The SELECT phase III study evaluated the TKI in patients with RAI-DTC and showed that lenvatinib prolonged the median PFS versus placebo (18.3 vs. 3.6 months) [28]. In an updated analysis of SELECT study, lenvatinib showed a significative duration of response (DOR) and confirmed its efficacy on PFS for the treatment of DTC-RAI [65]. The median DOR for all lenvatinib responders (patients with complete or partial responses) was 30 months and median PFS was 33.1 vs. 7.9 months in nonresponders. In 2018, the FDA also approved lenvatinib for first-line treatment of patients with unresectable HCC. Approval was based on REFLECT phase III trial where the TKI was compared with sorafenib. REFLECT demonstrated that lenvatinib was non-inferior but not statistically superior to sorafenib for OS (median OS of 13.6 and 12.3 months) [66]. REFLECT also demonstrated statistically significant improvements of lenvatinib vs. sorafenib in PSF (median PFS of 7.4 and 3.7 months) and ORR (24.1% vs. 9.2% per modified RECIST for HCC) [29,66]. Serious side effects were more common among lenvatinib-treated patients, and the most common observed reactions were hypertension, fatigue, diarrhea, decreased appetite, arthralgia/myalgia, decreased weight, abdominal pain, palmar-plantar erythrodysesthesia syndrome, proteinuria, dysphonia, hemorrhagic events, hypothyroidism, and nausea [29].

Nintedanib is a potent TKI that targets VEGFR-1, -2, and -3, FGFR-1, -2, and -3, and PDGFRα and β [30]. It has been approved for the treatment of idiopathic pulmonary fibrosis, and in addition, in European countries, nintedanib plus docetaxel is used for patients with advanced non-small cell lung cancer (NSCLC) after first-line chemotherapy [30]. The combination of nintedanib with docetaxel has been investigated in LUME-Lung 1, a phase III trial (see next paragraph).

Ponatinib is a third-generation TKI with potent activity against native Bcr-Abl1 and clinically relevant resistant mutants [31,32]. The TKI is currently approved in >30 countries for use in adults with refractory chronic myeloid leukemia (CML) or Philadelphia chromosome-positive acute lymphoblastic leukemia (Ph^+ ALL) and those harboring the resistant Bcr-Abl1^{T315I} mutant [33]. The PACE phase II trial showed responses to ponatinib of heavily pretreated patients with a median follow-up of 15 months [67]. Twelve-month estimates of PFS and OS in chronic phase CML patients were 80% and 94%, respectively. The final 5-year follow-up results of the PACE pivotal study were recently published and confirmed the durable and clinically meaningful responses obtained with ponatinib in a heavily pretreated chronic phase CML population [67]. In addition to its potent inhibition of Bcr-Abl, ponatinib also inhibits multiple other kinases, including kinases involved in angiogenesis, important for the pathogenesis of many malignancies, including FGFR-1–4, RET, PDGFR, c-KIT, FLT3, MEKK2, and VEGFR-1 and -2 [68–72]. Ponatinib has also demonstrated antitumor activity in preclinical models of a variety of human cancers including rhabdomyosarcoma, thyroid cancer, lung cancer, GIST, glioblastoma, endometrial cancer, and CRC [69,71,73–77].

Pazopanib is a multi-TK inhibitor that targets several RTKs, including VEGFR-1, -2, and -3 and PDGFRα. Pazopanib is FDA approved for metastatic non-adipocytic soft-tissue sarcoma (STS) after the first-line treatment [34,35]. The approval was a result of the PALETTE phase III trial of 372 patients with metastatic STS who had progressed on standard chemotherapy [35]. In this trial, a median PFS of 4.6 months in the pazopanib arm compared with 1.6 months in the placebo arm and OS of 12.5 and 10.7 months, respectively, were observed. The most common adverse effects of pazopanib include fatigue, diarrhea, nausea, weight loss, hypertension, and liver function test abnormalities. Further, in the non-inferiority EPAZ trail, pazopanib resulted as an acceptable alternative to cytotoxic chemotherapy in patients aged greater than 60 years, with non-resectable or metastatic STS that were noneligible for an anthracycline treatment [78]. Pazopanib is also approved as a first-line treatment for RCC, and a phase III study demonstrated similar efficacy of pazopanib versus sunitinib [36] and a preferential sequential therapy of pazopanib followed by sorafenib [37]. In the phase III trial NCT00334282, pazopanib significantly prolonged PFS compared with placebo (median PFS: 9.2 vs. 4.2 months), in both the treatment-naive subpopulation and the cytokine-pretreated subpopulation. The ORR was 30% with pazopanib compared with 3% with placebo ($P < .001$). The median DOR was longer than 1 year, and the common adverse events were diarrhea, hypertension, hair color changes, nausea, anorexia, and vomiting [79].

Regorafenib is a multi-TKI that targets VEGFR-1, -2, and -3, Tie2, PDGFR, FGFRs, and c-KIT, RET, RAF-1, BRAF, and BRAF^{V600E}, as well as the activity of Abl. It has shown to significantly prolong both OS and PFS in patients with refractory metastatic CRC, who progressed after all standard therapies, including Asian subpopulations in the CORRECT phase III and CONCUR trials, respectively [40–42]. In the recent open-label phase IIIb CONSIGN trial, carried out prospectively in more than 2800 patients across 25 countries, the TKI confirmed its safety profile, and the results reinforced the importance of using treatment modifications to manage adverse events [80]. Regorafenib also showed to improve PSF compared with placebo (median PFS: 4.8 vs. 0.9 months) in patients with metastatic GIST after progression on standard treatments [81]. A subsequent analysis suggested particular benefit of regorafenib treatment among patients with primary KIT exon 11 mutations and those with SDH-deficient GIST, and confirmed the safety profile, with hand-foot skin reaction and hypertension as the most common drug-related toxicities [82]. Further, the TKI is approved as a monotherapy for patients with HCC who progressed on sorafenib treatment (RESORSE, phase III trial) [83]. In the trail, regorafenib improved the OS vs. placebo (median OS: 10.6 vs. 7.8 months), with manageable adverse effects, and additional analysis confirmed its clinical benefit to patients with HCC regardless of the step of their disease progression during prior sorafenib treatment and regardless of their last sorafenib dose [84].

Sorafenib inhibits multiple targets including VEGFR-1, -2, and -3, as well as Raf and PDGFRβ. In patients with advanced or metastatic RCC, refractory to bevacizumab or sunitinib, sorafenib in the TARGET phase III trial significantly improved PFS from 2.8 to 5.6 months compared to that of placebo, with similar outcomes for both older (≥70) and younger patients (<70 years) [43,44]. Sorafenib or lenvatinib may be offered as first-line treatment of patients with advanced HCC, when there are contraindications to atezolizumab and/or bevacizumab, and as second-line therapy in patients progressed to atezolizumab and/or bevacizumab [45]. Monotherapy with sorafenib prolongs OS (median OS: 6.5 vs. 4.2 months) and delays the time to progression (2.8 vs. 1.4 months) compared to placebo in patients with advanced HCC who are not candidates for potentially curative treatment or trans-arterial chemoembolization, according to the results of two phase III trials (the SHARP trial and the Asia-Pacific trial) [46,47]. In the SHARP trial, drug-related adverse events of any grade occurring significantly more in sorafenib than placebo recipients included diarrhea, hand-foot skin reaction, anorexia, alopecia, weight loss, abdominal pain, voice changes, dry skin, and other dermatological events. A similar tolerability profile was seen in the Asia-Pacific trial. According to the DECISION phase III trial, sorafenib also significantly improved PFS compared with placebo (median PFS: 10.8 vs. 5.8 months) in patients with progressive RAI-DTC, irrespective of the mutation status. Adverse events were consistent with the known safety profile of sorafenib. Most adverse events were grade 1 or 2, and the most frequent treatment-emergent adverse events in the sorafenib group were hand-foot skin reaction, diarrhea, alopecia, and skin rash or desquamation [85]. Additional investigations confirmed the safety and tolerability profile of sorafenib in this carcinoma [86].

Sunitinib is a TKI that targets PDGFRα and PDGFRβ, VEGFR-1, -2, and -3, c-KIT, FLT3, CSF-1R, and RET. The clinical safety and efficacy of sunitinib have been studied in the treatment as a single agent of patients with GIST who were resistant to imatinib or intolerant to imatinib, patients with metastatic RCC, and patients with unresectable pancreatic neuroendocrine tumor (pNET). It was first approved by the FDA in 2006 as the treatment for GIST. In phase II trial number NCT00075218, median time to tumor progression was 27.3 weeks in patients receiving sunitinib and 6.4 weeks in those on placebo. Therapy was reasonably well tolerated; the most common treatment-related adverse events were fatigue, diarrhea, skin discoloration, and nausea [48]. In RCC sunitinib can increase PFS from 5.0 to 11.0 months compared to that of interferon (IFN) [49]. As a result, sunitinib is recommended in international treatment guidelines and is considered a reference standard of care in the first-line setting for patients at favorable or intermediate prognostic risk. Sunitinib is generally well tolerated with a predictable adverse-event profile. Sunitinib also showed an antitumor activity in pNET. In the phase III trial number NCT00428597, daily administration of sunitinib improved PFS (median PFS: 11.4 vs. 5.5 months), OS, and ORR (9.3% vs. 0%) as compared with placebo among patients with advanced pNET [50].

Tivozanib is a TKI targeting VEGFR-1, -2, and -3 [51,52]. This TKI received the EMA approval in 2017 and more recently in March 2021 the FDA approval for the treatment of adult patients with relapsed or refractory RCC following two or more prior systemic therapies. In the TIVO-3 phase III trial, tivozanib as third-line or fourth-line therapy improved PFS (median PSF: 5.6 vs. 3·9 months) and was better tolerated compared with sorafenib in patients with metastatic RCC, however, with no difference in global OS [53,54].

Vandetanib is a TKI targeting VEGFRs, EGFR, and RET signaling pathways, approved for the treatment of medullary thyroid cancer (MTC) with mutation in *RET* proto-oncogene. In the phase III trial number NCT00410761, the PFS at 6 months was 83% (vandetanib) and 63% (placebo) with a hazard ratio of 0.46 and an estimated 11-month prolongation of median PFS [55]. Common adverse events (any grade) occurred more frequently with vandetanib compared with placebo, including diarrhea, rash, nausea, hypertension, and headache.

Aflibercept (also known as VEGF Trap, AVE0005, or ziv-aflibercept in the United States) is a recombinant fusion protein containing VEGF-binding portions from the extracellular domains of human VEGFR-1 and -2, fused to the Fc portion of human immunoglobulin (Ig) G1. Aflibercept blocks the activity of VEGFA, VEGFB, and placental growth factor (PlGF) by acting as a high-affinity ligand trap to prevent these ligands from binding to their receptors. Aflibercept is approved by EMA in combination with fluorouracil, leucovorin, and irinotecan (FOLFIRI regimen) for the treatment of patients with metastatic CRC previously treated with an oxaliplatin-based regimen (see next paragraph) [56].

Approved combinatorial regimens with antiangiogenic drugs

Many clinical phase I/II trials prove that most antiangiogenic monotherapies failed to achieve any significant inhibition of tumor growth despite many successful preclinical results [23]. This ineffectiveness may be mainly attributed to (i) the occurrence of other mechanisms of vascularization in tumors such as vascular mimicry and vessel co-option; (ii) the abundance and heterogeneity of pro-angiogenic factors, in addition to VEGF, such as FGF, PDGF, and Ang-2 among others in human cancers; and (iii) the onset of antiangiogenic drug resistance induced by hypoxia associated with the activity of these drugs, which upregulates the production of growth factors and pro-angiogenic factors in the TME. Combinations of antiangiogenic therapies with standard chemo/targeted/immunotherapy or radiotherapy seem to overcome these difficulties. The decline of interstitial fluid pressure at tumor sites caused by antiangiogenic therapy should promote drug transport, contributing to successful combination therapies. Increased intra-tumoral tissue oxygenation obtained by vascular normalization and vessel maturation may significantly sensitize tumors to radiotherapy and immunotherapies. Recently, many studies in this field have appeared, and some different combination therapies already approved or still in clinical trial will be discussed in the following subsection. Specifically, we will focus on drug–drug combination, leaving out radiotherapy.

Antiangiogenic drugs and chemotherapy

Bevacizumab is not intrinsically cytotoxic in normal or tumor cells. When combined with chemotherapy, bevacizumab has been shown to prolong the survival of patients with cancer, especially for those with metastatic CRC and NSCLC. A series of phase III clinical studies have demonstrated the survival benefit of adding bevacizumab to chemotherapy in metastatic CRC [87]. The most common backbone regimen for combination is FOLFOX (fluorouracil-leucovorin-oxaliplatin) [88–90]. The AVEX phase III study, comparing capecitabine–bevacizumab with capecitabine monotherapy in elderly patients (age > 70 years) with untreated metastatic CRC, showed a significant bevacizumab-related benefit in median PFS (median PFS: 9.1 vs. 5.1 months) [91]. Several meta-analyses have confirmed the benefit of bevacizumab in terms of PFS and OS in the first-line treatment of metastatic CRC [92–95]. Nevertheless, it is widely accepted that the addition of bevacizumab to first-line chemotherapy offers a modest clinical benefit.

The therapeutic outcomes of bevacizumab combined with paclitaxel and carboplatin were also explored in lung cancer in phase II/III trials (AVF-0757g, JO19907, ECOG 4599, and AVAiL trials), where bevacizumab proved effective in improving outcomes [96]. The Center for Drug Evaluation of China approved bevacizumab, in combination with platinum-based doublet chemotherapy, for the first-line treatment of patients with unresectable, locally advanced, recurrent, or metastatic NSCLC [96–98]. Adding bevacizumab to platinum-based chemotherapy in advanced non-squamous NSCLC confers a significant benefit in terms of PFS (median PFS: 9.2 vs. 6.5 months) and OS (median OS: 24.3 vs. 17.7 months) compared with chemotherapy alone, while safety was like previous studies in NSCLC and no new safety signals were observed [97].

In a meta-analysis, the addition of bevacizumab to chemotherapy can significantly improve PFS and ORR both in first- and second-line treatments of advanced NSCLC; however, there was no benefit in terms of OS [99]. In the setting of maintenance therapy, PFS was significantly improved by bevacizumab combined with pemetrexed following first-line induction vs. bevacizumab alone (7.4 versus 3.7 months). Further, OS was prolonged upon combinatory maintenance therapy (17.1 versus 13.2 months), but the treatment led to an increase in adverse events, including anemia, thrombocytopenia, fatigue, neutropenia, febrile neutropenia, sensory neuropathy, and alopecia [100]. Some retrospective studies provided limitations regarding the benefit/risk profile in elderly patients; therefore, bevacizumab in combination with chemotherapy is not recommended for eligible patients aged >75 years [101–104].

First-line bevacizumab combined with chemotherapy significantly improves efficacy versus chemotherapy alone in HER2-negative locally recurrent or metastatic BC. The main clinical effectiveness data were derived from a single trial (E2100) that evaluated the addition of bevacizumab to weekly paclitaxel in patients who had not previously received chemotherapy for advanced disease. This trial reported significant increases in median PFS for the combination therapy (median PSF: 11.3 vs. 5.8 months), while median OS was not significantly different between the two groups [105,106]. The additional AVADO trial, evaluating the addition of bevacizumab to docetaxel, reported a markedly smaller benefit of adding bevacizumab to docetaxel than that reported for adding bevacizumab to paclitaxel in E2100 [107]. The combination of bevacizumab with commonly used chemotherapies, including capecitabine-based, taxane-based, or anthracycline-based chemotherapies, also improved PFS (5.1 to 7.2 months, RiBBON-2 phase III trial) in the second-line treatment of patients with HER2-negative metastatic BC, with a safety profile comparable with that in prior phase III studies. There was no statistically significant difference in OS, and the most common grade ≥3 adverse events related to bevacizumab treatment were hypertension and proteinuria [108]. FDA did not approve bevacizumab for BC treatment, while, based on E2100 and RiBBON-1 and -2 trials, EMA approved the combination of bevacizumab with paclitaxel and capecitabine in metastatic BC as first-line treatment.

Bevacizumab in combination with INFα-2a is active as first-line treatment in patients with metastatic RCC. AVOREN phase III trial demonstrated a median OS of 23.3 months with bevacizumab plus IFN and 21.3 months with IFN plus placebo, while the increase in the primary endpoint of OS by 2 months was not significant. Tolerability was similar to that reported previously [109].

Bevacizumab in combination with carboplatin and paclitaxel, in the first-line treatment, is approved for patients with epithelial ovarian, fallopian tube, or primary peritoneal cancer. The efficacy and safety of combination compared to chemotherapy alone were studied in two phase III trials (GOG-0218 and ICON7). In the GOG-0218 phase III trial, combining bevacizumab with chemotherapy, followed by maintenance of bevacizumab until progression, improved PFS (median PFS: 14.7 vs. 11.6 months), but not OS [110,111]. A retrospective analysis suggested a positive association between tumor microvessel density and tumor VEGF-A levels and magnitude of bevacizumab effect in ovarian cancer [112]. ICON7 trial showed that, in ovarian cancer, the benefits with respect to both PFS and OS were greater among those at high risk for disease progression [113].

Bevacizumab in combination with carboplatin and gemcitabine or in combination with carboplatin and paclitaxel is indicated for the treatment of patients with first recurrence of epithelial ovarian carcinoma, fallopian tube carcinoma, or primary peritoneal cancer. Three phase III trials studied the efficacy and safety of combinations. OCEANS trial evaluated the PFS of bevacizumab in combination with carboplatin and gemcitabine, followed by bevacizumab as a single agent in patients with platinum-sensitive recurrent epithelial ovarian, fallopian tube, or primary peritoneal cancer. The trial showed a statistically significant improvement in PFS (median PFS: 12.4 vs. 8.4 months) compared with chemotherapy, with grade 3 or higher hypertension and proteinuria, which occurred more frequently in the bevacizumab arm [114]. Results from final OS analysis of the study showed no significant difference [115]. GOG-0213 evaluated the combination of bevacizumab with carboplatin and paclitaxel in the same oncological setting and regimen. The trial showed that the addition of bevacizumab to standard chemotherapy, followed by maintenance therapy until progression, improved the median OS (median OS: 42.2 vs. 37.3 months) in patients with platinum-sensitive recurrent ovarian cancer [116]. AURELIA trial evaluated the efficacy and safety of bevacizumab in combination with paclitaxel, topotecan, or pegylated liposomal doxorubicin in patients with platinum-resistant recurrent epithelial ovarian, fallopian tube, or primary peritoneal cancer. The trial showed a significant improved PFS (median PSF: 6.7 vs. 3.4 months) and ORR (27.3% vs. 11.8%) in bevacizumab arm compared to chemotherapy-alone arm. In line with other studies, the OS trend was not significant and grade ≥2 hypertension and proteinuria were more common with bevacizumab [117]. In additional exploratory analyses, PFS and ORR improvement with bevacizumab was consistent in older and younger patients, while grade ≥3 hypertension was more common in elderly [118].

Finally, bevacizumab in combination with chemotherapy (paclitaxel and cisplatin or paclitaxel and topotecan) is also approved in the treatment for patients with persistent, recurrent, or metastatic carcinoma of the cervix. In GOG-0240 phase III trial, the addition of bevacizumab to combination chemotherapy was associated with an improvement of 3.7 months in median OS (median OS: 17.0 vs. 13.3 months) and 2.3 months in median PSF (median PSF: 6 vs. 8.3 months) and higher ORR (48% vs. 36%). Bevacizumab arm was associated with an increased incidence of hypertension of grade 2 or higher, thromboembolic events of grade 3 or higher, and gastrointestinal fistulas of grade 3 or higher [119].

Ramucirumab in combination with paclitaxel is indicated for the treatment of patients with advanced GC or GEJ adenocarcinoma with disease progression after prior platinum and fluoropyrimidine chemotherapy. RAINBOW phase III study investigated ramucirumab plus paclitaxel versus placebo plus paclitaxel in second-line treatment, following platinum- and fluoropyrimidine-containing chemotherapy, with or without anthracycline [120]. OS was improved in patients receiving ramucirumab plus paclitaxel compared with those receiving placebo plus paclitaxel (median OS: 9.63 vs. 7.36 months). Similarly, PSF was increased of 1.5 months in favor of the combination arm (4.4 months vs. 2.9 months in the placebo plus paclitaxel arm). The ORR in the Cyramza plus paclitaxel arm was 27.9% and in the placebo plus paclitaxel arm was 16.1%. Improvements in OS and PFS were consistently observed in pre-specified subgroups based on age, sex, and race and in most other pre-specified subgroups [121,122].

Ramucirumab in combination with irinotecan/5-fluorouracil/folinic acid (FOLFIRI) chemotherapy is indicated for the treatment of metastatic CRC with disease progression on or after prior therapy with bevacizumab, oxaliplatin, and fluoropyrimidine. RAISE phase III study investigated the efficacy and safety of ramucirumab plus FOLFIRI versus placebo plus FOLFIRI [123]. Both OS and PFS were improved in patients receiving the combination arm compared with chemotherapy alone (median OS: 13.3 vs. 11.7 months; median PSF: 5.7 vs. 4.5 months). The ORR was similar for both treatment arms (13.4% versus 12.5%, ramucirumab plus FOLFIRI versus placebo plus FOLFIRI, respectively), and patients in the ramucirumab plus FOLFIRI treatment arm reported a transient decrease in quality of life. Further, the addition of ramucirumab to FOLFIRI improved patient outcomes, regardless of RAS/RAF and BRAF mutation status, and tumor sidedness, while low baseline carcinoembryonic antigen (CEA) level might identify a subgroup of patients with metastatic CRC who obtain greater benefit from ramucirumab [124,125].

Ramucirumab in combination with docetaxel is indicated for the treatment of metastatic NSCLC with disease progression on or after platinum-based chemotherapy. REVEL phase III study demonstrated that OS was significantly improved in patients receiving ramucirumab plus docetaxel compared with those receiving placebo plus docetaxel ($P = .024$), with median OS of 10.5 in combination therapy arm vs. 9.1 months in the placebo plus docetaxel arm [126]. There was an increase in median PFS of 1.5 months in favor of the ramucirumab plus docetaxel arm and ORR (median PSF: 4.5 vs. 3 months; ORR: 22.9% vs. 13.6%). The most common grade 3 or worse adverse events were similar in the two arms. Improvement for PFS and OS for ramucirumab plus docetaxel vs. placebo plus docetaxel was observed in several subgroups, including non-squamous and squamous histology, time since start of prior therapy <9 months, and patient <65 years old. No differences in efficacy between treatment arms have been observed in the subgroups of patients ≥65 years old and those with time since start of prior therapy ≥9 months [127–129]. Finally, a recent exploratory analysis suggested that second-line ramucirumab plus docetaxel might be effective regardless of the prior treatment with platinum-based chemotherapy plus a taxane, pemetrexed, gemcitabine, or bevacizumab [126].

Efficacy and safety of nintedanib plus docetaxel have been assessed in the phase III LUME-Lung 1 study as second-line therapy for NSCLC. PFS was significantly improved in the combination arm compared with the docetaxel plus placebo arm (median PFS: 3.4 vs. 2.7 months). OS was significantly improved by the combination treatment for patients with adenocarcinoma histology (median OS: 12.6 vs. 10.3 months), but not in the total study population (median OS: 10.1 vs. 9.1 months). Grade 3 or worse adverse events were more common in the combination arm than in the docetaxel plus placebo group and included diarrhea and reversible increases in alanine aminotransferase and in aspartate aminotransferase [30].

Ziv-aflibercept in combination with FOLFIRI chemotherapy is indicated in adults with metastatic CRC that is resistant to or has progressed after an oxaliplatin-containing regimen. The combination of ziv-aflibercept with FOLFIRI in VELOUR phase III trial improved median OS (13.5 vs. 12.06 months), median PFS (6.90 vs. 4.67 months), and ORR (19.8% vs. 11.1%) [130]. Subgroup analyses for OS and PFS according to age, gender, presence of liver metastasis only, history of prior hypertension, and number of organs involved showed a better effect for ziv-aflibercept/FOLFIRI regimen than for placebo/FOLFIRI regimen [131,132].

Of note, except for nintedanib, all the other antiangiogenic multi-TKI drugs are currently not approved for use in combination with chemotherapy. As VEGF signaling has been shown to be necessary, but not sufficient, to control the tumor angiogenic process, we can expect interesting results from the many trials assessing the combination of antiangiogenic TKIs with chemotherapy. A list of ongoing trials on combination of antiangiogenic drugs combined with chemotherapy is reported in Table 2.

Antiangiogenic drugs and hormonal therapy

Endocrine therapy has become a key part in the pharmacotherapy of hormone-responsive cancers. Several preclinical models and clinical trials suggest that a combination of antiangiogenic drugs may improve the efficacy of hormonal therapy, particularly in hormone-sensitive BC [133,134]. The rationale of antiangiogenic drug use in endocrine-sensitive tumors is to delay the onset of resistance. In fact, the contribution of angiogenesis as the potential mechanism for endocrine therapy resistance has been extensively reported [135–137]. VEGF overexpression in drug-resistant cells has reported to increase the activation of MAPK, which is proven to contribute to anti-hormonal drug resistance [138,139]. Based on these observations, the combination of VEGF inhibitors with hormonal therapy was supposed to improve the efficacy of hormonal therapy and delay the emergence of resistance in patients who have hormone-sensitive BC. Surprisingly, many authors reported that the use of anti-VEGF was not observed to be useful in overcoming tamoxifen resistance, probably due to redundance of angiogenesis modulators [140]. The clinical efficacy of bevacizumab

TABLE 2 Clinical trials on combination of antiangiogenic drugs with conventional chemotherapy.

Clinical trials	Phase	Cancer type	Antiangiogenic agents	Chemotherapy	References
Approved					
NCT00484939 (AVEX)	III	mCRC	Bevacizumab	Capecitabine	[91]
(AVF-0757g)	II	NSCLC	Bevacizumab	Paclitaxel, carboplatin	[96–98]
CTI-060338 (JO19907)	II	NSCLC	Bevacizumab	Paclitaxel, carboplatin	[96–98]
NCT00021060 (ECOG 4599)	III	NSCLC	Bevacizumab	Paclitaxel, carboplatin	[96–98]
NCT00806923 (AVAiL)	III	NSCLC	Bevacizumab	Cisplatin, gemcitabine	[96–98]
NCT00961415 (AVAPERL (MO22089))	III	Advanced NSCLC	Bevacizumab	Pemetrexed	[100]
NCT00762034 (9707 H3E-MC-JMHD)	III	Advanced NSCLC	Bevacizumab	Pemetrexed, carboplatin	[101]
NCT01303926 (ERACLE)	III	Advanced NSCLC	Bevacizumab	Pemetrexed, carboplatin	[102]
NCT00451906 (MO19390)	IV	Non-squamous NSCLC	Bevacizumab	Platinum-based chemotherapy	[103]
NCT00028990 (E2100)	III	HER2-negative BC	Bevacizumab	Paclitaxel	[105,106]
NCT01200212 (AVADO)	III	BC	Bevacizumab	Taxane, xeloda	[107]
NCT00281697 (RiBBON-2)	III	BC	Bevacizumab	Taxane/gemcitabine/ vinorelbine/ capecitabine	[108]
NCT02056587 (AVOREN)	III	mRCC	Bevacizumab	INF-α	[109]
NCT00262847 (GOG-0218)	III	EOC, peritoneal, fallopian tube cancer	Bevacizumab	Carboplatin, paclitaxel	[110]
NCT00483782 (ICON7)	III	EOC, peritoneal, fallopian tube cancer	Bevacizumab	Carboplatin, paclitaxel	[113]
NCT00434642 (OCEANS)	III	EOC, peritoneal, fallopian tube cancer	Bevacizumab	Carboplatin, gemcitabine	[114,115]

TABLE 2 Clinical trials on combination of antiangiogenic drugs with conventional chemotherapy—cont'd

Clinical trials	Phase	Cancer type	Antiangiogenic agents	Chemotherapy	References
NCT00565851 (GOG-0213)	III	EOC, peritoneal, fallopian tube cancer	Bevacizumab	Paclitaxel, docetaxel, carboplatin	[116]
NCT00976911 (AURELIA)	III	EOC, peritoneal, fallopian tube cancer	Bevacizumab	Liposomal doxorubicin, paclitaxel, topotecan	[117,118]
NCT00803062 (GOG-0240)	III	CC	Bevacizumab	Paclitaxel, cisplatin/ paclitaxel, topotecan	[119]
NCT01170663 (RAINBOW)	III	Advanced GC or GEJ adenocarcinoma	Ramucirumab	Paclitaxel	[120–122]
NCT01183780 (RAISE)	III	mCRC	Ramucirumab	FOLFIRI	[123–125]
NCT01168973 (REVEL)	III	Metastatic NSCLC	Ramucirumab	Docetaxel	[126–129]
NCT00805194 (LUME-Lung 1)	III	NSCLC	Nintedanib	Docetaxel	[30]
NCT00561470 (VELOUR)	III	mCRC	Ziv-aflibercept	FOLFIRI	[130–132]
Ongoing					
NCT04733963 (FRUCA)	II	CRC	Fruquintinib/ bevacizumab	Capecitabine	
NCT01548924 (CNIO-BR-01-2011)	I	Solid tumor	Dovitinib	Paclitaxel	
NCT00219557 (A4061016)	II	Pancreatic cancer	Axitinib	Gemcitabine	
NCT00471146 (A4061028)	III	Pancreatic cancer	Axitinib	Gemcitabine	
NCT00870532 (08-3-LUN)	I	NSCLC	Sorafenib	Metronomic vinorelbine	
NCT00449033 (NExUS)	III	Advanced NSCLC	Sorafenib	Gemcitabine, cisplatin	
NCT03100955 (QU20170327)	III	SCLC	Apatinib	Cisplatin, etoposide	
NCT03775395 (HCC-S055)	III	HCC	Lenvatinib, sorafenib	Oxaliplatin, fluorouracil, leucovorin	
NCT05003700 (NFEC-2020-130)	III	HCC	Lenvatinib, camrelizumab	Oxaliplatin, raltitrexed	

Continued

TABLE 2 Clinical trials on combination of antiangiogenic drugs with conventional chemotherapy—cont'd

Clinical trials	Phase	Cancer type	Antiangiogenic agents	Chemotherapy	References
NCT02790580 (BR01/01/16)	II	BC	Sunitinib	Doxorubicin, cyclophosphamide	
NCT01803503 (BR01/08/13)	II	BC, NSCLC, GC, Prostate cancer	Sunitinib	Docetaxel	
NCT00457691 (A6181122)	III	mCRC	Sunitinib	FOLFIRI	
NCT01478594 (4130-CL-0201)	II	CRC	Tivozanib, bevacizumab	mFOLFOX6	
NCT01745367 (BATON-BC)	II	TNBC	Tivozanib	Paclitaxel	
NCT01834183 (13-084)	II	RCC	Tivozanib	Gemcitabine	

GC, *gastric cancer;* RCC, *renal cell carcinoma;* CC, *carcinoma of the cervix;* NSCLC, *non-small cell lung cancer;* SCLC, *small-cell lung cancer;* HCC, *hepatocellular carcinoma;* mCRC, *metastatic colorectal cancer;* GIST, *gastrointestinal stromal tumors;* EOC, *epithelial ovarian cancer;* BC, *breast cancer;* TNBC, *triple-negative breast cancer.*

addition to endocrine therapy as the first-line treatment for hormone receptor-positive metastatic BC is still debated. Recently, data from two phase III randomized trials showed that adding bevacizumab to first-line endocrine therapy of metastatic BC significantly improves PFS and ORR, but does not impact OS at the cost of significant toxicity [140]. The same trial showed that a combination of bevacizumab and letrozole, an aromatase inhibitor, improved PFS, but this benefit was associated with a markedly increased risk of grade 3–4 toxicities [141]. By contrast, a case-control study has showed a stable metastatic BC disease successfully treated with maintenance therapy using bevacizumab and tamoxifen [142]. Bevacizumab is currently approved with paclitaxel or capecitabine, but not hormone therapy, in metastatic BC in patients with HER-2-negative disease (for details see above paragraph).

Antiangiogenic drugs and target therapies

Targeted therapies interfere with specific proteins involved in tumor growth. The specific pharmacological inhibition of these oncogene products in tumor cells can restore blood vessels and potentially block the pro-angiogenic signaling activated by individual tumors. In this context, target tumor angiogenesis could rescue the equilibrium of angiogenic signals by targeting the mutated oncogenes, which play a central role in this process. To potentially reduce acquired resistance by tumor cells against target therapies, a combined strategy of antiangiogenic drugs and target therapies was evaluated in many clinical trials in the recent years [143]. Following is a list of targets with the related combination regimens.

EGFR

There are several ongoing or concluded clinical trials exploring the possibility of combining anti-EGFR drugs (e.g., erlotinib, cetuximab, and panitumumab) with different antiangiogenic drugs such as bevacizumab, ramucirumab, or sorafenib [143].

The randomized phase II study published by Seto et al. proved that the addition of bevacizumab to erlotinib as the first-line treatment confers a significant clinical improvement in patients with advanced NSCLC carrying activating EGFR mutations [144]. These results were confirmed in patients with EGFR-mutated NSCLC by another phase II, single-arm trial (BELIEF) that provided further evidence on the benefit of this drug association [145] and by a phase III study (NEJ026) demonstrating a significant longer PFS of this combination when compared to erlotinib alone (median PFS ad interim: 16.9 vs. 13.3 months) [146]. In April 2016, EMA approved bevacizumab in combination with erlotinib for the first-line treatment of adult patients with advanced, unresectable, metastatic, or recurrent NSCLC with activating mutations of EGFR [147].

Other anti-EGFR therapies (e.g., cetuximab and panitumumab) were evaluated in combination to bevacizumab in patients with metastatic CRC. Two large, randomized phase III studies have evaluated their efficacy to an oxaliplatin-containing chemotherapy doublet plus bevacizumab: the CAIRO-2 study (cetuximab) and PACCE study (panitumumab) [148,149]. Both studies reported that the addition of these anti-EGFR antibodies did not improve the clinical efficacy.

As for ramucirumab, a phase III, randomized clinical trial (RELAY) reported a superior PFS of ramucirumab + erlotinib vs. placebo + erlotinib in patients with untreated, advanced, EGFR-mutated, NSCLC (median PFS: 19.4 vs. 12.4 months) [150]. In 2020, both FDA and EMA approved ramucirumab in combination with erlotinib for the first-line treatment of metastatic NSCLC with activating EGFR mutations.

As for sorafenib, its use has been evaluated in several types of cancer in combination with an anti-EGFR therapy. The efficacy of sorafenib + erlotinib was evaluated in a phase III, randomized, double-blind, placebo-controlled trial in patients with advanced HCC (SEARCH trial) [151]. Also, the combination of sorafenib + cetuximab was investigated in different cancers such as CRC [152,153] and head and neck squamous cell carcinoma [154]. Unfortunately, for all these cancers, any meaningful benefit was observed.

A phase I study evaluating the antitumor property of the combination of regorafenib and cetuximab in patients with advanced cancers showed a clinical benefit in all patients and in particular the combination demonstrated early signals of activity in wild-type CRC [155].

In August 2021, only the combination of bevacizumab (EMA) and ramucirumab (EMA and FDA) with erlotinib has been approved for the treatment of NSCLC. Anyway, the association of other anti-VEGF therapies (such as cabozantinib) with anti-EGFR preliminarily demonstrated to be promising for different types of cancer (clincialtrials.gov Identifier NCT03667482, NCT01866410).

HER2

HER2 is a known target in the treatment of cancer and, in particular, in the treatment of advanced/metastatic BC [156]. A phase II study, evaluating the association of bevacizumab to trastuzumab and docetaxel in women with HER2-positive metastatic BC, reported PFS

(14.3 months) and ORR (46%) comparable to higher dose of docetaxel [157]. Two large phase III trials evaluated the efficacy of bevacizumab and trastuzumab with chemotherapy in HER2-positive BC (BETH and AVAREL). The results of these trials did not demonstrate that the association improved the outcome for these patients [158,159]. Moreover, a phase II trial evaluated the combination of pazopanib with lapatinib in HER2-positive BC showing no increase in PFS [160].

mTOR

Mechanistic target of rapamycin (mTOR) is a protein kinase regulating cell growth, survival, metabolism, and immunity, and many mTOR inhibitors have been developed to treat cancer [161]. In a phase II trial, the mTOR inhibitor everolimus has been associated with bevacizumab to treat patients with advanced papillary variant RCC. Results of this trial confirm robust activity of this association as a first-line treatment. Among 37 evaluable patients, the 6-month PFS rate was 78%, the median PFS was 13.7 months, and the ORR was 35%. With a median follow-up of 17.6 months, the median OS was 33.9 months and safety was consistent with prior reports for everolimus + bevacizumab [162]. Another phase II trial (RECORD-2) evaluated this association in patients with metastatic RCC. Anyway, in this trial, the efficacy of everolimus + bevacizumab association compared to INFα-2a + bevacizumab appeared similar (median PFS: 9.3 and 10.0 months) [163].

This association has also been tested in patients with refractory, progressive, and intracranial meningiomas [164], demonstrating a good tolerability and promoting stable disease in 88% of patients.

Other anti-VEGF therapies have been tested in combination with everolimus (e.g., sorafenib and lenvatinib). While sorafenib in combination with everolimus vs. sorafenib alone did not demonstrate an improved benefit in patients with advanced HCC [165], the association of lenvatinib + everolimus demonstrated to be effective compared to everolimus alone in patients with advanced RCC who have progressed after one previous VEGF-targeted therapy (median PFS: 14.6 vs. 5.5 months) [166]. Lenvatinib in combination with everolimus received the approval by FDA for the treatment of patients with advanced RCC following one prior antiangiogenic therapy.

BRCA

As for poly (ADP-ribose) polymerase (PARP) inhibitors, in 2019 AstraZeneca and MSD announced positive results from the phase III PAOLA-1 study. The study assessing olaparib in combination with bevacizumab demonstrated statistically significant and clinically relevant improvement in PFS in women with newly diagnosed advanced ovarian cancer [167]. Olaparib in combination with bevacizumab is approved by EMA for maintenance treatment of adult patients with advanced high-grade epithelial ovarian cancer, fallopian tube cancer, or peritoneal cancer in response (complete or partial) after completion of first-line platinum-based chemotherapy in combination with bevacizumab and whose tumor shows a deficiency of homologous recombination deficiency, defined by the presence of a BRCA1/2 mutation and/or genomic instability.

ALK

ALK gene has been found to be rearranged, mutated, or amplified in a series of tumors, including NSCLC. One prospective observation study reported that crizotinib, an ALK inhibitor, plus bevacizumab showed benefit in treating naive *ALK* rearrangement in NSCLC patients, and the toxicity was relatively acceptable (median PFS: 13.9 months; ORR: 58.3%) [168]. An ongoing clinical trial is evaluating the association of alectinib with bevacizumab (clinicaltrials.gov identifier NCT03779191) in patients with ALK-positive NSCLC.

MEK

Clinical trials are evaluating or have already evaluated the association of MEK inhibitors with antiangiogenic therapies and immunotherapy. One study is ongoing evaluating the possible association of bevacizumab + pembrolizumab + binimetinib in patients with refractory CRC (NCT03475004), while another is assessing the combination of bevacizumab + atezolizumab and cobimetinib in patients with gastrointestinal or other tumors (NCT02876224).

Antiangiogenic drugs and immune checkpoint inhibitors

For the last decade, immune checkpoint inhibitors (ICIs) have revolutionized the treatment landscape of advanced cancers. ICIs rejuvenate dysfunctional or exhausted cytotoxic TILs to exert potent antitumor effector functions, thereby enabling effective and durable control of previously refractory cancers. However, only 20%–30% of patients with cancer respond to ICI treatments, and it is difficult to predict these responders before treatment, because the immune system is too complex to decode with a single biomarker. Among the various determinants of cancer immunity, the TME functions as a major obstacle that severely impairs the efficacy of ICIs. Within the TME, the aberrant immune-vascular cross talk generates an endothelial barrier that hinders T-cell infiltration into the tumor and also impairs T-cell effector functions and even leads to T-cell apoptosis within the TME. VEGF, as reported above, negatively impacts cancer immunity at multiple steps. Thus, the inhibition of VEGF-induced signaling cascades could ideally suppress tumor growth through two mechanisms: by suppressing tumor angiogenesis characterized by aberrant and undifferentiated vessels and by exerting immunosupportive effects on tumor cells. Due to the dual effect of VEGF on angiogenesis and in the TME, several studies addressed the combinatory effect of antiangiogenic therapies with immune checkpoint blockade. However, to avoid the onset of extreme hypoxia in TME associated with antiangiogenic drugs and, on the contrary, to establish an immunosupportive TME, optimal dose of antiangiogenic treatment is required to achieve vessel normalization. It is not clear which dose corresponds to either the vascular normalizing dose or the vascular destructing dose in patients with cancer. In this context, metronomic chemotherapy could be a promising strategy [169,170] (see also Chapter 8 of this book).

There are currently few approved regimens of combining ICIs with antiangiogenic drugs by FDA, while EMA has not still approved any combination. In this paragraph we focus on most advanced clinical trials where this strategy is currently being investigated.

Cancer immunotherapy is based on the use of ICIs directed against immune checkpoint molecules, vital components of immune homeostasis. The most common targets of ICIs are the immune checkpoint molecules CTLA-4, PD-1, and its ligands PD-L1 and PD-L2. Their blockade results in the removal of inhibitory signals of T-cell activation, enhancing antitumor immune activity and leading to the inhibition of tumor growth, through CD8+ T-cell-mediated cancer cell death.

Anti-VEGFRs are the first-line therapy for patients with metastatic RCC. More recently, their use in combination with ICIs has been evaluated (Table 3). Two phase III clinical trials compared the effect of sunitinib versus the combinatory therapy of axitinib (VEGFR and PDGFR inhibitor) and pembrolizumab (anti-PD-1 monoclonal antibody) (KEYNOTE-426) or combined with avelumab (anti-PD-L1 monoclonal antibody) (JAVELIN Renal 101) [175,177]. In both cases, the combination induced a significant increase in PFS of cancer patients (median PFS: 16.6 vs. 11.2 months in KEYNOTE-426 trial for combination treatment arm vs. sunitinib arm, and 15.1 vs. 11.1 months in JAVELIN Renal 101 trial for combination treatment arm vs. sunitinib). In KEYNOTE-426 trial, patients with previously untreated metastatic RCC were treated with either pembrolizumab and axitinib combination therapy or sunitinib monotherapy, and the combination group had a 47% reduced risk of death, and the ORR of the combination was 59.3% compared with 35.7% in the sunitinib group. Although the incidence of hepatic toxicity was higher in the combination group, no relevant lethal event occurred. Based on the significative efficacy and acceptable toxicity profile, combination therapy with pembrolizumab and axitinib was approved by the FDA for treatment-naïve patients with metastatic RCC. Additionally, the combination of axitinib and pembrolizumab prolonged the OS of cancer patients [177,179].

JAVELIN Renal 101 (NCT02684006) is a phase III clinical trial that evaluated the efficacy of avelumab and axitinib combination therapy against sunitinib monotherapy in patients with metastatic RCC in a first-line setting. In addition to PFS reported earlier, the ORR was 51.4% vs. 25.7%, showing a significant efficacy of the combination compared to the monotherapy.

TABLE 3 Clinical trials investigating the combination effect of antiangiogenic agents and ICIs.

Clinical trial	Phase	Cancer type	Antiangiogenic agents	ICI	References
NCT02366143 (IMpower150)	III	NSCLC	Bevacizumab	Atezolizumab	[171]
NCT03434379 (IMbrave 150)	III	HCC	Bevacizumab	Atezolizumab	[172]
NCT03475953 (REGOMUNE)	II	CRC	Regorafenib	Avelumab	[173]
NCT02420821 (IMmotion151)	III	RCC	Bevacizumab	Atezolizumab	[174]
NCT02684006 (JAVELIN Renal 101)	III	RCC	Axitinib	Avelumab	[175]
NCT02501096	II	RCC	Lenvatinib	Pembrolizumab	[176]
NCT02853331 (KEYNOTE-426)	III	RCC	Axitinib	Pembrolizumab	[177]
NCT02501096	II	EC	Lenvatinib	Pembrolizumab	[178]

EC, *endometrial cancer;* CRC, *colorectal cancer;* HCC, *hepatocellular carcinoma;* NSCLC, *non-small cell lung cancer;* RCC, *renal cell cancer.*
From http://clinicaltrials.gov/.

Grade $\geq$3 toxicity was comparable (71.2% vs. 71.5%) between the two groups. Based on this study, the FDA approved avelumab for use in combination with axitinib as first-line treatment for patients with advanced RCC [180]. Additional clinical trials are ongoing in patients with advanced RCC based on preexisting studies that showed promising results with PD-1/PD-L1 inhibitor and antiangiogenic agent combination therapy.

For the treatment of RCC, another combination of ICIs and antiangiogenic drugs has been shown to be very effective and indicated for the first-line treatment of advanced RCC in adults: cabozantinib plus nivolumab. The phase III NCT03141177 trial showed that nivolumab plus cabozantinib had significant benefits over sunitinib with respect to PFS (PFS at 18 months: 16.6 vs. 8.3 months), OS (probability of OS at 12 months: 85.7% vs. 75.6%), and ORR (75.3% vs. 70.6%) in first-line setting for advanced RCC [181]. In addition, in combination with atezolizumab, cabozantinib was investigated in the first-line setting versus sorafenib for advanced HCC in the phase III COSMIC-312 trial [182].

In HCC, the phase III trial IMBRAVE 150 demonstrated significant improvements in PFS and OS, using the combination of atezolizumab and bevacizumab compared with sorafenib. This was the first study to propose a new first-line treatment option that is superior to sorafenib, which has been the standard of care for a decade. The median PFS was 6.8 versus 4.5 months, and the ORR was 27% versus 12%, respectively. In addition, the grade 3–4 adverse event rate was comparable (57% vs. 55% of patients, in combination therapy arm and in the sorafenib arm, respectively). Further, the combination group exhibited delayed deterioration of quality of life compared with sorafenib [183]. The safety and efficacy of the combination of pembrolizumab and lenvatinib as a potential first-line treatment for patients with advanced HCC are now under evaluation in LEPP-002, a phase III trial [172,183].

The IMpower150 phase III trial reported a significant improvement of PFS and OS in metastatic non-squamous subset of NSCLC patients after the combined administration of atezolizumab, bevacizumab, and chemotherapy (carboplatin and paclitaxel) (median PFS: 8.3 vs. 6.8 months; median OS: 19.2 vs. 14.7 months) [171]. The ORR was significantly higher in the ICI combination group than in the standard combination group (ORR: 63.5% vs. 48.0%), whereas the adverse event rate was comparable. Based on these results, atezolizumab was approved by the FDA for use in combination with bevacizumab, paclitaxel, and carboplatin as first-line treatment for patients with metastatic non-squamous NSCLC. Of note, atezolizumab combinatory usage in NSCLC treatment is effective regardless of PD-L1 expression.

Recently, the FDA granted accelerated approval for the use of a combination of pembrolizumab and lenvatinib in patients with advanced endometrial cancer who have experienced disease progression after systemic therapy. The approval was based on the results of the single-arm, multicenter, open-label, multicohort phase Ib/II KEYNOTE-146 trial. In the study, interim analysis showed that the ORR was 39.6% and 45.3% at 24 weeks by investigator review and independent review, respectively. The most common treatment-related adverse events were hypertension, fatigue, diarrhea, and hypothyroidism. No cases of grade 4 adverse events were reported [178].

Finally, an interesting phase II trial, REGOMUNE, investigated the efficacy of regorafenib plus avelumab in a subset of patients with microsatellite stable advanced CRC. The study showed that outcomes (median PFS and OS were 3.6 months and 10.8 months, respectively) were positively correlated with tumor infiltration of macrophages and cytotoxic CD8+ T cells [173].

In addition to the abovementioned clinical trials, numerous phase I studies have been concluded or are ongoing in various malignancies to prove the efficacy of combining PD-1/PD-L1 inhibitors and anti-VEGF agents. We expect that these ongoing trials generate consistent results, which will evolve the therapeutic landscape of advanced cancers.

Conclusions

Years of research have demonstrated the potential of antiangiogenic drugs as well as their limitations, which have led to further attempts to overcome their limits by combination therapies. The combination of antiangiogenic agents with chemotherapies and, more recently, with ICIs is yielding promising results in clinical studies, changing the treatment landscape for several solid tumors (Fig. 1).

Nonetheless, several issues remain to optimize the efficacy of this combination therapy. First, predictive biomarkers must be developed to identify the subset of patients who will

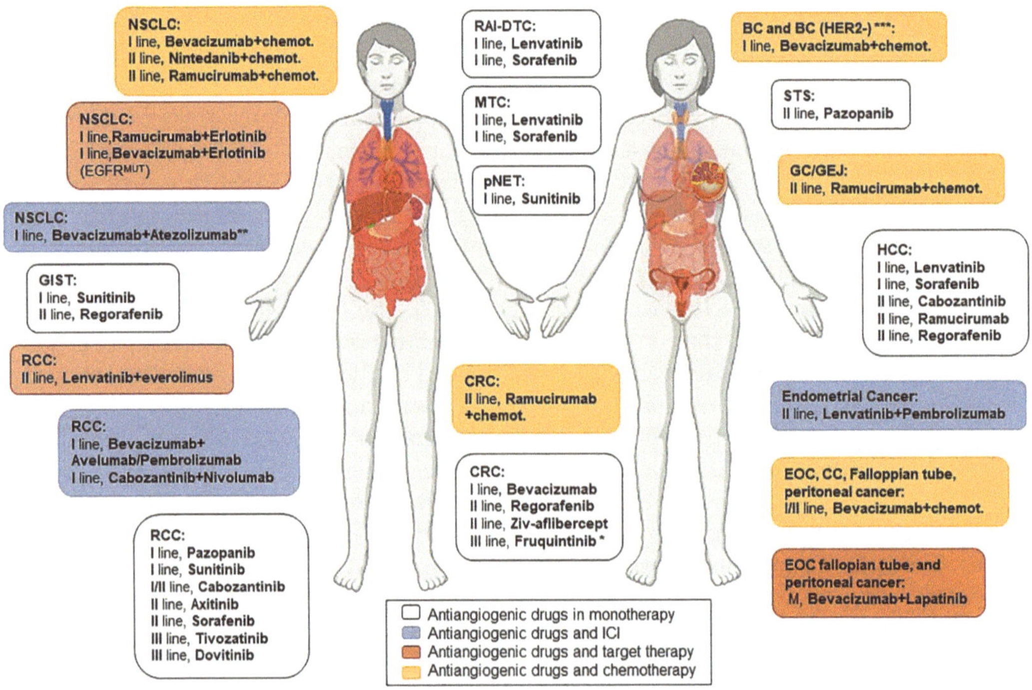

FIG. 1 Approved antiangiogenic regimens for the treatment of cancer. Graphic summary of the approved use of antiangiogenic drugs in therapeutic regimes against cancer. *GC*, gastric cancer; *RCC*, renal cell carcinoma; *NSCLC*, non-small cell lung cancer; *HCC*, hepatocellular carcinoma; *CRC*, colorectal cancer; *GIST*, gastrointestinal stromal tumors; *GC*, gastric cancer; *GEJ*, gastroesophageal junction adenocarcinoma; *EOC*, epithelial ovarian cancer; *CC*, carcinoma of the cervix; *BC*, breast cancer; *STS*, soft-tissue sarcoma, *RAI-DTC*, radioactive iodine-refractory differentiated thyroid cancer; *MTC*, medullary thyroid cancer; *pNET*, pancreatic neuroendocrine tumor; *M*, maintenance; *chemot.*, chemotherapy. *Approved only by CFDA; **approved only by the FDA; ***approved only by the EMA.

benefit from these combination treatments (see Chapter 10 specifically for CRC). For tumor angiogenesis and the immune status of TME there are still not clear and validated biomarkers. Second, the focus on anti-VEGF/VEGFR agents as the main antiangiogenic drugs should be expanded to agents targeting other angiogenic mediators, such as FGF/FGFR, PDGF/PDGFR, and Ang2, among others. Finally, the angiogenic phenotype differs according to organ, as well as the modality of neovessel formation in tumors; thus, more in-depth analyses must be performed to expand our knowledge of the response to antiangiogenic treatment and its combination at the organ level.

Acknowledgment

This work was sponsored by Regione Toscana-Bando Ricerca Salute 2018_CORELAB.

References

[1] Nagl L, Horvath L, Pircher A, Wolf D. Tumor endothelial cells (TECs) as potential immune directors of the tumor microenvironment – new findings and future perspectives. Front Cell Dev Biol 2020;8:766. https://doi.org/10.3389/fcell.2020.00766.

[2] Finetti F, Donnini S, Giachetti A, Morbidelli L, Ziche M. Prostaglandin E(2) primes the angiogenic switch via a synergic interaction with the fibroblast growth factor-2 pathway. Circ Res 2009;105(7):657–66. https://doi.org/10.1161/CIRCRESAHA.109.203760.

[3] Finetti F, Travelli C, Ercoli J, Colombo G, Buoso E, Trabalzini L. Prostaglandin E2 and cancer: insight into tumor progression and immunity. Biology (Basel) 2020;9(12):434. https://doi.org/10.3390/biology9120434.

[4] Motz GT, Santoro SP, Wang LP, Garrabrant T, Lastra RR, Hagemann IS, Lal P, Feldman MD, Benencia F, Coukos G. Tumor endothelium FasL establishes a selective immune barrier promoting tolerance in tumors. Nat Med 2014;20(6):607–15. https://doi.org/10.1038/nm.3541.

[5] Dirkx AE, Oude Egbrink MG, Kuijpers MJ, van der Niet ST, Heijnen VV, Bouma-ter Steege JC, Wagstaff J, Griffioen AW. Tumor angiogenesis modulates leukocyte-vessel wall interactions in vivo by reducing endothelial adhesion molecule expression. Cancer Res 2003;63(9):2322–9.

[6] Shrimali RK, Yu Z, Theoret MR, Chinnasamy D, Restifo NP, Rosenberg SA. Antiangiogenic agents can increase lymphocyte infiltration into tumor and enhance the effectiveness of adoptive immunotherapy of cancer. Cancer Res 2010;70(15):6171–80. https://doi.org/10.1158/0008-5472.CAN-10-0153.

[7] Millán J, Hewlett L, Glyn M, Toomre D, Clark P, Ridley AJ. Lymphocyte transcellular migration occurs through recruitment of endothelial ICAM-1 to caveola- and F-actin-rich domains. Nat Cell Biol 2006;8(2):113–23. https://doi.org/10.1038/ncb1356.

[8] Chouaib S, Kieda C, Benlalam H, Noman MZ, Mami-Chouaib F, Rüegg C. Endothelial cells as key determinants of the tumor microenvironment: interaction with tumor cells, extracellular matrix and immune killer cells. Crit Rev Immunol 2010;30(6):529–45. https://doi.org/10.1615/critrevimmunol.v30.i6.30.

[9] Fukumura D, Kloeppe J, Amoozgar Z, Duda DG, Jain RK. Enhancing cancer immunotherapy using antiangiogenics: opportunities and challenges. Nat Rev Clin Oncol 2018;15(5):325–40.

[10] De Sanctis F, Ugel S, Facciponte J, Facciabene A. The dark side of tumor-associated endothelial cells. Semin Immunol 2018;35:35–47. https://doi.org/10.1016/j.smim.2018.02.002.

[11] LaGory EL, Giaccia AJ. The ever-expanding role of HIF in tumour and stromal biology. Nat Cell Biol 2016;18(4):356–65. https://doi.org/10.1038/ncb3330.

[12] Végran F, Boidot R, Michiels C, Sonveaux P, Feron O. Lactate influx through the endothelial cell monocarboxylate transporter MCT1 supports an NF-κB/IL-8 pathway that drives tumor angiogenesis. Cancer Res 2011;71(7):2550–60. https://doi.org/10.1158/0008-5472.CAN-10-2828.

[13] Rodig N, Ryan T, Allen JA, Pang H, Grabie N, Chernova T, Greenfield EA, Liang SC, Sharpe AH, Lichtman AH, Freeman GJ. Endothelial expression of PD-L1 and PD-L2 down-regulates CD8+ T cell activation and cytolysis. Eur J Immunol 2003;33(11):3117–26. https://doi.org/10.1002/eji.200324270.

[14] Geindreau M, Ghiringhelli F, Bruchard M. Vascular endothelial growth factor, a key modulator of the antitumor immune response. Int J Mol Sci 2021;22(9):4871. https://doi.org/10.3390/ijms22094871.
[15] Oyama T, Ran S, Ishida T, Nadaf S, Kerr L, Carbone DP, Gabrilovich DI. Vascular endothelial growth factor affects dendritic cell maturation through the inhibition of nuclear factor-kappa B activation in hemopoietic progenitor cells. J Immunol 1998;160(3):1224–32.
[16] Almand B, Resser JR, Lindman B, Nadaf S, Clark JI, Kwon ED, Carbone DP, Gabrilovich DI. Clinical significance of defective dendritic cell differentiation in cancer. Clin Cancer Res 2000;6(5):1755–66.
[17] Hicklin DJ, Ellis LM. Role of the vascular endothelial growth factor pathway in tumor growth and angiogenesis. J Clin Oncol 2005;23(5):1011–27. https://doi.org/10.1200/JCO.2005.06.081.
[18] Carmeliet P, Jain RK. Molecular mechanisms and clinical applications of angiogenesis. Nature 2011;473(7347):298–307. https://doi.org/10.1038/nature10144.
[19] Rini BI, Escudier B, Tomczak P, Kaprin A, Szczylik C, Hutson TE, Michaelson MD, Gorbunova VA, Gore ME, Rusakov IG, Negrier S, Ou YC, Castellano D, Lim HY, Uemura H, Tarazi J, Cella D, Chen C, Rosbrook B, Kim S, Motzer RJ. Comparative effectiveness of axitinib versus sorafenib in advanced renal cell carcinoma (AXIS): a randomised phase 3 trial. Lancet 2011;378(9807):1931–9. https://doi.org/10.1016/S0140-6736(11)61613-9 [Erratum in: Lancet. 2012;380(9856):1818].
[20] Ferrara N, Hillan KJ, Gerber HP, Novotny W. Discovery and development of bevacizumab, an anti-VEGF antibody for treating cancer. Nat Rev Drug Discov 2004;3(5):391–400. https://doi.org/10.1038/nrd1381.
[21] Anon. Avastin [prescribing information]. South San Francisco, CA: Genentech; 2014.
[22] Mortimer J, Zonder HB, Pal SK. Lessons learned from the bevacizumab experience. Cancer Control 2012;19(4):309–16. https://doi.org/10.1177/107327481201900407.
[23] Oguntade AS, Al-Amodi F, Alrumayh A, Alobaida M, Bwalya M. Anti-angiogenesis in cancer therapeutics: the magic bullet. J Egypt Natl Canc Inst 2021;33(1):15. https://doi.org/10.1186/s43046-021-00072-6.
[24] Motzer RJ, Porta C, Vogelzang NJ, Sternberg CN, Szczylik C, Zolnierek J, Kollmannsberger C, Rha SY, Bjarnason GA, Melichar B, De Giorgi U, Grünwald V, Davis ID, Lee JL, Esteban E, Urbanowitz G, Cai C, Squires M, Marker M, Shi MM, Escudier B. Dovitinib versus sorafenib for third-line targeted treatment of patients with metastatic renal cell carcinoma: an open-label, randomised phase 3 trial. Lancet Oncol 2014;15(3):286–96. https://doi.org/10.1016/S1470-2045(14)70030-0.
[25] Li J, Qin S, Xu RH, Shen L, Xu J, Bai Y, Yang L, Deng Y, Chen ZD, Zhong H, Pan H, Guo W, Shu Y, Yuan Y, Zhou J, Xu N, Liu T, Ma D, Wu C, Cheng Y, Chen D, Li W, Sun S, Yu Z, Cao P, Chen H, Wang J, Wang S, Wang H, Fan S, Hua Y, Su W. Effect of Fruquintinib vs placebo on overall survival in patients with previously treated metastatic colorectal cancer: the FRESCO randomized clinical trial. JAMA 2018;319(24):2486–96. https://doi.org/10.1001/jama.2018.7855.
[26] Okamoto K, Kodama K, Takase K, Sugi NH, Yamamoto Y, Iwata M, Tsuruoka A. Antitumor activities of the targeted multi-tyrosine kinase inhibitor lenvatinib (E7080) against RET gene fusion-driven tumor models. Cancer Lett 2013;340(1):97–103. https://doi.org/10.1016/j.canlet.2013.07.007.
[27] Tohyama O, Matsui J, Kodama K, Hata-Sugi N, Kimura T, Okamoto K, Minoshima Y, Iwata M, Funahashi Y. Antitumor activity of lenvatinib (e7080): an angiogenesis inhibitor that targets multiple receptor tyrosine kinases in preclinical human thyroid cancer models. J Thyroid Res 2014;2014:638747. https://doi.org/10.1155/2014/638747.
[28] Schlumberger M, Tahara M, Wirth LJ, Robinson B, Brose MS, Elisei R, Habra MA, Newbold K, Shah MH, Hoff AO, Gianoukakis AG, Kiyota N, Taylor MH, Kim SB, Krzyzanowska MK, Dutcus CE, de las Heras B, Zhu J, Sherman SI. Lenvatinib versus placebo in radioiodine-refractory thyroid cancer. N Engl J Med 2015;372(7):621–30. https://doi.org/10.1056/NEJMoa1406470.
[29] Hilberg F, Roth GJ, Krssak M, Kautschitsch S, Sommergruber W, Tontsch-Grunt U, Garin-Chesa P, Bader G, Zoephel A, Quant J, Heckel A, Rettig WJ. BIBF 1120: triple angiokinase inhibitor with sustained receptor blockade and good antitumor efficacy. Cancer Res 2008;68(12):4774–82. https://doi.org/10.1158/0008-5472.CAN-07-6307.
[30] Reck M, Kaiser R, Mellemgaard A, Douillard JY, Orlov S, Krzakowski M, von Pawel J, Gottfried M, Bondarenko I, Liao M, Gann CN, Barrueco J, Gaschler-Markefski B, Novello S, LUME-Lung 1 Study Group. Docetaxel plus nintedanib versus docetaxel plus placebo in patients with previously treated non-small-cell lung cancer (LUME-lung 1): a phase 3, double-blind, randomised controlled trial. Lancet Oncol 2014;15(2):143–55. https://doi.org/10.1016/S1470-2045(13)70586-2.

[31] O'Hare T, Shakespeare WC, Zhu X, Eide CA, Rivera VM, Wang F, Adrian LT, Zhou T, Huang WS, Xu Q, Metcalf 3rd CA, Tyner JW, Loriaux MM, Corbin AS, Wardwell S, Ning Y, Keats JA, Wang Y, Sundaramoorthi R, Thomas M, Zhou D, Snodgrass J, Commodore L, Sawyer TK, Dalgarno DC, Deininger MW, Druker BJ, Clackson T. AP24534, a pan-BCR-ABL inhibitor for chronic myeloid leukemia, potently inhibits the T315I mutant and overcomes mutation-based resistance. Cancer Cell 2009;16(5):401–12. https://doi.org/10.1016/j.ccr.2009.09.028.

[32] Baccarani M, Deininger MW, Rosti G, Hochhaus A, Soverini S, Apperley JF, Cervantes F, Clark RE, Cortes JE, Guilhot F, Hjorth-Hansen H, Hughes TP, Kantarjian HM, Kim DW, Larson RA, Lipton JH, Mahon FX, Martinelli G, Mayer J, Müller MC, Niederwieser D, Pane F, Radich JP, Rousselot P, Saglio G, Saußele S, Schiffer C, Silver R, Simonsson B, Steegmann JL, Goldman JM, Hehlmann R. European LeukemiaNet recommendations for the management of chronic myeloid leukemia: 2013. Blood 2013;122(6):872–84. https://doi.org/10.1182/blood-2013-05-501569.

[33] Anon. Iclusig [summary of product characteristics]. Leatherhead, United Kingdom: ARIAD Pharmaceuticals, Inc, a wholly owned subsidiary of Takeda Pharmaceutical Company Limited; 2017.

[34] Sleijfer S, Ray-Coquard I, Papai Z, Le Cesne A, Scurr M, Schöffski P, Collin F, Pandite L, Marreaud S, De Brauwer A, van Glabbeke M, Verweij J, Blay JY. Pazopanib, a multikinase angiogenesis inhibitor, in patients with relapsed or refractory advanced soft tissue sarcoma: a phase II study from the European organisation for research and treatment of cancer-soft tissue and bone sarcoma group (EORTC study 62043). J Clin Oncol 2009;27(19):3126–32. https://doi.org/10.1200/JCO.2008.21.3223.

[35] van der Graaf WT, Blay JY, Chawla SP, Kim DW, Bui-Nguyen B, Casali PG, Schöffski P, Aglietta M, Staddon AP, Beppu Y, Le Cesne A, Gelderblom H, Judson IR, Araki N, Ouali M, Marreaud S, Hodge R, Dewji MR, Coens C, Demetri GD, Fletcher CD, Dei Tos AP, Hohenberger P, EORTC Soft Tissue and Bone Sarcoma Group; PALETTE study group. Pazopanib for metastatic soft-tissue sarcoma (PALETTE): a randomised, double-blind, placebo-controlled phase 3 trial. Lancet 2012;379(9829):1879–86. https://doi.org/10.1016/S0140-6736(12)60651-5.

[36] Escudier B, Porta C, Bono P, Powles T, Eisen T, Sternberg CN, Gschwend JE, De Giorgi U, Parikh O, Hawkins R, Sevin E, Négrier S, Khan S, Diaz J, Redhu S, Mehmud F, Cella D. Randomized, controlled, double-blind, cross-over trial assessing treatment preference for pazopanib versus sunitinib in patients with metastatic renal cell carcinoma: PISCES study. J Clin Oncol 2014;32(14):1412–8. https://doi.org/10.1200/JCO.2013.50.8267.

[37] Retz M, Bedke J, Bögemann M, Grimm MO, Zimmermann U, Müller L, Leiber C, Teber D, Wirth M, Bolenz C, van Alphen R, De Santis M, Beeker A, Lehmann J, Indorf M, Frank M, Bokemeyer C, Gschwend JE. SWITCH II: Phase III randomized, sequential, open-label study to evaluate the efficacy and safety of sorafenib-pazopanib versus pazopanib-sorafenib in the treatment of advanced or metastatic renal cell carcinoma (AUO AN 33/11). Eur J Cancer 2019;107:37–45. https://doi.org/10.1016/j.ejca.2018.11.001.

[38] Fuchs CS, Tomasek J, Yong CJ, Dumitru F, Passalacqua R, Goswami C, Safran H, Dos Santos LV, Aprile G, Ferry DR, Melichar B, Tehfe M, Topuzov E, Zalcberg JR, Chau I, Campbell W, Sivanandan C, Pikiel J, Koshiji M, Hsu Y, Liepa AM, Gao L, Schwartz JD, Tabernero J, REGARD Trial Investigators. Ramucirumab monotherapy for previously treated advanced gastric or gastro-oesophageal junction adenocarcinoma (REGARD): an international, randomised, multicentre, placebo-controlled, phase 3 trial. Lancet 2014;383(9911):31–9. https://doi.org/10.1016/S0140-6736(13)61719-5.

[39] Zhu AX, Kang YK, Yen CJ, Finn RS, Galle PR, Llovet JM, Assenat E, Brandi G, Pracht M, Lim HY, Rau KM, Motomura K, Ohno I, Merle P, Daniele B, Shin DB, Gerken G, Borg C, Hiriart JB, Okusaka T, Morimoto M, Hsu Y, Abada PB, Kudo M, REACH-2 study investigators. Ramucirumab after sorafenib in patients with advanced hepatocellular carcinoma and increased α-fetoprotein concentrations (REACH-2): a randomised, double-blind, placebo-controlled, phase 3 trial. Lancet Oncol 2019;20(2):282–96. https://doi.org/10.1016/S1470-2045(18)30937-9.

[40] Grothey A, Van Cutsem E, Sobrero A, Siena S, Falcone A, Ychou M, Humblet Y, Bouché O, Mineur L, Barone C, Adenis A, Tabernero J, Yoshino T, Lenz HJ, Goldberg RM, Sargent DJ, Cihon F, Cupit L, Wagner A, Laurent D, CORRECT Study Group. Regorafenib monotherapy for previously treated metastatic colorectal cancer (CORRECT): an international, multicentre, randomised, placebo-controlled, phase 3 trial. Lancet 2013;381(9863):303–12. https://doi.org/10.1016/S0140-6736(12)61900-X.

[41] Yoshino T, Komatsu Y, Yamada Y, Yamazaki K, Tsuji A, Ura T, Grothey A, Van Cutsem E, Wagner A, Cihon F, Hamada Y, Ohtsu A. Randomized phase III trial of regorafenib in metastatic colorectal cancer: analysis of the CORRECT Japanese and non-Japanese subpopulations. Invest New Drugs 2015;33(3):740–50. https://doi.org/10.1007/s10637-014-0154-x.

[42] Li J, Qin S, Xu R, Yau TC, Ma B, Pan H, Xu J, Bai Y, Chi Y, Wang L, Yeh KH, Bi F, Cheng Y, Le AT, Lin JK, Liu T, Ma D, Kappeler C, Kalmus J, Kim TW, CONCUR Investigators. Regorafenib plus best supportive care versus placebo plus best supportive care in Asian patients with previously treated metastatic colorectal cancer (CONCUR): a randomised, double-blind, placebo-controlled, phase 3 trial. Lancet Oncol 2015;16(6):619–29. https://doi.org/10.1016/S1470-2045(15)70156-7.

[43] Eisen T, Oudard S, Szczylik C, Gravis G, Heinzer H, Middleton R, Cihon F, Anderson S, Shah S, Bukowski R, Escudier B, TARGET Study Group. Sorafenib for older patients with renal cell carcinoma: subset analysis from a randomized trial. J Natl Cancer Inst 2008;100(20):1454–63. https://doi.org/10.1093/jnci/djn319.

[44] Escudier B, Eisen T, Stadler WM, Szczylik C, Oudard S, Siebels M, Negrier S, Chevreau C, Solska E, Desai AA, Rolland F, Demkow T, Hutson TE, Gore M, Freeman S, Schwartz B, Shan M, Simantov R, Bukowski RM, TARGET Study Group. Sorafenib in advanced clear-cell renal-cell carcinoma. N Engl J Med 2007;356(2):125–34. https://doi.org/10.1056/NEJMoa060655 [Erratum in: N Engl J Med. 2007;357(2):203].

[45] Gordan JD, Kennedy EB, Abou-Alfa GK, Beg MS, Brower ST, Gade TP, Goff L, Gupta S, Guy J, Harris WP, Iyer R, Jaiyesimi I, Jhawer M, Karippot A, Kaseb AO, Kelley RK, Knox JJ, Kortmansky J, Leaf A, Remak WM, Shroff RT, Sohal DPS, Taddei TH, Venepalli NK, Wilson A, Zhu AX, Rose MG. Systemic therapy for advanced hepatocellular carcinoma: ASCO guideline. J Clin Oncol 2020;38(36):4317–45. https://doi.org/10.1200/JCO.20.02672.

[46] Cheng AL, Kang YK, Chen Z, Tsao CJ, Qin S, Kim JS, Luo R, Feng J, Ye S, Yang TS, Xu J, Sun Y, Liang H, Liu J, Wang J, Tak WY, Pan H, Burock K, Zou J, Voliotis D, Guan Z. Efficacy and safety of sorafenib in patients in the Asia-Pacific region with advanced hepatocellular carcinoma: a phase III randomised, double-blind, placebo-controlled trial. Lancet Oncol 2009;10(1):25–34. https://doi.org/10.1016/S1470-2045(08)70285-7.

[47] Llovet JM, Ricci S, Mazzaferro V, Hilgard P, Gane E, Blanc JF, de Oliveira AC, Santoro A, Raoul JL, Forner A, Schwartz M, Porta C, Zeuzem S, Bolondi L, Greten TF, Galle PR, Seitz JF, Borbath I, Häussinger D, Giannaris T, Shan M, Moscovici M, Voliotis D, Bruix J, SHARP Investigators Study Group. Sorafenib in advanced hepatocellular carcinoma. N Engl J Med 2008;359(4):378–90. https://doi.org/10.1056/NEJMoa0708857.

[48] Demetri GD, van Oosterom AT, Garrett CR, Blackstein ME, Shah MH, Verweij J, McArthur G, Judson IR, Heinrich MC, Morgan JA, Desai J, Fletcher CD, George S, Bello CL, Huang X, Baum CM, Casali PG. Efficacy and safety of sunitinib in patients with advanced gastrointestinal stromal tumour after failure of imatinib: a randomised controlled trial. Lancet 2006;368(9544):1329–38. https://doi.org/10.1016/S0140-6736(06)69446-4.

[49] Escudier B. Sunitinib for the management of advanced renal cell carcinoma. Expert Rev Anticancer Ther 2010;10(3):305–17. https://doi.org/10.1586/era.10.26.

[50] Raymond E, Dahan L, Raoul JL, Bang YJ, Borbath I, Lombard-Bohas C, Valle J, Metrakos P, Smith D, Vinik A, Chen JS, Hörsch D, Hammel P, Wiedenmann B, Van Cutsem E, Patyna S, Lu DR, Blanckmeister C, Chao R, Ruszniewski P. Sunitinib malate for the treatment of pancreatic neuroendocrine tumors. N Engl J Med 2011;364(6):501–13. https://doi.org/10.1056/NEJMoa1003825 [Erratum in: N Engl J Med. 2011;364(11):1082].

[51] Fotivda. (Tivozanib) SmPC. August; 2017.

[52] Motzer RJ, Nosov D, Eisen T, Bondarenko I, Lesovoy V, Lipatov O, Tomczak P, Lyulko O, Alyasova A, Harza M, Kogan M, Alekseev BY, Sternberg CN, Szczylik C, Cella D, Ivanescu C, Krivoshik A, Strahs A, Esteves B, Berkenblit A, Hutson TE. Tivozanib versus sorafenib as initial targeted therapy for patients with metastatic renal cell carcinoma: results from a phase III trial. J Clin Oncol 2013;31(30):3791–9. https://doi.org/10.1200/JCO.2012.47.4940.

[53] Rini BI, Pal SK, Escudier BJ, Atkins MB, Hutson TE, Porta C, Verzoni E, Needle MN, McDermott DF. Tivozanib versus sorafenib in patients with advanced renal cell carcinoma (TIVO-3): a phase 3, multicentre, randomised, controlled, open-label study. Lancet Oncol 2020;21(1):95–104. https://doi.org/10.1016/S1470-2045(19)30735-1.

[54] Pal SK, Escudier BJ, Atkins MB, Hutson TE, Porta C, Verzoni E, Needle MN, Powers D, McDermott DF, Rini BI. Final overall survival results from a phase 3 study to compare Tivozanib to Sorafenib as third- or fourth-line therapy in subjects with metastatic renal cell carcinoma. Eur Urol 2020;78(6):783–5. https://doi.org/10.1016/j.eururo.2020.08.007.

[55] Wells Jr SA, Robinson BG, Gagel RF, Dralle H, Fagin JA, Santoro M, Baudin E, Elisei R, Jarzab B, Vasselli JR, Read J, Langmuir P, Ryan AJ, Schlumberger MJ. Vandetanib in patients with locally advanced or metastatic medullary thyroid cancer: a randomized, double-blind phase III trial. J Clin Oncol 2012;30(2):134–41. https://doi.org/10.1200/JCO.2011.35.5040 [Erratum in: J Clin Oncol. 2013;31(24):3049].

[56] Van Cutsem E, Tabernero J, Lakomy R, Prenen H, Prausová J, Macarulla T, Ruff P, van Hazel GA, Moiseyenko V, Ferry D, McKendrick J, Polikoff J, Tellier A, Castan R, Allegra C. Addition of aflibercept to fluorouracil,

leucovorin, and irinotecan improves survival in a phase III randomized trial in patients with metastatic colorectal cancer previously treated with an oxaliplatin-based regimen. J Clin Oncol 2012;30(28):3499–506. https://doi.org/10.1200/JCO.2012.42.8201.

[57] Han B, Li K, Wang Q, Zhang L, Shi J, Wang Z, Cheng Y, He J, Shi Y, Zhao Y, Yu H, Zhao Y, Chen W, Luo Y, Wu L, Wang X, Pirker R, Nan K, Jin F, Dong J, Li B, Sun Y. Effect of anlotinib as a third-line or further treatment on overall survival of patients with advanced non-small cell lung cancer: the alter 0303 phase 3 randomized clinical trial. JAMA Oncol 2018;4(11):1569–75. https://doi.org/10.1001/jamaoncol.2018.3039 [Erratum in: JAMA Oncol. 2018;4(11):1625].

[58] Chen D, Xu J, Zhao Y, Chu T, Zhong H, Han B, Zhong R. Prognostic value of tumor cavitation in extensive-stage small-cell lung cancer patients treated with anlotinib. J Cancer Res Clin Oncol 2020;146(2):401–6. https://doi.org/10.1007/s00432-019-03064-1.

[59] Chi Y, Shu Y, Ba Y, Bai Y, Qin B, Wang X, Xiong J, Xu N, Zhang H, Zhou J, Xu J, Cheng Y, Feng J, Hu C, Chen Y, Chen Z, Wang J, Dang C, Wang J, Wan Y, Tang Y, Wang D, Liu J, Wu M, Deng Y, Li X, Li Y, Dong J, Jiang D, Li G, Wu Q, Li J, Qi Y, Sun Y, Cai J. Anlotinib monotherapy for refractory metastatic colorectal cancer: a double-blinded, placebo-controlled, randomized phase III trial (ALTER0703). Oncologist 2021. https://doi.org/10.1002/onco.13857.

[60] Li J, Qin S, Xu J, Xiong J, Wu C, Bai Y, Liu W, Tong J, Liu Y, Xu R, Wang Z, Wang Q, Ouyang X, Yang Y, Ba Y, Liang J, Lin X, Luo D, Zheng R, Wang X, Sun G, Wang L, Zheng L, Guo H, Wu J, Xu N, Yang J, Zhang H, Cheng Y, Wang N, Chen L, Fan Z, Sun P, Yu H. Randomized, double-blind, placebo-controlled phase III trial of apatinib in patients with chemotherapy-refractory advanced or metastatic adenocarcinoma of the stomach or gastroesophageal junction. J Clin Oncol 2016;34(13):1448–54. https://doi.org/10.1200/JCO.2015.63.5995.

[61] Choueiri TK, Escudier B, Powles T, Tannir NM, Mainwaring PN, Rini BI, Hammers HJ, Donskov F, Roth BJ, Peltola K, Lee JL, Heng DYC, Schmidinger M, Agarwal N, Sternberg CN, McDermott DF, Aftab DT, Hessel C, Scheffold C, Schwab G, Hutson TE, Pal S, Motzer RJ, METEOR investigators. Cabozantinib versus everolimus in advanced renal cell carcinoma (METEOR): final results from a randomised, open-label, phase 3 trial. Lancet Oncol 2016;17(7):917–27. https://doi.org/10.1016/S1470-2045(16)30107-3.

[62] George DJ, Hessel C, Halabi S, Michaelson MD, Hahn O, Walsh M, Picus J, Small EJ, Dakhil S, Feldman DR, Mangeshkar M, Scheffold C, Morris MJ, Choueiri TK. Cabozantinib versus Sunitinib for untreated patients with advanced renal cell carcinoma of intermediate or poor risk: subgroup analysis of the Alliance A031203 CABOSUN trial. Oncologist 2019;24(11):1497–501. https://doi.org/10.1634/theoncologist.2019-0316.

[63] Kelley RK, Meyer T, Rimassa L, Merle P, Park JW, Yau T, Chan SL, Blanc JF, Tam VC, Tran A, Dadduzio V, Markby DW, Kaldate R, Cheng AL, El-Khoueiry AB, Abou-Alfa GK. Serum alpha-fetoprotein levels and clinical outcomes in the phase III CELESTIAL study of cabozantinib versus placebo in patients with advanced hepatocellular carcinoma. Clin Cancer Res 2020;26(18):4795–804. https://doi.org/10.1158/1078-0432.CCR-19-3884.

[64] Cao J, Zhang J, Peng W, Chen Z, Fan S, Su W, Li K, Li J. A phase I study of safety and pharmacokinetics of fruquintinib, a novel selective inhibitor of vascular endothelial growth factor receptor-1, -2, and -3 tyrosine kinases in Chinese patients with advanced solid tumors. Cancer Chemother Pharmacol 2016;78(2):259–69. https://doi.org/10.1007/s00280-016-3069-8.

[65] Gianoukakis AG, Dutcus CE, Batty N, Guo M, Baig M. Prolonged duration of response in lenvatinib responders with thyroid cancer. Endocr Relat Cancer 2018;25(6):699–704. https://doi.org/10.1530/ERC-18-0049.

[66] Nair A, Reece K, Donoghue MB, Yuan WV, Rodriguez L, Keegan P, Pazdur R. FDA supplemental approval summary: lenvatinib for the treatment of unresectable hepatocellular carcinoma. Oncologist 2021;26(3):e484–91. https://doi.org/10.1002/onco.13566.

[67] Cortes JE, Kim DW, Pinilla-Ibarz J, le Coutre P, Paquette R, Chuah C, Nicolini FE, Apperley JF, Khoury HJ, Talpaz M, DiPersio J, DeAngelo DJ, Abruzzese E, Rea D, Baccarani M, Müller MC, Gambacorti-Passerini C, Wong S, Lustgarten S, Rivera VM, Clackson T, Turner CD, Haluska FG, Guilhot F, Deininger MW, Hochhaus A, Hughes T, Goldman JM, Shah NP, Kantarjian H, PACE Investigators. A phase 2 trial of ponatinib in Philadelphia chromosome-positive leukemias. N Engl J Med 2013;369(19):1783–96. https://doi.org/10.1056/NEJMoa1306494.

[68] Gozgit JM, Wong MJ, Moran L, Wardwell S, Mohemmad QK, Narasimhan NI, Shakespeare WC, Wang F, Clackson T, Rivera VM. Ponatinib (AP24534), a multitargeted pan-FGFR inhibitor with activity in multiple FGFR-amplified or mutated cancer models. Mol Cancer Ther 2012;11(3):690–9. https://doi.org/10.1158/1535-7163.MCT-11-0450.

[69] Mologni L, Redaelli S, Morandi A, Plaza-Menacho I, Gambacorti-Passerini C. Ponatinib is a potent inhibitor of wild-type and drug-resistant gatekeeper mutant RET kinase. Mol Cell Endocrinol 2013;377(1–2):1–6. https://doi.org/10.1016/j.mce.2013.06.025.

[70] Sadovnik I, Lierman E, Peter B, Herrmann H, Suppan V, Stefanzl G, Haas O, Lion T, Pickl W, Cools J, Vandenberghe P, Valent P. Identification of Ponatinib as a potent inhibitor of growth, migration, and activation of neoplastic eosinophils carrying FIP1L1-PDGFRA. Exp Hematol 2014;42(4):282–293.e4. https://doi.org/10.1016/j.exphem.2013.12.007.

[71] Garner AP, Gozgit JM, Anjum R, Vodala S, Schrock A, Zhou T, Serrano C, Eilers G, Zhu M, Ketzer J, Wardwell S, Ning Y, Song Y, Kohlmann A, Wang F, Clackson T, Heinrich MC, Fletcher JA, Bauer S, Rivera VM. Ponatinib inhibits polyclonal drug-resistant KIT oncoproteins and shows therapeutic potential in heavily pretreated gastrointestinal stromal tumor (GIST) patients. Clin Cancer Res 2014;20(22):5745–55. https://doi.org/10.1158/1078-0432.CCR-14-1397.

[72] Ahmad S, Johnson GL, Scott JE. Identification of ponatinib and other known kinase inhibitors with potent MEKK2 inhibitory activity. Biochem Biophys Res Commun 2015;463(4):888–93. https://doi.org/10.1016/j.bbrc.2015.06.029.

[73] De Falco V, Buonocore P, Muthu M, Torregrossa L, Basolo F, Billaud M, Gozgit JM, Carlomagno F, Santoro M. Ponatinib (AP24534) is a novel potent inhibitor of oncogenic RET mutants associated with thyroid cancer. J Clin Endocrinol Metab 2013;98(5):E811–9. https://doi.org/10.1210/jc.2012-2672.

[74] Li SQ, Cheuk AT, Shern JF, Song YK, Hurd L, Liao H, Wei JS, Khan J. Targeting wild-type and mutationally activated FGFR4 in rhabdomyosarcoma with the inhibitor ponatinib (AP24534). PLoS One 2013;8(10). https://doi.org/10.1371/journal.pone.0076551, e76551.

[75] Gozgit JM, Squillace RM, Wongchenko MJ, Miller D, Wardwell S, Mohemmad Q, Narasimhan NI, Wang F, Clackson T, Rivera VM. Combined targeting of FGFR2 and mTOR by ponatinib and ridaforolimus results in synergistic antitumor activity in FGFR2 mutant endometrial cancer models. Cancer Chemother Pharmacol 2013;71(5):1315–23. https://doi.org/10.1007/s00280-013-2131-z.

[76] Wynes MW, Hinz TK, Gao D, Martini M, Marek LA, Ware KE, Edwards MG, Böhm D, Perner S, Helfrich BA, Dziadziuszko R, Jassem J, Wojtylak S, Sejda A, Gozgit JM, Bunn Jr PA, Camidge DR, Tan AC, Hirsch FR, Heasley LE. FGFR1 mRNA and protein expression, not gene copy number, predict FGFR TKI sensitivity across all lung cancer histologies. Clin Cancer Res 2014;20(12):3299–309. https://doi.org/10.1158/1078-0432.CCR-13-3060.

[77] Zhang J, Zhou Q, Gao G, Wang Y, Fang Z, Li G, Yu M, Kong L, Xing Y, Gao X. The effects of ponatinib, a multi-targeted tyrosine kinase inhibitor, against human U87 malignant glioblastoma cells. Onco Targets Ther 2014;7:2013–9. https://doi.org/10.2147/OTT.S67556.

[78] Karch A, Koch A, Grünwald V. A phase II trial comparing pazopanib with doxorubicin as first-line treatment in elderly patients with metastatic or advanced soft tissue sarcoma (EPAZ): study protocol for a randomized controlled trial. Trials 2016;17(1):312. https://doi.org/10.1186/s13063-016-1434-x.

[79] Sternberg CN, Davis ID, Mardiak J, Szczylik C, Lee E, Wagstaff J, Barrios CH, Salman P, Gladkov OA, Kavina A, Zarbá JJ, Chen M, McCann L, Pandite L, Roychowdhury DF, Hawkins RE. Pazopanib in locally advanced or metastatic renal cell carcinoma: results of a randomized phase III trial. J Clin Oncol 2010;28(6):1061–8. https://doi.org/10.1200/JCO.2009.23.9764.

[80] Van Cutsem E, Martinelli E, Cascinu S, Sobrero A, Banzi M, Seitz JF, Barone C, Ychou M, Peeters M, Brenner B, Hofheinz RD, Maiello E, André T, Spallanzani A, Garcia-Carbonero R, Arriaga YE, Verma U, Grothey A, Kappeler C, Miriyala A, Kalmus J, Falcone A, Zaniboni A. Regorafenib for patients with metastatic colorectal cancer who progressed after standard therapy: results of the large, single-arm, open-label phase IIIb CONSIGN study. Oncologist 2019;24(2):185–92. https://doi.org/10.1634/theoncologist.2018-0072.

[81] Demetri GD, Reichardt P, Kang YK, Blay JY, Rutkowski P, Gelderblom H, Hohenberger P, Leahy M, von Mehren M, Joensuu H, Badalamenti G, Blackstein M, Le Cesne A, Schöffski P, Maki RG, Bauer S, Nguyen BB, Xu J, Nishida T, Chung J, Kappeler C, Kuss I, Laurent D, Casali PG, GRID Study Investigators. Efficacy and safety of regorafenib for advanced gastrointestinal stromal tumours after failure of imatinib and sunitinib (GRID): an international, multicentre, randomised, placebo-controlled, phase 3 trial. Lancet 2013;381(9863):295–302. https://doi.org/10.1016/S0140-6736(12)61857-1.

[82] Ben-Ami E, Barysauskas CM, von Mehren M, Heinrich MC, Corless CL, Butrynski JE, Morgan JA, Wagner AJ, Choy E, Yap JT, Van den Abbeele AD, Solomon SM, Fletcher JA, Demetri GD, George S. Long-term follow-up results of the multicenter phase II trial of regorafenib in patients with metastatic and/or unresectable GI stromal

tumor after failure of standard tyrosine kinase inhibitor therapy. Ann Oncol 2016;27(9):1794–9. https://doi.org/10.1093/annonc/mdw228.

[83] Bruix J, Qin S, Merle P, Granito A, Huang YH, Bodoky G, Pracht M, Yokosuka O, Rosmorduc O, Breder V, Gerolami R, Masi G, Ross PJ, Song T, Bronowicki JP, Ollivier-Hourmand I, Kudo M, Cheng AL, Llovet JM, Finn RS, MA LB, Baumhauer A, Meinhardt G, Han G, RESORCE Investigators. Regorafenib for patients with hepatocellular carcinoma who progressed on sorafenib treatment (RESORCE): a randomised, double-blind, placebo-controlled, phase 3 trial. Lancet 2017;389(10064):56–66. https://doi.org/10.1016/S0140-6736(16)32453-9.

[84] Finn RS, Merle P, Granito A, Huang YH, Bodoky G, Pracht M, Yokosuka O, Rosmorduc O, Gerolami R, Caparello C, Cabrera R, Chang C, Sun W, LeBerre MA, Baumhauer A, Meinhardt G, Bruix J. Outcomes of sequential treatment with sorafenib followed by regorafenib for HCC: additional analyses from the phase III RESORCE trial. J Hepatol 2018;69(2):353–8. https://doi.org/10.1016/j.jhep.2018.04.010.

[85] Brose MS, Nutting CM, Jarzab B, Elisei R, Siena S, Bastholt L, de la Fouchardiere C, Pacini F, Paschke R, Shong YK, Sherman SI, Smit JW, Chung J, Kappeler C, Peña C, Molnár I, Schlumberger MJ, DECISION investigators. Sorafenib in radioactive iodine-refractory, locally advanced or metastatic differentiated thyroid cancer: a randomised, double-blind, phase 3 trial. Lancet 2014;384(9940):319–28. https://doi.org/10.1016/S0140-6736(14)60421-9.

[86] Worden F, Fassnacht M, Shi Y, Hadjieva T, Bonichon F, Gao M, Fugazzola L, Ando Y, Hasegawa Y, Park DJ, Shong YK, Smit JW, Chung J, Kappeler C, Meinhardt G, Schlumberger M, Brose MS. Safety and tolerability of sorafenib in patients with radioiodine-refractory thyroid cancer. Endocr Relat Cancer 2015;22(6):877–87. https://doi.org/10.1530/ERC-15-0252.

[87] Lai E, Cascinu S, Scartozzi M. Are all anti-Angiogenic drugs the same in the treatment of second-line metastatic colorectal cancer? Expert opinion on clinical practice. Front Oncol 2021;11:637823. https://doi.org/10.3389/fonc.2021.637823.

[88] Hurwitz H, Fehrenbacher L, Novotny W, Cartwright T, Hainsworth J, Heim W, Berlin J, Baron A, Griffing S, Holmgren E, Ferrara N, Fyfe G, Rogers B, Ross R, Kabbinavar F. Bevacizumab plus irinotecan, fluorouracil, and leucovorin for metastatic colorectal cancer. N Engl J Med 2004;350(23):2335–42. https://doi.org/10.1056/NEJMoa032691.

[89] Saltz LB, Clarke S, Díaz-Rubio E, Scheithauer W, Figer A, Wong R, Koski S, Lichinitser M, Yang TS, Rivera F, Couture F, Sirzén F, Cassidy J. Bevacizumab in combination with oxaliplatin-based chemotherapy as first-line therapy in metastatic colorectal cancer: a randomized phase III study. J Clin Oncol 2008;26(12):2013–9. https://doi.org/10.1200/JCO.2007.14.9930 [Erratum in: J Clin Oncol. 2008;26(18):3110. Erratum in: J Clin Oncol. 2009;27(4):653].

[90] Bennouna J, Sastre J, Arnold D, Österlund P, Greil R, Van Cutsem E, von Moos R, Viéitez JM, Bouché O, Borg C, Steffens CC, Alonso-Orduña V, Schlichting C, Reyes-Rivera I, Bendahmane B, André T, Kubicka S, ML18147 Study Investigators. Continuation of bevacizumab after first progression in metastatic colorectal cancer (ML18147): a randomised phase 3 trial. Lancet Oncol 2013;14(1):29–37. https://doi.org/10.1016/S1470-2045(12)70477-1.

[91] Cunningham D, Lang I, Marcuello E, Lorusso V, Ocvirk J, Shin DB, Jonker D, Osborne S, Andre N, Waterkamp D, Saunders MP, AVEX Study Investigators. Bevacizumab plus capecitabine versus capecitabine alone in elderly patients with previously untreated metastatic colorectal cancer (AVEX): an open-label, randomised phase 3 trial. Lancet Oncol 2013;14(11):1077–85. https://doi.org/10.1016/S1470-2045(13)70154-2.

[92] Petrelli F, Borgonovo K, Cabiddu M, Ghilardi M, Lonati V, Maspero F, Sauta MG, Beretta GD, Barni S. FOLFIRI-bevacizumab as first-line chemotherapy in 3500 patients with advanced colorectal cancer: a pooled analysis of 29 published trials. Clin Colorectal Cancer 2013;12(3):145–51. https://doi.org/10.1016/j.clcc.2013.04.006.

[93] Macedo LT, da Costa Lima AB, Sasse AD. Addition of bevacizumab to first-line chemotherapy in advanced colorectal cancer: a systematic review and meta-analysis, with emphasis on chemotherapy subgroups. BMC Cancer 2012;12:89. https://doi.org/10.1186/1471-2407-12-89.

[94] Meyerhardt JA, Li L, Sanoff HK, Carpenter 4th W, Schrag D. Effectiveness of bevacizumab with first-line combination chemotherapy for Medicare patients with stage IV colorectal cancer. J Clin Oncol 2012;30(6):608–15. https://doi.org/10.1200/JCO.2011.38.9650.

[95] Hurwitz HI, Lyman GH. Registries and randomized trials in assessing the effects of bevacizumab in colorectal cancer: is there a common theme? J Clin Oncol 2012;30(6):580–1. https://doi.org/10.1200/JCO.2011.40.7031.

[96] Soria JC, Mauguen A, Reck M, Sandler AB, Saijo N, Johnson DH, Burcoveanu D, Fukuoka M, Besse B, Pignon JP, Meta-analysis of bevacizumab in advanced NSCLC collaborative group. Systematic review and meta-analysis of randomised, phase II/III trials adding bevacizumab to platinum-based chemotherapy as first-line treatment in patients with advanced non-small-cell lung cancer. Ann Oncol 2013;24(1):20–30. https://doi.org/10.1093/annonc/mds590 [Erratum in: Ann Oncol. 2013;24(4):1133. Erratum in: Ann Oncol. 2013;24(4):1133].
[97] Zhou C, Wu YL, Chen G, Liu X, Zhu Y, Lu S, Feng J, He J, Han B, Wang J, Jiang G, Hu C, Zhang H, Cheng G, Song X, Lu Y, Pan H, Zheng W, Yin AY. BEYOND: a randomized, double-blind, placebo-controlled, multicenter, phase III study of first-line carboplatin/paclitaxel plus bevacizumab or placebo in Chinese patients with advanced or recurrent nonsquamous non-Small-cell lung cancer. J Clin Oncol 2015;33(19):2197–204. https://doi.org/10.1200/JCO.2014.59.4424.
[98] Ellis PM. Anti-angiogenesis in personalized therapy of lung cancer. Adv Exp Med Biol 2016;893:91–126. https://doi.org/10.1007/978-3-319-24223-1_5.
[99] Sun L, Ma JT, Zhang SL, Zou HW, Han CB. Efficacy and safety of chemotherapy or tyrosine kinase inhibitors combined with bevacizumab versus chemotherapy or tyrosine kinase inhibitors alone in the treatment of non-small cell lung cancer: a systematic review and meta-analysis. Med Oncol 2015;32:473. https://doi.org/10.1007/s12032-014-0473-y.
[100] Barlesi F, Scherpereel A, Gorbunova V, Gervais R, Vikström A, Chouaid C, Chella A, Kim JH, Ahn MJ, Reck M, Pazzola A, Kim HT, Aerts JG, Morando C, Loundou A, Groen HJ, Rittmeyer A. Maintenance bevacizumab-pemetrexed after first-line cisplatin-pemetrexed-bevacizumab for advanced nonsquamous nonsmall-cell lung cancer: updated survival analysis of the AVAPERL (MO22089) randomized phase III trial. Ann Oncol 2014;25(5):1044–52. https://doi.org/10.1093/annonc/mdu098.
[101] Patel JD, Socinski MA, Garon EB, Reynolds CH, Spigel DR, Olsen MR, Hermann RC, Jotte RM, Beck T, Richards DA, Guba SC, Liu J, Frimodt-Moller B, John WJ, Obasaju CK, Pennella EJ, Bonomi P, Govindan R. PointBreak: a randomized phase III study of pemetrexed plus carboplatin and bevacizumab followed by maintenance pemetrexed and bevacizumab versus paclitaxel plus carboplatin and bevacizumab followed by maintenance bevacizumab in patients with stage IIIB or IV nonsquamous non-small-cell lung cancer. J Clin Oncol 2013;31(34):4349–57. https://doi.org/10.1200/JCO.2012.47.9626.
[102] Galetta D, Cinieri S, Pisconti S, Gebbia V, Morabito A, Borsellino N, Maiello E, Febbraro A, Catino A, Rizzo P, Montrone M, Misino A, Logroscino A, Rizzi D, Di Maio M, Colucci G. Cisplatin/pemetrexed followed by maintenance pemetrexed versus carboplatin/paclitaxel/bevacizumab followed by maintenance bevacizumab in advanced nonsquamous lung cancer: the GOIM (Gruppo Oncologico Italia Meridionale) ERACLE phase III randomized trial. Clin Lung Cancer 2015;16(4):262–73. https://doi.org/10.1016/j.cllc.2014.12.002.
[103] Laskin J, Crinò L, Felip E, Franke F, Gorbunova V, Groen H, Jiang GL, Reck M, Schneider CP. Safety and efficacy of first-line bevacizumab plus chemotherapy in elderly patients with advanced or recurrent nonsquamous non-small cell lung cancer: safety of avastin in lung trial (MO19390). J Thorac Oncol 2012;7(1):203–11. https://doi.org/10.1097/JTO.0b013e3182370e02.
[104] Langer CJ, Socinski MA, Patel JD, Sandler AB, Schiller JH, Leon L, Hazard SJ, Ramalingam SS. Isolating the role of bevacizumab in elderly patients with previously untreated nonsquamous non-small cell lung cancer: secondary analyses of the ECOG 4599 and PointBreak trials. Am J Clin Oncol 2016;39(5):441–7. https://doi.org/10.1097/COC.0000000000000163.
[105] Miller KD. E2100: a phase III trial of paclitaxel versus paclitaxel/bevacizumab for metastatic breast cancer. Clin Breast Cancer 2003;3(6):421–2. https://doi.org/10.3816/CBC.2003.n.007.
[106] Gray R, Bhattacharya S, Bowden C, Miller K, Comis RL. Independent review of E2100: a phase III trial of bevacizumab plus paclitaxel versus paclitaxel in women with metastatic breast cancer. J Clin Oncol 2009;27(30):4966–72. https://doi.org/10.1200/JCO.2008.21.6630.
[107] Pivot X, Schneeweiss A, Verma S, Thomssen C, Passos-Coelho JL, Benedetti G, Ciruelos E, von Moos R, Chang HT, Duenne AA, Miles DW. Efficacy and safety of bevacizumab in combination with docetaxel for the first-line treatment of elderly patients with locally recurrent or metastatic breast cancer: results from AVADO. Eur J Cancer 2011;47(16):2387–95. https://doi.org/10.1016/j.ejca.2011.06.018.
[108] Brufsky AM, Hurvitz S, Perez E, Swamy R, Valero V, O'Neill V, Rugo HS. RIBBON-2: a randomized, double-blind, placebo-controlled, phase III trial evaluating the efficacy and safety of bevacizumab in combination with chemotherapy for second-line treatment of human epidermal growth factor receptor 2-negative metastatic breast cancer. J Clin Oncol 2011;29(32):4286–93. https://doi.org/10.1200/JCO.2010.34.1255.

[109] Escudier B, Bellmunt J, Négrier S, Bajetta E, Melichar B, Bracarda S, Ravaud A, Golding S, Jethwa S, Sneller V. Phase III trial of bevacizumab plus interferon alfa-2a in patients with metastatic renal cell carcinoma (AVOREN): final analysis of overall survival. J Clin Oncol 2010;28(13):2144–50. https://doi.org/10.1200/JCO.2009.26.7849.

[110] Burger RA, Brady MF, Rhee J, Sovak MA, Kong G, Nguyen HP, Bookman MA. Independent radiologic review of the gynecologic oncology group study 0218, a phase III trial of bevacizumab in the primary treatment of advanced epithelial ovarian, primary peritoneal, or fallopian tube cancer. Gynecol Oncol 2013;131(1):21–6. https://doi.org/10.1016/j.ygyno.2013.07.100.

[111] Tewari KS, Burger RA, Enserro D, Norquist BM, Swisher EM, Brady MF, Bookman MA, Fleming GF, Huang H, Homesley HD, Fowler JM, Greer BE, Boente M, Liang SX, Ye C, Bais C, Randall LM, Chan JK, Ferriss JS, Coleman RL, Aghajanian C, Herzog TJ, DiSaia PJ, Copeland LJ, Mannel RS, Birrer MJ, Monk BJ. Final overall survival of a randomized trial of bevacizumab for primary treatment of ovarian cancer. J Clin Oncol 2019;37(26):2317–28. https://doi.org/10.1200/JCO.19.01009.

[112] Bais C, Mueller B, Brady MF, Mannel RS, Burger RA, Wei W, Marien KM, Kockx MM, Husain A, Birrer MJ, NRG Oncology/Gynecologic Oncology Group. Tumor microvessel density as a potential predictive marker for bevacizumab benefit: GOG-0218 biomarker analyses. J Natl Cancer Inst 2017;109(11). https://doi.org/10.1093/jnci/djx066, djx066.

[113] Perren TJ, Swart AM, Pfisterer J, Ledermann JA, Pujade-Lauraine E, Kristensen G, Carey MS, Beale P, Cervantes A, Kurzeder C, du Bois A, Sehouli J, Kimmig R, Stähle A, Collinson F, Essapen S, Gourley C, Lortholary A, Selle F, Mirza MR, Leminen A, Plante M, Stark D, Qian W, Parmar MK, Oza AM, ICON7 Investigators. A phase 3 trial of bevacizumab in ovarian cancer. N Engl J Med 2011;365(26):2484–96. https://doi.org/10.1056/NEJMoa1103799 [Erratum in: N Engl J Med. 2012;366(3):284].

[114] Aghajanian C, Blank SV, Goff BA, Judson PL, Teneriello MG, Husain A, Sovak MA, Yi J, Nycum LR. OCEANS: a randomized, double-blind, placebo-controlled phase III trial of chemotherapy with or without bevacizumab in patients with platinum-sensitive recurrent epithelial ovarian, primary peritoneal, or fallopian tube cancer. J Clin Oncol 2012;30(17):2039–45. https://doi.org/10.1200/JCO.2012.42.0505.

[115] Aghajanian C, Goff B, Nycum LR, Wang YV, Husain A, Blank SV. Final overall survival and safety analysis of OCEANS, a phase 3 trial of chemotherapy with or without bevacizumab in patients with platinum-sensitive recurrent ovarian cancer. Gynecol Oncol 2015;139(1):10–6. https://doi.org/10.1016/j.ygyno.2015.08.004.

[116] Coleman RL, Brady MF, Herzog TJ, Sabbatini P, Armstrong DK, Walker JL, Kim BG, Fujiwara K, Tewari KS, O'Malley DM, Davidson SA, Rubin SC, DiSilvestro P, Basen-Engquist K, Huang H, Chan JK, Spirtos NM, Ashfaq R, Mannel RS. Bevacizumab and paclitaxel-carboplatin chemotherapy and secondary cytoreduction in recurrent, platinum-sensitive ovarian cancer (NRG Oncology/Gynecologic Oncology Group study GOG-0213): a multicentre, open-label, randomised, phase 3 trial. Lancet Oncol 2017;18(6):779–91. https://doi.org/10.1016/S1470-2045(17)30279-6.

[117] Pujade-Lauraine E, Hilpert F, Weber B, Reuss A, Poveda A, Kristensen G, Sorio R, Vergote I, Witteveen P, Bamias A, Pereira D, Wimberger P, Oaknin A, Mirza MR, Follana P, Bollag D, Ray-Coquard I. Bevacizumab combined with chemotherapy for platinum-resistant recurrent ovarian cancer: the AURELIA open-label randomized phase III trial. J Clin Oncol 2014;32(13):1302–8. https://doi.org/10.1200/JCO.2013.51.4489.

[118] Sorio R, Roemer-Becuwe C, Hilpert F, Gibbs E, García Y, Kaern J, Huizing M, Witteveen P, Zagouri F, Coeffic D, Lück HJ, González-Martín A, Kristensen G, Levaché CB, Lee CK, Gebski V, Pujade-Lauraine E, AURELIA Investigators. Safety and efficacy of single-agent bevacizumab-containing therapy in elderly patients with platinum-resistant recurrent ovarian cancer: subgroup analysis of the randomised phase III AURELIA trial. Gynecol Oncol 2017;144(1):65–71. https://doi.org/10.1016/j.ygyno.2016.11.006.

[119] Tewari KS, Sill MW, Long 3rd HJ, Penson RT, Huang H, Ramondetta LM, Landrum LM, Oaknin A, Reid TJ, Leitao MM, Michael HE, Monk BJ. Improved survival with bevacizumab in advanced cervical cancer. N Engl J Med 2014;370(8):734–43. https://doi.org/10.1056/NEJMoa1309748 [Erratum in: N Engl J Med. 2017;377(7):702].

[120] Wilke H, Muro K, Van Cutsem E, Oh SC, Bodoky G, Shimada Y, Hironaka S, Sugimoto N, Lipatov O, Kim TY, Cunningham D, Rougier P, Komatsu Y, Ajani J, Emig M, Carlesi R, Ferry D, Chandrawansa K, Schwartz JD, Ohtsu A, RAINBOW Study Group. Ramucirumab plus paclitaxel versus placebo plus paclitaxel in patients with previously treated advanced gastric or gastro-oesophageal junction adenocarcinoma (RAINBOW): a double-blind, randomised phase 3 trial. Lancet Oncol 2014;15(11):1224–35. https://doi.org/10.1016/S1470-2045(14)70420-6.

[121] Muro K, Oh SC, Shimada Y, Lee KW, Yen CJ, Chao Y, Cho JY, Cheng R, Carlesi R, Chandrawansa K, Orlando M, Ohtsu A. Subgroup analysis of East Asians in RAINBOW: a phase 3 trial of ramucirumab plus paclitaxel for advanced gastric cancer. J Gastroenterol Hepatol 2016;31(3):581–9. https://doi.org/10.1111/jgh.13153.

[122] Cascinu S, Bodoky G, Muro K, Van Cutsem E, Oh SC, Folprecht G, Ananda S, Girotto G, Wainberg ZA, Miron MLL, Ajani J, Wei R, Liepa AM, Carlesi R, Emig M, Ohtsu A. Tumor response and symptom palliation from RAINBOW, a phase III trial of ramucirumab plus paclitaxel in previously treated advanced gastric cancer. Oncologist 2021;26(3):e414–24. https://doi.org/10.1002/onco.13623.

[123] Tabernero J, Yoshino T, Cohn AL, Obermannova R, Bodoky G, Garcia-Carbonero R, Ciuleanu TE, Portnoy DC, Van Cutsem E, Grothey A, Prausová J, Garcia-Alfonso P, Yamazaki K, Clingan PR, Lonardi S, Kim TW, Simms L, Chang SC, Nasroulah F, RAISE Study Investigators. Ramucirumab versus placebo in combination with second-line FOLFIRI in patients with metastatic colorectal carcinoma that progressed during or after first-line therapy with bevacizumab, oxaliplatin, and a fluoropyrimidine (RAISE): a randomised, double-blind, multicentre, phase 3 study. Lancet Oncol 2015;16(5):499–508. https://doi.org/10.1016/S1470-2045(15)70127-0 [Erratum in: Lancet Oncol. 2015:16(6):e262].

[124] Yoshino T, Portnoy DC, Obermannová R, Bodoky G, Prausová J, Garcia-Carbonero R, Ciuleanu T, García-Alfonso P, Cohn AL, Van Cutsem E, Yamazaki K, Lonardi S, Muro K, Kim TW, Yamaguchi K, Grothey A, O'Connor J, Taieb J, Wijayawardana SR, Hozak RR, Nasroulah F, Tabernero J. Biomarker analysis beyond angiogenesis: RAS/RAF mutation status, tumour sidedness, and second-line ramucirumab efficacy in patients with metastatic colorectal carcinoma from RAISE-a global phase III study. Ann Oncol 2019;30(1):124–31. https://doi.org/10.1093/annonc/mdy461.

[125] Yoshino T, Obermannová R, Bodoky G, Garcia-Carbonero R, Ciuleanu T, Portnoy DC, Kim TW, Hsu Y, Ferry D, Nasroulah F, Tabernero J. Baseline carcinoembryonic antigen as a predictive factor of ramucirumab efficacy in RAISE, a second-line metastatic colorectal carcinoma phase III trial. Eur J Cancer 2017;78:61–9. https://doi.org/10.1016/j.ejca.2017.03.007.

[126] Garon EB, Ciuleanu TE, Arrieta O, Prabhash K, Syrigos KN, Goksel T, Park K, Gorbunova V, Kowalyszyn RD, Pikiel J, Czyzewicz G, Orlov SV, Lewanski CR, Thomas M, Bidoli P, Dakhil S, Gans S, Kim JH, Grigorescu A, Karaseva N, Reck M, Cappuzzo F, Alexandris E, Sashegyi A, Yurasov S, Pérol M. Ramucirumab plus docetaxel versus placebo plus docetaxel for second-line treatment of stage IV non-small-cell lung cancer after disease progression on platinum-based therapy (REVEL): a multicentre, double-blind, randomised phase 3 trial. Lancet 2014;384(9944):665–73. https://doi.org/10.1016/S0140-6736(14)60845-X.

[127] Paz-Ares LG, Pérol M, Ciuleanu TE, Kowalyszyn RD, Reck M, Lewanski CR, Syrigos K, Arrieta O, Prabhash K, Park K, Pikiel J, Göksel T, Lee P, Zimmermann A, Carter GC, Alexandris E, Garon EB. Treatment outcomes by histology in REVEL: a randomized phase III trial of ramucirumab plus docetaxel for advanced non-small cell lung cancer. Lung Cancer 2017;112:126–33. https://doi.org/10.1016/j.lungcan.2017.05.021.

[128] Ramalingam SS, Pérol M, Reck M, Kowalyszyn RD, Gautschi O, Kimmich M, Cho EK, Czyzewicz G, Grigorescu A, Karaseva N, Dakhil S, Lee P, Zimmerman A, Sashegyi A, Alexandris E, Carter GC, Winfree KB, Garon EB. Efficacy and safety of ramucirumab with docetaxel versus placebo with docetaxel as second-line treatment of advanced non-small-cell lung cancer: a subgroup analysis according to patient age in the REVEL trial. Clin Lung Cancer 2018;19(3):270–279.e3. https://doi.org/10.1016/j.cllc.2017.12.011.

[129] Smit EF, Garon EB, Reck M, Cappuzzo F, Bidoli P, Cohen RB, Gao L, O'Brien LM, Lee P, Zimmermann A, Ferry DR, Melemed AS, Pérol M. Exposure-response relationship for ramucirumab from the randomized, double-blind, phase 3 REVEL trial (docetaxel versus docetaxel plus ramucirumab) in second-line treatment of metastatic non-small cell lung cancer. Cancer Chemother Pharmacol 2018;82(1):77–86. https://doi.org/10.1007/s00280-018-3560-5.

[130] Van Cutsem E, Tabernero J, Lakomy R, Prenen H, Prausová J, Macarulla T, Ruff P, van Hazel GA, Moiseyenko V, Ferry D, McKendrick J, Polikoff J, Tellier A, Castan R, Allegra C. Addition of aflibercept to fluorouracil, leucovorin, and irinotecan improves survival in a phase III randomized trial in patients with metastatic colorectal cancer previously treated with an oxaliplatin-based regimen. J Clin Oncol 2012;30(28):3499–506. https://doi.org/10.1200/JCO.2012.42.8201.

[131] Tabernero J, Van Cutsem E, Lakomý R, Prausová J, Ruff P, van Hazel GA, Moiseyenko VM, Ferry DR, McKendrick JJ, Soussan-Lazard K, Chevalier S, Allegra CJ. Aflibercept versus placebo in combination with fluorouracil, leucovorin and irinotecan in the treatment of previously treated metastatic colorectal cancer: prespecified subgroup analyses from the VELOUR trial. Eur J Cancer 2014;50(2):320–31. https://doi.org/10.1016/j.ejca.2013.09.013.

[132] Ruff P, Van Cutsem E, Lakomy R, Prausova J, van Hazel GA, Moiseyenko VM, Soussan-Lazard K, Dochy E, Magherini E, Macarulla T, Papamichael D. Observed benefit and safety of aflibercept in elderly patients with metastatic colorectal cancer: an age-based analysis from the randomized placebo-controlled phase III VELOUR trial. J Geriatr Oncol 2018;9(1):32–9. https://doi.org/10.1016/j.jgo.2017.07.010.

[133] Yardley DA, Burris 3rd HA, Clark BL, Shipley D, Rubin M, Barton Jr J, Arrowsmith E, Hainsworth JD. Hormonal therapy plus bevacizumab in postmenopausal patients who have hormone receptor-positive metastatic breast cancer: a phase II trial of the Sarah Cannon oncology research consortium. Clin Breast Cancer 2011;11 (3):146–52. https://doi.org/10.1016/j.clbc.2011.03.010.

[134] Manders P, Beex LV, Tjan-Heijnen VC, Span PN, Sweep CG. Vascular endothelial growth factor is associated with the efficacy of endocrine therapy in patients with advanced breast carcinoma. Cancer 2003;98(10):2125–32. https://doi.org/10.1002/cncr.11764.

[135] Massarweh S, Schiff R. Unraveling the mechanisms of endocrine resistance in breast cancer: new therapeutic opportunities. Clin Cancer Res 2007;13(7):1950–4. https://doi.org/10.1158/1078-0432.CCR-06-2540.

[136] Osborne CK, Shou J, Massarweh S, Schiff R. Crosstalk between estrogen receptor and growth factor receptor pathways as a cause for endocrine therapy resistance in breast cancer. Clin Cancer Res 2005;11(2 Pt 2):865s–70s.

[137] Yao J, Deng K, Huang J, Zeng R, Zuo J. Progress in the understanding of the mechanism of tamoxifen resistance in breast cancer. Front Pharmacol 2020;11:592912. https://doi.org/10.3389/fphar.2020.592912.

[138] Peng WX, Huang JG, Yang L, Gong AH, Mo YY. Linc-RoR promotes MAPK/ERK signaling and confers estrogen-independent growth of breast cancer. Mol Cancer 2017;16(1):161. https://doi.org/10.1186/s12943-017-0727-3.

[139] Heckler MM, Thakor H, Schafer CC, Riggins RB. ERK/MAPK regulates ERRγ expression, transcriptional activity and receptor-mediated tamoxifen resistance in ER+ breast cancer. FEBS J 2014;281(10):2431–42. https://doi.org/10.1111/febs.12797.

[140] Martín M, Loibl S, Hyslop T, De la Haba-Rodríguez J, Aktas B, Cirrincione CT, Mehta K, Barry WT, Morales S, Carey LA, Garcia-Saenz JA, Partridge A, Martinez-Jañez N, Hahn O, Winer E, Guerrero-Zotano A, Hudis C, Casas M, Rodriguez-Martin C, Furlanetto J, Carrasco E, Dickler MN, GEICAM Spanish Breast Cancer Group; GBG (German Breast Group); Alliance for Clinical Trials in Oncology (Alliance). Evaluating the addition of bevacizumab to endocrine therapy as first-line treatment for hormone receptor-positive metastatic breast cancer: a pooled analysis from the LEA (GEICAM/2006-11_GBG51) and CALGB 40503 (Alliance) trials. Eur J Cancer 2019;117:91–8. https://doi.org/10.1016/j.ejca.2019.06.002.

[141] Dickler MN, Barry WT, Cirrincione CT, Ellis MJ, Moynahan ME, Innocenti F, Hurria A, Rugo HS, Lake DE, Hahn O, Schneider BP, Tripathy D, Carey LA, Winer EP, Hudis CA. Phase III trial evaluating Letrozole as first-line endocrine therapy with or without bevacizumab for the treatment of postmenopausal women with hormone receptor-positive advanced-stage breast cancer: CALGB 40503 (Alliance). J Clin Oncol 2016;34 (22):2602–9. https://doi.org/10.1200/JCO.2015.66.1595.

[142] Roviello G, Francini E, Perrella A, Laera L, Mazzei MA, Guerrini S, Roviello F, Marrelli D, Petrioli R. Five years of stable disease with maintenance therapy using bevacizumab and tamoxifen in a patient with metastatic breast cancer. Cancer Biol Ther 2015;16(4):493–7. https://doi.org/10.1080/15384047.2015.1017692.

[143] Comunanza V, Bussolino F. Therapy for cancer: strategy of combining anti-angiogenic and target therapies. Front Cell Dev Biol 2017;5:101. https://doi.org/10.3389/fcell.2017.00101.

[144] Seto T, Kato T, Nishio M, Goto K, Atagi S, Hosomi Y, Yamamoto N, Hida T, Maemondo M, Nakagawa K, Nagase S, Okamoto I, Yamanaka T, Tajima K, Harada R, Fukuoka M, Yamamoto N. Erlotinib alone or with bevacizumab as first-line therapy in patients with advanced non-squamous non-small-cell lung cancer harbouring EGFR mutations (JO25567): an open-label, randomised, multicentre, phase 2 study. Lancet Oncol 2014;15(11):1236–44. https://doi.org/10.1016/S1470-2045(14)70381-X [Erratum in: Lancet Oncol. 2014;15(11): e475].

[145] Rosell R, Dafni U, Felip E, Curioni-Fontecedro A, Gautschi O, Peters S, Massutí B, Palmero R, Aix SP, Carcereny E, Früh M, Pless M, Popat S, Kotsakis A, Cuffe S, Bidoli P, Favaretto A, Froesch P, Reguart N, Puente J, Coate L, Barlesi F, Rauch D, Thomas M, Camps C, Gómez-Codina J, Majem M, Porta R, Shah R, Hanrahan E, Kammler R, Ruepp B, Rabaglio M, Kassapian M, Karachaliou N, Tam R, Shames DS, Molina-Vila MA, Stahel RA, BELIEF collaborative group. Erlotinib and bevacizumab in patients with advanced non-small-cell lung cancer and activating EGFR mutations (BELIEF): an international, multicentre, single-arm, phase 2 trial. Lancet Respir Med 2017;5(5):435–44. https://doi.org/10.1016/S2213-2600(17)30129-7 [Erratum in: Lancet Respir Med. 2018;6(12):e57].

[146] Saito H, Fukuhara T, Furuya N, Watanabe K, Sugawara S, Iwasawa S, Tsunezuka Y, Yamaguchi O, Okada M, Yoshimori K, Nakachi I, Gemma A, Azuma K, Kurimoto F, Tsubata Y, Fujita Y, Nagashima H, Asai G, Watanabe S, Miyazaki M, Hagiwara K, Nukiwa T, Morita S, Kobayashi K, Maemondo M. Erlotinib plus bevacizumab versus erlotinib alone in patients with EGFR-positive advanced non-squamous non-small-cell lung cancer (NEJ026): interim analysis of an open-label, randomised, multicentre, phase 3 trial. Lancet Oncol 2019;20(5):625–35. https://doi.org/10.1016/S1470-2045(19)30035-X.

[147] Anon, https://www.ema.europa.eu/en/documents/smop/chmp-post-authorisation-summary-positive-opinion-avastin_en.pdf.

[148] Hecht JR, Mitchell E, Chidiac T, Scroggin C, Hagenstad C, Spigel D, Marshall J, Cohn A, McCollum D, Stella P, Deeter R, Shahin S, Amado RG. A randomized phase IIIB trial of chemotherapy, bevacizumab, and panitumumab compared with chemotherapy and bevacizumab alone for metastatic colorectal cancer. J Clin Oncol 2009;27(5):672–80. https://doi.org/10.1200/JCO.2008.19.8135.

[149] Tol J, Koopman M, Cats A, Rodenburg CJ, Creemers GJ, Schrama JG, Erdkamp FL, Vos AH, van Groeningen CJ, Sinnige HA, Richel DJ, Voest EE, Dijkstra JR, Vink-Börger ME, Antonini NF, Mol L, van Krieken JH, Dalesio O, Punt CJ. Chemotherapy, bevacizumab, and cetuximab in metastatic colorectal cancer. N Engl J Med 2009;360 (6):563–72. https://doi.org/10.1056/NEJMoa0808268 [Erratum in: N Engl J Med. 2010 Dec 23;363(26):2573].

[150] Nakagawa K, Garon EB, Seto T, Nishio M, Ponce Aix S, Paz-Ares L, Chiu CH, Park K, Novello S, Nadal E, Imamura F, Yoh K, Shih JY, Au KH, Moro-Sibilot D, Enatsu S, Zimmermann A, Frimodt-Moller B, Visseren-Grul C, Reck M, RELAY Study Investigators. Ramucirumab plus erlotinib in patients with untreated, EGFR-mutated, advanced non-small-cell lung cancer (RELAY): a randomised, double-blind, placebo-controlled, phase 3 trial. Lancet Oncol 2019;20(12):1655–69. https://doi.org/10.1016/S1470-2045(19)30634-5.

[151] Zhu AX, Rosmorduc O, Evans TR, Ross PJ, Santoro A, Carrilho FJ, Bruix J, Qin S, Thuluvath PJ, Llovet JM, Leberre MA, Jensen M, Meinhardt G, Kang YK. SEARCH: a phase III, randomized, double-blind, placebo-controlled trial of sorafenib plus erlotinib in patients with advanced hepatocellular carcinoma. J Clin Oncol 2015;33(6):559–66. https://doi.org/10.1200/JCO.2013.53.7746.

[152] Do K, Cao L, Kang Z, Turkbey B, Lindenberg ML, Larkins E, Holkova B, Steinberg SM, Raffeld M, Peer CJ, Figg WD, Eugeni M, Jacobs P, Choyke P, Wright JJ, Doroshow JH, Kummar S. A phase II study of Sorafenib combined with cetuximab in EGFR-expressing, KRAS-mutated metastatic colorectal cancer. Clin Colorectal Cancer 2015;14(3):154–61. https://doi.org/10.1016/j.clcc.2015.02.007.

[153] Galal KM, Khaled Z, Mourad AM. Role of cetuximab and sorafenib in treatment of metastatic colorectal cancer. Indian J Cancer 2011;48(1):47–54. https://doi.org/10.4103/0019-509X.75825.

[154] Gilbert J, Schell MJ, Zhao X, Murphy B, Tanvetyanon T, Leon ME, Neil Hayes D, Haigentz Jr M, Saba N, Nieva J, Bishop J, Sidransky D, Ravi R, Bedi A, Chung CH. A randomized phase II efficacy and correlative studies of cetuximab with or without sorafenib in recurrent and/or metastatic head and neck squamous cell carcinoma. Oral Oncol 2015;51(4):376–82. https://doi.org/10.1016/j.oraloncology.2014.12.011.

[155] Subbiah V, Khawaja MR, Hong DS, Amini B, Yungfang J, Liu H, Johnson A, Schrock AB, Ali SM, Sun JX, Fabrizio D, Piha-Paul S, Fu S, Tsimberidou AM, Naing A, Janku F, Karp DD, Overman M, Eng C, Kopetz S, Meric-Bernstam F, Falchook GS. First-in-human trial of multikinase VEGF inhibitor regorafenib and anti-EGFR antibody cetuximab in advanced cancer patients. JCI Insight 2017;2(8). https://doi.org/10.1172/jci.insight.90380, e90380.

[156] Iqbal N, Iqbal N. Human epidermal growth factor receptor 2 (HER2) in cancers: overexpression and therapeutic implications. Mol Biol Int 2014;2014:852748. https://doi.org/10.1155/2014/852748.

[157] Zhao M, Pan X, Layman R, Lustberg MB, Mrozek E, Macrae ER, Wesolowski R, Carothers S, Puhalla S, Shapiro CL, Ramaswamy B. A phase II study of bevacizumab in combination with trastuzumab and docetaxel in HER2 positive metastatic breast cancer. Invest New Drugs 2014;32(6):1285–94. https://doi.org/10.1007/s10637-014-0122-5.

[158] Gianni L, Romieu GH, Lichinitser M, Serrano SV, Mansutti M, Pivot X, Mariani P, Andre F, Chan A, Lipatov O, Chan S, Wardley A, Greil R, Moore N, Prot S, Pallaud C, Semiglazov V. AVEREL: a randomized phase III trial evaluating bevacizumab in combination with docetaxel and trastuzumab as first-line therapy for HER2-positive locally recurrent/metastatic breast cancer. J Clin Oncol 2013;31(14):1719–25. https://doi.org/10.1200/JCO.2012.44.7912.

[159] Slamon DJ, Swain SM, Buyse M, Martin M, Geyer CE, Im Y-H, Pienkowski T, Kim S-B, Robert NJ, Steger G, Crown J, Verma S, Eiermann W, Costantino J, Im S-A, Mamounas E, Schwartzberg L, Paterson A, Mackey J, Provencher L, Press M, Thirlwell M, Bee-Munteanu V, Henschel V, Crepelle-Flechais A, Wolmark

N. Primary results from BETH, a phase 3 controlled study of adjuvant chemotherapy and trastuzumab ± bevacizumab in patients with HER2-positive, node-positive or high risk node-negative breast cancer. In: 36th Annual San Antonio Breast Cancer Symposium (SABCS)/AACR American Association for Cancer Research:S1-03.2013; 2013.

[160] Cristofanilli M, Johnston SR, Manikhas A, Gomez HL, Gladkov O, Shao Z, Safina S, Blackwell KL, Alvarez RH, Rubin SD, Ranganathan S, Redhu S, Trudeau ME. A randomized phase II study of lapatinib + pazopanib versus lapatinib in patients with HER2+ inflammatory breast cancer. Breast Cancer Res Treat 2013;137(2):471–82. https://doi.org/10.1007/s10549-012-2369-x.

[161] Hua H, Kong Q, Zhang H, Wang J, Luo T, Jiang Y. Targeting mTOR for cancer therapy. J Hematol Oncol 2019;12 (1):71. https://doi.org/10.1186/s13045-019-0754-1.

[162] Feldman DR, Ged Y, Lee CH, Knezevic A, Molina AM, Chen YB, Chaim J, Coskey DT, Murray S, Tickoo SK, Reuter VE, Patil S, Xiao H, Aghalar J, Apollo AJ, Carlo MI, Motzer RJ, Voss MH. Everolimus plus bevacizumab is an effective first-line treatment for patients with advanced papillary variant renal cell carcinoma: final results from a phase II trial. Cancer 2020;126(24):5247–55. https://doi.org/10.1002/cncr.33148.

[163] Ravaud A, Barrios CH, Alekseev B, Tay MH, Agarwala SS, Yalcin S, Lin CC, Roman L, Shkolnik M, Anak O, Gogov S, Pelov D, Louveau AL, Melichar B. RECORD-2: phase II randomized study of everolimus and bevacizumab versus interferon α-2a and bevacizumab as first-line therapy in patients with metastatic renal cell carcinoma. Ann Oncol 2015;26(7):1378–84. https://doi.org/10.1093/annonc/mdv170.

[164] Shih KC, Chowdhary S, Rosenblatt P, Weir 3rd AB, Shepard GC, Williams JT, Shastry M, Burris 3rd HA, Hainsworth JD. A phase II trial of bevacizumab and everolimus as treatment for patients with refractory, progressive intracranial meningioma. J Neurooncol 2016;129(2):281–8. https://doi.org/10.1007/s11060-016-2172-3.

[165] Koeberle D, Dufour JF, Demeter G, Li Q, Ribi K, Samaras P, Saletti P, Roth AD, Horber D, Buehlmann M, Wagner AD, Montemurro M, Lakatos G, Feilchenfeldt J, Peck-Radosavljevic M, Rauch D, Tschanz B, Bodoky G, Swiss Group for Clinical Cancer Research (SAKK). Sorafenib with or without everolimus in patients with advanced hepatocellular carcinoma (HCC): a randomized multicenter, multinational phase II trial (SAKK 77/08 and SASL 29). Ann Oncol 2016;27(5):856–61. https://doi.org/10.1093/annonc/mdw054.

[166] Motzer RJ, Hutson TE, Glen H, Michaelson MD, Molina A, Eisen T, Jassem J, Zolnierek J, Maroto JP, Mellado B, Melichar B, Tomasek J, Kremer A, Kim HJ, Wood K, Dutcus C, Larkin J. Lenvatinib, everolimus, and the combination in patients with metastatic renal cell carcinoma: a randomised, phase 2, open-label, multicentre trial. Lancet Oncol 2015;16(15):1473–82. https://doi.org/10.1016/S1470-2045(15)00290-9 [Erratum in: Lancet Oncol. 2016;17(7):e270. Erratum in: Lancet Oncol. 2018;19(10):e509].

[167] Anon, https://www.astrazeneca.it/area-stampa/press-releases/2018/OLAPARIB_PAOLA_ESMO2019.html.

[168] Huang Z, Xiong Q, Cui Z, Tao H, Zhang S, Wang L, Cui P, Chen S, Huang D, Yang B, Hu Y. Efficacy and safety of crizotinib plus bevacizumab in *ALK/ROS-1/c-MET* positive non-small cell lung cancer: an open-label, single-arm, prospective observational study. Am J Transl Res 2021;13(3):1526–34.

[169] Kareva I. A combination of immune checkpoint inhibition with metronomic chemotherapy as a way of targeting therapy-resistant cancer cells. Int J Mol Sci 2017;18(10):2134. https://doi.org/10.3390/ijms18102134.

[170] Skavatsou E, Semitekolou M, Morianos I, Karampelas T, Lougiakis N, Xanthou G, Tamvakopoulos C. Immunotherapy combined with metronomic dosing: an effective approach for the treatment of NSCLC. Cancers (Basel) 2021;13(8):1901. https://doi.org/10.3390/cancers13081901.

[171] Reck M, TSK M, Nishio M, Jotte RM, Cappuzzo F, Orlandi F, Stroyakovskiy D, Nogami N, Rodríguez-Abreu D, Moro-Sibilot D, Thomas CA, Barlesi F, Finley G, Lee A, Coleman S, Deng Y, Kowanetz M, Shankar G, Lin W, Socinski MA, IMpower150 Study Group. Atezolizumab plus bevacizumab and chemotherapy in non-small-cell lung cancer (IMpower150): key subgroup analyses of patients with EGFR mutations or baseline liver metastases in a randomised, open-label phase 3 trial. Lancet Respir Med 2019;7(5):387–401. https://doi.org/10.1016/S2213-2600(19)30084-0.

[172] Galle PR, Finn RS, Qin S, Ikeda M, Zhu AX, Kim TY, Kudo M, Breder V, Merle P, Kaseb A, Li D, Mulla S, Verret W, Xu DZ, Hernandez S, Ding B, Liu J, Huang C, Lim HY, Cheng AL, Ducreux M. Patient-reported outcomes with atezolizumab plus bevacizumab versus sorafenib in patients with unresectable hepatocellular carcinoma (IMbrave150): an open-label, randomised, phase 3 trial. Lancet Oncol 2021;22(7):991–1001. https://doi.org/10.1016/S1470-2045(21)00151-0.

[173] Cousin S, Cantarel C, Guegan JP, Gomez-Roca C, Metges JP, Adenis A, Pernot S, Bellera C, Kind M, Auzanneau C, Le Loarer F, Soubeyran I, Bessede A, Italiano A. Regorafenib-avelumab combination in patients with

microsatellite stable colorectal cancer (REGOMUNE): a single-arm, open-label, phase II trial. Clin Cancer Res 2021;27(8):2139–47. https://doi.org/10.1158/1078-0432.CCR-20-3416.

[174] Jonasch E, Slack RS, Geynisman DM, Hasanov E, Milowsky MI, Rathmell WK, Stovall S, Juarez D, Gilchrist TR, Pruitt L, Ornstein MC, Plimack ER, Tannir NM, Rini BI. Phase II study of two weeks on, one week off sunitinib scheduling in patients with metastatic renal cell carcinoma. J Clin Oncol 2018;36(16):1588–93. https://doi.org/10.1200/JCO.2017.77.1485.

[175] Uemura M, Tomita Y, Miyake H, Hatakeyama S, Kanayama HO, Numakura K, Takagi T, Kato T, Eto M, Obara W, Uemura H, Choueiri TK, Motzer RJ, Fujii Y, Kamei Y, Umeyama Y, di Pietro A, Oya M. Avelumab plus axitinib vs sunitinib for advanced renal cell carcinoma: Japanese subgroup analysis from JAVELIN renal 101. Cancer Sci 2020;111(3):907–23. https://doi.org/10.1111/cas.14294.

[176] Bedke J, Albiges L, Capitanio U, Giles RH, Hora M, Lam TB, Ljungberg B, Marconi L, Klatte T, Volpe A, Abu-Ghanem Y, Dabestani S, Pello SF, Hofmann F, Kuusk T, Tahbaz R, Powles T, Bex A. The 2021 updated european association of urology guidelines on renal cell carcinoma: immune checkpoint inhibitor-based combination therapies for treatment-naive metastatic clear-cell renal cell carcinoma are standard of care. Eur Urol 2021;80(4):393–7. https://doi.org/10.1016/j.eururo.2021.04.042.

[177] Rini BI, Plimack ER, Stus V, Gafanov R, Hawkins R, Nosov D, Pouliot F, Alekseev B, Soulières D, Melichar B, Vynnychenko I, Kryzhanivska A, Bondarenko I, Azevedo SJ, Borchiellini D, Szczylik C, Markus M, McDermott RS, Bedke J, Tartas S, Chang YH, Tamada S, Shou Q, Perini RF, Chen M, Atkins MB, Powles T, KEYNOTE-426 Investigators. Pembrolizumab plus axitinib versus sunitinib for advanced renal-cell carcinoma. N Engl J Med 2019;380(12):1116–27. https://doi.org/10.1056/NEJMoa1816714.

[178] Makker V, Rasco D, Vogelzang NJ, Brose MS, Cohn AL, Mier J, Di Simone C, Hyman DM, Stepan DE, Dutcus CE, Schmidt EV, Guo M, Sachdev P, Shumaker R, Aghajanian C, Taylor M. Lenvatinib plus pembrolizumab in patients with advanced endometrial cancer: an interim analysis of a multicentre, open-label, single-arm, phase 2 trial. Lancet Oncol 2019;20(5):711–8. https://doi.org/10.1016/S1470-2045(19)30020-8.

[179] Atkins MB, Plimack ER, Puzanov I, Fishman MN, McDermott DF, Cho DC, Vaishampayan U, George S, Olencki TE, Tarazi JC, Rosbrook B, Fernandez KC, Lechuga M, Choueiri TK. Axitinib in combination with pembrolizumab in patients with advanced renal cell cancer: a non-randomised, open-label, dose-finding, and dose-expansion phase 1b trial. Lancet Oncol 2018;19(3):405–15. https://doi.org/10.1016/S1470-2045(18)30081-0.

[180] Motzer RJ, Penkov K, Haanen J, Rini B, Albiges L, Campbell MT, Venugopal B, Kollmannsberger C, Negrier S, Uemura M, Lee JL, Vasiliev A, Miller Jr WH, Gurney H, Schmidinger M, Larkin J, Atkins MB, Bedke J, Alekseev B, Wang J, Mariani M, Robbins PB, Chudnovsky A, Fowst C, Hariharan S, Huang B, di Pietro A, Choueiri TK. Avelumab plus Axitinib versus Sunitinib for advanced renal-cell carcinoma. N Engl J Med 2019;380(12):1103–15. https://doi.org/10.1056/NEJMoa1816047.

[181] Choueiri TK, Powles T, Burotto M, Escudier B, Bourlon MT, Zurawski B, Oyervides Juárez VM, Hsieh JJ, Basso U, Shah AY, Suárez C, Hamzaj A, Goh JC, Barrios C, Richardet M, Porta C, Kowalyszyn R, Feregrino JP, Żołnierek J, Pook D, Kessler ER, Tomita Y, Mizuno R, Bedke J, Zhang J, Maurer MA, Simsek B, Ejzykowicz F, Schwab GM, Apolo AB, Motzer RJ, CheckMate 9ER Investigators. Nivolumab plus cabozantinib versus sunitinib for advanced renal-cell carcinoma. N Engl J Med 2021;384(9):829–41. https://doi.org/10.1056/NEJMoa2026982.

[182] Kelley RK, Oliver JW, Hazra S, Benzaghou F, Yau T, Cheng AL, Rimassa L. Cabozantinib in combination with atezolizumab versus sorafenib in treatment-naive advanced hepatocellular carcinoma: COSMIC-312 phase III study design. Future Oncol 2020;16(21):1525–36. https://doi.org/10.2217/fon-2020-0283.

[183] Finn RS, Qin S, Ikeda M, Galle PR, Ducreux M, Kim TY, Kudo M, Breder V, Merle P, Kaseb AO, Li D, Verret W, Xu DZ, Hernandez S, Liu J, Huang C, Mulla S, Wang Y, Lim HY, Zhu AX, Cheng AL, IMbrave150 Investigators. Atezolizumab plus bevacizumab in unresectable hepatocellular carcinoma. N Engl J Med 2020;382(20):1894–905. https://doi.org/10.1056/NEJMoa1915745.

CHAPTER

3

Antiangiogenic agents in the treatment of colorectal, gastric, and gastroesophageal junction adenocarcinoma

Roberto Petrioli, Salvatora Tindara Miano, and Ignazio Martellucci

Medical Oncology Unit, Medical Hospital and University of Siena, Siena, Italy

Abstract

Angiogenesis supplies for increased request of oxygen and nutrients which is typical of the fast-growing microenvironment of solid tumors. Tumor angiogenesis provides for the formation of new blood vessels in an uncontrolled and not-organized fashion, ultimately resulting in enhanced tumor growth and increased metastatic potential. The angiogenetic mechanism represents an appealing target for novel anticancer therapies and several antiangiogenic drugs are available for the treatment of some tumor types.

Herein, we summarize the clinical evidence supporting the use of antiangiogenic agents in patients with colorectal cancer and the promising activity of antiangiogenic drugs, alone or combined with conventional therapy, exhibited in patients with advanced gastric cancer (GC) or gastroesophageal junction adenocarcinoma (GEJ).

The antiangiogenic drug bevacizumab combined with conventional chemotherapy was effective in increasing the median OS of patients with metastatic colorectal cancer. Nevertheless, although the inhibition of angiogenesis became an interesting area of GC research, the role of antiangiogenic agents remains controversial in this tumor type. The antiangiogenic agent ramucirumab was the first and unique antiangiogenic drug which was approved for the treatment of metastatic GC patients after failure of first-line treatment. Several oral multitarget tyrosine kinase inhibitors (TKI) were investigated in advanced GC and GEJ adenocarcinoma, and some of these agents (sunitinib, sorafenib, apatinib, trebananib, and regorafenib) achieved encouraging results in terms of PFS and OS. Recently, the combination of antiangiogenic agents plus immunotherapy exhibited promising results in patients with advanced GC or GEJ adenocarcinoma. The ongoing trials will help to clarify the real efficacy of the combination of antiangiogenic agents with conventional anticancer treatment and with immune checkpoint inhibitors (ICIs) in advanced GC and GEJ adenocarcinoma, and the role of predictive and prognostic biomarkers.

https://doi.org/10.1016/B978-0-323-90190-1.00007-X

Abbreviations

5FU/FA	5-fluorouracil/folinic acid
CAPEOX	capecitabine and oxaliplatin
DFS	disease-free survival
FOLFIRI	irinotecan, fluorouracil, and leucovorin
FOLFOX	5-fluorouracil, leucovorin, and oxaliplatin
GC	gastric cancer
GEJ	gastroesophageal junction adenocarcinoma
ICI	immune checkpoint inhibitors
PD-L1	programmed death ligand-1
PlGF	placental growth factor
TKI	tyrosine kinase inhibitors
TME	tumor microenvironment
VEGF	vascular endothelial growth factor
XELOX	xeloda and oxaliplatin

Conflict of interest

No potential conflicts of interest were disclosed.

Introduction

Angiogenesis is a biological process in which novel capillary blood vessels grow from preexisting vasculature, providing tissues with oxygen and nutrients. It is correlated with numerous complicated interactions between various biological components, such as several cell types, soluble angiogenic factors, and extracellular matrix components. The process of angiogenesis is complex and consists of four distinct sequential steps: (1) degradation of basement membrane glycoproteins and other components of the extracellular matrix surrounding the blood vessels by proteolytic enzymes; (2) endothelial cell activation and migration; (3) endothelial cell proliferation; and (4) endothelial cells transforming into tubelike structures and then forming capillary tubes with a novel basement membrane. Aberrant angiogenesis is a key feature and a major process in cancer development.

Angiogenesis supplies for increased request of oxygen and nutrients which is typical of the fast-growing microenvironment of solid tumors. Tumor angiogenesis provides for the formation of new blood vessels in an uncontrolled and not-organized fashion, ultimately resulting in enhanced tumor growth and increased metastatic potential. The angiogenic process is mainly regulated by the interactions between tissue vascular endothelial growth factor receptors (VEGFRs) and their soluble ligands (vascular endothelial growth factors, VEGFs), which are kept under a finely regulated control in healthy conditions. VEGF molecules consist of several sub-variants, including VEGF-A, VEGF-B, VEGF-C, VEGF-D, and placental growth factor (PIGF). The receptor family involves three molecular subtypes (VEGFR-1, VEGFR-2, and VEGFR-3) which are type II transmembrane proteins characterized by a tyrosine kinase activity [1]. Among these receptors, VEGFR-2 has been found to be the one which is majorly implicated in the pathological over-formation of blood vessels in the context of several solid tumors, and therefore VEGFR-2 is considered as the main factor responsible for tumor angiogenesis. Although associated with overall worse clinical outcomes, VEGFR-2 represents an appealing target for novel anticancer therapies. Today, it is possible to inhibit the signal arising from the activation of VEGFR-2 through several pharmacodynamics approaches, including seizure of the ligand (bevacizumab), receptor blockade (ramucirumab), and small-molecule inhibition (sorafenib, sunitinib, apatinib, cediranib, telatinib, regorafenib, cabozantinib) [2]. Although malignant cells continue to grow in the absence of VEGF, they exhibit reduced invasion ability resulting in reduced metastatic activity. In addition, their reduced intracellular pressure makes them more vulnerable to chemotherapy and radiotherapy.

Herein, we summarize the clinical evidence supporting the use of antiangiogenic agents in patients with colorectal cancer and the promising activity of antiangiogenic drugs, alone or combined with conventional therapy, exhibited in patients with advanced gastric cancer (GC) or gastroesophageal junction adenocarcinoma (GEJ).

Colorectal cancer

Bevacizumab is a humanized monoclonal antibody that targets VEGF, thus inhibiting tumor vascularization. Bevacizumab combined with conventional chemotherapy was effective in increasing the median OS of patients with metastatic colorectal cancer [3,4] and, therefore, the use of bevacizumab combined with irinotecan-based or oxaliplatin-based chemotherapy was approved by the US Food and Drug Administration in 2013.

Bevacizumab combined with conventional chemotherapy (5-fluorouracil, leucovorin, and oxaliplatin—FOLFOX; capecitabine and oxaliplatin—CAPEOX; irinotecan, fluorouracil, and leucovorin—FOLFIRI) is commonly used as the first-line treatment for metastatic colorectal cancer. In addition, for patients with favorable clinical conditions who require rapid and strong tumor shrinkage, FOLFOXIRI, triple combination of bevacizumab plus oxaliplatin, irinotecan, and 5-fluorouracil/folinic acid (5FU/FA), is an effective treatment option [5]. It is important to consider that patients undergoing bevacizumab treatment can experience hypertension, nephrotic syndrome, gastrointestinal perforation, delayed wound healing, arterial and venous thromboembolic disease, hemorrhage, and proteinuria. Therefore, patient characteristics and clinical criteria should be evaluated in order to establish who will benefit of bevacizumab treatment.

The antiangiogenic drug aflibercept is a fusion protein composed of extracellular domains of VEGFR-1 and VEGFR-2 and inhibits tumor angiogenesis by targeting VEGF-A, VEGF-B, and PIGF. Aflibercept combined with FOLFIRI significantly improved OS in patients with metastatic colorectal cancer compared with FOLFIRI alone, after failure of an oxaliplatin-based regimen [6].

Some studies showed that both patients with KRAS mutations and those with a wild-type genotype may benefit from bevacizumab and both left- and right-sided colon tumors respond well to this antiangiogenic drug [7,8]. Nevertheless, a recent meta-analysis suggested that the addition of bevacizumab to conventional chemotherapy optimizes outcome in RAS wild-type patients, but not necessarily in RAS-mutated metastatic colorectal cancer patients [9]. Other trials validated the efficacy of bevacizumab as second-line treatment, and a better response rate as well as a longer PFS were observed with the combination of conventional FOLFOX chemotherapy with bevacizumab with respect to FOLFOX alone for patients with metastatic colorectal cancer who progressed after FOLFOX treatment [10]. Interesting data were also reported with bevacizumab as maintenance after first-line chemotherapy in patients with metastatic colorectal cancer who achieved stable disease. Maintenance with bevacizumab significantly improved the OS in patients with metastatic colorectal cancer compared with no maintenance. Similarly, trends of longer OS for maintenance bevacizumab plus xeloda (capecitabine) and oxaliplatin (XELOX) chemotherapy versus bevacizumab alone (MACRO trial) and for maintenance bevacizumab compared with no treatment (SAKK trial) were reported [11–13].

In addition, bevacizumab combined with oxaliplatin-based chemotherapy was investigated as adjuvant treatment. However, bevacizumab-avastin adjuvant (AVANT) phase III trial failed to demonstrate the superiority of conventional chemotherapy FOLFOX4 or XELOX combined with bevacizumab compared with FOLFOX4 alone in terms of disease-free survival (DFS) in patients with stage III C colon [14]. Similarly, the National Surgical Adjuvant Breast and Bowel Project c-08 phase III trial reported that bevacizumab combined with adjuvant oxaliplatin-based chemotherapy did not increase DFS in patients with stage II or III colon cancer [15]. The addition of bevacizumab to capecitabine achieved no DFS benefit in patients with high-risk stage II or stage III colorectal cancer [16].

Gastric cancer and gastroesophageal junction adenocarcinoma

A few randomized trials investigated the efficacy and safety of antiangiogenic agent bevacizumab associated with chemotherapy in patients with GC and GEJ cancer. The Avastin for Advanced Gastric Cancer trial (AVAGAST) investigated the combination of bevacizumab with cisplatin and capecitabine in patients with locally advanced or metastatic GC or GEJ cancer. Although the median OS was better in the bevacizumab group (12.1 months) with respect to placebo group (10.1 months), the difference was not statistically significant (HR = 0.87; 95% CI, 0.3–1.03; $P = .1002$). Conversely, both the overall RR (46% vs. 37.4%; HR = 8.61; CI, 0.6–16.6, $P = .0315$) and the median PFS (6.7 vs. 5.3 months; HR = 0.80; 95% CI 0.68–0.93; $P = .0037$) were significantly improved with bevacizumab versus placebo [17]. A smaller phase III trial (AVATAR) investigated the combination of bevacizumab with chemotherapy in 202 Chinese patients. The difference in median OS was not statistically different for bevacizumab plus chemotherapy with respect to chemotherapy alone in the AVATAR trial (OS: 10.5 months vs. 11.4 months; HR = 1.11; 95% CI, 0.79–1.56; $P = .56$) [18]. In these trials, with the exception of the incidence of grade > 3 hypertension, the incidence of grade > 3 adverse events of special interest for bevacizumab was similar between the two treatment arms.

The main results of AVAGAST and AVATAR trials are summarized in Table 1.

Although the inhibition of angiogenesis has become an interesting area of GC research, the role of antiangiogenic agents remains controversial in GC. Interestingly, a meta-analysis of 22 trials exploring targeted therapy for a total of 7022 advanced GC patients demonstrated promising efficacy results for antiangiogenic therapy [19]. However, based on published results, the combination of bevacizumab with chemotherapy cannot be recommended for patients with locally advanced or metastatic GC or GEJ cancer.

Tumor site, histology, and ethnic differences may influence the expression of potential antiangiogenic targets and the chosen backbone of chemotherapy may further influence

TABLE 1 Randomized clinical trials on first-line treatment of patients with metastatic gastric cancer or gastroesophageal junction cancer with bevacizumab combined with conventional chemotherapy.

Trial	Patients (n)	Regimen	ORR	PFS (months)	OS (months)
AVAGAST	387	Cis/Cap + Bev	46%	6.7	12.1
	vs				
	387	Cis/Cap + Placebo	37.4%	5.3	10.1
			$P=.03$	$P=.03$	$P=.1$
AVATAR	100	Cis/Cap + Bev	41%	6.3	10.5
	vs				
	102	Cis/Cap + Bev	34%	6.0	11.4
			$P=.34$	$P=.47$	$P=.56$

the response to antiangiogenic treatment [20]. Unfortunately, the lack of specific predictive biomarkers will not allow an appropriate selection of treatment based on the peculiar patient characteristics.

Ramucirumab

Antiangiogenic agent ramucirumab was the first and unique antiangiogenic drug which was approved for the treatment of metastatic GC patients after failure of first-line treatment; in fact most of the antiangiogenic drugs failed to show an improvement in terms of OS. Two phase III clinical trials (REGARD and RAINBOW) demonstrated that ramucirumab is a valuable therapeutic option in GC [21,22]. The REGARD trial enrolled patients with advanced GC or GEJ adenocarcinoma and disease progression after first-line platinum-containing or fluoropyrimidine-containing chemotherapy.

Patients were randomized to receive ramucirumab or ramucirumab plus placebo. The median OS was significantly better in the ramucirumab group than in the placebo group (5.2 months vs. 3.8 months, HR = 0.776, $P=.047$), and the rate of disease control was significantly higher in the ramucirumab group than in placebo group (49% vs. 23%, $P<.00001$). Although the rates of serious adverse events were similar between the two randomized groups, the rates of hypertension were higher in the ramucirumab group than in patients given placebo (16% vs. 8%).

The RAINBOW trial investigated ramucirumab or placebo plus paclitaxel in patients with advanced GC or GEJ adenocarcinoma and disease progression on or within 4 months after first-line chemotherapy. Overall survival was significantly longer in the ramucirumab plus paclitaxel group than in the placebo plus paclitaxel group (median, 9.6 vs. 7.4 months: HR = 0.807, $P=.0169$). PFS was significantly better in ramucirumab group than in the placebo

TABLE 2 Randomized clinical trials on second-line treatment of patients with metastatic gastric cancer or gastroesophageal junction cancer with ramucirumab alone or combined with paclitaxel.

Trial	Patients (n)	Regimen	ORR/DCR	PFS (months)	OS (months)
REGARD	238	Ram	DCR 49%	2.1	5.2
	vs				
	217	Placebo	DCR 23%	1.3	3.8
			$P=.0001$	$P=.0001$	$P=.047$
RAINBOW	330	Ram/Pac	RR 28%	4.4	9.6
	vs				
	335	Pac	RR 16%	2.8	7.4
			$P=.0001$	$P=.0001$	$P=.016$

plus paclitaxel group (4.40 vs. 2.86 months, HR = 0.635; $P=.0001$). At 6 months, the PFS rate was 36% versus 17% and at 9 months was 22% versus 10%.

Based on the data from the RAINBOW trial, in November 2014, the FDA approved the combination of ramucirumab and paclitaxel as a treatment for patients with previously treated advanced GC or GEJ adenocarcinoma. The main results of REGARD and RAINBOW trials are summarized in Table 2.

Unfortunately, other studies failed to demonstrate the superiority of ramucirumab over chemotherapy in the first-line setting. A large phase II trial enrolled 168 patients with previously untreated unresectable locally advanced or metastatic esophageal, GC, or GEJ adenocarcinoma who were randomized to receive ramucirumab plus standard chemotherapy or placebo [23]. PFS and OS were similar (6.4 vs. 6.7 months and 11.7 vs. 11.5 months, respectively). Negative results of this study may be explained by the high rate of patients with esophageal cancer (>45%) and the higher rate of treatment discontinuation before progression in the ramucirumab group.

Tyrosine kinase inhibitors (TKIs)

Several oral multi-target TKI inhibitors were investigated in advanced GC and GEJ adenocarcinoma and some of these agents (sunitinib, sorafenib, apatinib, trebananib, and regorafenib) achieved encouraging results in terms of PFS and OS in patients with advanced GC [24,25]. However, the reported results do not actually support the role of these oral antiangiogenic agents in this setting.

As single agent, sunitinib was investigated in a phase II study involving 78 patients as second-line therapy: the median PFS was 2.3 months, and the median OS was 6.8 months. In another phase II study sunitinib achieved a median OS of 5.8 months in patients with chemoresistant advanced GC [26].

The TKI sorafenib combined with docetaxel and cisplatin was evaluated in a large phase II study in patients with metastatic or advanced GC or GEJ adenocarcinoma. The RR was 41%, PFS was 5.8 months, and OS was 13.6 months. Another phase II study investigated the efficacy and toxicity of sorafenib combined with oxaliplatin as second-line therapy for GC after failure to first-line therapy with cisplatin and fluoropyrimidine. The median PFS was 3 months, and the median OS was 6.5 months. The association of sorafenib plus oxaliplatin resulted in a good safety profile [27]. The STARGATE trial compared capecitabine plus cisplatin with capecitabine plus cisplatin with sorafenib. The overall RR was 54% in the experimental arm and 52% in the chemotherapy arm ($P = .8$). The median PFS was 5.6 months in the sorafenib group and 5.3 months in the chemotherapy-alone group ($P = .6$). There was no statistically significant difference in OS between the two treatment groups (median 11.7 months vs. 10.8 months, $P = .6$). Although the addition of sorafenib to chemotherapy was well tolerated, it did not result more effective than chemotherapy alone [28].

Another phase II study investigated the safety and efficacy of apatinib as a treatment option for heavily pretreated patients with metastatic GC. Toxicity was tolerable and the dosing regimen of oral 850 mg once daily was recommended for the following studies [29].

A randomized phase III trial randomized patients with metastatic GC to receive apatinib or placebo as third-line treatment. The median OS was significantly longer in the apatinib group compared with that in the placebo group (195 days vs. 140 days, $P < .01$). The objective RR was 2.84% in the apatinib group and 0.00% in the placebo group, respectively. The grade 3/4 adverse events that occurred in more than 2% of patients were hand-and-foot syndrome, hypertension, proteinuria, anorexia, fatigue, and elevated aminotransferase [30]. Promising results were recently reported with the combination of apatinib and the chemotherapeutic agents paclitaxel and S1 as preoperative treatment in 30 evaluable patients with locally advanced GC. After treatment, the RR was 73.3%: 18 patients underwent surgery and R0 resection was achieved in 17 patients. The patients who underwent the conversion surgery had a better OS than patients who did not (3 years OS: 52.9% vs. 8.3%, $P = .001$) [31].

An unsatisfactory activity was reported with the TKI cediranib in a small group of patients, while promising results in terms of disease control rate were described with the use of telatinib associated with chemotherapy, with a good toxicity profile.

Interesting data were reported with the TKI regorafenib in a randomized phase II study: patients with advanced GC who progressed after a first and second line of therapy were randomized to receive regorafenib or placebo. Median PFS was 11.1 weeks and 3.9 weeks for regorafenib and placebo, respectively ($P = .0001$); median OS was 25 weeks and 19.4 weeks, respectively ($P = .11$) [32].

Everolimus is an oral inhibitor of the mammalian target of rapamycin serine–threonine kinase (mTOR), inhibiting the PI3K/Akt/mTOR pathway, which is important in angiogenesis. A randomized phase II study enrolled patients with advanced GC that progressed after one or two lines of systemic chemotherapy to receive everolimus or placebo. The median OS was 5.4 months with everolimus and 4.3 months with placebo ($P = .124$). The median PFS was 1.7 months with everolimus and 1.4 months with placebo ($P = .001$). However, formal statistical significance could not be declared per the hierarchical testing strategy [33].

Angiogenesis and immune checkpoint inhibitors (ICIs)

Treatment with ICIs exhibits benefit in patients with mismatch-repair/microsatellite instability high (MMR/MSI-H) and in patients with high programmed death ligand-1 (PD-L1) expression as third-line therapy [34,35]. Resistance to immunotherapy may be due to the immune-suppressive nature of the tumor microenvironment (TME) which is characterized by increased infiltration of regulatory T cells, tumor-associated macrophages, and myeloid-derived suppressor cells, with a consequent decrease in the activity of immune effector cells such as cytotoxic T cells and helper T cells [36]. VEGF, which is induced by local tissue hypoxia and acidosis, has direct local and systemic immune-suppressive mechanisms, and therefore, anti-VEGF therapy might "normalize" the TME and potentially reverse the resistance to ICIs. The combination of ICIs with antiangiogenic agents is a promising new treatment option in cancer treatment.

The combination of the antiangiogenic agent ramucirumab plus immunotherapy (nivolumab, pembrolizumab, durvalumab) exhibited promising results in recent studies performed in patients with advanced GC or GEJ adenocarcinoma [37–39]. A recent phase I/II study involved patients with advanced GC who received ramucirumab combined with nivolumab; the 6-month PFS was 37.4% (90% confidential intervals: 25.7%–49.2%) and met the primary endpoint [37]. The combination of ramucirumab plus pembrolizumab showed manageable safety profile with RR of 7% and DCR of 44% in 41 pretreated patients with advanced GC or GEJ adenocarcinoma [38].

An encouraging RR of 21%, PFS of 2.6 months, and OS of 12.4 months were recently reported with the combination of ramucirumab and durvalumab in a group of cancer patients including GC and GEJ adenocarcinoma [39].

Some VEGFR TKIs demonstrated immunomodulatory activity, suggesting a potential role in the combination with ICIs in order to overcome the resistance to immunotherapy. In this setting, the combination of the VEGFR multi-TKI regorafenib combined with nivolumab in chemo-refractory patients with advanced GC obtained 44% RR with 5.6 months PFS and 12.3 months OS. Grade 3 toxicity was observed in about 40% of patients, with palmoplantar erythrodysesthesia, rash, and proteinuria the most common adverse events [40].

The VEGFR2 inhibitor apatinib combined with anti-PD-1 antibody achieved 16% RR and 78% DCR in patients with advanced GC or GEJ adenocarcinoma [41].

Interesting results in terms of RR (25%), DCR (85%), and PFS (4 months) were recently observed with the VEGFR TKI cabozantinib combined with durvalumab in pretreated patients with advanced GC and GEJ adenocarcinoma [42].

Promising results were reported with the combination of the VEGFR multi-TKI lenvatinib with pembrolizumab in chemo-naive and chemo-refractory gastric and GEJ adenocarcinoma patients, with 69% RR, 100% DCR, and 6.9-month PFS [43]. Grade 3 toxicity was observed in 48% of patients, and proteinuria, thrombocytopenia, and hypertension were the most common adverse events.

To date, although the MSI-H status, PD-L1 expression, and tumor molecular burden correlate with treatment response from monotherapy with PD-1/PD-L1 blockade, the role of predictive biomarkers for the efficacy of the combination of VEGFR with ICIs is unknown. The ongoing trials will help to clarify the real efficacy of the combination of antiangiogenic agents with ICIs in advanced GC and GEJ adenocarcinoma, and the role of predictive and prognostic biomarkers.

Conclusions

The inhibition of pro-angiogenic growth factors is an interesting research field in cancer treatment, and the inhibition of VEGF has entered clinical practice in some tumor types including colorectal cancer. The combination of the antiangiogenic drug bevacizumab can improve OS in patients with advanced or metastatic colorectal cancer when combined with conventional chemotherapy. Although the inhibition of angiogenesis has become a promising area of research in GC and GEJ adenocarcinoma, to date no antiangiogenic drug, alone or combined with chemotherapy, can be recommended as first-line therapy in these tumors. Ramucirumab, alone or combined with paclitaxel, is the only antiangiogenic drug which is approved as second-line treatment in advanced GC. Other VEGF-targeted therapies, such as the several available oral TKIs, are currently being studied for the treatment of GC and encouraging results are reported.

Antiangiogenic therapy with VEGFR inhibitors combined with PD-1/PD-L1 inhibitors is a promising approach in patients with advanced GC cancer and GEJ adenocarcinoma, and preliminary data are very promising. Ongoing studies and well-designed clinical trials will clarify the role of antiangiogenic therapy in gastrointestinal tumors, and some predictive biomarkers will be available to predict the benefit of antiangiogenic agents in these solid tumors.

References

[1] Ferrara N, Gerber HP, LeCouter J. The biology of VEGF and its receptors. J Nat Med 2003;9:669–76.

[2] Roviello G, Petrioli R, Marano L, Polom K, Marrelli D, Perrella A, Roviello F. Angiogenesis inhibitors in gastric and gastroesophageal junction cancer. Gastric Cancer 2016;19:31–41.

[3] Hurwitz H, Fehrenbacher L, Novotny W, Cartwright T, Hainsworth J, Heim W, Berlin J, Baron A, Griffing S, Holmgren E, Ferrara N, Fyfe G, Rogers B, Ross R, Kabbinavar F. Bevacizumab plus irinotecan, fluorouracil, and leucovorin for metastatic colorectal cancer. N Engl J Med 2004;3:2335–42.

[4] Golfinopoulos V, Salanti G, Pavlidis N, Ioannidis JPA. Survival and disease-progression benefits with treatment regimens for advanced colorectal cancer: a meta-analysis. Lancet Oncol 2007;8:898–911.

[5] Loupakis F, Cremolini C, Masi G, Lonardi S, Zagonel V, Salvatore L, Cortesi E, Tomasello G, Ronzoni M, Spadi R, Zaniboni A, Tonini G, Buonadonna A, Amoroso D, Chiara S, Carlomagno C, Boni C, Allegrini G, Boni L, Falcone A. Initial therapy with FOLFOXIRI and bevacizumab for metastatic colorectal cancer. N Engl J Med 2014;371:1609–18.

[6] Van Cutsem E, Tabernero J, Lakomy R, Prenen H, Prausova J, Macarulla T, Ruff P, van Hazel GA, Moiseyenko V, Ferry D, McKendrick J, Polikoff J, Tellier A, Castan R, Allegra C. Addition of aflibercept to fluorouracil, leucovorin, and irinotecan improves survival in a phase III randomized trial in patients with metastatic colorectal cancer previously treated with an oxaliplatin-based regimen. J Clin Oncol 2012;30:3499–506.

[7] Venook AP, Niedzwiecki D, Lenz HJ, Innocenti F, Fruth B, Meyerhardt JA, Schrag D, Greene C, O'Neil BH, Atkins JN, Berry S, Polite BN, O'Reilly EM, Goldberg RM, Hochster HS, Schilsky RL, Bertagnolli MM, El-Khoueiry AB, Watson P, Benson 3rd AB, Mulkerin DL, Mayer RJ, Blanke C. Effect of first-line chemotherapy combined with cetuximab or bevacizumab on overall survival in patients with KRAS wild-type advanced or metastatic colorectal cancer: a randomized clinical trial. JAMA 2017;317:2392–401.

[8] Heinemann V, von Weikersthal LF, Decker T, Kiani A, Vehling-Kaiser U, Al-Batran SE, Heintges T, Lerchenmüller C, Kahl C, Seipelt G, Kullmann F, Stauch M, Scheithauer W, Hielscher J, Scholz M, Muller S, Link H, Niederle N, Rost A, Hoffkes HG, Moehler M, Lindig RU, Modest DP, Rossius L, Kirchner T, Jung A, Stintzing S. FOLFIRI plus cetuximab versus FOLFIRI plus bvacizumab as first-line treatment for patients with metastatic colorectal cancer (FIRE-3): a randomised, open-label, phase 3 trial. Lancet Oncol 2014;10:1065–75.

[9] Stahler A, Heinemann V, Ricard I, von Einem JC, Giessen-Jung C, Westphalen CB, Michl M, Heinrich K, Miller-phillips L, Jelas I, Stintzing S, Dominik PD. Modest current treatment options in RAS mutant metastatic colorectal cancer patients: a meta-analysis of 14 randomized phase III trials. J Cancer Res Clin Oncol 2020;146: 2077–87.

[10] Cassidy J, Saltz LB, Giantonio BJ, Kabbinavar FF, Hurwitz HI, Rohr UP. Effect of bevacizumab in older patients with metastatic colorectal cancer: pooled analysis of four randomized studies. J Cancer Res Clin Oncol 2010;136:737–43.

[11] Salvatore L, Bria E, Sperduti I, Hinke A, Hegewisch-Becker S, Aparicio T, Le Malicot K, Boige V, Koeberle D, Baertschi D, Dietrich D, Tortora G, Arnold D. Bevacizumab as maintenance therapy in patients with metastatic colorectal cancer: a meta-analysis of individual patients'data from 3 phase III studies. Cancer Treat Rev 2021;97:102202.
[12] Diaz-Rubio E, Gomez-Espana A, Massuti B, Sastre J, Abad A, Valladares M, Rivera F, Safont MJ, Martinez de Prado P, Gallen M, Gonzalez E, Marcuello E, Benavides V, Fernandez-Martos C, Losa F, Escudero P, Arrivi A, Cervantes A, Duenas R, Lopez-Ladron A, Lacasta A, Llanos M, Tabernero JM, Anton A, Aranda E, Spanish Cooperative Group for the Treatment Digestive Tumors. First-line XELOX plus bevacizumab followed by XELOX plus bevacizumab or single-agent bevacizumab as maintenance therapy in patients with metastatic colorectal cancer: the phase III MACRO TTD study. Oncologist 2012;17:15–25.
[13] Koeberle D, Betticher DC, von Moos R, Dietrich D, Brauchli P, Baertschi D, Matter K, Winterhalder R, Borner M, Anchisi S, Moosmann P, Kollar A, Saletti P, Roth A, Frueth M, Kueng M, Popescu RA, Schaler S, Hess V, Herrmann R. Bevacizumab continuation versus no continuation after first-line chemotherapy plus bevacizumab in patients with metastatic colorectal cancer: a randomized phase III non-inferiority trial (SAKK 41/06). Ann Oncol 2015;26:709–14.
[14] de Gramont A, Van Cutsem E, Schmoll HJ, Tabernero J, Clarke S, Moore MJ, Cunningham D, Cartwright TH, Hecht JR, Rivera F, Im SA, Bodoky G, Salazar R, Maindrault-Goebel F, Shacham-Shmueli E, Bajetta E, Makrutzki M, Shang A, André T, Hoff PM. Bevacizumab plus oxaliplatin-based chemotherapy as adjuvant treatment for colon cancer (AVANT): a phase 3 randomised controlled trial. Lancet Oncol 2012;13:1225–33.
[15] Shah MA, Renfro LA, Allegra CJ, André T, de Gramont A, Schmoll HJ, Haller DG, Alberts SR, Yothers G, Sargent DJ. Impact of patient factors on recurrence risk and time dependency of oxaliplatin benefit in patients with colon cancer: analysis from modern-era adjuvant studies in the adjuvant colon cancer end points (ACCENT) database. J Clin Oncol 2016;34:843–53.
[16] Kerr RS, Love S, Sagelov E, Johnstone E, Falcon B, Hewett P, Weaver A, Church D, Scudder C, Pearson S, Julier P, Pezzella F, Tomlinson I, Domingo E, Kerr DJ. Adjuvant capecitabine plus bevacizumab versus capecitabine alone in patients with colorectal cancer (QUASAR 2): an open-label, randomised phase 3 trial. Lancet Oncol 2016;17:1543–57.
[17] Ohtsu A, Shah MA, Van Cutsem E, Rha SY, Sawaki A, Park SR, Lim HY, Yamada Y, Wu J, Langer B, Stamawski M, Kang YK. Bevacizumab in combination with chemotherapy as first-line therapy in advanced gastric cancer: a randomized, double-blind, placebo-controlled phase III study. J Clin Oncol 2011;29:3968–76.
[18] Shen L, Li J, Xu J, Pan H, Dai G, Qin S, Wang L, Wang J, Yang Z, Shu Y, Xu R, Chen L, Liu Y, Yu S, Bu L, Piao Y. Bevacizumab plus capecitabine and cisplatin in Chinese patients with inoperable locally advanced or metastatic gastric or gastroesophageal junction cancer: randomized, double-blind, phase III study (AVATAR study). Gastric Cancer 2015;18:168–76.
[19] Ciliberto D, Staropoli N, Caglioti F, Gualtieri S, Fiorillo L, Chiellino S, De Angelis AM, Mendicino F, Bottoa C, Craglia M, Tassone P, Tagliaferri P. A systematic review and meta-analysis of randomized trials on the role of targeted therapy in the management of advanced gastric cancer: evidence does not translate? Cancer Biol Ther 2015;16:1148–59.
[20] Roviello G, Ravelli A, Fiaschi AI, Cappelletti MR, Gobbi A, Senti C, Zanotti L, Polom K, Reynolds AR, Fox SB, Generali D. Apatinib for the treatment of gastric cancer. Expert Rev Gastroenterol Hepatol 2016;10:887–92.
[21] Fuchs CS, Tomasek J, Yong CJ, Dumitru F, Passalacqua R, Goswami C, Safran H, Dos Santos LV, Aprile G, Ferry DR, Melichar B, Tehfe M, Topuzov E, Zalcberg JR, Chau J, Campbell W, Sivanandan C, Pikiel J, Koshiji M, Hsu Y, Liepa AM, Gao L, Schawartz JD, Tabernero J. REGARD Trial Investigators Ramucirumab monotherapy for previously treated advanced gastric or gastro-oesophageal junction adenocarcinoma (REGARD): an international, randomised, multicentre, placebo-controlled, phase 3 trial. Lancet 2014;383:31–9.
[22] Wilke H, Muro K, Van Cutsem E, Oh SC, Bodoky G, Shimada Y, Hironaka S, Sugimoto N, Lipatov O, Kim TY, Cunningham D, Rougier P, Komatsu Y, Ajani J, Emig M, Carlesi R, Ferry D, Chandrawansa K, Schwartz JD, Ohtsu A. Ramucirumab plus paclitaxel versus placebo plus paclitaxel in patients with previously treated advanced gastric or gastro-oesophageal junction adenocarcinoma (RAINBOW): a double-blind, randomised phase 3 trial. RAINBOW Study Group. Lancet Oncol 2014;15:1224–35.
[23] Yoon HH, Bendell JC, Braiteh FS, Firdaus I, Philip AP, Al C, Lewis N, Anderson DM, Arrowsmith E, Schwartz JD, Gao L, Hsu Y, Xu Y, Ferry D, Alberts SR, Wainberg ZA. Ramucirumab combined with FOLFOX as front-line therapy for advanced esophageal, gastroesophageal junction, or gastric adenocarcinoma: a randomized, double-blind, multicenter Phase II trial. Ann Oncol 2016;27:2196–203.

[24] Yu J, Zhang Y, Leung LH, Liu L, Yang F, Yao X. Efficacy and safety of angiogenesis inhibitors in advanced gastric cancer: a systematic review and meta-analysis. J Hematol Oncol 2016;9:111.

[25] Mawalla B, Yuan X, Luo X, Chalya PL. Treatment outcome of anti-angiogenesis through VEGF-pathway in the management of gastric cancer: a systematic review of phase II and III clinical trials. BMC Res Notes 2018;11(1):21.

[26] Bang YJ, Kang YK, Kang WK, Boku N, Chung HC, Chen JS, Doi T, Sun Y, Shen L, Qin S, Ng WT, Tursi JM, Lechuga MJ, Lu DR, Ruiz-Garcia A, Sobrero A. Phase II study of sunitinib as second-line treatment for advanced gastric cancer. Invest New Drugs 2011;29:1449–58.

[27] Martin-Richard M, Gallego R, Pericay C, Foncillas JG, Queralt B, Casado E, Barriuso J, Iranzo V, Juez I, Visa L, Saigi E, Barnadas A, Garcia-Albeniz X, Maurel J. Multicenter phase II study of oxaliplatin and sorafenib in advanced gastric adenocarcinoma after failure of cisplatin and fluoropyrimidine treatment: a GEMCAD study. Invest New Drugs 2013;31:1573–9.

[28] Kang Y, Lee KH, Shen L, Yeh K, Hong YS. Randomized phase II study of capecitabine and cisplatin with or without sorafenib in patients with metastatic gastric cancer: STARGATE study. Ann Oncol 2014;25(Suppl. 4): iv210–53.

[29] Li J, Qin S, Xu J, Guo WJ, Xiong JP, Bai Y, San G, Yang Y, Wang L, Xu N, Cheng Y, Zhe-Hai W, Zheng L, Tao M. A randomized, double-blind, multicenter, phase II, three-arm, placebo-control study of apatinib as third-line treatment in patients with metastatic gastric carcinoma. J Clin Oncol 2011;29(suppl):4019.

[30] Qin S. Phase III study of apatinib in advanced gastric cancer: a randomized, double-blind, placebo-controlled trial. J Clin Oncol 2014;32(5s):4003.

[31] Xu Z, Hu C, Yu J, Du Y, Hu P, Yu G, Hu C, Zhang Y, Mao W, Chen S, Cheng X. Efficacy of conversion surgery following apatinib plus paclitaxel/S1 for advanced gastric cancer with unresectable factors: a multicenter, single-arm, phase II trial. Front Pharmacol 2021;12:642511.

[32] Sjoquist KM, Martin AJ, Tsobanis E, Yip S, Kang YK, Bang YJ, Alcindor T, O'Callaghan CJ, Burnell MJ, Tebbutt NC, Rha SY, Lee J, Cho JY, Lipton LR, Wong M, Strickland A, Kim JW. Regorafenib for the treatment of advanced gastric cancer (INTEGRATE): a multinational placebo-controlled phase II trial. Clin Oncol 2016;34:2728–35.

[33] Ohtsu A, Ajani JA, Bai YX, Bang YJ, Chung HC, Pan HM, Sahmoud T, Shen L, Yeh KH, Chin K, Muro K, Kim YH, Ferry D, Tebbutt NC, Al-Batran SE, Smith H, Costantini C, Rzvi S, Lebwohl D, Van Cutsem E. Everolimus for previously treated advanced gastric cancer: results of the randomized, double-blind, phase III GRANITE-1 study. J Clin Oncol 2013;31:3935–43.

[34] Le DT, Uram JN, Wang H, Bartlett BR, Kemberling H, Eyring AD, Skora AD, Luber BS, Azad NS, Laheru D, Biedrzycki B, Donehower RC, Zaheer A, Fisher GA, Crocenzi TS, Lee JJ, Duffy SM, Goldberg RM, de la Chapelle A, Koshiji M, Bhaijee F, Huebner T, Hruban RH, Wood LD, Cuka N, Pardoll DM, Papadopoulos N, Kinzler KW, Zhou S, Cornish TC, Taube JM, Anders RA, Eshleman JR, Vogelstein B, Diaz Jr LA. PD-1 blockade in tumors with mismatch-repair deficiency. N Engl J Med 2015;372:2509–20.

[35] Fuchs CS, Doi T, Jang RW, Muro K, Satoh T, Machado M, Sun W, Jalal SI, Shah MA, Metges J-P, Garrido M, Golan T, Mandala M, Wainberg ZA, Catenacci DV, Ohtsu A, Shitara K, Geva R, Bleeker J, Ko AH, Ku G, Philip P, Enzinger P, Bang YJ, Levitan D, Wang J, Rosales M, Dalal RP, Yoon HH. Safety and efficacy pf pembrolizumab monotherapy in patients with previously treated advanced gastric and gastroesophageal junction cancer: phase 2 clinical keynote-059 trial. JAMA Oncol 2018;4, e180013.

[36] Binnewies M, Roberts EW, Kersten K, Chan V, Fearon DF, Merad M, Coussens LM, Gabrilovich DI, Ostrand-Rosenberg S, Hedrick CC, Vonderheide RH, Pittet MJ, Jain RK, Zou W, Howcroft TK, Woodhouse EC, Weinberg RA, Krummel MF. Understanding the tumor immune microenvironment (TIME) for effective therapy. Nat Med 2018;24:541–50.

[37] Fukuoka S, Hara H, Takahashi N, Kojima T, Kawazoe A, Asayama M, Yoshii T, Kotani D, Tamura H, Mikamoto Y, Hirano N, Wakabayashi M, Nomura S, Sato A, Kuwata T, Togashi Y, Nishikawa H, Shitara K. Regorafenib plus nivolumab in patients with advanced gastric or colorectal cancer: an open-label, dose-escalation, and dose-expansion phase Ib trial (REGONIVO, EPOC1603). J Clin Oncol 2020;38(18):2053–61.

[38] Cahu I, Penel N, Soriano AO, Arkenau H-T, Cultrera J, Santana-Davila R, Calvo E, Le Tourneau C, Zender L, Bendell JC, Mi G, Gao L, McNeely SC, Oliveira JM, Ferry D, Herbst RS, Fuchs CS. Ramucirumab in combiantion with pembrolizumab in treatment-naive advanced gastric or GEJ adenocarcinoma: safety and antitumor activity from the phase 1a/b JVDF trial. Cancer 2020;12:2985.

[39] Bang YJ, Golan T, Dahan L, Fu S, Moreno V, Park K, Geva R, De Braud F, Wainberg ZA, Reck M, Goff L, Laing N, Mi G, Oliveira JM, Wasserstrom H, Lin CC. Ramucirumab and durvalumab for previously treated, advanced

non-small-cell lung cancer, gastric/gastro-oesophageal junction adenocarcinoma, or hepatocellular carcinoma: an open-label, phase Ia/b study (JVDJ). Eur J Cancer 2020;137:272–84.

[40] Fukuoka S, Hara H, Takahashi N, Kojima T, Kawazoe A, Asayama M, Yoshii T, Kotani D, Tamura H, Mikamoto Y, Hirano N, Wakabayashi M, Nomura S, Sato A, Kuwata T, Togashi Y, Nishikawa H. Shitara K regorafenib plus nivolumab in patients with advanced gastric or colorectal cancer: an open-label, dose-escalation, and dose-expansion phase Ib trial (REGONIVO, EPOC1603). J Clin Oncol 2020;38:2053–61.

[41] Xu J, Zhang Y, Jia R, Yue C, Chang L, Liu R, Zhang G, Zhao C, Zhang Y, Chen C, Wang Y, Yi X, Hu Z, Zou J, Wang Q. Anti-PD-1 antibody SHR-1210 combined with apatinib for advanced hepatocellular carcinoma, gastric, or esophagogastric junction cancer: an open-label, dose escalation and expansion study. Clin Cancer Res 2019;25(2):515–23.

[42] Massard C, Gordon MS, Sharma S, Rafii S, Wainberg ZA, Luke J, Curiel TJ, Colon-Otero G, Hamid O, Sanborn RE, O'Donnell PH, Drakaki A, Tan W, Kurland JF, Rebelatto MC, Jin X, Blake-Haskins JA, Gupta A, Segal NH. Safety and efficacy of durvalumab (MEDI4736), an anti-programmed cell death ligand-1 immune checkpoint inhibitor, in patients with advanced urothelial bladder cancer. J Clin Oncol 2016;34:3119–25.

[43] Kawazoe A, Fukuoka S, Nakamura Y, Kuboki Y, Wakabayashi M, Nomura S, Mikamoto Y, Shima H, Fujishiro N, Higuchi T, Sato A, Kuwata T, Shitara K. Lenvatinib plus pembrolizumab in patients with advanced gastric cancer in the first-line or second-line setting (EPOC1706): an open-label, single-arm, phase 2 trial. Lancet Oncol 2020;21:1057–65.

CHAPTER

4

Antiangiogenic drugs as chemosensitizer in renal cell cancer

Giandomenico Roviello[a], Pietro Spatafora[b], Francesco Sessa[b,c], Claudio Bisegna[b], Martina Catalano[a], Andrea Liaci[b], Lorenzo Verdelli[b], Ester Illiano[d], Ilaria Camilla Galli[e], Beatrice Detti[f], Gabriella Nesi[e], Graziano Vignolini[b], and Donata Villari[b,c]

[a]Department of Medical Oncology, Careggi Hospital, Florence, Italy [b]Unit of Urological Robotic Surgery and Renal Transplantation, Careggi Hospital, Florence, Italy [c]Department of Experimental and Clinical Medicine, University of Florence, Florence, Italy [d]Department of Surgical and Biomedical Science, Andrological and Urogynecological Clinic, Santa Maria Terni Hospital, University of Perugia, Terni, Italy [e]Division of Pathological Anatomy, University of Florence, Florence, Italy [f]Radiation Oncology Unit - Oncology Department, Careggi Hospital, Florence, Italy

Abstract

Although surgical resection of localized renal cell carcinoma (RCC) can be curative for localized disease, some patients may relapse. In addition, RCCs can be clinically silent during their course, and the initial diagnosis can be delayed until disease is either locally advanced and unresectable or metastatic. The natural history of disease for patients with advanced or metastatic RCC can widely be different from few months to many years depending on the clinical, pathologic, laboratory, and radiographic features of disease, as well as therapy response. The understanding of the molecular pathogenesis of RCC allowed to identify targets for therapeutic intervention with the development of multiple agents, having an important role in the management of patients with advanced disease. Among them, antiangiogenic drugs have changed the management of these patients considering the central role of angiogenesis in the metastatic RCC tumorigenesis and progression. In this chapter, an overview on the use of antiangiogenic drugs in the treatment of RCC, with regard to their chemosensitizer aspect, is provided.

Abbreviations

AEs	adverse events
CI	confidence interval
FDA	Food and Drug Administration

https://doi.org/10.1016/B978-0-323-90190-1.00005-6

HR	hazard ratio
ICI	immune checkpoint inhibitor
IMDC	International Metastatic RCC Database Consortium
mRCC	metastatic renal cell carcinoma
mTOR	mammalian target of rapamycin
ORR	objective response rate
OS	overall survival
PD-1	programmed cell death protein 1
PD-L1	programmed cell death ligand 1
PFS	progression-free survival
RCC	renal cell carcinoma
TKI	tyrosine kinase inhibitors
VEGF	vascular endothelial growth factor
VEGFR1-3	vascular endothelial growth factor receptor 1–3

Conflict of interest

No potential conflicts of interest were disclosed.

Introduction

Renal cell cancer (RCC) is not sensitive to radiotherapy and chemotherapy. For early-stage RCC, surgical resection is the most effective treatment, but nearly one-third of patients have had distant metastases at the time of diagnosis, and even if they have undergone surgery, 30% of patients still have tumor recurrence or metastasis [1]. Over the past two decades, the introduction of antiangiogenic drugs such as bevacizumab and tyrosine kinase inhibitors (TKIs) has transformed the management of the metastatic renal cell cancer (mRCC). In fact, angiogenesis plays a central role in the mRCC tumorigenesis and progression, regulating the immune landscape through abnormal tumor vessel formation, dysregulation of various immune cells, and promotion of an immunosuppressive tumor microenvironment. Nowadays, many agents are being developed to focus on vascular endothelial growth factor (VEGF)/VEGFR pathway such as pazopanib, sunitinib, axitinib, bevacizumab, and mammalian target of rapamycin (mTOR) pathway, e.g., everolimus and temsirolimus [2–5].

By the use of these drugs, survival rates of patients with mRCC have been remarkably enhanced and the current approvals as first-line treatment for these patients are numerous [6].

Clinical applications in renal cell cancer

The therapeutic scenario of mRCC has markedly increased, ranging from the most studied molecular target therapies (e.g., sunitinib, pazopanib) to those most recently introduced (e.g., cabozantinib, lenvatinib), up to immune checkpoint inhibitors (ICIs) (Fig. 1).

The most recent clinical trials with the combination of molecular targeted agents and ICIs show how, by restoring an efficient immune response against cancer cells and by establishing an immunological memory, it is possible to obtain not only a better radiological response but also a longer progression-free survival (PFS) and overall survival (OS). Several clinical studies combining antiangiogenic drugs with checkpoint therapies have reported improvements in

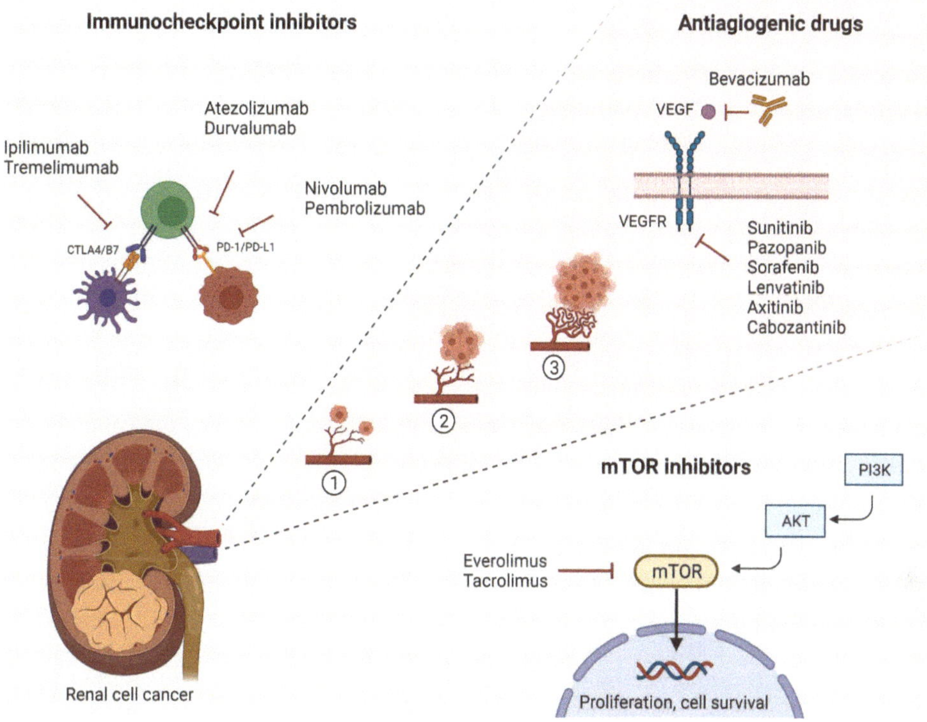

FIG. 1 Drug categories approved in the treatment of advanced renal cell cancer (RCC). *Image created with BioRender.com.*

tumor immune responses with associated clinical benefit. A phase III, open-label study, IMmotion151, randomized untreated patients with mRCC stratified according to programmed death ligand 1 (PD-L1) expression to receive atezolizumab plus bevacizumab [7,8]. In the PD-L1-positive group, PFS was longer with the combination of atezolizumab plus bevacizumab compared to sunitinib (11.2 versus 7.7 months, HR: 0.74, 95% CI: 0.57–0.96) [7]. The objective response rate (ORR) among PD-L1-positive patients was 43% for the combination group versus 35% in the sunitinib group. Treatment-related grade 3 and higher adverse events (AEs) were observed in 40% of patients receiving atezolizumab plus bevacizumab versus 54% of patients receiving sunitinib. Overall, the study met one but not both co-primary endpoints, and the combination has not been approved for first-line use in mRCC.

The phase III study, JAVELIN Renal 1010, randomized patients with untreated mRCC to receive either avelumab plus axitinib or sunitinib [9]. Patients with PD-L1-positive tumors represented 63.2% of the patients; in this population, PFS was 13.8 months with the combination versus 7.2 months with sunitinib (HR: 0.61, 95% CI: 0.47–0.79). There was not a significant difference in OS among these patients, although a higher ORR of 55.2% with avelumab plus axitinib versus 25.5% with sunitinib was recorded. Across the complete study population, median PFS was longer with avelumab plus axitinib (13.8 months) versus sunitinib (8.4months) (HR: 0.69, 95% CI: 0.56–0.84). Grade 3 or higher treatment-related AEs were observed in 71.2% of patients receiving avelumab plus axitinib versus 71.5% of patients receiving sunitinib. Based on these results, FDA approved this combination for the first-line treatment of patients with mRCC. The phase III KEYNOTE-426 trial randomized treatment-naïve mRCC patients to receive either pembrolizumab plus axitinib or sunitinib [10]. Median PFS was of 15.1months in the pembrolizumab plus axitinib group versus 11.1months in the sunitinib group (HR: 0.69, 95% CI: 0.57–0.84). Median OS was not reached in either group, but the risk of death was 47% lower in the pembrolizumab plus axitinib group (HR: 0.53, 95% CI: 0.38–0.74). There was also a higher ORR of 59.3% in the pembrolizumab plus axitinib group compared to 35.7% in the sunitinib group, which was statistically significant. The combination of pembrolizumab plus axitinib has also been approved by the FDA for the first-line treatment of mRCC.

Recent data from CheckMate 9ER document the superiority of cabozantinib plus nivolumab versus sunitinib in previously untreated advanced RCC [11]. After a median follow-up of 18.1months, the median PFS was 16.6months with nivolumab plus cabozantinib and 8.3months with sunitinib (HR: 0.51; 95% CI: 0.41–0.64). The experimental treatment was also significantly better in terms of OS: the probability of survival at 12months was 85.7% with nivolumab plus cabozantinib compared to 75.6% with sunitinib (HR: 0.60; 98.89% CI: 0.40–0.89). The experimental treatment also resulted in a significant increase in OR (55.7% versus 27.1%). Grade 3 or worse AEs occurred in 75.3% with nivolumab plus cabozantinib and 70.6% with sunitinib.

In addition, a phase III trial, COSMIC-313, evaluating the combination of cabozantinib with nivolumab and ipilimumab versus nivolumab and ipilimumab in patients with international metastatic RCC database consortium (IMDC)-defined intermediate and poor-risk mRCC [12], has been initiated.

However, the role of tyrosine kinase inhibitors remains of paramount importance, especially in patients that, for clinical characteristics, tumor burden, and comorbidity, could have greater benefit from the use of TKIs in monotherapy rather than in combination with other therapies.

The combination of VEGFR and mTOR inhibitors is recognized as an effective therapeutic strategy in the management of RCC [13]. Several combinations of mTOR-targeted agents (e.g., everolimus and temsirolimus) and VEGF-targeted agents (e.g., bevacizumab and sorafenib) have been investigated as first-line therapies showing only modest clinical activity and greater toxic effects compared to single-agent targeted treatments [14].

Compared to everolimus alone, lenvatinib, a VEGFR1–3 multi-TKI, plus everolimus significantly prolonged PFS (median 14.6months versus 5.5months; HR: 0.40; 95% CI: 0.24–0.68) in patients with mRCC who have progressed after one previous VEGF-targeted therapy [15]. Grade 3 and 4 events occurred in fewer patients allocated single-agent everolimus (50%) compared with those assigned lenvatinib plus everolimus (71%). Based on these results, this combination has been approved for RCC patients after one prior VEGF-targeted therapy [16].

In another phase III trial, CLEAR, lenvatinib plus everolimus and lenvatinib plus pembrolizumab showed longer PFS than sunitinib (median, 23.9 versus 9.2months; HR: 0.39; 95% CI: 0.32–0.49), while only lenvatinib plus pembrolizumab was associated with significant OS than sunitinib (HR: 0.66; 95% CI: 0.49–0.88). Grade 3 or higher AEs emerged in 82.4% of the patients who received lenvatinib plus pembrolizumab, 83.1% of those who received lenvatinib plus everolimus, and 71.8% of those who received sunitinib [16].

However, although antiangiogenic drugs have been moderately successful in the treatment of mRCC, a significant problem for these patients is represented by the development of resistance to target therapy.

Most resistance mechanisms are biologically mediated, such as by alteration of the target gene itself or by activation of bypass pathways [17]. Pharmacokinetic features, involving poor penetration of the drug or activation of cellular pumps that drive the drug from the cell, and phenotypic transformations such as epithelial-mesenchymal transition represent other resistance mechanisms [18].

Therefore, patients who progressed to VEGF-target therapy could benefit from a combination approach; moreover, the use of new multitarget VEGFR inhibitors such as lenvatinib combined with other therapies could correlate with better efficacy results compared to bevacizumab, largely tested in previous studies.

Conclusions

At present time, there are no definitive data supporting one specific treatment sequence over the others, with multiple drugs recommended in the first-, second-, and third-line settings. Antiangiogenic therapy thus remains a valid option in selected patients (VEGF-dependent, IMDC favorable risk group first-line mRCC) and enhances the activity of immunotherapy, modulating immune response as demonstrated in several clinical trials. Prospective trial will be needed to reassess the new sequence of anti-PD-1/PD-L1 and VEGF/VEGFR inhibitors and their combination, particularly in the second-line setting, mostly after the impressive results of the combination trials of ICIs and VEGF/VEGFR inhibitor in the first-line therapy. Moreover, a better identification of efficacy and toxicity predictive biomarkers of antiangiogenic agents could improve therapeutic indications.

References

[1] Chen W, Zheng R, Baade PD, Zhang S, Zeng H, Bray F, Jemal A, Yu XQ, He J. Cancer statistics in China, 2015. CA Cancer J Clin 2016;66(2):115–32.

[2] Aglietta M, Barone C, Sawyer MB, Moore MJ, Miller WH, Bagalà C, Colombi F, Cagnazzo C, Gioeni L, Wang E, Huang B, Fly KD, Leone F. A phase I dose escalation trial of tremelimumab (CP-675,206) in combination with gemcitabine in chemotherapy-naive patients with metastatic pancreatic cancer. Ann Oncol 2014;25(9):1750–5.

[3] Yang JC, Hughes M, Kammula U, Royal R, Sherry RM, Topalian SL, Suri KB, Levy C, Allen T, Mavroukakis S, Lowy I, White DE, Rosenberg SA. Ipilimumab (anti-CTLA4 antibody) causes regression of metastatic renal cell cancer associated with enteritis and hypophysitis. J Immunother 2007;30(8):825–30.

[4] Rowshanravan B, Halliday N, Sansom DM. CTLA-4: a moving target in immunotherapy. Blood 2018;131:58–67.

[5] Amin A, Plimack ER, Infante JR, Ernstoff MS, Rini BI, McDermott DF, Knox JJ, Kumar Pal S, Voss MH, Sharma P, Kollmannsberger CK, Chin Heng DY, Spratlin JL, Shen Y, Kurland JF, Gagnier P, Hammers HJ. Nivolumab (anti-PD-1; BMS-936558, ONO-4538) in combination with sunitinib or pazopanib in patients (pts) with metastatic renal cell carcinoma (mRCC). J Clin Oncol 2014;32(15 suppl):5010.

[6] Kreamer CRNP, MSN KM. Immune checkpoint blockade: a new paradigm in treating advanced cancer. J Adv Pract Oncol 2014;5(6):418–31.

[7] Rini BI, Powles T, Atkins MB, Escudier B, DF MD, Suarez C, Bracarda S, Stadler WM, Donskov F, Lee JL, Hawkins R, Ravaud A, Alekseev B, Staehler M, Uemura M, De Giorgi U, Mellado B, Porta C, Melichar B, Gurney H, Bedke J, Choueiri TK, Parnis F, Khaznadar T, Thobhani A, Li S, Piault-Louis E, Frantz G, Huseni M, Schiff C, Green MC, Motzer RJ, IMmotion151 Study Group. Atezolizumab plus bevacizumab versus sunitinib in patients with previously untreated metastatic renal cell carcinoma (IMmotion151): a multicentre, open-label, phase 3, randomised controlled trial. Lancet 2019;393(10189):2404–15.

[8] McDermott DF, Huseni MA, Atkins MB, Motzer RJ, Rini BI, Escudier B, Fong L, Joseph RW, Pal SK, Reeves JA, Sznol M, Hainsworth J, Rathmell WK, Stadler WM, Hutson T, Gore ME, Ravaud A, Bracarda S, Suárez C, Danielli R, Gruenwald V, Choueiri TK, Nickles D, Jhunjhunwala S, Piault-Louis E, Thobhani A, Qiu J, Chen DS, Hegde PS, Schiff C, Fine GD, Powles T. Clinical activity and molecular correlates of response to atezolizumab alone or in combination with bevacizumab versus sunitinib in renal cell carcinoma. Nat Med 2018;24(6):749–57.

[9] Motzer RJ, Penkov K, Haanen J, Rini B, Albiges L, Campbell MT, Venugopal B, Kollmannsberger C, Negrier S, Uemura M, Lee JL, Vasiliev A, Miller Jr WH, Gurney H, Schmidinger M, Larkin J, Atkins MB, Bedke J, Alekseev B, Wang J, Mariani M, Robbins PB, Chudnovsky A, Fowst C, Hariharan S, Huang B, di Pietro A, Choueiri TK. Avelumab plus axitinib versus sunitinib for advanced renal-cell carcinoma. N Engl J Med 2019;380(12):1103–15.

[10] Rini BI, Plimack ER, Stus V, Gafanov R, Hawkins R, Nosov D, Pouliot F, Alekseev B, Soulières D, Melichar B, Vynnychenko I, Kryzhanivska A, Bondarenko I, Azevedo SJ, Borchiellini D, Szczylik C, Markus M, McDermott RS, Bedke J, Tartas S, Chang YH, Tamada S, Shou Q, Perini RF, Chen M, Atkins MB, Powles T, KEYNOTE-426 Investigators. Pembrolizumab plus axitinib versus sunitinib for advanced renal-cell carcinoma. N Engl J Med 2019;380(12):1116–27.

[11] Choueiri TK, Powles T, Burotto M, Escudier B, Bourlon MT, Zurawski B, Oyervides Juárez VM, Hsieh JJ, Basso U, Shah AY, Suárez C, Hamzaj A, Goh JC, Barrios C, Richardet M, Porta C, Kowalyszyn R, Feregrino JP, Żołnierek J, Pook D, Kessler ER, Tomita Y, Mizuno R, Bedke J, Zhang J, Maurer MA, Simsek B, Ejzykowicz F, Schwab GM, Apolo AB, Motzer RJ, CheckMate 9ER Investigators. Nivolumab plus cabozantinib versus sunitinib for advanced renal-cell carcinoma. N Engl J Med 2021;384(9):829–41.

[12] ClinicalTrials.gov. Study of cabozantinib in combination with nivolumab and ipilimumab in patients with previously untreated advanced or metastatic renal cell carcinoma [Internet]. Available from: https://clinicaltrials.gov/ct2/show/NCT03937219?term=NCT03937219&draw=2&rank=1; 2021.

[13] Courtney KD, Corcoran RB, Engelman JA. The PI3K pathway as drug target in human cancer. J Clin Oncol 2010;28:1075–83.

[14] Battelli C, Cho DC. MTOR inhibitors in renal cell carcinoma. Therapy 2011;8:359–67.

[15] Motzer RJ, Hutson TE, Glen H, Michaelson MD, Molina A, Eisen T, Jassem J, Zolnierek J, Maroto JP, Mellado B, Melichar B, Tomasek J, Kremer A, Kim HJ, Wood K, Dutcus C, Larkin J. Lenvatinib, everolimus, and the combination in patients with metastatic renal cell carcinoma: a randomised, phase 2, open-label, multicentre trial. Lancet Oncol 2015;16(15):1473–82.

[16] Motzer R, Alekseev B, Rha S-Y, Porta C, Eto M, Powles T, Grünwald V, Hutson TE, Kopyltsov E, Méndez-Vidal MJ, Kozlov V, Alyasova A, Hong SH, Kapoor A, Alonso Gordoa T, Merchan JR, Winquist E, Maroto P, Goh JC, Kim M, Gurney H, Patel V, Peer A, Procopio G, Takagi T, Melichar B, Rolland F, De Giorgi U, Wong S, Bedke J, Schmidinger M, Dutcus CE, Smith AD, Dutta L, Mody K, Perini RF, Xing D, Choueiri TK, CLEAR Trial Investigators. Lenvatinib plus pembrolizumab or everolimus for advanced renal cell carcinoma. N Engl J Med 2021;384(14):1289–300.

[17] Kamli H, Li L, Gobe GC. Limitations to the therapeutic potential of tyrosine kinase inhibitors and alternative therapies for kidney cancer. Ochsner J 2019;19(2):138–51.

[18] Piva F, Giulietti M, Santoni M, Occhipinti G, Scarpelli M, Lopez-Beltran A, Cheng L, Principato G, Montironi R. Epithelial to mesenchymal transition in renal cell carcinoma: implications for cancer therapy. Mol Diagn Ther 2016;20:111–7.

CHAPTER

5

Antiangiogenic drugs as chemosensitizer in genitourinary cancer

Francesco Sessa[a,b], Pietro Spatafora[a], Giandomenico Roviello[c], Claudio Bisegna[a], Martina Catalano[c], Andrea Liaci[a], Lorenzo Verdelli[a], Ester Illiano[d], Ilaria Camilla Galli[e], Beatrice Detti[f], Gabriella Nesi[e], Graziano Vignolini[a], and Donata Villari[a,b]

[a]Unit of Urological Robotic Surgery and Renal Transplantation, Careggi Hospital, Florence, Italy [b]Department of Experimental and Clinical Medicine, University of Florence, Florence, Italy [c]Department of Medical Oncology, Careggi Hospital, Florence, Italy [d]Department of Surgical and Biomedical Science, Andrological and Urogynecological Clinic, Santa Maria Terni Hospital, University of Perugia, Terni, Italy [e]Division of Pathological Anatomy, University of Florence, Florence, Italy [f]Radiation Oncology Unit - Oncology Department, Careggi Hospital, Florence, Italy

Abstract

Tumor angiogenesis is a crucial factor associated with tumor growth, progression, and metastasis. Therefore, there has been a great deal of interest in the development of antiangiogenesis strategies that inhibit tumor vascularization. Conventional approaches include administering antiangiogenic drugs that target and block the activity of proangiogenic factors with good results in some cancers, although their efficacy as single agents is still debated. Exploiting their chemosensitizer feature, new strategies have focused on combining antiangiogenic agents with chemotherapy or immunotherapy. In this chapter, the role of antiangiogenic agents investigated as a therapeutic approach in genitourinary cancers and their promising future directions are described.

Abbreviations

CI confidence interval
c-KIT tyrosine-protein kinase Kit
DFS disease-free survival

https://doi.org/10.1016/B978-0-323-90190-1.00015-9

FGFR	fibroblast growth factor receptor
HDCT	high-dose chemotherapy
ICI	immune checkpoint inhibitor
LH-RH	luteinizing hormone-releasing hormone
mTOR	mammalian target of rapamycin
NGS	next-generation sequencing
ORR	overall respond rate
OS	overall survival
PDGFR	platelet-derived growth factor receptor
PD-L1	programmed death ligand 1
PFS	progression-free survival
PTEN	phosphatase and tensin homolog
RCC	renal cell carcinoma
TGCT	testicular germ cell tumors
TSC1	tuberous sclerosis complex 1
TSC2	tuberous sclerosis complex 2
VEGF	vascular endothelial growth factor
VEGFR	vascular endothelial growth factor receptor

Conflict of interest

No potential conflicts of interest were disclosed.

Introduction

As angiogenesis is an essential process for tumor growth, progression, and metastasis, identifying proangiogenic molecules and blocking their activity through various approaches represent a potential anticancer therapy. Although the use of antiangiogenic drugs has proved to effectively inhibit tumor progression, it cannot eradicate the tumor as a stand-alone strategy; on the other hand, the combination of antiangiogenesis agents and chemotherapy or immunotherapy seems to provide an effective cancer treatment [1]. Approved combination regimens with antiangiogenic drugs are currently limited to renal cell cancer (RCC), although it appears that angiogenesis inhibitors may represent a promising new therapeutic approach in different types of genitourinary cancer. Fig. 1 reports all investigated drugs in genitourinary cancer.

Urothelial cancer

Current guidelines for the treatment of advanced urothelial cancer suggest combination chemotherapy in the front-line setting, immunotherapy for those that relapse after chemotherapy or are cisplatin-ineligible with tumors expressing programmed death ligand 1 (PD-L1), and the FGFR inhibitor erdafitinib in the third-line setting after the failure of these prior treatments in eligible patients.

Several promising investigational studies on drugs with antiangiogenic activity in combination with traditional chemotherapy as an alternative to chemotherapy alone for advanced bladder cancers have been conducted.

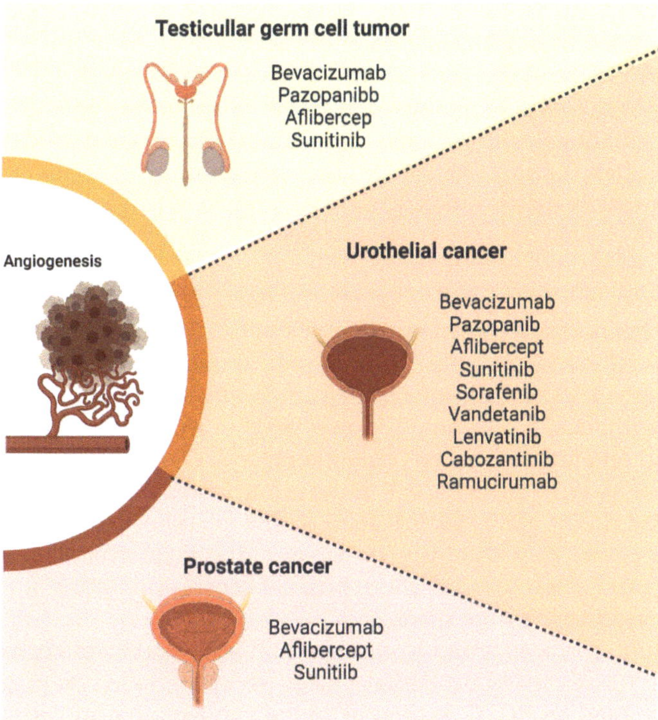

FIG. 1 Relevant investigated agents in genitourinary cancer. *Image created with BioRender.com.*

In 2013, a phase II trial examined the response to gemcitabine, carboplatin, and bevacizumab in patients with visceral metastatic disease finding an overall survival (OS) of 13.9 months (95% CI 11.9–18.1), compared to 10.3 months of the control population that received gemcitabine and carboplatin alone. The disease-free survival (DFS) recorded was 6.5 months (95% CI 4.7–7.8), representing a 35% improvement over the documented 4.8-month DFS for gemcitabine and cisplatin alone [2].

Another phase II single-arm trial evaluated gemcitabine, cisplatin, and bevacizumab in patients with metastatic transitional cell carcinoma, recording an improvement in DFS of 8.2 months (95% CI 6.8–10.3) compared to previous studies without bevacizumab, although not statistically significant [3]. However, the observed median OS of 19.1 months (95% CI 12.4–22.7) was a favorable improvement over chemotherapy alone.

A phase III trial has been conducted on patients with metastatic urothelial cancer receiving gemcitabine and cisplatin with either bevacizumab or placebo in the first-line setting. While the final results of a phase III trial are still pending, trial data suggested, thus far, that the addition of bevacizumab leads to an improvement in progression-free survival (PFS) but not in OS [4]. Encouraging response rates with the use of bevacizumab compared to chemotherapy alone have emerged from these single-arm trials and preliminary abstract data, but further studies are still needed.

Pazopanib, VEGFR, PDGFR, c-KIT, and FGFR inhibitor, in combination with paclitaxel, have been tested in the refractory setting demonstrating an impressive overall response rate (ORR) of 54% (95% CI 33.9–72.5) and a PFS and an OS of 6.2 and 10 months, respectively [5]. Of note, a phase I trial found that response to a combination of pazopanib and everolimus was significantly improved in patients with metastatic urothelial carcinoma who also had specific mutations in tuberous sclerosis complex 1 and 2 (TSC1/TSC2) or mTOR [6].

A phase II trial was conducted to evaluate the efficacy of ramucirumab (VEGFR2 antagonist) or icrucumab (VEGFR1 inhibitor) plus docetaxel versus docetaxel monotherapy in the second-line setting after first-line platinum-based therapy. PFS was significantly longer in the ramucirumab group (5.4 months; 95% CI 3.1–6.9) compared to chemotherapy alone (2.8 months; 95% CI 1.9–3.6), whereas no significant improvement in OS was demonstrated [7]. Similar results have been observed in a larger phase III randomized double-blinded trial, RANGE, which compared the safety and efficacy of ramucirumab and docetaxel versus docetaxel and placebo in patients with locally advanced, unresectable, or metastatic bladder cancer who had disease progression on platinum-based chemotherapy. A PFS of 4.07 months (95% CI 2.96–4.47) and an ORR of 24.5% (95% CI 18.8–30.3) in the ramucirumab group versus 2.76 months (95% CI 2.60–2.96) and 14.0% (95% CI 9.4–18.6) in the placebo group were recorded [8]. This study shows superior PFS with ramucirumab and docetaxel over chemotherapy alone in patients with platinum-refractory advanced bladder cancer, validating anti-VEGF mechanisms as effective in treating invasive disease.

Finally, vandetanib, a VEGFR inhibitor, was tested in combination with docetaxel versus placebo and docetaxel in patients with advanced urothelial cancer who progressed on prior platinum-based chemotherapy. The median PFS (2.56 months in the vandetanib arm versus 1.02 months in the placebo arm) was not found to be statistically significant, and no difference in OS and ORR was recorded between two groups in light of higher toxicity for the combination regimen [9]. Negative results also emerged from another phase II trial comparing carboplatin, gemcitabine, and vandetanib to chemotherapy alone [10].

Given the potential synergistic effect of VEGF inhibition and immune checkpoint inhibition being shown in other malignancies such as renal cell cancer, this combination should be further studied in urothelial cancer [11]. As antiangiogenic agents may prime the immune system for an enhanced response to immune checkpoint inhibitors (ICIs), future studies should evaluate how the specific sequence and combination of treatments play a role in survival benefit. Multiple ongoing clinical trials investigating the use of ICIs with angiogenesis inhibitors such as atezolizumab with bevacizumab, and pembrolizumab with ramucirumab, are ongoing [12–14].

Testicular germ cell tumor

Despite their high sensitivity to cisplatin-based chemotherapy, 20% of patients with testicular germ cell tumors (TGCT) progress or relapse after first-line chemotherapy [15]. Of these patients, 25%–70% can be rescued with salvage chemotherapy, with or without additional surgery post-chemotherapy. The prognosis of patients is far less favorable when further relapse occurs, and durable responses can be achieved in <10% of cases [16]. Therapeutic options for patients who have failed multiple chemotherapy regimens, possibly including high-dose chemotherapy (HDCT), are limited thus far. As demonstrated in other cancer types, antiangiogenic agents can be useful when used in a combinatorial or sequential strategy by relying on their potential for short-term activity, coupled with their good toxicity profile. Previous studies suggested that VEGF and its receptors have an important role in the development and progression of testicular germ cell tumors [17]. Adam et al. described increased expression of VEGF and VEGFR2 in patients with TGCT especially in non-seminoma [18]. Nitzsche et al. showed that blocking VEGFR2 inhibited the growth of platinum-sensible and -resistant TGCT cells and suppressed tumor angiogenesis [19]. In Jennewein et al.'s study, VEGFR2 is significantly upregulated in testicular tumor as compared to normal tissue, with a higher expression in non-seminoma tumor than in seminoma [20]. Moreover, Jones et al. have shown that non-seminomatous tumors have a higher microvascular density and higher expression of VEGF [21].

Based on these results, sunitinib, an orally applicable VEGFR2 inhibitor, has been evaluated in a phase II study in five platinum-refractory advanced germ cell tumor patients reporting a clinical benefit in one of those.

Most recently, the activity of pazopanib as sequential or combination strategy for salvage therapy of TGCT was tested in another larger study [22].

In this study, pazopanib salvage therapy in patients receiving three or more regimens did not improve the 3-month PFS, as reported by previous results. However, given its potent short-term activity and good tolerability, pazopanib has been proposed as an optimal bridging therapy preceding or combined with standard salvage chemotherapy.

In fact, administering pazopanib before HDCT may have the advantage of improving tolerability compared to chemotherapy, with no additional side effects beyond those from subsequent therapies. In the context of maintenance palliative therapy, the combination of pazopanib with well-tolerated single-agent therapies like paclitaxel may result in increased efficacy, as it has been achieved for other tumor types, therefore representing one possible next step [19]. Finally, pazopanib course may anticipate salvage surgery by potentially facilitating prolonged systemic disease control while allowing aggressive surgical removal of residual, chemoresistant tissue. These data suggest that antiangiogenic therapy might be a promising option in the treatment of testicular cancer, although other large studies are needed to evaluate whether this therapy should be administered individually, sequentially, or in combination.

Prostate cancer

Prostate cancer is the most diagnosed cancer among men in the Western world. Localized disease can be effectively treated with established surgical (radical prostatectomy) and radiopharmaceutical treatment options, and hormonal manipulation options include surgical castration (orchidectomy) or medical castration (LH-RH antagonists). Treatment options for high-stage metastatic castration-resistant cancer include chemotherapy, hormonotherapy, active cellular immunotherapy with sipuleucel-T, and radium-223. Among the emerging therapeutic options, angiogenesis appears to have an important role in prostate cancer, but the results of antiangiogenic therapy in castration-resistant refractory disease have hitherto been disappointing. Possible explanations for this lack of efficacy in castration-resistant refractory disease are due to redundancy of angiogenic pathways, molecular heterogeneity of the disease, loss of tumor-suppressor protein phosphatase and tensin homolog (PTEN) expression, as well as various VEGF-A splicing isoforms with pro- and antiangiogenic activity. A better understanding of the molecular mechanisms of angiogenesis may help to develop effective antiangiogenic therapy in prostate cancer.

VEGF-A isoform has been extensively studied demonstrating an important role in prostate cancer angiogenesis [23]. Moreover, VEGF-A has been found to be overexpressed in prostate cancer and a high level of VEGF-A is associated with distant metastasis and a poorer prognosis [24–26].

Based on these findings, several antiangiogenic clinical studies in prostate cancer predominantly targeting VEGF-A have begun. A randomized phase II trial on bevacizumab, involving patients with hormone-sensitive prostate cancer, showed improved relapse-free survival when bevacizumab was used alongside hormone-deprivation therapy [27].

On the other hand, another randomized phase III clinical study showed some improvement in PFS but found no significant improvement in OS in metastatic, castration-resistant prostate cancer, when bevacizumab was used with docetaxel and prednisone therapy [28]. Furthermore, bevacizumab resulted in increased toxicity and a greater incidence of treatment-related deaths. As well as bevacizumab, aflibercept combined with docetaxel chemotherapy and hormonal therapy did not show any improvement in OS in a phase III multicenter, randomized study in patients with castration-resistant refractory tumors [29].

Promising results have emerged from a phase II study that combined bevacizumab and lenalidomide therapy to docetaxel and prednisone chemotherapy and hormonal therapy in patients with metastatic castration-resistant prostate cancer. This study showed that combined antiangiogenic therapy can be safely administered, but further randomized trials are required to confirm clinical benefits [30].

Therefore, administration of hormonal therapy and chemotherapy with antiangiogenic drugs, mainly bevacizumab, in hormone-sensitive prostate cancer showed some positive clinical effects, further supporting the hypothesis that angiogenesis is an important factor in prostate cancer.

Conclusions

Next-generation sequencing (NGS) of tumor tissue and circulating tumor cells has provided new insights into the biological makeup of genitourinary carcinoma, mainly urothelial cancer.

The identification of specific genomic alterations may have important implications for prognosis and predicted response to therapy representing the next frontier for patients with urothelial and other cancer types.

Analyzing the molecular profile of tumors from patients with urothelial cancer achieving groundbreaking therapeutic responses has allowed to identify the association of gene mutations such as TSC1 mutation or TSC1/TSC2 mutations to the activity of everolimus and everolimus plus pazopanib, respectively.

Additionally, mutation and amplification of the FGFR3 gene were found in the same tumor type in one patient responding to pazopanib treatment [31]. Another study on TGCT patients found a correlation between new missense mutations of genes, including TSC2, and responders to pazopanib [32].

Future investigations to confirm and clarify the role of these and other mutations are needed in a context of combined therapeutic approaches.

References

[1] Rajabi M, Mousa SA. The role of angiogenesis in cancer treatment. Biomedicine 2017;5(2):34. https://doi.org/10.3390/biomedicines5020034.

[2] Balar AV, Apolo AB, Ostrovnaya I, Mironov S, Iasonos A, Trout A, Regazzi AM, Garcia-Grossman IR, Gallagher DJ, Milowsky MI, Bajorin DF. Phase II study of gemcitabine, carboplatin, and bevacizumab in patients with advanced unresectable or metastatic urothelial cancer. J Clin Oncol 2013;31(6):724–30.

[3] Hahn NM, Stadler WM, Zon RT, Waterhouse D, Picus J, Nattam S, Johnson CS, Perkins SM, Waddell MJ, Sweeney CJ, Hoosier Oncology Group. Phase II trial of cisplatin, gemcitabine, and bevacizumab as first-line therapy for metastatic urothelial carcinoma: Hoosier oncology group GU 04-75. J Clin Oncol 2011;29(12):1525–30.

[4] Rosenberg JE, Ballman KV, Halabi S, Watt C, Hahn OM, Steen PD, Dreicer R, Flaig TW, Stadler WM, Sweeney C, Mortazavi A, Morris MJ. CALGB 90601 (Alliance): randomized, double-blind, placebo-controlled phase III trial comparing gemcitabine and cisplatin with bevacizumab or placebo in patients with metastatic urothelial carcinoma. J Clin Oncol 2019;37(15_suppl):4503.

[5] Narayanan S, Lam A, Vaishampayan U, Harshman L, Fan A, Pachynski R, Poushnejad S, Haas D, Li S, Srinivas S. Phase II study of pazopanib and paclitaxel in patients with refractory urothelial cancer. Clin Genitourin Cancer 2016;14(5):432–7.

[6] Bellmunt J, Lalani AKA, Jacobus S, Wankowicz SA, Polacek L, Takeda DY, Harshman LC, Wagle N, Moreno I, Lundgren K, Bossé D, Van Allen EM, Choueiri TK, Rosenberg JE. Everolimus and pazopanib (E/P) benefit genomically selected patients with metastatic urothelial carcinoma. Br J Cancer 2018;119(6):707–12.

[7] Petrylak DP, Tagawa ST, Kohli M, Eisen A, Canil C, Sridhar SS, Spira A, Yu EY, Burke JM, Shaffer D, Pan CX, Kim JJ, Aragon-Ching JB, Quinn DI, Vogelzang NJ, Tang S, Zhang H, Cavanaugh CT, Gao L, Kauh JS, Walgren RA, Chi KN. Docetaxel as monotherapy or combined with ramucirumab or icrucumab in second-line treatment for locally advanced or metastatic urothelial carcinoma: an open-label, three-arm, randomized controlled phase II trial. J Clin Oncol 2016;34(13):1500–9.

[8] Petrylak D, de Wit R, Chi KN, Drakaki A, Sternberg CN, Nishiyama H, Castellano D, Hussain SA, Fléchon A, Bamias A, Yu EY, van der Heijden MS, Matsubara N, Alekseev B, Necchi A, Géczi L, Ou YC, Coskun HS, Su WP, Bedke J, Gakis G, Percent IJ, Lee JL, Tucci M, Semenov A, Laestadius F, Peer A, Tortora G, Safina S, Garcia Del Muro X, Rodriguez-Vida A, Cicin I, Harputluoglu H, Tagawa ST, Vaishampayan U, Aragon-Ching JB, Hamid O, Liepa AM, Wijayawardana S, Russo F, Walgren RA, Zimmermann AH, Hozak RR, Bell-McGuinn KM, Powles T, RANGE study investigators. Ramucirumab plus docetaxel versus placebo plus docetaxel in patients with locally advanced or metastatic urothelial carcinoma after platinum-based therapy (RANGE): a randomised, double-blind, phase 3 trial. Lancet 2017;390(10109):2266–77.

[9] Choueiri TK, Ross RW, Jacobus S, Vaishampayan U, Yu EY, Quinn DI, Hahn NM, Hutson TE, Sonpavde G, Morrissey SC, Buckle GC, Kim WY, Petrylak DP, Ryan CW, Eisenberger MA, Mortazavi A, Bubley GJ, Taplin ME, Rosenberg JE, Kantoff PW. Double-blind, randomized trial of docetaxel plus vandetanib versus docetaxel plus placebo in platinum-pretreated metastatic urothelial cancer. J Clin Oncol 2012;30(5):507–12.

[10] Jones RJ, Crabb SJ, Chester JD, Elliott T, Huddart RA, Birtle AJ, Evans L, Lester J, Jagdev S, Casbard A, Huang C, Madden TA, Griffiths G. TOUCAN: a randomised phase II trial of carboplatin and gemcitabine +/− vandetanib in first line treatment of advanced urothelial cell cancer in patients who are not suitable to receive cisplatin. J Clin Oncol 2016;34(2_suppl):448.

[11] Atkins MB, Tannir NM. Current and emerging therapies for first-line treatment of metastatic clear cell renal cell carcinoma. Cancer Treat Rev 2018;70:127–37. https://doi.org/10.1016/j.ctrv.2018.07.009.

[12] ClinicalTrials.gov. Search of: NCT03272217—list results. [Internet]. Available from: https://clinicaltrials.gov/ct2/results?cond=&term=NCT03272217&cntry=&state=&city=&dist; 2021.

[13] ClinicalTrials.gov. Search of: NCT02443324—list results. [Internet]. Available from: https://clinicaltrials.gov/ct2/results?cond=&term=NCT02443324&cntry=&state=&city=&dist; 2021.

[14] ClinicalTrials.gov. Search of: NCT04179110—list results. [Internet]. Available from: https://clinicaltrials.gov/ct2/results?cond=&term=NCT04179110&cntry=&state=&city=&dist; 2021.

[15] Horwich A, Shipley J, Huddart R. Testicular germ-cell cancer. Lancet 2006;367(9512):754–65. https://doi.org/10.1016/S0140-6736(06)68305-0.

[16] Oing C, Kollmannsberger C, Oechsle K, Bokemeyer C. Investigational targeted therapies for the treatment of testicular germ cell tumors. Expert Opin Investig Drugs 2016;25(9):1033–43. https://doi.org/10.1080/13543784.2016.1195808.

[17] Bentas W, Beecken WD, Glienke W, Binder J, Schuldes H. Serum levels of basic fibroblast growth factor reflect disseminated disease in patients with testicular germ cell tumors. Urol Res 2003;30(6):390–3.

[18] Adam M, Schmidt D, Wardelmann E, Wernert N, Albers P. Angiogenetic protooncogene ets-1 induced neovascularization is involved in the metastatic process of testicular germ cell tumors. Eur Urol 2003;44(3):329–36.

[19] Nitzsche B, Gloesenkamp C, Schrader M, Ocker M, Preissner R, Lein M, Zakrzewicz A, Hoffmann B, Höpfner M. Novel compounds with antiangiogenic and antiproliferative potency for growth control of testicular germ cell tumours. Br J Cancer 2010;103(1):18–28.

[20] Jennewein L, Bartsch G, Gust K, Kvasnicka HM, Haferkamp A, Blaheta R, Mittelbronn M, Harter PN, Mani J. Increased tumor vascularization is associated with the amount of immune competent PD-1 positive cells in testicular germ cell tumors. Oncol Lett 2018;15(6):9852–60.

[21] Jones A, Fujiyama C, Turner K, Fuggle S, Cranston D, Turley H, Valtola R, Bicknell R, Harris AL. Angiogenesis and lymphangiogenesis in stage 1 germ cell tumours of the testis. BJU Int 2000;86(1):80–6.

[22] Subbiah V, Meric-Bernstam F, Mills GB, Shaw KRM, Bailey AM, Rao P, Ward JF, Pagliaro LC. Next generation sequencing analysis of platinum refractory advanced germ cell tumor sensitive to Sunitinib (Sutent®) a VEGFR2/PDGFRβ/c-kit/ FLT3/RET/CSF1R inhibitor in a phase II trial. J Hematol Oncol 2014;7(1):1–10.

[23] Wong SY, Haack H, Crowley D, Barry M, Bronson RT, Hynes RO. Tumor-secreted vascular endothelial growth factor-C is necessary for prostate cancer lymphangiogenesis, but lymphangiogenesis is unnecessary for lymph node metastasis. Cancer Res 2005;65(21):9789–98.

[24] Duque JLF, Loughlin KR, Adam RM, Kantoff PW, Zurakowski D, Freeman MR. Plasma levels of vascular endothelial growth factor are increased in patients with metastatic prostate cancer. Urology 1999;54(3):523–7.

[25] Green MML, Hiley CT, Shanks JH, Bottomley IC, West CML, Cowan RA, Stratford IJ. Expression of vascular endothelial growth factor (VEGF) in locally invasive prostate cancer is prognostic for radiotherapy outcome. Int J Radiat Oncol Biol Phys 2007;67(1):84–90.

[26] Wegiel B, Bjartell A, Ekberg J, Gadaleanu V, Brunhoff C, Persson JL. A role for cyclin A1 in mediating the autocrine expression of vascular endothelial growth factor in prostate cancer. Oncogene 2005;24(42):6385–93.

[27] McKay RR, Zurita AJ, Werner L, Bruce JY, Carducci MA, Stein MN, Heath EI, Hussain A, Tran HT, Sweeney CJ, Ross RW, Kantoff PW, Slovin SF, Taplin ME. A randomized phase II trial of short-course androgen deprivation therapy with or without bevacizumab for patients with recurrent prostate cancer after definitive local therapy. J Clin Oncol 2016;34(16):1913–20.

[28] Kelly WK, Halabi S, Carducci M, George D, Mahoney JF, Stadler WM, Morris M, Kantoff P, Monk JP, Kaplan E, Vogelzang NJ, Small EJ. Randomized, double-blind, placebo-controlled phase III trial comparing docetaxel and prednisone with or without bevacizumab in men with metastatic castration-resistant prostate cancer: CALGB 90401. J Clin Oncol 2012;30(13):1534–40.

[29] Tannock IF, Fizazi K, Ivanov S, Karlsson CT, Fléchon A, Skoneczna I, Orlandi F, Gravis G, Matveev V, Bavbek S, Gil T, Viana L, Arén O, Karyakin O, Elliott T, Birtle A, Magherini E, Hatteville L, Petrylak D, Tombal B, Rosenthal M, VENICE investigators. Aflibercept versus placebo in combination with docetaxel and prednisone for treatment of men with metastatic castration-resistant prostate cancer (VENICE): a phase 3, double-blind randomised trial. Lancet Oncol 2013;14(8):760–8.

[30] Madan RA, Karzai FH, Ning YM, Adesunloye BA, Huang X, Harold N, Couvillon A, Chun G, Cordes L, Sissung T, Beedie SL, Dawson NA, Theoret MR, McLeod DG, Rosner I, Trepel JB, Lee MJ, Tomita Y, Lee S, Chen C, Steinberg SM, Arlen PM, Gulley JL, Figg WD, Dahut WL. Phase II trial of docetaxel, bevacizumab, lenalidomide and prednisone in patients with metastatic castration-resistant prostate cancer. BJU Int 2016;118(4):590–7.

[31] Palma N, Morris JC, Ali SM, Ross JS, Pal SK. Exceptional response to pazopanib in a patient with urothelial carcinoma harboring FGFR3 activating mutation and amplification. Eur Urol 2015;68:168–70.

[32] Necchi A, Lo Vullo S, Giannatempo P, Raggi D, Calareso G, Togliardi E, Crippa F, Pennati M, Zaffaroni N, Perrone F, Busico A, Colecchia M, Nicolai N, Mariani L, Salvioni R. Pazopanib in advanced germ cell tumors after chemotherapy failure: results of the open-label, single-arm, phase 2 Pazotest trial. Ann Oncol 2017;28(6):1346–51.

CHAPTER

6

Antiangiogenic drugs as chemosensitizer in melanoma

Davide Quaresmini and Michele Guida

Rare Tumors and Melanoma Unit, IRCCS Istituto Oncologico "Giovanni Paolo II", Bari, Italy

Abstract

Neoangiogenesis is the capacity of cancer to generate a new vascular network to supply metabolic substrates to cancer cells. As a universal property of all cancers, angiogenesis is a recognized hallmark of cancer and has also been evidenced in melanoma, where vascular proliferation correlates with clinical aggressiveness and poorer prognosis. Melanoma can induce neoangiogenesis in multiple ways including vascular mimicry, consisting in the capacity of melanoma cells to assume endothelial-like characteristics and functionally express pro-angiogenic mediators and receptors that participate in the interplay with extracellular matrix components and are even potentiated by matrix remodeling and alterations that are peculiar to the metastatic phase. Despite the unavoidable biological role of angiogenesis and its association with worse outcomes, in melanoma combinations of antiangiogenic and chemotherapeutic agents have been investigated in clinical trials with only poor improvement of the already unsatisfactory outcomes. In recent years, immunotherapy has induced previously unexpected results in the treatment of melanoma, yet efforts have been made to furtherly increase the results possibly by new schedules of association therapies. Immune checkpoint inhibitory treatment acts on the interplay between immune responses and vascular normalization, as an independent factor antagonizing neoplastic angiogenesis. Furthermore, recent clinical trials have been investigating the efficacy of the association between antiangiogenic agents and immune-checkpoint inhibitors in advanced melanoma, with interesting outcomes. Early data from a recent trial, in particular, show the possible role of antiangiogenic treatment in overcoming the resistance to immune checkpoint inhibitors. This chapter reviews the biological bases of angiogenesis in melanoma and the currently available data on the efficacy of antiangiogenic drugs in the treatment of advanced melanoma.

Abbreviations

ANG-2 angiopoietin-2
BRAF V-Raf murine sarcoma viral oncogene homolog B1
CTLA-4 common T leukocyte antigen 4
DC dendritic cell
HIF-1α hypoxia inducible factor 1 α

Antiangiogenic Drugs as Chemosensitizers in Cancer Therapy
https://doi.org/10.1016/B978-0-323-90190-1.00003-2

MDSC myeloid-derived stromal cell
MEK mitogen-activated protein kinase kinase 1
PD-1 programmed death 1
PD-L-1 programmed death ligand 1
TAM tumor associated macrophage
TLR toll-like receptor
VEGF vascular endothelial growth factor

Conflict of interest

No potential conflicts of interest were disclosed.

Introduction

Melanoma is an aggressive cancer that originates from melanocytic cells with elevated potential of locoregional and metastatic diffusion. During tumor growth, simple diffusion of oxygen and nutrients is sufficient only initially and when the dimensions of the neoplastic tissue exceed the threshold of 100 to 200 μm diffusion is insufficient, hence a cancer-specific net of new blood vessels is essential to grant tumor survival [1]. The newly generated vascular network is aberrant with dilated, distorted, and leaky vessels with incomplete pericytic coverage, abnormal proliferation of endothelial cells, an irregular distribution within cancer tissue, and wide fenestrations that ultimately contribute to the immission of cancer cells in the circulation [2–5]. The activation of quiescent vasculature by cancer cells is called "angiogenic switch" and is induced by a molecular disequilibrium between activators and inhibitors of angiogenesis, in favor of angiogenesis [6]. Angiogenesis is essential in the growth, progression, and metastasization of cancers, including melanoma [7]. Warren first described the angiogenetic capacity of melanoma in an animal model where a graft of melanoma tissue on a hamster cheek pouch showed a florid and rich neovascularization [8]. Since then, further investigations have contributed to building a more complete picture of angiogenesis in melanoma which is far more complex than initially imagined, made of multiple mechanisms of manipulation of blood vessels [9]. First, melanoma cells induce the spread of new blood vessels from pre-existing ones; secondly, melanoma cells recruit bone marrow progenitors to the hypoxic areas within the tumor microenvironment where they can induce vascular formation; finally, melanoma cells themselves can acquire an endothelial-like phenotype for the phenomenon of the so-called vascular mimicry, in which melanoma cells take direct part in the structure of newborn vessels [9]. To date, antiangiogenic drugs have been widely used in oncology in many cancers including renal, colorectal, and ovarian neoplasms, while data from clinical research in melanoma are poor. The current treatment for advanced-stage disease in melanoma is based on the inhibition of either the aberrantly activated pathway of BRAF and MEK, when V600 BRAF mutation is identified, or the immune checkpoints PD-1 and CTLA-4. Such treatments have provided previously unexpected outcomes in advanced disease, yet there is still a compelling need for further improvements. This chapter describes the biological bases of angiogenesis in melanoma, also in correlation with the most recent treatments, and summarizes the currently available clinical data on the use of antiangiogenic compounds in melanoma.

Insights concerning the molecular mechanisms of angiogenesis

The recent evolution of melanoma treatment has led to a primary role of targeted and immune therapy both in the adjuvant and the metastatic setting. Interestingly, the molecular architecture of the melanoma microenvironment is far more complex than previously thought, and a strong cross-talk has been clearly demonstrated between the angiogenic and the immune components of cancer stroma with mutual influences in the molecular compartments that are involved in response to systemic treatments. Angiogenic factors are produced by tumor cells, stromal cells including cancer-associated fibroblasts (CAFs), and inflammatory cells like lymphocytes and macrophages. The hyperproduction of pro-angiogenetic factors is induced by several mechanisms, primarily the activation of the hypoxia-induced HIF-1α pathway, the oncogene-induced transcription of Vascular Endothelial Growth Factor (VEGF), and the loss of oncosuppressor genes including p53, which both stimulates the production of antiangiogenic factors like thrombospondin-1 and inhibits the expression of proangiogenic factors [10,11].

Vascular endothelial growth factor (VEGF)

VEGF-A is the archetype and most biologically relevant among proangiogenic factors, with a strong effect on endothelial cells survival and migration [12], and on vasculogenic mimicking properties in melanoma cells [13]. VEGF-A is upregulated by HIF-1α and oncogene signaling pathways [14–16]. Remarkably, VEGF-A has also shown immunosuppressive capacity. In particular, the increased production of VEGF-A in cancer leads to the inhibition of T cells in several ways: (1) reducing the activity of functional T cells both directly [17] and indirectly through the endothelial PGE2-mediated suppressive action on T cells [18]; (2) decreasing neoantigen presentation to lymphocytes by inhibiting the maturation of dendritic cells (DCs) [19], mainly interfering with NF-kB activation [20]; (3) recruiting immunosuppressive T regulatory cells (Tregs) into the tumor microenvironment [21,22]; (4) limiting endothelial cytokine response and adhesion molecule expression, hence affecting vascular functional permeability to leukocytes and their peripheral recruitment to cancer microenvironment [23,24]. Relevantly for immunotherapy, VEGF pathway activation also enhances T cell exhaustion mediated by immune checkpoints like PD-L-1, CTLA-4, TIM-3, and LAG-3 [25]. While lowering the immunogenic compartment of immunity, VEGF potentiates the counteracting immunopermissive microenvironment [26] both by recruiting immune-suppressive Tregs [27] and myeloid-derived stromal cells (MDSCs) [28] and activating tumor-associated macrophages (TAMs) at the tumor site [29]. To confirm this role of VEGF, the anti-VEGF antibody bevacizumab induces DCs maturation and a reduction in Tregs and MDSCs recruitment to cancer sites [30–32]. VEGF can be produced by cancer cells and immune cells from the tumor microenvironment, mostly from Tregs and in smaller proportions from TAMs, MDSCs, and DCs [33], creating cellular communications that either directly or indirectly convey on the inhibition of cytotoxic T lymphocytes [34].

Angiopoietin-2 (ANG-2)

ANG-2 is an antagonist cytokine of the Angiopoietin-1/Tie2 pathway that acts as a facilitator of VEGF-dependent angiogenesis [35]. ANG-2 has long been considered an exclusive product of endothelium but more recently Pari and colleagues demonstrated that it can also be produced by melanoma cells [36]. ANG-2 levels in sera are increased in stage III and IV melanoma patients but not in stage I and II [37]. Consistently with this evidence, ANG-2 is produced by melanoma cells themselves, especially by metastatic sites [36]. In a mouse model, melanoma-derived ANG-2, differently from the stromal-derived ANG-2, was not shown to increase the microvessel density of melanoma microenvironment, but rather showed protection of tumor cells from oxidative stress and a role in reactive oxygen species associated metastases to the lungs [36]. High serum ANG-2 levels were correlated with a worse response to immunotherapy in melanoma [38]. ANG-2 contributes to immune microenvironment composition by acting on the Tie-2 expressing subpopulation of circulating monocytes that are recruited by ANG-2 and converted to M2-like macrophages [39] and secrete IL-10, which is a known promoter of Treg expansion and inhibitor of effector T cell activity [40].

Toll-like receptors (TLRs)

TLRs are a family of pattern recognition receptors involved in anti-microbial immunity, apoptotic cell clearance, and cancer. Among all family members, TLR-4 is expressed in 90% of primary and 93% of metastatic melanomas where it plays a role in the aggressive behavior of cancer cells [41]. TLR-4 signaling involves the activation of signal transducer and activator of transcription 3 (STAT3) which in turn promotes melanoma growth and aggressiveness associated features including angiogenesis and epithelial to mesenchymal transition [42]. During melanomagenesis, ultraviolet radiation recruits and activates neutrophils in a TLR4-mediated mechanism, inducing an inflammation that facilitates angiogenesis and favors melanoma angiotropism [41]. Moreover, STAT3 has also been associated with immunosuppression in melanoma [43].

Immune cells in microenvironment and angiogenesis

In the cancer microenvironment, there is a constant dynamic cross-talk between all resident cells which is far beyond the mere activity of cancer cells alone: the dynamic interaction between all cell components is responsible for the biological behavior of cancer. Accordingly, angiogenesis in cancer is not only induced by cancer cells themselves: the immune cells in the tumor microenvironment can sustain angiogenesis in cancer [44]. Globally, tumor cells can influence immune infiltrates toward an immune permissive phenotype. VEGF, for example, is produced by tumor-associated macrophages (TAMs), tumor-associated neutrophils, regulatory DCs, myeloid-derived suppressor cells, NK cells, and γδT17 cells [44]. VEGF-Receptor-1 and -2 are expressed on DCs which can promote angiogenesis [45]. Neutrophils and TAMs secrete pro-angiogenic factors including VEGF, TNFα, IL-8, and chemokines [46], together with matrix metalloproteases, which are essential to remodel the extracellular matrix during angiogenesis and metastases [47]. In an analysis on mouse models of breast cancers, extended to some of TCGA databases excluding melanoma, Tian et al. [48] demonstrated that the activation of vessel normalization (pericyte coverage, reduced vascular leakage, improved blood perfusion) is one of the effects of immune checkpoint inhibitors (ICIs) mediated by the activation of CD4+ T lymphocytes, in particular IFNγ producing Th1 cells [48]. Specifically, the authors postulated the existence of a positive feedback loop involving Th1 cells that localize proximal to tumoral vessels and change the local tumor microenvironment via cytokines like IFNγ. IFNγ itself then induces a reduction in VEGF-A production and an increase in CXCL9, CXCL10, CXCL11 that recruit pericytes and Th1 cells, self-sustaining the process [48]. Such findings would confirm the evidence that inhibition of IFNγ signaling is associated with secondary resistance to immune checkpoint inhibitors (ICIs) [49]. Furthermore, after interacting with DCs, T cells can acquire neurophilin-1 which is a ligand of VEGF-A that promotes angiogenesis [50]. Endothelial cells, formerly considered passive lining cells of blood vessels, are actively responsible for the intense reciprocal cellular interaction consisting in the so-called angiocrine signaling that is essential in normal organ development [51] and can be exacerbated in cancer. Tumor cells can then induce endothelial activation mediated by cytokines including ANG-2, which are responsible for the autocrine induction of STAT3 signaling in the endothelium, followed by the expression of chemokines (CCL2) and adhesion molecules (ICAM1) that recruit CCR2+ macrophages to the cancer site [38]. Endothelial cells take part in granulocyte differentiation in physiology and pathology, given the common developmental origin between endothelial cells and hematopoietic cells [38]: the endothelium secretes cytokines (SCF, CXCL12) that contribute to the quiescence of hematopoietic cells in the bone marrow, but can also promote granulopoiesis in case of inflammation [52]. The hypoxia-regulated Endothelin B receptor on tumor endothelium acts as an obstacle to T cell adhesion and has been identified in some cases of resistance to immune therapy [53].

Manipulation of angiogenesis by therapeutic agents in melanoma

Antiangiogenics are a class of kinase inhibitors that bind either angiogenic factors or their receptors. The first antiangiogenic agent to be developed was bevacizumab, an anti-VEGF monoclonal antibody, which is still indicated in the treatment of cancers including colorectal, ovarian, or uterine carcinomas. Many other agents were synthesized subsequently, presenting a wider spectrum of pharmacodynamic targeting, including aflibercept, pazopanib, ramucirumab, regorafenib, sorafenib, and sunitinib, being so far approved in daily practice either alone or in association with chemotherapy. Antiangiogenic agents shape the irregular and leaky vessels of cancer to create an almost normalized intratumoral vascular network, at least transiently. Such improved vascular efficiency of cancer microcirculation is thus mainly responsible for more efficient transport of chemotherapeutic agents to cancer cells, and also the molecular background to the association of antiangiogenic therapies with traditional chemotherapies [54]. Part of the effects of antiangiogenics in cancer may also be attributed to the modulation of immune cell composition in the tumor microenvironment, triggered by the reduction in the tissue hypoxia that is associated with the immature cancer vasculature. The response to hypoxia favors the polarization of the tumor microenvironment towards an immune-suppressive phenotype in terms of increase in Tregs and M2-TAMs, reduction of DC activity, and increase in PD-L1 expression on endothelial cells, TAMs, DCs, and cytotoxic lymphocytes [34]. Antiangiogenics can then interfere with both CD8+ T cells trafficking and TAMs repolarization, inducing an immunostimulatory milieu [38]. Anti-VEGF-A agents also improve immune responses [55]. In a preclinical model of antiangiogenic-driven vascular normalization in melanoma and other primary cancers, Schmittnaegel et al. demonstrated that the administration of a bispecific anti-VEGF-ANG-2 antibody was associated with increased recruitment and activation of CD4+ and CD8+ T cells at the areas with normalized vessels [56]. In the cross-talk between adaptive and innate immunity, IFNγ from CD4+ and CD8+ T cells then stimulates M1-TAMs to angiostatic activity and antagonize endothelial cell proliferation [57]. In a preclinic mouse model with multiple cancer cell lines including melanoma, DeAlmeida and colleagues demonstrated that therapy with an anti-VEGF-A targeting agent induces the HIF-1α-mediated activation of intratumoral CD8+ cells resulting in an increase in IFNγ and TNFα production [58]. Unfortunately, the response is usually transient [59] and short in duration [60], due to mechanisms of escape that include the activation of metabolic stress responses in cancer cells, the activation of alternative angiogenic pathways like the ANG-2 pathway, the participation in new vessel growth according to the vascular co-option mechanism, the normalization of cancer vessels [57]. In renal carcinoma, two different molecular subsets have been identified: the angiogenic subset characterized by upregulation of angiogenesis-associated genes and responsiveness to antiangiogenic therapy, and the inflammatory subset presenting an up-regulation in immune-related genes and refractoriness to antiangiogenic therapy [61]. In contrast, no similar dichotomy has been identified in melanoma, for which resistance to antiangiogenic therapies may be intrinsically associated with its biology: as evidenced by Donnem, melanoma primary and metastatic lesions strongly rely on vessel co-option for their vascular supply as an alternative to angiogenesis [62]. In melanoma, the mutation of the BRAF gene has primary importance since mutations at the codon 600 are druggable and clinically associated with significant responses. The constitutive activation of BRAF kinase is not only a driver mutation in melanoma but also has effects on melanoma microenvironment composition [63]. In particular, BRAF mutations are associated with a double density of Treg infiltrate [64], and a reduction in T cell activity by the indirect secretion of IL-1α and IL-1β from CAFs [65]. PD-L1 expression in immune infiltrate is also more prevalent in BRAF mutated tumor specimens [66]. Finally, BRAF mutated cells also have the angiogenic capacity [67], since they can induce other cancer cells and microenvironmental cells to secrete CXCL8 and CCL2, two pro-tumorigenic chemokines, leading to cancer cell proliferation and macrophage-mediated angiogenesis [63]. Furthermore, the pathway of VEGFR-1 has recently been identified as an escape mechanism to the BRAF-inhibitor vemurafenib [68]. As evidenced by Aztori and colleagues, melanoma cell lines express higher VEGF-receptors when transformed into vemurafenib-resistant, and silencing of such receptors can prolong the maintenance of sensitivity to vemurafenib [68]. The backbone of treatment in advanced melanoma is represented by the immune therapy with either anti-CTLA-4 or anti-PD-1 agents, that remove immune checkpoint inhibition to potentiate the immune response to melanoma. Despite an overall improvement in the outcomes of treatment, still immune therapy is often associated with secondary resistance and progression and, yet more rarely, to early resistance. Among the possible mechanisms underpinning these resistances, recent evidence also identified some vascular-related mechanisms. In particular, high serum ANG-2 levels correlated with a worse overall response rate to ICI therapy in melanoma, possibly because ANG-2 can recruit monocytes and induce PD-L1 expression in M2-macrophages [38]. Wu and colleagues identified a subset of melanoma patients characterized by significant tumor infiltration of CD68+ macrophages that particularly responded to treatment with ipilimumab and bevacizumab with a neat decrease in ANG-2 expression [69]. Allen and colleagues demonstrated that the combination therapy of antiangiogenics and anti-PD-1 agents induced an increase in intratumoral high endothelial venules responsible for selective leukocyte infiltration and for the switch of microenvironment toward immunosensitive features [70]. A bispecific anti-VEGFA and anti-ANG-2 inhibitor was also shown to potentiate the efficacy of an anti-PD-1 treatment [56].

Evidence from clinical trials and possible perspectives

The first clinical trials investigating the efficacy of antiangiogenic compounds in melanoma date back to early experiences in the 1990s when chemotherapy was still the only option for the treatment of the advanced stages of the disease, at an exclusive palliative intent due to its dramatically poor outcomes. Apart from isolated publications on small cohorts of patients, most of the clinical studies performed were phase 1 or 2 trials treating patients with stage III unresectable or stage IV melanoma, either in a single arm or with a control arm to compare, enrolling small cohorts of patients, usually 20–30 in number (Table 1). Evidence from preclinical data suggested that the concomitant antiangiogenic treatment can induce a more regular vascular network which in turn can distribute chemotherapies to cancer cells with a more even, and possibly more cytotoxic, distribution. Yet, such advantage was never observed in these trials, and no improvement was shown by the association of antiangiogenics with chemotherapies (Table 1) therefore they have never been tested in wider trials, excepting for an adjuvant phase III clinical trial in high-risk patients that showed no significant results of bevacizumab versus placebo [79].

More recently, immunotherapy became one of the mainstays of systemic treatment for advanced-stage melanoma, together with the anti-BRAF and anti-MEK combinations for BRAF-V600 mutated patients. ICIs have clearly demonstrated a remarkable efficacy when combined with antiangiogenic compounds in other cancers including renal clear cell cancer or non-small cell lung cancer. To investigate such associations also in melanoma, Hodi and colleagues performed an investigational phase I trial enrolling 46 patients affected by advanced melanoma without brain metastases, receiving a first (37% of patients) or second (63% of patients) line of treatment with ipilimumab and the antiangiogenic bevacizumab: the best overall response rate (ORR) was 19.6%, the disease control rate (DCR) was 67.4% and the median time to progression was 9 months [80]. Serial biopsies of target lesions performed on these patients showed changes in melanoma-associated endothelium, presenting an increased expression of E-selectin and an association with an extensive CD8+ T cell and macrophage infiltration, consistent with preclinical data [80]. Unfortunately, even considering the possible confounding factor of the small number of treated patients, the clinical results of

TABLE 1 Clinical trials of the association of antiangiogenic treatments with old drugs in advanced melanoma.

Reference	Phase	Clinical setting	Line of treatment in metastatic setting	Arm 1	Arm 2	Primary endpoints	Secondary endpoints	Furhter analyses
[71]	II	Stage IV cutaneous melanoma	1st line	Bevacizumab + Fotemustine (20 pts)	None	CR 1/20 PR 2/20 SD 10/20	TTP 8 m OS 20 m G3 toxicity 14/20 pts	Reduction of VEGF levels post-therapy
[72]	II	Stage III unresectable or stage IV cutaneous melanoma No active brain metastases	1st line or further	Aflibercept (40 pts)	None	ORR 7.5%	PFS 4 m Os 16 m	Hypertension correlated with OS
[73]	II	Stage IV cutaneous melanoma No brain metastases	1st line	Bevacizumab + Temozolomide (62 pts)	None	SD 52%	ORR 16% PFS 4 m OS 10 m	OS higher in BRAF wt 12 vs 9 m
[74]	II Randomized	Stage IV melanoma No brain metastases	1st line	Carboplatin + Paclitaxel + Placebo (71 pts)	Carboplatin + Paclitaxel + Bevacizumab (143 pts)	PFS 4 vs 5 m	OS 9 vs 12 m OR 11/67 vs 36/141 DOR 8 vs 7 m G3-G5 toxicity 57 vs 45%	
[75]	II	Stage IV melanoma. No brain metastases	2nd line	Bevacizumab (35 pts)	None	DCR 31%	PFS 2 m OS 9 m Toxicity 7/11 pts. who had disease control developed hypertension	

[76]	II	Stage IV melanoma No active brain metastases cKit mutated	2nd line or further. No prior immunotherapy.	Sunitinib (10 pts)	None	ORR 3/4 in mutated cKIT pts.; 1/6 in amplified or overexpressed cKIT pts.		
[77]	II	Stage III unresectable or stage IV cutaneous melanoma No active brain metastases	2nd or 3rd line	Bevacizumab + Sorafenib (14 pts)	None	ORR 0% SD 21%	PFS 8 m G3-G4 toxicity 43%	Low VEGF values correlated with longer PFS
[78]	II	Stage IV cutaneous melanoma. No brain metastases	1st line	Bevacizumab + Dacarbazine (40 pts)	None	ORR 19%	TTP 5 m Discontinuation 92% G3-G4 toxicity 22%	
NCT02158520	II Randomized	Stage IV melanoma No brain metastases	1st line or further	Nab-Paclitaxel + Bevacizumab (12 pts)	Ipilimumab (12 pts)	PFS 129 vs 94 days	OS 18 vs 27 m ORR (2 vs 0 CR; 1 vs 1 PR) G3-G4 Toxicity (9 vs 7)	Recruitment completed. (24 enrolled pts. vs 176 initially designed)

these associations were too poor and did not improve the known outcomes of immunotherapy, so they were not furtherly explored.

A more promising antiangiogenic compound is lenvatinib, a tyrosine kinase inhibitor that inhibits VEGF-receptors (VEGFR-1, VEGFR-2, and VEGFR-3), fibroblast growth factor receptors (FGFR-1, FGFR-2, FGFR-3, FGFR-4), platelet-derived growth factor receptor alpha (PDGFR-α), and other kinases (c-KIT, RET). In a recent phase IB/II trial, Taylor and colleagues treated patients with advanced solid tumors including melanoma with the association of pembrolizumab and lenvatinib. Among the 21 patients of the melanoma cohort, the combination treatment demonstrated a manageable safety profile, and promising antitumor activity with a 33% ORR also including 1 complete response, while the overall DCR was 81% [81]. These results led researchers to design the phase II LEAP-004 trial investigating the association of pembrolizumab and lenvatinib in an exclusive cohort of 103 advanced melanoma patients, progressing on immunotherapy with either anti-PD-1 or anti-CTLA-4 immunotherapy. Globally, the cohort of patients included a significant proportion of high-risk patients: 68% had stage IV with either visceral or brain secondary lesions; 55.3% had increased LDH levels; 58% had received at least 2 prior lines of systemic therapy. From the preliminary data that have been recently published [82], ORR was 21.4%, including 3 complete responses. DCR was 66% and the median duration of response was 8.2months with an estimated 37% duration of response at 9months. Progression-free survival was 4.2months and overall survival was 14months. Interestingly, a 22% ORR was observed in patients progressing on immunotherapy, including cases of secondary and primary resistance [73], suggesting that antiangiogenic treatment can overcome resistance to immunotherapy in a significant proportion of progressing patients. As for side effects, 45% of patients had grade 3 or 4 toxicity, that led to a reduction in the dose of lenvatinib (56%) or treatment discontinuation (7%. The encouraging data of this trial are now being furtherly investigated in a randomized phase III trial (Table 2). Moreover, other studies are currently ongoing, mostly still at the recruitment phase, for treatment associations of antiangiogenics with either anti-PD-1 (nivolumab or pembrolizumab) or anti-PD-L1 (avelumab) inhibitors (Table 2).

Conclusion

As described above, angiogenesis plays a major role in the natural history of melanoma, from its intrinsic aggressiveness to some forms of resistance to systemic therapy. Despite widely intertwined mechanisms between angiogenesis and immunity, the efficacy of antiangiogenic therapies is currently insufficient. Hence, much interest is addressed to the ongoing clinical trials of combined antiangiogenic and immune therapies, to pursue better outcomes in the therapy of advanced melanoma.

TABLE 2 Clinical trials on the association of antiangiogenic with immunotherapy or anti-BRAF/anti-MEK targeted therapy in advanced melanoma.

Reference	Phase	Clinical setting	Line of treatment in metastatic setting	Arm 1	Arm 2	Primary endpoints	Secondary endpoints	Further details
[80]	I	Stage III unresectable or stage IV melanoma. No brain metastases	1st or 2nd line (46 pts)	Ipilimumab + Bevacizumab	None	ORR 8 PR, 22 SD	DCR 67% OS 25 m	
NCT01950390	II randomized	Stage III unresectable or IV cutaneous melanoma No brain metastases	1st or 2nd line (168 pts)	Ipilimumab	Ipilimumab+ Bevacizumab	OS	PFS ORR	Active, not recruiting. Results pending
[81]	IB/II	Advanced solid tumors including melanoma	2nd or 3rd line (21 pts. with melanoma)	Pembrolizumab + Lenvatinib	None	Safety	ORR at week 24 48% in melanoma (1 CR and 9 PR)	
[82]	II	Stage III unresectable or IV cutaneous melanoma	2nd or further line (103 pts)	Pembrolizumab + Lenvatinib	None	ORR 21% DCR 66%	PFS 4.2 m OS 14 m	ORR 22% in progressing on or early after immune therapy
NCT03820986	III Randomized	Stage III unresectable or IV cutaneous melanoma No active brain metastases	1st or 2nd line (660 pts)	Pembrolizumab + Lenvatinib	Pembrolizumab + Placebo	PRS OS	ORR DOR Toxicity	Active, not recruiting

Continued

TABLE 2 Clinical trials on the association of antiangiogenic with immunotherapy or anti-BRAF/anti-MEK targeted therapy in advanced melanoma—cont'd

Reference	Phase	Clinical setting	Line of treatment in metastatic setting	Arm 1	Arm 2	Primary endpoints	Secondary endpoints	Further details
NCT01495988	II	Stage IIIC unresectable or stage IV melanoma. BRAF V600E/ V600K positive No active brain metastases	1st or further line (10 pts. enrolled, vs initially designed 180 pts)	Vemurafenib + Cobimetinib + Bevacizumab (arm 2)	Vemurafenib + Cobimetinib (arm 1) Vemurafenib (arm 3) Vemurafenib + Bevacizumab (arm 4)	PFS	OS RR Toxicity (2 pts. in V+C+B arm. 1 of them, multi-organ failure)	Slow accrural, toxicity, change in priorities
NCT04356729	II	Stage III unresectable or IV cutaneous melanoma PDL1 negative No active brain metastases	Any line (no prior immunotherapy)	Atezolizumab + Bevacizumab (30 estimated pts)	None	ORR	OS TTP DOR Safety Change in TILs	Recruiting
NCT03175432	II	Stage IV melanoma with brain metastases. BRAF wt	Progression after anti-PD1	Atezolizumab + Bevacizumab + Cobimetinib	None	Intracranial ORR Safety	ORR DOR Neurocognitive function	Recruiting. Estimated 60 pts
NCT02681549	II	Stage IV melanoma or NSCLC with brain metastases	2nd or further line	Pembrolizumab + Bevacizumab	None	Intracranial ORR	Need for steroids ORR PFS Safety Biomarkers	Recruiting. Estimated 53 pts
NCT03239145	I	Advanced solid tumors No active brain metastases	2nd or further line	Pembrolizumab + Trebananib	None	Maximum dose	ORR PFS OS	Active, not recruiting

References

[1] Hanahan D, Weinberg RA. Hallmarks of cancer: the next generation. Cell 2011;144(5):646–74. https://doi.org/10.1016/j.cell.2011.02.013.

[2] Cooke VG, LeBleu VS, Keskin D, Khan Z, O'Connell JT, Teng Y, Duncan MB, Xie L, Maeda G, Vong S, Sugimoto H, Rocha RM, Damascena A, Brentani RR, Kalluri R. Pericyte depletion results in hypoxia-associated epithelial-to-mesenchymal transition and metastasis mediated by met signaling pathway. Cancer Cell 2012;21(1): 66–81. https://doi.org/10.1016/j.ccr.2011.11.024.

[3] Nagy JA, Chang SH, Shih SC, Dvorak AM, Dvorak HF. Heterogeneity of the tumor vasculature. Semin Thromb Hemost 2010;36(3):321–31. https://doi.org/10.1055/s-0030-1253454.

[4] Baluk P, Hashizume H, McDonald DM. Cellular abnormalities of blood vessels as targets in cancer. Curr Opin Genet Dev 2005;15(1):102–11. https://doi.org/10.1016/j.gde.2004.12.005.

[5] Jain RK. Antiangiogenesis strategies revisited: from starving tumors to alleviating hypoxia. Cancer Cell 2014; 26(5):605–22. https://doi.org/10.1016/j.ccell.2014.10.006.

[6] Hanahan D, Folkman J. Patterns and emerging mechanisms of the angiogenic switch during tumorigenesis. Cell 1996;86(3):353–64. https://doi.org/10.1016/s0092-8674(00)80108-7.

[7] Cho WC, Jour G, Aung PP. Role of angiogenesis in melanoma progression: update on key angiogenic mechanisms and other associated components. Semin Cancer Biol 2019;59:175–86. https://doi.org/10.1016/j.semcancer.2019.06.015.

[8] Warren BA, Shubik P. The growth of the blood supply to melanoma transplants in the hamster cheek pouch. Lab Invest 1966;15(2):464–78.

[9] Jour G, Ivan D, Aung PP. Angiogenesis in melanoma: an update with a focus on current targeted therapies. J Clin Pathol 2016;69(6):472–83. https://doi.org/10.1136/jclinpath-2015-203482.

[10] Baeriswyl V, Christofori G. The angiogenic switch in carcinogenesis. Semin Cancer Biol 2009;19(5): 329–37. https://doi.org/10.1016/j.semcancer.2009.05.003.

[11] Bergers G, Benjamin LE. Tumorigenesis and the angiogenic switch. Nat Rev Cancer 2003;3(6):401–10. https://doi.org/10.1038/nrc1093.

[12] Apte RS, Chen DS, Ferrara N. VEGF in signaling and disease: beyond discovery and development. Cell 2019; 176(6):1248–64. https://doi.org/10.1016/j.cell.2019.01.021.

[13] Vartanian A, Stepanova E, Grigorieva I, Solomko E, Baryshnikov A, Lichinitser M. VEGFR1 and PKCα signaling control melanoma vasculogenic mimicry in a VEGFR2 kinase-independent manner. Melanoma Res 2011; 21(2):91–8. https://doi.org/10.1097/CMR.0b013e328343a237.

[14] Ferrara N. Vascular endothelial growth factor. Arterioscler Thromb Vasc Biol 2009;29(6):789–91. https://doi.org/10.1161/ATVBAHA.108.179663.

[15] Mac Gabhann F, Popel AS. Systems biology of vascular endothelial growth factors. Microcirculation 2008; 15(8):715–38. https://doi.org/10.1080/10739680802095964.

[16] Carmeliet P. VEGF as a key mediator of angiogenesis in cancer. Oncology 2005;69(Suppl 3):4–10. https://doi.org/10.1159/000088478.

[17] Ziogas AC, Gavalas NG, Tsiatas M, Tsitsilonis O, Politi E, Terpos E, Rodolakis A, Vlahos G, Thomakos N, Haidopoulos D, Antsaklis A, Dimopoulos MA, Bamias A. VEGF directly suppresses activation of T cells from ovarian cancer patients and healthy individuals via VEGF receptor type 2. Int J Cancer 2012;130(4):857–64. https://doi.org/10.1002/ijc.26094.

[18] Mulligan JK, Rosenzweig SA, Young MR. Tumor secretion of VEGF induces endothelial cells to suppress T cell functions through the production of PGE2. J Immunother 2010;33(2):126–35. https://doi.org/10.1097/CJI.0b013e3181b91c9c.

[19] Gabrilovich DI, Chen HL, Girgis KR, Cunningham HT, Meny GM, Nadaf S, Kavanaugh D, Carbone DP. Production of vascular endothelial growth factor by human tumors inhibits the functional maturation of dendritic cells. Nat Med 1996;2(10):1096–103. https://doi.org/10.1038/nm1096-1096.

[20] Oyama T, Ran S, Ishida T, Nadaf S, Kerr L, Carbone DP, Gabrilovich DI. Vascular endothelial growth factor affects dendritic cell maturation through the inhibition of nuclear factor-kappa B activation in hemopoietic progenitor cells. J Immunol 1998;160(3):1224–32.

[21] Li B, Lalani AS, Harding TC, Luan B, Koprivnikar K, Huan Tu G, Prell R, VanRoey MJ, Simmons AD, Jooss K. Vascular endothelial growth factor blockade reduces intratumoral regulatory T cells and enhances the efficacy

of a GM-CSF-secreting cancer immunotherapy. Clin Cancer Res 2006;12(22):6808–16. https://doi.org/10.1158/1078-0432.CCR-06-1558.

[22] Terme M, Pernot S, Marcheteau E, Sandoval F, Benhamouda N, Colussi O, Dubreuil O, Carpentier AF, Tartour E, Taieb J. VEGFA-VEGFR pathway blockade inhibits tumor-induced regulatory T-cell proliferation in colorectal cancer. Cancer Res 2013;73(2):539–49. https://doi.org/10.1158/0008-5472.CAN-12-2325.

[23] Griffioen AW, Damen CA, Blijham GH, Groenewegen G. Tumor angiogenesis is accompanied by a decreased inflammatory response of tumor-associated endothelium. Blood 1996;88(2):667–73.

[24] Dirkx AE, Oude Egbrink MG, Kuijpers MJ, van der Niet ST, Heijnen VV, Bouma-ter Steege JC, Wagstaff J, Griffioen AW. Tumor angiogenesis modulates leukocyte-vessel wall interactions in vivo by reducing endothelial adhesion molecule expression. Cancer Res 2003;63(9):2322–9.

[25] Voron T, Colussi O, Marcheteau E, Pernot S, Nizard M, Pointet AL, Latreche S, Bergaya S, Benhamouda N, Tanchot C, Stockmann C, Combe P, Berger A, Zinzindohoue F, Yagita H, Tartour E, Taieb J, Terme M. VEGF-A modulates expression of inhibitory checkpoints on CD8+ T cells in tumors. J Exp Med 2015;212(2):139–48. https://doi.org/10.1084/jem.20140559.

[26] Rahma OE, Hodi FS. The intersection between tumor angiogenesis and immune suppression. Clin Cancer Res 2019;25(18):5449–57. https://doi.org/10.1158/1078-0432.CCR-18-1543.

[27] Wada J, Suzuki H, Fuchino R, Yamasaki A, Nagai S, Yanai K, Koga K, Nakamura M, Tanaka M, Morisaki T, Katano M. The contribution of vascular endothelial growth factor to the induction of regulatory T-cells in malignant effusions. Anticancer Res 2009;29(3):881–8.

[28] Huang Y, Chen X, Dikov MM, Novitskiy SV, Mosse CA, Yang L, Carbone DP. Distinct roles of VEGFR-1 and VEGFR-2 in the aberrant hematopoiesis associated with elevated levels of VEGF. Blood 2007;110(2):624–31. https://doi.org/10.1182/blood-2007-01-065714.

[29] Varney ML, Johansson SL, Singh RK. Tumour-associated macrophage infiltration, neovascularization and aggressiveness in malignant melanoma: role of monocyte chemotactic protein-1 and vascular endothelial growth factor-A. Melanoma Res 2005;15(5):417–25. https://doi.org/10.1097/00008390-200510000-00010.

[30] Osada T, Chong G, Tansik R, Hong T, Spector N, Kumar R, Hurwitz HI, Dev I, Nixon AB, Lyerly HK, Clay T, Morse MA. The effect of anti-VEGF therapy on immature myeloid cell and dendritic cells in cancer patients. Cancer Immunol Immunother 2008;57(8):1115–24. https://doi.org/10.1007/s00262-007-0441-x.

[31] Terme M, Pernot S, Marcheteau E, Sandoval F, Benhamouda N, Colussi O, Dubreuil O, Carpentier AF, Tartour E, Taieb J. VEGFA-VEGFR pathway blockade inhibits tumor-induced regulatory T-cell proliferation in colorectal cancer. Cancer Res 2013;73(2):539–49. https://doi.org/10.1158/0008-5472.CAN-12-2325.

[32] Kusmartsev S, Eruslanov E, Kübler H, Tseng T, Sakai Y, Su Z, Kaliberov S, Heiser A, Rosser C, Dahm P, Siemann D, Vieweg J. Oxidative stress regulates expression of VEGFR1 in myeloid cells: link to tumor-induced immune suppression in renal cell carcinoma. J Immunol 2008;181(1):346–53. https://doi.org/10.4049/jimmunol.181.1.346.

[33] Facciabene A, Peng X, Hagemann IS, Balint K, Barchetti A, Wang LP, Gimotty PA, Gilks CB, Lal P, Zhang L, Coukos G. Tumour hypoxia promotes tolerance and angiogenesis via CCL28 and T(reg) cells. Nature 2011;475(7355):226–30. https://doi.org/10.1038/nature10169.

[34] Fukumura D, Kloepper J, Amoozgar Z, Duda DG, Jain RK. Enhancing cancer immunotherapy using antiangiogenics: opportunities and challenges. Nat Rev Clin Oncol 2018;15(5):325–40. https://doi.org/10.1038/nrclinonc.2018.29.

[35] Augustin HG, Koh GY, Thurston G, Alitalo K. Control of vascular morphogenesis and homeostasis through the angiopoietin-Tie system. Nat Rev Mol Cell Biol 2009;10(3):165–77. https://doi.org/10.1038/nrm2639.

[36] Abdul Pari AA, Singhal M, Hübers C, Mogler C, Schieb B, Gampp A, Gengenbacher N, Reynolds LE, Terhardt D, Géraud C, Utikal J, Thomas M, Goerdt S, Hodivala-Dilke KM, Augustin HG, Felcht M. Tumor cell-derived angiopoietin-2 promotes metastasis in melanoma. Cancer Res 2020;80(12):2586–98. https://doi.org/10.1158/0008-5472.CAN-19-2660.

[37] Helfrich I, Edler L, Sucker A, Thomas M, Christian S, Schadendorf D, Augustin HG. Angiopoietin-2 levels are associated with disease progression in metastatic malignant melanoma. Clin Cancer Res 2009;15(4):1384–92. https://doi.org/10.1158/1078-0432.CCR-08-1615.

[38] Singhal M, Augustin HG. Beyond angiogenesis: exploiting angiocrine factors to restrict tumor progression and metastasis. Cancer Res 2020;80(4):659–62. https://doi.org/10.1158/0008-5472.CAN-19-3351.

[39] Coffelt SB, Tal AO, Scholz A, De Palma M, Patel S, Urbich C, Biswas SK, Murdoch C, Plate KH, Reiss Y, Lewis CE. Angiopoietin-2 regulates gene expression in TIE2-expressing monocytes and augments their inherent proangiogenic functions. Cancer Res 2010;70(13):5270–80. https://doi.org/10.1158/0008-5472.CAN-10-0012.

[40] Coffelt SB, Chen YY, Muthana M, Welford AF, Tal AO, Scholz A, Plate KH, Reiss Y, Murdoch C, De Palma M, Lewis CE. Angiopoietin 2 stimulates TIE2-expressing monocytes to suppress T cell activation and to promote regulatory T cell expansion. J Immunol 2011;186(7):4183–90. https://doi.org/10.4049/jimmunol.1002802.

[41] Bald T, Quast T, Landsberg J, Rogava M, Glodde N, Lopez-Ramos D, Kohlmeyer J, Riesenberg S, van den Boorn-Konijnenberg D, Hömig-Hölzel C, Reuten R, Schadow B, Weighardt H, Wenzel D, Helfrich I, Schadendorf D, Bloch W, Bianchi ME, Lugassy C, Barnhill RL, Koch M, Fleischmann BK, Förster I, Kastenmüller W, Kolanus W, Hölzel M, Gaffal E, Tüting T. Ultraviolet-radiation-induced inflammation promotes angiotropism and metastasis in melanoma. Nature 2014;507(7490):109–13. https://doi.org/10.1038/nature13111.

[42] Fu XQ, Liu B, Wang YP, Li JK, Zhu PL, Li T, Tse KW, Chou JY, Yin CL, Bai JX, Liu YX, Chen YJ, Yu ZL. Activation of STAT3 is a key event in TLR4 signaling-mediated melanoma progression. Cell Death Dis 2020;11(4):246. https://doi.org/10.1038/s41419-020-2440-1.

[43] Kortylewski M, Jove R, Yu H. Targeting STAT3 affects melanoma on multiple fronts. Cancer Metastasis Rev 2005;24(2):315–27. https://doi.org/10.1007/s10555-005-1580-1.

[44] Albini A, Bruno A, Noonan DM, Mortara L. Contribution to tumor angiogenesis from innate immune cells within the tumor microenvironment: implications for immunotherapy. Front Immunol 2018;9:527. https://doi.org/10.3389/fimmu.2018.00527.

[45] Riboldi E, Musso T, Moroni E, Urbinati C, Bernasconi S, Rusnati M, Adorini L, Presta M, Sozzani S. Cutting edge: proangiogenic properties of alternatively activated dendritic cells. J Immunol 2005;175(5):2788–92. https://doi.org/10.4049/jimmunol.175.5.2788.

[46] Owen JL, Mohamadzadeh M. Macrophages and chemokines as mediators of angiogenesis. Front Physiol 2013;4:159. https://doi.org/10.3389/fphys.2013.00159.

[47] Murdoch C, Muthana M, Coffelt SB, Lewis CE. The role of myeloid cells in the promotion of tumour angiogenesis. Nat Rev Cancer 2008;8(8):618–31. https://doi.org/10.1038/nrc2444.

[48] Tian L, Goldstein A, Wang H, Ching Lo H, Sun Kim I, Welte T, Sheng K, Dobrolecki LE, Zhang X, Putluri N, Phung TL, Mani SA, Stossi F, Sreekumar A, Mancini MA, Decker WK, Zong C, Lewis MT, Zhang XH. Mutual regulation of tumour vessel normalization and immunostimulatory reprogramming. Nature 2017;544(7649): 250–4. https://doi.org/10.1038/nature21724.

[49] Gao J, Shi LZ, Zhao H, Chen J, Xiong L, He Q, Chen T, Roszik J, Bernatchez C, Woodman SE, Chen PL, Hwu P, Allison JP, Futreal A, Wargo JA, Sharma P. Loss of IFN-γ pathway genes in tumor cells as a mechanism of resistance to anti-CTLA-4 therapy. Cell 2016;167(2):397–404.e9. https://doi.org/10.1016/j.cell.2016.08.069.

[50] Bourbié-Vaudaine S, Blanchard N, Hivroz C, Roméo PH. Dendritic cells can turn CD4+T lymphocytes into vascular endothelial growth factor-carrying cells by intercellular neuropilin-1 transfer. J Immunol 2006;177(3): 1460–9. https://doi.org/10.4049/jimmunol.177.3.1460.

[51] Rafii S, Butler JM, Ding BS. Angiocrine functions of organ-specific endothelial cells. Nature 2016;529(7586): 316–25. https://doi.org/10.1038/nature17040.

[52] Boettcher S, Gerosa RC, Radpour R, Bauer J, Ampenberger F, Heikenwalder M, Kopf M, Manz MG. Endothelial cells translate pathogen signals into G-CSF-driven emergency granulopoiesis. Blood 2014;124(9):1393–403. https://doi.org/10.1182/blood-2014-04-570762.

[53] Buckanovich RJ, Facciabene A, Kim S, Benencia F, Sasaroli D, Balint K, Katsaros D, O'Brien-Jenkins A, Gimotty PA, Coukos G. Endothelin B receptor mediates the endothelial barrier to T cell homing to tumors and disables immune therapy. Nat Med 2008;14(1):28–36. https://doi.org/10.1038/nm1699.

[54] Dewhirst MW, Secomb TW. Transport of drugs from blood vessels to tumour tissue. Nat Rev Cancer 2017; 17(12):738–50. https://doi.org/10.1038/nrc.2017.93.

[55] Yang J, Yan J, Liu B. Targeting VEGF/VEGFR to modulate antitumor immunity. Front Immunol 2018;9:978. https://doi.org/10.3389/fimmu.2018.00978.

[56] Schmittnaegel M, Rigamonti N, Kadioglu E, Cassará A, Wyser Rmili C, Kiialainen A, Kienast Y, Mueller HJ, Ooi CH, Laoui D, De Palma M. Dual angiopoietin-2 and VEGFA inhibition elicits antitumor immunity that is enhanced by PD-1 checkpoint blockade. Sci Transl Med 2017;9(385). https://doi.org/10.1126/scitranslmed.aak9670, eaak9670.

[57] De Palma M, Biziato D, Petrova TV. Microenvironmental regulation of tumour angiogenesis. Nat Rev Cancer 2017;17(8):457–74. https://doi.org/10.1038/nrc.2017.51.

[58] De Almeida PE, Mak J, Hernandez G, Jesudason R, Herault A, Javinal V, Borneo J, Kim JM, Walsh KB. Anti-VEGF treatment enhances CD8+ T-cell antitumor activity by amplifying hypoxia. Cancer Immunol Res 2020;8(6):806–18. https://doi.org/10.1158/2326-6066.CIR-19-0360.

[59] Winkler F, Kozin SV, Tong RT, Chae SS, Booth MF, Garkavtsev I, Xu L, Hicklin DJ, Fukumura D, di Tomaso E, Munn LL, Jain RK. Kinetics of vascular normalization by VEGFR2 blockade governs brain tumor response to radiation: role of oxygenation, angiopoietin-1, and matrix metalloproteinases. Cancer Cell 2004;6(6):553–63. https://doi.org/10.1016/j.ccr.2004.10.011.

[60] Huang Y, Yuan J, Righi E, Kamoun WS, Ancukiewicz M, Nezivar J, Santosuosso M, Martin JD, Martin MR, Vianello F, Leblanc P, Munn LL, Huang P, Duda DG, Fukumura D, Jain RK, Poznansky MC. Vascular normalizing doses of antiangiogenic treatment reprogram the immunosuppressive tumor microenvironment and enhance immunotherapy. Proc Natl Acad Sci U S A 2012;109(43):17561–6. https://doi.org/10.1073/pnas.1215397109.

[61] Şenbabaoğlu Y, Gejman RS, Winer AG, Liu M, Van Allen EM, de Velasco G, Miao D, Ostrovnaya I, Drill E, Luna A, Weinhold N, Lee W, Manley BJ, Khalil DN, Kaffenberger SD, Chen Y, Danilova L, Voss MH, Coleman JA, Russo P, Reuter VE, Chan TA, Cheng EH, Scheinberg DA, Li MO, Choueiri TK, Hsieh JJ, Sander C, Hakimi AA. Tumor immune microenvironment characterization in clear cell renal cell carcinoma identifies prognostic and immunotherapeutically relevant messenger RNA signatures. Genome Biol 2016;17(1):231. https://doi.org/10.1186/s13059-016-1092-z.

[62] Donnem T, Reynolds AR, Kuczynski EA, Gatter K, Vermeulen PB, Kerbel RS, Harris AL, Pezzella F. Non-angiogenic tumours and their influence on cancer biology. Nat Rev Cancer 2018;18(5):323–36. https://doi.org/10.1038/nrc.2018.14.

[63] Croce L, Coperchini F, Magri F, Chiovato L, Rotondi M. The multifaceted anti-cancer effects of BRAF-inhibitors. Oncotarget 2019;10(61):6623–40. https://doi.org/10.18632/oncotarget.27304.

[64] Leslie C, Bowyer SE, White A, Grieu-Iacopetta F, Trevenen M, Iacopetta B, Amanuel B, Millward M. FOXP3+ T regulatory lymphocytes in primary melanoma are associated with BRAF mutation but not with response to BRAF inhibitor. Pathology 2015;47(6):557–63. https://doi.org/10.1097/PAT.0000000000000314.

[65] Khalili JS, Liu S, Rodríguez-Cruz TG, Whittington M, Wardell S, Liu C, Zhang M, Cooper ZA, Frederick DT, Li Y, Zhang M, Joseph RW, Bernatchez C, Ekmekcioglu S, Grimm E, Radvanyi LG, Davis RE, Davies MA, Wargo JA, Hwu P, Lizée G. Oncogenic BRAF(V600E) promotes stromal cell-mediated immunosuppression via induction of interleukin-1 in melanoma. Clin Cancer Res 2012;18(19):5329–40. https://doi.org/10.1158/1078-0432.CCR-12-1632.

[66] Kim K, Jeon S, Kim TM, Jung CK. Immune gene signature delineates a subclass of papillary thyroid cancer with unfavorable clinical outcomes. Cancers (Basel) 2018;10(12):494. https://doi.org/10.3390/cancers10120494.

[67] Sharma A, Trivedi NR, Zimmerman MA, Tuveson DA, Smith CD, Robertson GP. Mutant V599EB-Raf regulates growth and vascular development of malignant melanoma tumors. Cancer Res 2005;65(6):2412–21. https://doi.org/10.1158/0008-5472.CAN-04-2423.

[68] Atzori MG, Ceci C, Ruffini F, Trapani M, Barbaccia ML, Tentori L, D'Atri S, Lacal PM, Graziani G. Role of VEGFR-1 in melanoma acquired resistance to the BRAF inhibitor vemurafenib. J Cell Mol Med 2020;24(1): 465–75. https://doi.org/10.1111/jcmm.14755.

[69] Wu X, Giobbie-Hurder A, Liao X, Connelly C, Connolly EM, Li J, Manos MP, Lawrence D, McDermott D, Severgnini M, Zhou J, Gjini E, Lako A, Lipschitz M, Pak CJ, Abdelrahman S, Rodig S, Hodi FS. Angiopoietin-2 as a biomarker and target for immune checkpoint therapy. Cancer Immunol Res 2017;5(1): 17–28. https://doi.org/10.1158/2326-6066.CIR-16-0206.

[70] Allen E, Jabouille A, Rivera LB, Lodewijckx I, Missiaen R, Steri V, Feyen K, Tawney J, Hanahan D, Michael IP, Bergers G. Combined antiangiogenic and anti-PD-L1 therapy stimulates tumor immunity through HEV formation. Sci Transl Med 2017;9(385). https://doi.org/10.1126/scitranslmed.aak9679, eaak9679.

[71] Del Vecchio M, Mortarini R, Canova S, Di Guardo L, Pimpinelli N, Sertoli MR, Bedognetti D, Queirolo P, Morosini P, Perrone T, Bajetta E, Anichini A. Bevacizumab plus fotemustine as first-line treatment in metastatic melanoma patients: clinical activity and modulation of angiogenesis and lymphangiogenesis factors. Clin Cancer Res 2010;16(23):5862–72. https://doi.org/10.1158/1078-0432.CCR-10-2363.

[72] Tarhini AA, Frankel P, Margolin KA, Christensen S, Ruel C, Shipe-Spotloe J, Gandara DR, Chen A, Kirkwood JM. Aflibercept (VEGF trap) in inoperable stage III or stage iv melanoma of cutaneous or uveal origin. Clin Cancer Res 2011;17(20):6574–81. https://doi.org/10.1158/1078-0432.CCR-11-1463.

[73] von Moos R, Seifert B, Simcock M, Goldinger SM, Gillessen S, Ochsenbein A, Michielin O, Cathomas R, Schläppi M, Moch H, Schraml PH, Mjhic-Probst D, Mamot C, Schönewolf N, Dummer R, Swiss Group for Clinical Cancer Research (SAKK). First-line temozolomide combined with bevacizumab in metastatic melanoma: a multicentre phase II trial (SAKK 50/07). Ann Oncol 2012;23(2):531–6. https://doi.org/10.1093/annonc/mdr126.

[74] Kim KB, Sosman JA, Fruehauf JP, Linette GP, Markovic SN, McDermott DF, Weber JS, Nguyen H, Cheverton P, Chen D, Peterson AC, Carson 3rd WE, O'Day SJ. BEAM: a randomized phase II study evaluating the activity of bevacizumab in combination with carboplatin plus paclitaxel in patients with previously untreated advanced melanoma. J Clin Oncol 2012;30(1):34–41. https://doi.org/10.1200/JCO.2011.34.6270.

[75] Schuster C, Eikesdal HP, Puntervoll H, Geisler J, Geisler S, Heinrich D, Molven A, Lønning PE, Akslen LA, Straume O. Clinical efficacy and safety of bevacizumab monotherapy in patients with metastatic melanoma: predictive importance of induced early hypertension. PLoS One 2012;7(6). https://doi.org/10.1371/journal.pone.0038364, e38364.

[76] Minor DR, Kashani-Sabet M, Garrido M, O'Day SJ, Hamid O, Bastian BC. Sunitinib therapy for melanoma patients with KIT mutations. Clin Cancer Res 2012;18(5):1457–63. https://doi.org/10.1158/1078-0432.CCR-11-1987.

[77] Mahalingam D, Malik L, Beeram M, Rodon J, Sankhala K, Mita A, Benjamin D, Ketchum N, Michalek J, Tolcher A, Wright J, Sarantopoulos J. Phase II study evaluating the efficacy, safety, and pharmacodynamic correlative study of dual antiangiogenic inhibition using bevacizumab in combination with sorafenib in patients with advanced malignant melanoma. Cancer Chemother Pharmacol 2014;74(1):77–84. https://doi.org/10.1007/s00280-014-2479-8.

[78] Ferrucci PF, Minchella I, Mosconi M, Gandini S, Verrecchia F, Cocorocchio E, Passoni C, Pari C, Testori A, Coco P, Munzone E. Dacarbazine in combination with bevacizumab for the treatment of unresectable/metastatic melanoma: a phase II study. Melanoma Res 2015;25(3):239–45. https://doi.org/10.1097/CMR.0000000000000146.

[79] Corrie PG, Marshall A, Nathan PD, Lorigan P, Gore M, Tahir S, Faust G, Kelly CG, Marples M, Danson SJ, Marshall E, Houston SJ, Board RE, Waterston AM, Nobes JP, Harries M, Kumar S, Goodman A, Dalgleish A, Martin-Clavijo A, Westwell S, Casasola R, Chao D, Maraveyas A, Patel PM, Ottensmeier CH, Farrugia D, Humphreys A, Eccles B, Young G, Barker EO, Harman C, Weiss M, Myers KA, Chhabra A, Rodwell SH, Dunn JA, Middleton MR, AVAST-M Investigators, Nathan P, Lorigan P, Dziewulski P, Holikova S, Panwar U, Tahir S, Faust G, Thomas A, Corrie P, Sirohi B, Kelly C, Middleton M, Marples M, Danson S, Lester J, Marshall E, Ajaz M, Houston S, Board R, Eaton D, Waterston A, Nobes J, Loo S, Gray G, Stubbings H, Gore M, Harries M, Kumar S, Goodman A, Dalgleish A, Martin-Clavijo A, Marsden J, Westwell S, Casasola R, Chao D, Maraveyas A, Marshall E, Patel P, Ottensmeier C, Farrugia D, Humphreys A, Eccles B, Dega R, Herbert C, Price C, Brunt M, Scott-Brown M, Hamilton J, Hayward RL, Smyth J, Woodings P, Nayak N, Burrows L, Wolstenholme V, Wagstaff J, Nicolson M, Wilson A, Barlow C, Scrase C, Podd T, Gonzalez M, Stewart J, Highley M, Wolstenholme V, Grumett S, Goodman A, Talbot T, Nathan K, Coltart R, Gee B, Gore M, Farrugia D, Martin-Clavijo A, Marsden J, Price C, Farrugia D, Nathan K, Coltart R, Nathan K, Coltart R. Adjuvant bevacizumab for melanoma patients at high risk of recurrence: survival analysis of the AVAST-M trial. Ann Oncol 2018;29(8):1843–52. https://doi.org/10.1093/annonc/mdy229.

[80] Hodi FS, Lawrence D, Lezcano C, Wu X, Zhou J, Sasada T, Zeng W, Giobbie-Hurder A, Atkins MB, Ibrahim N, Friedlander P, Flaherty KT, Murphy GF, Rodig S, Velazquez EF, Mihm Jr MC, Russell S, DiPiro PJ, Yap JT, Ramaiya N, Van den Abbeele AD, Gargano M, McDermott D. Bevacizumab plus ipilimumab in patients with metastatic melanoma. Cancer Immunol Res 2014;2(7):632–42. https://doi.org/10.1158/2326-6066.CIR-14-0053.

[81] Taylor MH, Lee CH, Makker V, Rasco D, Dutcus CE, Wu J, Stepan DE, Shumaker RC, Motzer RJ. Phase IB/II trial of lenvatinib plus pembrolizumab in patients with advanced renal cell carcinoma, endometrial Cancer, and other selected advanced solid tumors. J Clin Oncol 2020;38(11):1154–63. https://doi.org/10.1200/JCO.19.01598.

[82] Arance AM, De la Cruz-Merino L, Petrella TM, Jamal R, Ny L, Carneiro A, Berrocal A, Marquez-Rodas I, Spreafico A, Atkinson V, Costa Svedman F, Mant A, Smith AD, Chen K, Diede SJ, Krepler C, Long GV. Lenvatinib (len) plus pembrolizumab (pembro) for patients (pts) with advanced melanoma and confirmed progression on a PD-1 or PD-L1 inhibitor: Updated findings of LEAP-004. J Clin Oncol 2021;39(15 suppl):9504.

CHAPTER

7

Antiangiogenic drugs as chemosensitizers in hematological tumors

Vanessa Desantis[a,b], *Aurelia Lamanuzzi*[a], *Ilaria Saltarella*[a], *Maria Antonia Frassanito*[c], *Angelo Vacca*[a], *Monica Montagnani*[b], *and Domenico Ribatti*[d]

[a]Department of Biomedical Sciences and Human Oncology, Unit of Internal Medicine and Clinical Oncology, University of Bari Medical School, Bari, Italy [b]Department of Biomedical Sciences and Human Oncology, Pharmacology Section, Medical School, University of Bari "Aldo Moro", Bari, Italy [c]Department of Biomedical Sciences and Human Oncology, General Pathology Unit, University of Bari Medical School, Bari, Italy [d]Department of Basic Medical Sciences, Neurosciences and Sensory Organs, University of Bari Medical School, Bari, Italy

Abstract

Angiogenesis, the formation of new capillaries from preexisting blood vessels, plays an important role in cancer progression. When the tumor mass expands, the balance is shifted toward a pro-angiogenic milieu to maintain sustainable angiogenic processes. In this context, there is an up-regulation of several pro-angiogenic factors, including vascular endothelial growth factor (VEGF), fibroblast growth factor-2 (FGF-2), placental growth factor (PlGF), platelet derived endothelial cell growth factor (PD-ECGF), angiopoietins (Angs), transforming growth factors (TGFs) -α and -β, and epidermal cell growth factor (EGF), which collectively activate the proliferation of circulating endothelial progenitor cells (EPCs) able to enter in the peripheral blood circulation, migrating to sites of angiogenesis. Hence, the number of antiangiogenic agents developed for cancer treatment has risen over the past years. To date, the most common approaches to the inhibition of the VEGF axis include the blockade of VEGF receptors (VEGFRs) or ligands by neutralizing antibodies, as well as the tyrosine kinase inhibitors (TKIs), immunomodulatory drugs (IMiDs), and monoclonal antibodies (mAbs). Here, we focus on the role of circulating EPCs, which mediate the cross-talk between cancer angiogenesis and neoplastic clone, as potential novel targets for antiangiogenic drugs with particular relevance for hematological malignancies.

Antiangiogenic Drugs as Chemosensitizers in Cancer Therapy
https://doi.org/10.1016/B978-0-323-90190-1.00010-X

Abbreviations

3′-UTR	3′-untranslated region
AI	angiogenesis inhibitor
ALL	acute lymphoblastic leukemia
AML	acute myeloid leukemia
Ang	angiopoietin
BM	bone marrow
BMDCs	bone marrow-derived cells
BMSCs	bone marrow-derived stromal cells
CAFs	cancer associated fibroblasts
CAM	chorioallantoic membrane
CLL	chronic lymphocytic leukemia
CRC	colorectal cancer
DARPin	designed ankyrin repeat protein
DCs	dendritic cells
ECM	extracellular matrix
ECs	endothelial cells
EGF	epidermal growth factor
EGFR	epidermal growth factor receptor
EMT	epithelial mesenchymal transition
EPCs	endothelial progenitor cells
FDA	Food and Drug Administration
FGF-2	fibroblast growth factor-2
GIST	gastrointestinal stromal tumor
HCC	hepatocellular carcinoma
HGF	hepatocyte growth factor
HGF/cMET	hepatocyte growth factor (HGF) and its receptor c-MET
HIF-1α	hypoxia-inducible factor 1-alpha
HSCs	hematopoietic stromal cells
Ig	immunoglobulin
IGF	insulin-like growth factor
IL	interleukin
IMiDs	immunomodulatory drugs
mAbs	monoclonal antibodies
MDS	myelodysplastic syndromes
MDV	microvessel density
MGUS	gammopathy of undetermined significance
MM	multiple myeloma
MSCs	mesenchymal stromal cells
mTOR	mammalian target of rapamycin
NF-κB	nuclear factor-κB
PD-ECGF	platelet derived endothelial growth factor
PFS	progression-free survival
PlGF	placental growth factor
PNET	pancreatic neuroendocrine tumor
RCC	renal cell carcinoma
SCs	stromal cells
SDF1	stromal cell-derived factor 1
TGF	transforming growth factor
TKIs	tyrosine kinase inhibitors
TNF-α	tumor necrosis factor-alpha
TTP	time-to-progression

VDAs vascular disrupting agents
VEGF vascular endothelial growth factor

Conflict of interest

No potential conflicts of interest were disclosed.

Introduction

Angiogenesis, the formation of new blood vessels, is a tightly balanced mechanism regulated by both pro-angiogenic and antiangiogenic factors [1,2] and a constant hallmark of tumor progression and metastasis [3].

In most cancers, this balance is shifted toward a pro-angiogenic milieu to maintain sustainable angiogenic processes [4]. When the tumor mass expands beyond the support capacity of the existing vasculature, the resulting decreased levels of oxygen and nutrients and the accumulation of metabolic wastes prompt tumor cells to up-regulate several pro-angiogenic factors, including vascular endothelial growth factor (VEGF), fibroblast growth factor-2 (FGF-2), placental growth factor (PlGF), platelet derived endothelial cell growth factor (PD-ECGF), angiopoietins (Angs), transforming growth factors (TGFs) -α and - β, and epidermal growth factors (EGF), which collectively activate quiescent endothelial cells (ECs) and promote their migration into the tumor with growth and survival of new blood vessels [5]. The cancer's ability to induce the formation of new capillaries from preexisting host vessels is named "angiogenic switch" and is a turning point from the avascular phase of a dormant tumor to the vascular phase, in which changes of the tumor microenvironment ensure an exponential tumor growth [6]. Inhibition of angiogenesis, rather than simply blocking the delivery of oxygen and micronutrients to cancer cells, may also help to disrupt the interdependency of these various cellular players and the paracrine effects they exert to maintain the malignant phenotype.

The survival benefits obtained with angiogenesis inhibitors (AIs) in combination with cytotoxic agents undoubtedly indicate angiogenesis as a key target for novel cancer therapies. However, treatment with antiangiogenic drugs alone often results in modest clinical benefits; in part, this might be related to alternative mechanisms that sustain tumor vascularization, or be secondary to acquired resistance to the drugs. Thus, a deeper understanding on alternative molecular mechanisms of blood vessel formation remains a priority to identify additional targets and provide more effective therapeutic strategies to fight cancer.

Angiogenesis can occur locally by the proliferation of differentiated ECs, or systemically by mobilization of non-hematopoietic bone marrow (BM) populations, CD45-, named endothelial progenitor cells (EPCs), which enter the peripheral blood circulation, migrate to sites of angiogenesis, and incorporate into growing vessels [7,8] (Fig. 1). Accumulating evidence indicates that tumor angiogenesis is supported by the mobilization and functional incorporation of different cells, including EPCs, whose highly proliferative ratio distinguishes them from circulating ECs [9]. By promoting the recruitment of EPCs, hematopoietic stem and progenitor cells, [8] tumor-derived angiogenic factors enhance angiogenesis and neovessel formation [10]. Importantly, circulating EPCs express the VEGF receptor (VEGFR)-2, whereas subsets of hematopoietic cells express VEGFR-1; thus, cooperation between co-recruited

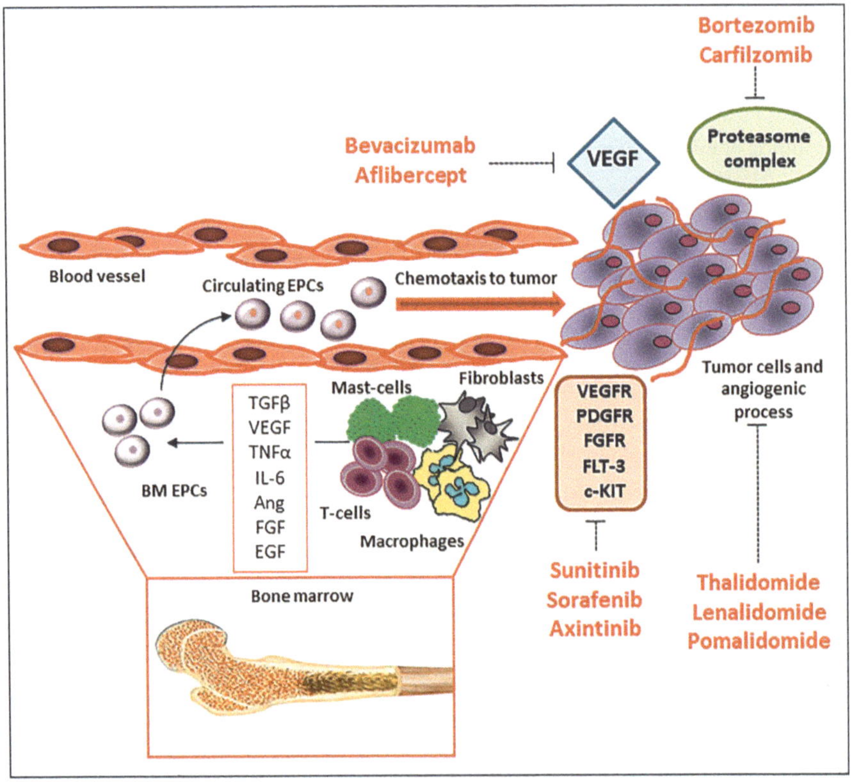

FIG. 1 Mobilization of endothelial progenitor cells (EPCs) from bone marrow (BM) to peripheral blood and homing to tumor to perform angiogenesis. In BM microenvironment different cell types (fibroblasts, mast cells, T-cells, macrophages) release tumor cytokines (TGFβ, VEGF, TNFα, IL-6, Ang, FGF, EGF) resulting in the mobilization of EPCs from the BM to blood vessels. EPCs homing to the tumor bed importantly contribute to subsequent tumor growth and angiogenesis. These processes are targeted by antiangiogenic drugs on distinct levels: on VEGF ligand, as for VEGF inhibitors (bevacizumab and aflibercept), or on tyrosine kinase (TK) receptors, as for TK inhibitors (sunitinib, sorafenib, and axintinib). Other drugs work as immunomodulatory agents (thalidomide, lenalidomide, and pomalidomide) or with a proteasome inhibiting activity (bortezomib and carfilzomib).

EPCs and hematopoietic cells facilitates the differentiation and integration of EPCs into rapidly expanding tumor vasculature [11].

Due to its ubiquitous and increased expression in most human cancers, the VEGF pathway is currently the dominant target for antiangiogenic drug therapy [12,13].

Besides bevacizumab (Avastin®), the first humanized anti-VEGF monoclonal antibody (mAb) approved by the Food and Drug Administration (FDA), the antiangiogenic drugs available incorporate newer players acting on distinct molecular targets, such as tyrosine kinase inhibitors (TKIs) (sunitinib, sorafenib), anti-cytokine drugs (thalidomide, lenalidomide), and proteasome inhibitors (bortezomib, carfilzomib), whose clinical effectiveness has been observed in relevant tumor types including hematological malignancies [14].

The following paragraphs recapitulate the current knowledge on players involved in tumor angiogenesis, focusing on the role and importance of circulating EPCs as potential novel targets for antiangiogenic drugs in the context of a global, clinically effective anti-cancer approach.

The cross-talk between EPCs and other players in tumor angiogenesis enlights effective drug targets

The complex microenvironment milieu surrounding malignant cells plays a fundamental role in tumor progression. It is composed of a non-cellular compartment formed by extracellular matrix (ECM) proteins (laminin, fibronectin, and collagen) and soluble factors (cytokines, growth factors, chemokines), and by a rich cellular compartment consisting of hematopoietic cells [myeloid cells, T and B cells, natural killer (NK) cells] and non-hematopoietic cells (fibroblasts, osteoblasts, osteoclasts, ECs, EPCs, pericytes, mesenchymal stem cells, mesenchymal stromal cells) [15]. The reciprocal and dynamic interplay among all components of the tumor microenvironment contributes to creating a continuously evolving setting, deeply involved in cell survival, cell proliferation, and, potentially, drug resistance [16].

During tumor angiogenesis, ECs are highly proliferative and express a number of plasma membrane proteins characteristic of the activated endothelium, such as growth factor receptors and adhesion molecules. In this context, a pivotal role has been ascribed to those anti-angiogenic drugs able to target molecular pathways involved in ECs proliferation, migration, and/or survival.

Nevertheless, clinical benefits obtained with antiangiogenic agents in monotherapy are minimal, suggesting the existence of contra-regulatory mechanisms. For example, long-term administration of AIs has been shown to up-regulate the expression of the hypoxia-induced factor-1α (HIF-1α) in the tumor microenvironment [17]. In turn, the activation of HIF-1α enhances the release of the stromal cell-derived factor 1 (SDF1) and VEGF, among the main chemotactic factors for EPCs [18]. Upon stimulation by these agents, EPCs secrete proangiogenic cytokines and promote angiogenesis [19], thereby limiting the overall effect of anti-angiogenic drugs.

Interestingly, this hypoxic condition also results in the expansion and recruitment of bone marrow-derived cells (BMDCs) and of cancer-associated fibroblasts (CAFs) into the tumor microenvironment, leading to a weakened antitumor response on one side [20] and further promoting angiogenesis, tumor growth, epithelial-mesenchymal transition (EMT), and metastasis [21].

In this vicious circle, the role played by CAFs seems to be particularly interesting: when activated by growth factors released from tumor and inflammatory cells, including TGF-β, PD-ECGF, and FGF-2 [22], CAFs secrete several pro-angiogenic growth factors [EGF, hepatocyte growth factor (HGF), and FGF-2] and chemokines as SDF1, which directly stimulate tumor cells and recruit more EPCs and further BMDCs into the tumor tissue [23] (Fig. 1). The interplay between EPCs and BMDCs and/or CAFs with the subsequent release of additional pro-angiogenic factors reinforces the key role of angiogenesis in tumor growth and survival. Although the use of AIs has been shown to reduce tumor mass, prolong disease

TABLE 1 Antiangiogenic therapies and relative targets, and pathologic indications.

Drug	Brand name	Drug classification	Target	Pathologic indication	Reference
Bevacizumab	Avastin®	Anti-VEGF strategy	VEGF-A, FGF	metastatic CRC, ovarian, RCC, NSCLC, GBM, AML	[14,26]
Aflibercept	Zaltrap®	Anti-VEGF strategy	VEGF-A/B	CRC	[27,28]
Sunitinib	Sutent®	TKI	VEGFR, FGFR, PDGFR, c-KIT,	RCC, GIST	[29,30]
Sorafenib	Nexavar®	TKI	VEGFR, PDGFR, FGFR, FLT3	RCC, HCC	[14,30]
Axintinib	Inlyta®	TKI	VEGFR, PDGFR, c-KIT	RCC	[31]
Thalidomide	Thalidomid®	IMiDs	VEGF, FGF-2, TNF-α, IL-6	MM	[14,32,33]
Lenalidomide	Revlimid®	IMiDs	VEGF, FGF-2, TNF-α, PI3K-Akt	MM, MDS	[14,34,35]
Pomalidomide	Actimid®	IMiDs	VEGF, FGF-2, TNF-α	MM	[36,37]
Bortezomib	Velcade®	Proteasome inhibitor	26S proteasome, NF-κB	MM, MDS, ALL, AML	[14,35,38,39]
Carfilzomib	Kyprolis®	Proteasome inhibitor	20S proteasome	MM	[14,40]

VEGF, vascular endothelial growth factor; *FGF*, fibroblast growth factor; *CRC*, metastatic colorectal cancer; *RCC*, renal cell carcinoma; *NSCLC*, non-small cell lung cancer; *GBM*, glioblastoma mutliforme; *AML*, acute myeloid leukemia; *TKI*, tyrosine kinase inhibitor; *VEGFR*, vascular endothelial growth factor receptor; *FGFR*, fibroblast growth factor receptor; *PDGFR*, platelet-derived growth factor receptor; *GIST*, gastrointestinal stromal tumors; *FLT3*, fms-like tyrosine kinase 3; *HCC*, hepatocellular carcinoma; *IMiDs*, immunomodulatory drugs; *TNF-α*, tumor necrosis factor-alpha; *IL-6*, interleukin-6; *MM*, multiple myeloma; *MDS*, myelodysplastic syndromes; *NF-κB*, nuclear factor κB; *ALL*, acute lymphocytic leukemia.

stabilization and increase patient survival rates [24], responses to AIs are variable and largely dependent on the treatment type and dose. The very same intricate and mutual relationship between various players of the tumor microenvironment might help to explain controversial and unexpected effects of AIs under specific conditions [25].

Among the currently available antiangiogenic therapies, bevacizumab (Table 1) is the most extensively characterized. It is approved for the treatment of metastatic colorectal cancer (CRC), ovarian, renal cell carcinoma (RCC), non-squamous cell lung cancer, and glioblastoma multiforme [26]. Unfortunately, bevacizumab in monotherapy did not obtain the expected results. [41]. However, when used in combination with other cancer drugs, it enhances the overall therapeutic potential. It has been hypothesized that, by reducing the tortuous and dysfunctional tumor vasculature induced by VEGF, bevacizumab may contribute to normalize the blood vessel texture and increase the delivery of the chemotherapeutic agent to the cancer tissue [42].

Aflibercept (Zaltrap®) (Table 1) is a soluble VEGF decoy receptor, able to neutralize both VEGF and PlGF because it consists of the extracellular domains of VEGFR-1 and -2 and the Fc portion of human immunoglobulin G1 (IgG1) [27]. FDA approved aflibercept for the treatment of metastatic CRC in combination with other chemotherapies; compared with bevacizumab treatment, the use of aflibercept demonstrates a higher tumor-suppressive activity, at least in patient-derived xenograft models of CRC [28].

Besides mAbs, several TKIs are able to counteract VEGF-induced angiogenesis by interacting with the kinase domain on the VEGFRs; interestingly, for some of these small molecules, a remarkable clinical benefit has been observed even when they are used as single antiangiogenic agents for the treatment of solid tumors, including RCC, hepatocellular carcinoma (HCC), thyroid cancer, gastrointestinal stromal tumor (GIST), and pancreatic neuroendocrine tumor (PNET) [43].

For sunitinib (Sutent®) (Table 1) an antiangiogenic TKI targeting VEGFRs and several other receptor tyrosine kinases, superior effectiveness with respect to the standard of care interferon-α treatment has been observed in patients with advanced RCC [29]. However, conflicting data exist regarding the effects of this drug on the growth of metastases in preclinical models. The phenotypic vascular changes observed under sunitinib treatment result in a decreased expression of adherent junction proteins and a reduced basement membrane and pericyte coverage, with subsequently increased leakiness. Regrettably, these alterations may involve both tumor vessels and normal organ vessels, and enhance local intra- and extravasation of tumor cells that potentially favor metastatic colonization. [44]. Notably, Singh et al. demonstrated that sunitinib, in contrast to the anti-VEGF antibody monotherapy, enhanced the aggressiveness of tumor cells [45]. Similarly, pretreatment with sunitinib and other TKIs such as imatinib (Glivec®), or sorafenib (Nexavar®) (Table 1) have been shown to promote lung metastasis following the injection of 66c14 cells, while the use of an anti-VEGFR-2 antibody inhibited the formation of lung nodules [30]. These evidences demonstrate that TKIs differ from the anti-VEGF antibody bevacizumab in their antiangiogenic actions and/or in their ability to facilitate cytotoxic drug delivery [8] and drive attention to the contribution of circulating ECs and EPCs on increased metastasis and enhanced invasiveness in tumors.

BM-derived EPCs have been detected in tumor blood vessels: although their precise contribution to tumor angiogenesis is uncertain [9], results from preclinical studies underline their participation in tumor growth and survival.

Treatment of tumor-bearing mice with vascular disrupting agents (VDAs) leads to an acute mobilization of EPCs, which home to the viable tumor rim persisting after such therapy. Disruption of this EPCs spike either by antiangiogenic drugs or by genetic manipulation markedly reduces tumor rim size and blood flow, and concomitantly improves VDAs antitumor activity [46]. These findings underline the involvement of circulating EPCs on tumor angiogenesis and provide evidence that a significant reduction in EPCs recruitment may be achieved when treatment with antiangiogenesis drugs is combined with VDAs.

Consistent with this, the decreased chemotherapy-induced vasculogenesis observed under pretreatment with DC101 or axintinib (Inlyta®) (Table 1) has been ascribed to the ability of these drugs to limit EPCs mobilization [31]. On this basis, inhibition of circulating EPCs has been proposed as an additional target to foster the efficacy of combined treatments in controlling tumor angiogenesis, cancer growth, and metastatic infiltration.

The antiangiogenic drugs in hematological malignancies

Targeting angiogenesis is a fundamental aspect in the pathophysiology of hematopoietic malignancies [47], and antiangiogenic drugs such as thalidomide (Thalomid®), lenalidomide (Revlimid®), bortezomib (Velcade®), and bevacizumab, have demonstrated an important clinical activity [14] (Table 1).

The VEGF neutralizing effects of bevacizumab previously described make this drug highly active in the cancer milieu and its use is effective also for the treatment of hematologic malignancies [48].

In terms of signaling involved, thalidomide, as an immunomodulatory antagonist of tumor necrosis factor-alpha (TNF-α), inhibits ECs activation by VEGF and other pro-angiogenic factors [49]. Lenalidomide is a potent thalidomide analog whose complex mechanism of action includes antiangiogenic activity [50]. Bortezomib, a proteasome inhibitor, counteracts the anti-apoptotic nuclear factor κ-B (NF-κB) pathway inducing ECs apoptosis and represses HIF-1α and VEGF production [38].

The role of circulating EPCs, which mediate the cross-talk between cancer angiogenesis and neoplastic clone, might be especially relevant in hematological cancers. Thus, deepening the description of neoplastic-infiltrating EPCs in hematological malignancies may help to shed light on additional targets for a more detailed and effective antiangiogenic strategy, with the advantage of tipping the balance of critical phases in the disease progression [51]. The physiological angiogenesis in the BM is induced and maintained by the balanced interplay of cells, cytokines, and growth factors within a complex microenvironment where the proliferation of ECs and pericytes is closely related to dendritic cells (DCs), inflammatory cells, and hematopoietic stem cells. BM malignant cells evolve and proliferate by producing and responding to these same factors, thereby stimulating angiogenesis and increasing vascularity [52].

The first and most convincing evidence of angiogenesis in hematologic malignancies was observed by Vacca and colleagues in multiple myeloma (MM) [53]. MM is characterized by a multistep natural history, as well as by variable pre-neoplastic stages preceding full-blown disease [54]. The substantial increase of microvessel density (MDV) observed in the BM of active MM patients compared to those with smoldering MM and monoclonal gammopathy of undetermined significance (MGUS) was indicative of a significant correlation between the extent of BM angiogenesis and plasma cell proliferation, as well as between angiogenesis and disease progression [53]. Following these seminal findings, more recent literature data demonstrating that EPCs are associated with disease stage, prognosis, and response to treatment have proposed vasculogenesis as an early pathogenic event that contributes to the "angiogenic switch" in MM [54]. More specifically, the deregulated interactions between MM cells and different cell types in the BM microenvironment might represent the foundation of the MM clinical manifestations, including osteolytic bone lesions, hypercalcemia, and suppressed hematopoietic functions. A specialized microenvironment, termed "niche" is required to support cell subpopulations characterized by stem cell potential [55]. This is a permissive microenvironment capable of supporting metastatic cells where the migration of macrophages and mast cells is associated with the recruitment of circulating EPCs, mesenchymal stromal cells (MSCs), and hematopoietic stromal cells (HSCs) and with the production of angiogenic factors stimulating angiogenesis process, cell migration, survival, and proliferation [55].

As a consequence of the important role of angiogenesis in the MM pathogenesis, a number of antiangiogenic therapeutic strategies have been evaluated; indeed, the use of thalidomide, lenalidomide, and bortezomib have changed clinical practice in MM patients with both a newly active and/or a relapsed stage of the disease.

Thalidomide was the first angiogenesis inhibitor used in MM treatment [32]. Besides the drug's ability to down-regulate VEGF and interleukin (IL)-6 productions within the BM microenvironment in patients with active disease, thalidomide has been shown to enhance T cell- and NK cell-mediated immunological responses and to induce caspase-8 mediated apoptosis [33]. When used in combination with dexamethasone, thalidomide has been shown to ameliorate long-term outcomes, including time-to-progression (TTP) and progression-free survival (PFS), in patients with newly diagnosed MM [56].

Based on these promising results, a series of immunomodulatory drugs (IMiDs), including lenalidomide and pomalidomide (Actimid®) (Table 1) has been developed. These drugs inhibit VEGF and FGF-2 secretion from both MM and BM stromal cells (BMSCs) and block ECs migration and proliferation, with an in vitro inhibitory activity toward TNF-α much greater than thalidomide [36]. In particular, lenalidomide-mediated inhibition of angiogenesis seems to involve VEGF-induced PI3K-Akt pathway and HIF-1α expression; this drug exerts an antitumor TNF-α activity, modulates the immune response stimulating T cells and NK cells activities, induces apoptosis of tumor cells, and decreases the binding of MM cells to BMSCs [34]. In addition, by down-regulating key genes and proteins related to MM angiogenesis, lenalidomide has been shown to inhibit MM plasma cells-induced angiogenesis in vivo in the chorioallantoic membrane (CAM) assay, and ECs-induced angiogenesis in vitro in the Matrigel® assay [35]. When used in combination with dexamethasone, lenalidomide induces 90% of therapeutic response in newly diagnosed MM patients [57]. Interestingly, lenalidomide sensitizes MM plasma cells to bortezomib, and a combination treatment with lenalidomide/bortezomib/dexamethasone elicited a positive response in 84% of relapsed/refractory patients, achieving a complete or nearly complete response in 21% of patients [58].

Pomalidomide anti-MM activity in vitro has been related to inhibition of angiogenesis by targeting VEGF and HIF-1α, but also to several other distinct mechanisms spanning from induction of apoptosis via caspase-8 to decrease proliferation, inhibition of NF-κB activation, and reduction of SCs stimulatory cytokine secretion. Currently, pomalidomide is used as second-line treatment when MM cells are resistant to bortezomib and lenalidomide [37] and is associated with a low-dose dexamethasone in patients with MM refractory to other IMiDs and/or to bortezomib [59].

To date, bortezomib is used in the treatment of MM patients in the relapsed setting post-transplant or as a second-line treatment in patients unsuitable for transplantation. In combination with dexamethasone, bortezomib works with a synergistic effect to sensitize MM cells to other chemotherapeutic agents [35]. Main beneficial activities for this proteasome inhibitor have been related to its ability to inhibit ECs proliferation and migration, to modulate VEGF and Ang expression, and to promote ECs apoptosis by downregulating the VEGF, IL-6, Ang-1 and -2, and insulin-like growth factor (IGF)-1 secretion [60]. A second-generation proteasome inhibitor, carfilzomib (Kyprolis®) (Table 1), is used to treat refractory MM, in newly diagnosed MM patients and in patients resistant to bortezomib [40].

Recently, new players with potential importance for MM treatment have been proposed. Among them, a significant role has been ascribed to microRNAs. These endogenous single-stranded non-coding RNA molecules regulate gene expression by targeting the 3′-untranslated region (3′-UTR) of mRNAs and are involved in a wide range of physiologic and pathologic cellular processes such as proliferation, differentiation, metabolism, angiogenesis, and cell death [15,61]. For example, a down-regulation of microRNA-15a/-16 in MM plasma cells, compared to normal cells, has been related to an antiangiogenic activity characterized by a reduced VEGF secretion from MM cells [62].

Similarly, in vitro and in vivo studies on HGF/cMET axis revealing the increased expression of VEGF/VEGFR-2 in MM ECs have also shown that the use of SU11274, a cMET inhibitor, significantly inhibits angiogenesis and decreases ECs activities including chemotaxis, motility, adhesion, and spreading, either alone or in combination with bortezomib, lenalidomide, and dexamethasone [63].

Even more interesting is the promising activity of MP0250, a multi-domain designed ankyrin repeat protein (DARPin) drug, which simultaneously binds and neutralizes VEGF and HGF with high specificity and affinity [64]. In vitro, MP0250 influences the secretory profile of MM ECs, inhibits their angiogenic activities, and strongly reduces capillary network formation and vessel-sprouting on a Matrigel® angiogenesis assay. In vivo, preclinical evaluation of MP0250 activity in syngeneic 5T33MM tumor model results in decreased MVD and, when combined with bortezomib, provides a supplementary decrease in angiogenesis by reducing the percentage of idiotype positive cells and the serum levels of M-protein [64].

An additional target for angiogenesis in MM is the mammalian target of rapamycin (mTOR), whose activation has been demonstrated in ECs isolated from MGUS and MM patients [65]. ECs from MM show a significantly higher expression of mTOR and RICTOR than MGUS-ECs; this finding, accompanied by higher activation of mTORC2 downstream effectors, suggests a major role of mTORC2 in the MM angiogenic switch. Consistent with this, a specific inhibition of mTOR activity obtained with both siRNA targeting RICTOR and dual mTOR inhibitor PP242, resulted in the reduced pro-angiogenic potential of MM ECs in vitro as a result of decreased cell migration, chemotaxis, adhesion, invasion, and cytoskeleton reorganization. In vivo, treatment with PP242 showed an antiangiogenic activity in the CAM assay, and in Matrigel® plug assays, highly enhanced when PP242 was combined with lenalidomide and bortezomib in MM patients [65].

Besides MM, increased angiogenesis is observed in myelodysplastic syndromes (MDS) and is associated with a BM dysfunction that impairs the production of erythrocytes, leucocytes, and platelets and results in anemia, infection, or bleeding. Conversion of MDS to acute myeloid leukemia (AML) occurs in up to 40% of patients and is associated with a poor prognosis. In the BM of patients with MDS, immature myeloid cells overexpress VEGF that is concomitantly associated with an increase of BM vascularity and with the presence of circulating EPCs [14]. In MDS patients, thalidomide treatment induces downregulation of BM microcirculation coupled to the plasma concentration of endostatin, a predictor of overall survival. This suggests that improper angiogenesis stimulation may participate in tumorigenesis of hematologic disease in a manner similar to the angiogenic switch proposed for solid tumors [66,67]. Analogous results have been obtained with lenalidomide used as an immunomodulatory agent to control clinical activity in MDS, since this drug is also able to interfere with hematopoiesis and retains antiangiogenic effects. Conversely, direct antiangiogenic agents, including SU5416 and AG-013736 (small molecules that inhibit the VEGFR tyrosine kinases) and bevacizumab have limited clinical activity in patients with MDS [68].

Literature data support the idea that angiogenesis is involved in leukemia pathophysiology, since an increase of MVD is reported in BM from patients with AML and with pediatric acute lymphoblastic leukemia (ALL) [69]. Angiogenesis in leukemia is characterized by increased expression of several key mediators including HIF-1α, VEGF, FGF-2, Ang-2, and soluble VEGFRs. Different antiangiogenesis drugs, implicated in the modulation of the VEGF pathway, have been used in adult leukemia patients, and encompass anti-cytokine agents, proteasome inhibitors, multikinase inhibitors, mAbs, and mTOR inhibitors [14]. This last class of drugs includes deforolimus (Ridaforolimus®; MK-8669), everolimus (Afinitor®), rapamycin (Sirolimus®), and temsirolimus (Torisel®) and in particular rapamycin, that inhibits VEGF expression and cell proliferation [70].

Thalidomide has been used in advanced AML patients, at first as monotherapy with modest clinical efficacy and then in combination with chemotherapy [71]; parental drug lenalidomide is indicated in monotherapy for the treatment of patients with relapsed/refractory chronic lymphocytic leukemia (CLL) [72]. In patients with untreated AML, bortezomib has shown an encouraging remission rate when proposed in combination with daunorubicin and cytarabine to maintain chemotherapy effects with a tolerable toxicity profile [39]. Finally, in patients with heavily-treated refractory AML, monotherapy with bevacizumab has demonstrated a time-dependent reduction of VEGF expression [73], while in combination with mitoxantrone, a type II topoisomerase inhibitor, bevacizumab improves the overall response rate [74].

Concluding remarks

The concept of targeting tumor angiogenesis is an important advancement in the treatment of hematological malignancies. However, the use of different therapeutic agents enrolled as monotherapy or in combination with other drugs has shown both clinical benefits and/or unexpected side effects. Indeed, despite promising results in terms of PFS and overall survival, the treatment with antiangiogenic drugs may give rise to more resistant tumors with higher patient relapse rates. The development of drug resistance to antiangiogenic agents is still a clinically significant problem (see Chapter 9 of the present book). This should encourage researchers to identify new players as target therapy to optimize the antiangiogenic treatments and to overcome resistance mechanisms. In this context, novel literature studies focusing on nanocarriers for microRNA delivery are regarded as a great promise for the treatment of hematological diseases. This potential therapeutic option, in addition to targeting angiogenesis pathways, should consider different approaches to elude the problem of resistance, for example by taking advantage of endogenous angiogenic inhibitors which are able to hinder the EPCs activity to the new vessel formation in a neoplastic microenvironment.

Overall, multiple therapeutic approaches have shown an efficient profile against the angiogenesis process. The association between existing antiangiogenic drugs, cancer immunotherapy, and the use of innovative technologies could help re-programming the tumor milieu in terms of new vessel formation and significantly advance our understanding of progression and treatment of tumor angiogenesis.

Acknowledgments

This work was supported by INNOLABS-POR Puglia FESR-FSE 2014-2020 (CITEL-Telemedicine Research Center) to A.V. and "Progetto Regione Puglia - Fondo europeo di sviluppo regionale e Fondo sociale europeo (FESR e FSE)" to A. V. The sponsors of this study are public or non-profit organizations that support science in general.

References

[1] Kazerounian S, Lawler J. Integration of pro- and anti-angiogenic signals by endothelial cells. J Cell Commun Signal 2018;12:171–9.
[2] Folkman J. Angiogenesis. Annu Rev Med 2006;57:1–18.
[3] Hanahan D, Weinberg RA. Hallmarks of cancer: the next generation. Cell 2011;144:646–74.
[4] El-Kenawi AE, El-Remessy AB. Angiogenesis inhibitors in cancer therapy: mechanistic perspective on classification and treatment rationales. Br J Pharmacol 2013;170:712–29.
[5] Ma J, Waxman DJ. Combination of antiangiogenesis with chemotherapy for more effective cancer treatment. Mol Cancer Ther 2008;7:3670–84.
[6] Ribatti D, Nico B, Crivellato E, Roccaro AM, Vacca A. The history of the angiogenic switch concept. Leukemia 2007;21:44–52.
[7] Hill JM, Zalos G, Halcox JP, Schenke WH, Waclawiw MA, Quyyumi AA, Finkel T. Circulating endothelial progenitor cells, vascular function, and cardiovascular risk. N Engl J Med 2003;348:593–600.
[8] Rafii S, Lyden D, Benezra R, Hattori K, Heissig B. Vascular and haematopoietic stem cells: novel targets for antiangiogenesis therapy? Nat Rev Cancer 2002;2:826–35.
[9] Lyden D, Hattori K, Dias S, Costa C, Blaikie P, Butros L, Chadburn A, Heissig B, Marks W, Witte L, Wu Y, Hicklin D, Zhu Z, Hackett NR, Crystal RG, Moore MA, Hajjar KA, Manova K, Benezra R, Rafii S. Impaired recruitment of bone-marrow-derived endothelial and hematopoietic precursor cells blocks tumor angiogenesis and growth. Nat Med 2001;7:1194–201.
[10] Shaked Y, Ciarrocchi A, Franco M, Lee CR, Man S, Cheung AM, Hicklin DJ, Chaplin D, Foster FS, Benezra R, Kerbel RS. Therapy-induced acute recruitment of circulating endothelial progenitor cells to tumors. Science 2006;313:1785–7.
[11] Carmeliet P, Jain RK. Molecular mechanisms and clinical applications of angiogenesis. Nature 2011;473:298–307.
[12] Jayson GC, Kerbel R, Ellis LM, Harris AL. Antiangiogenic therapy in oncology: current status and future directions. Lancet 2016;388:518–29.
[13] Dahan N, Magidey K, Shaked Y, editors. Resistance to inhibitors of angiogenesis. In: Resistance to anti-cancer therapeutics targeting receptor tyrosine kinases and downstream pathways. Cham: Springer; 2018.
[14] Li WW, Hutnik M, Gehr G. Antiangiogenesis in haematological malignancies. Br J Haematol 2008;143:622–31.
[15] Di Marzo L, Desantis V, Solimando AG, Ruggieri S, Annese T, Nico B, Fumarulo R, Vacca A, Frassanito MA. Microenvironment drug resistance in multiple myeloma: emerging new players. Oncotarget 2016;7:60698–711.
[16] Manier S, Sacco A, Leleu X, Ghobrial IM, Roccaro AM. Bone marrow microenvironment in multiple myeloma progression. J Biomed Biotechnol 2012;2012:157496.
[17] Goel S, Duda DG, Xu L, Munn LL, Boucher Y, Fukumura D, et al. Normalization of the vasculature for treatment of cancer and other diseases. Physiol Rev 2011;91:1071–121.
[18] Ceradini DJ, Kulkarni AR, Callaghan MJ, Tepper OM, Bastidas N, Kleinman ME, Capla JM, Galiano RD, Levine JP, Gurtner GC. Progenitor cell trafficking is regulated by hypoxic gradients through HIF-1 induction of SDF-1. Nat Med 2004;10:858–64.
[19] Crawford Y, Ferrara N. Tumor and stromal pathways mediating refractoriness/resistance to anti-angiogenic therapies. Trends Pharmacol Sci 2009;30:624–30.
[20] Sica A, Allavena P, Mantovani A. Cancer related inflammation: the macrophage connection. Cancer Lett 2008;267:204–15.
[21] Frassanito MA, Rao L, Moschetta M, Ria R, Di Marzo L, De Luisi A, Racanelli V, Catacchio I, Berardi S, Basile A, Menu E, Ruggieri S, Nico B, Ribatti D, Fumarulo R, Dammacco F, Vanderkerken K, Vacca A. Bone marrow fibroblasts parallel multiple myeloma progression in patients and mice: in vitro and in vivo studies. Leukemia 2014;28:904–16.

[22] Turner N, Grose R. Fibroblast growth factor signalling: from development to cancer. Nat Rev Cancer 2010;10:116–29.

[23] Orimo A, Gupta PB, Sgroi DC, Arenzana-Seisdedos F, Delaunay T, Naeem R, Carey VJ, Richardson AL, Weinberg RA. Stromal fibroblasts present in invasive human breast carcinomas promote tumor growth and angiogenesis through elevated SDF-1/CXCL12 secretion. Cell 2005;121:335–48.

[24] Joyce JA, Fearon DT. T cell exclusion, immune privilege, and the tumor microenvironment. Science 2015;348: 74–80.

[25] Ellis LM, Reardon DA. Is there really a yin and yang to VEGF-targeted therapies? Lancet Oncol 2010;11:809–11. https://doi.org/10.1016/S1470-2045(10)70161-3.

[26] Garcia J, Hurwitz HI, Sandler AB, Miles D, Coleman RL, Deurloo R, Chinot OL. Bevacizumab (Avastin®) in cancer treatment: a review of 15 years of clinical experience and future outlook. Cancer Treat Rev 2020;86:102017.

[27] Ciombor KK, Berlin J, Chan E. Aflibercept. Clin Cancer Res 2013;19:1920–5.

[28] Chiron M, Bagley RG, Pollard J, Mankoo PK, Henry C, Vincent L, Geslin C, Baltes N, Bergstrom DA. Differential antitumor activity of aflibercept and bevacizumab in patient-derived xenograft models of colorectal cancer. Mol Cancer Ther 2014;13:1636–44.

[29] Motzer RJ, Hutson TE, Tomczak P, Michaelson MD, Bukowski RM, Rixe O, Oudard S, Negrier S, Szczylik C, Kim ST, Chen I, Bycott PW, Baum CM, Figlin RA. Sunitinib versus interferon alfa in metastatic renal-cell carcinoma. N Engl J Med 2007;356:115–24.

[30] Chung AS, Kowanetz M, Wu X, Zhuang G, Ngu H, Finkle D, Komuves L, Peale F, Ferrara N. Differential drug class-specific metastatic effects following treatment with a panel of angiogenesis inhibitors. J Pathol 2012;227:404–16.

[31] Paul S, Foutz T, Calleri A, et al. AG-013736, a potent VEGF/PDGF receptor tyrosine kinase inhibitor, is active against lymphoma growth and chemotherapy-induced vasculogenesis. Blood 2003;102:649a.

[32] D'Amato RJ, Loughnan MS, Flynn E, Folkman J. Thalidomide is an inhibitor of angiogenesis. Proc Natl Acad Sci U S A 1994;91:4082–5.

[33] Davies FE, Raje N, Hideshima T, Lentzsch S, Young G, Tai YT, Lin B, Podar K, Gupta D, Chauhan D, Treon SP, Richardson PG, Schlossman RL, Morgan GJ, Muller GW, Stirling DI, Anderson KC. Thalidomide and immunomodulatory derivatives augment natural killer cell cytotoxicity in multiple myeloma. Blood 2001;98:210–6.

[34] Ribatti D, Vacca A. New insights in anti-angiogenesis in multiple myeloma. Int J Mol Sci 2018;19:2031.

[35] De Luisi A, Ferrucci A, Coluccia AM, Ria R, Moschetta M, de Luca E, Pieroni L, Maffia M, Urbani A, Di Pietro G, Guarini A, Ranieri G, Ditonno P, Berardi S, Caivano A, Basile A, Cascavilla N, Capalbo S, Quarta G, Dammacco F, Ribatti D, Vacca A. Lenalidomide restrains motility and overangiogenic potential of bone marrow endothelial cells in patients with active multiple myeloma. Clin Cancer Res 2011;17:1935–46.

[36] Vacca A, Ria R, Reale A, Ribatti D. Angiogenesis in multiple myeloma. Chem Immunol Allergy 2014;99:180–96.

[37] Chanan-Khan AA, Swaika A, Paulus A, Kumar SK, Mikhael JR, Rajkumar SV, Dispenzieri A, Lacy MQ. Pomalidomide: the new immunomodulatory agent for the treatment of multiple myeloma. Blood Cancer J 2013;3, e143.

[38] Shin DH, Chun YS, Lee DS, Huang LE, Park JW. Bortezomib inhibits tumor adaptation to hypoxia by stimulating the FIH-mediated repression of hypoxia-inducible factor-1. Blood 2008;111:3131–6.

[39] Attar EC, Johnson JL, Amrein PC, Lozanski G, Wadleigh M, DeAngelo DJ, Kolitz JE, Powell BL, Voorhees P, Wang ES, Blum W, Stone RM, Marcucci G, Bloomfield CD, Moser B, Larson RA. Bortezomib added to daunorubicin and cytarabine during induction therapy and to intermediate-dose cytarabine for consolidation in patients with previously untreated acute myeloid leukemia age 60 to 75 years: CALGB (Alliance) study 10502. J Clin Oncol 2013;31:923–9.

[40] Ziogas DC, Terpos E, Kastritis E, Dimopoulos MA. An overview of the role of carfilzomib in the treatment of multiple myeloma. Expert Opin Pharmacother 2017;18:1883–97.

[41] Jain RK. Normalization of tumor vasculature: an emerging concept in antiangiogenic therapy. Science 2005;307:58–62.

[42] Jain RK. Antiangiogenesis strategies revisited: from starving tumors to alleviating hypoxia. Cancer Cell 2014;26:605–22.

[43] Ebos JM, Kerbel RS. Antiangiogenic therapy: impact on invasion, disease progression, and metastasis. Nat Rev Clin Oncol 2011;8:210–21.

[44] Welti JC, Powles T, Foo S, Gourlaouen M, Preece N, Foster J, Frentzas S, Bird D, Sharpe K, van Weverwijk A, Robertson D, Soffe J, Erler JT, Pili R, Springer CJ, Mather SJ, Reynolds AR. Contrasting effects of sunitinib within in vivo models of metastasis. Angiogenesis 2012;15:623–41.
[45] Singh M, Couto SS, Forrest WF, Lima A, Cheng JH, Molina R, Long JE, Hamilton P, McNutt A, Kasman I, Nannini MA, Reslan HB, Cao TC, Ho CC, Barck KH, Carano RA, Foreman O, Eastham-Anderson J, Jubb AM, Ferrara N, Johnson L. Anti-VEGF antibody therapy does not promote metastasis in genetically engineered mouse tumour models. J Pathol 2012;227:417–30.
[46] Shaked Y, Ciarrocchi A, Franco M, Lee CR, Man S, Cheung AM, Hicklin DJ, Chaplin D, Foster FS, Benezra R, Kerbel RS. Therapy-induced acute recruitment of circulating endothelial progenitor cells to tumors. Science 2006;313:1785–7.
[47] Vacca A, Ribatti D, Roncali L, Dammacco F. Angiogenesis in B cell lymphoproliferative diseases. Biological and clinical studies. Leuk Lymphoma 1995;20:27–38.
[48] Ferrara N, Hillan KJ, Gerber HP, Novotny W. Discovery and development of bevacizumab, an anti-VEGF antibody for treating cancer. Nat Rev Drug Discov 2004;3:391–400.
[49] D'Amato RJ, Loughnan MS, Flynn E, Folkman J. Thalidomide is an inhibitor of angiogenesis. Proc Natl Acad Sci U S A 1994;91:4082–5.
[50] Kotla V, Goel S, Nischal S, Heuck C, Vivek K, Das B, Verma A. Mechanism of action of lenalidomide in hematological malignancies. J Hematol Oncol 2009;2:36.
[51] Solimando AG, Summa S, Vacca A, Ribatti D. Cancer-associated angiogenesis: the endothelial cell as a checkpoint for immunological patrolling. Cancers (Basel) 2020;12:3380.
[52] Medinger M, Passweg J. Role of tumour angiogenesis in haematological malignancies. Swiss Med Wkly 2014;144:w14050.
[53] Vacca A, Ribatti D, Roncali L, Ranieri G, Serio G, Silvestris F, Dammacco F. Bone marrow angiogenesis and progression in multiple myeloma. Br J Haematol 1994;87:503–8.
[54] Moschetta M, Mishima Y, Sahin I, Manier S, Glavey S, Vacca A, Roccaro AM, Ghobrial IM. Role of endothelial progenitor cells in cancer progression. Biochim Biophys Acta 1846;2014:26–39.
[55] Ribatti D, Nico B, Vacca A. Multiple myeloma as a model for the role of bone marrow niches in the control of angiogenesis. Int Rev Cell Mol Biol 2015;314:259–82.
[56] Rajkumar SV, Rosiñol L, Hussein M, Catalano J, Jedrzejczak W, Lucy L, Olesnyckyj M, Yu Z, Knight R, Zeldis JB, Bladé J. Multicenter, randomized, double-blind, placebo-controlled study of thalidomide plus dexamethasone compared with dexamethasone as initial therapy for newly diagnosed multiple myeloma. J Clin Oncol 2008;26:2171–7.
[57] Rajkumar SV. Combination therapy with lenalidomide plus dexamethasone (Rev/Dex) for newly diagnosed myeloma. Blood 2005;106:4050–3.
[58] Dimopoulos MA, Kastritis E, Christoulas D, Migkou M, Gavriatopoulou M, Gkotzamanidou M, Iakovaki M, Matsouka C, Mparmparoussi D, Roussou M, Efstathiou E, Terpos E. Treatment of patients with relapsed/refractory multiple myeloma with lenalidomide and dexamethasone with or without bortezomib: prospective evaluation of the impact of cytogenetic abnormalities and of previous therapies. Leukemia 2010;24:1769–78.
[59] Lacy MQ, Allred JB, Gertz MA, Hayman SR, Short KD, Buadi F, Dispenzieri A, Kumar S, Greipp PR, Lust JA, Russell SJ, Dingli D, Zeldenrust S, Fonseca R, Bergsagel PL, Roy V, Stewart AK, Laumann K, Mandrekar SJ, Reeder C, Rajkumar SV, Mikhael JR. Pomalidomide plus low-dose dexamethasone in myeloma refractory to both bortezomib and lenalidomide: comparison of 2 dosing strategies in dual-refractory disease. Blood 2011;118:2970–5.
[60] Roccaro AM, Hideshima T, Raje N, Kumar S, Ishitsuka K, Yasui H, Shiraishi N, Ribatti D, Nico B, Vacca A, Dammacco F, Richardson PG, Anderson KC. Bortezomib mediates antiangiogenesis in multiple myeloma via direct and indirect effects on endothelial cells. Cancer Res 2006;66:184–91.
[61] Dimopoulos K, Gimsing P, Grønbæk K. Aberrant microRNA expression in multiple myeloma. Eur J Haematol 2013;91:95–105.
[62] Roccaro AM, Sacco A, Thompson B, Leleu X, Azab AK, Azab F, Runnels J, Jia X, Ngo HT, Melhem MR, Lin CP, Ribatti D, Rollins BJ, Witzig TE, Anderson KC, Ghobrial IM. MicroRNAs 15a and 16 regulate tumor proliferation in multiple myeloma. Blood 2009;113:6669–80.
[63] Ferrucci A, Moschetta M, Frassanito MA, Berardi S, Catacchio I, Ria R, Racanelli V, Caivano A, Solimando AG, Vergara D, Maffia M, Latorre D, Rizzello A, Zito A, Ditonno P, Maiorano E, Ribatti D, Vacca A. A HGF/cMET autocrine loop is operative in multiple myeloma bone marrow endothelial cells and may represent a novel therapeutic target. Clin Cancer Res 2014;20:5796–807.

[64] Rao L, De Veirman K, Giannico D, Saltarella I, Desantis V, Frassanito MA, Solimando AG, Ribatti D, Prete M, Harstrick A, Fiedler U, De Raeve H, Racanelli V, Vanderkerken K, Vacca A. Targeting angiogenesis in multiple myeloma by the VEGF and HGF blocking DARPin® protein MP0250: a preclinical study. Oncotarget 2018;9:13366–81.

[65] Lamanuzzi A, Saltarella I, Desantis V, Frassanito MA, Leone P, Racanelli V, Nico B, Ribatti D, Ditonno P, Prete M, Solimando AG, Dammacco F, Vacca A, Ria R. Inhibition of mTOR complex 2 restrains tumor angiogenesis in multiple myeloma. Oncotarget 2018;9:20563–77.

[66] Moehler TM, Ho AD, Goldschmidt H, Barlogie B. Angiogenesis in hematologic malignancies. Crit Rev Oncol Hematol 2003;45:227–44.

[67] Fuchs O. Treatment of lymphoid and myeloid malignancies by immunomodulatory drugs. Cardiovasc Hematol Disord Drug Targets 2019;19:51–78.

[68] Savic A, Cemerikic-Martinovic V, Dovat S, Rajic N, Urosevic I, Sekulic B, Kvrgic V, Popovic S. Angiogenesis and survival in patients with myelodysplastic syndrome. Pathol Oncol Res 2012;18:681–90.

[69] Keyhani A, Jendiroba DB, Freireich EJ. Angiogenesis and leukemia. Leuk Res 2001;25:639–45.

[70] Dormond O, Madsen JC, Briscoe DM. The effects of mTOR-Akt interactions on anti-apoptotic signaling in vascular endothelial cells. J Biol Chem 2007;282:23679–86.

[71] Barr P, Fu P, Lazarus H, Kane D, Meyerson H, Hartman P, Reyes R, Creger R, Stear K, Laughlin M, Tse W, Cooper B. Antiangiogenic activity of thalidomide in combination with fludarabine, carboplatin, and topotecan for high-risk acute myelogenous leukemia. Leuk Lymphoma 2007;48:1940–9.

[72] Wiernik PH. Lenalidomide in lymphomas and chronic lymphocytic leukemia. Expert Opin Pharmacother 2013;14:475–88.

[73] Zahiragic L, Schliemann C, Bieker R, Thoennissen NH, Burow K, Kramer C, Zühlsdorf M, Berdel WE, Mesters RM. Bevacizumab reduces VEGF expression in patients with relapsed and refractory acute myeloid leukemia without clinical antileukemic activity. Leukemia 2007;21:1310–2.

[74] Karp JE, Gojo I, Pili R, Gocke CD, Greer J, Guo C, Qian D, Morris L, Tidwell M, Chen H, Zwiebel J. Targeting vascular endothelial growth factor for relapsed and refractory adult acute myelogenous leukemias: therapy with sequential 1-beta-d-arabinofuranosylcytosine, mitoxantrone, and bevacizumab. Clin Cancer Res 2004;10:3577–85.

CHAPTER

8

Metronomic chemotherapy and antiangiogenic drugs: Preclinical and clinical data

Guido Bocci[a], *Marta Banchi*[a], *Diana Gonzalez Garcia*[b], *Serina Batson*[b], *and Giulio Francia*[b]

[a]Dipartimento di Medicina Clinica e Sperimentale, Università di Pisa, Pisa, Italy [b]Department of Biological Sciences/BBRC, University of Texas at El Paso (UTEP), El Paso, TX, United States

Abstract

Metronomic chemotherapy (MC) is the frequent, regular administration of chemotherapeutic drugs that maintain a low, prolonged and active range of plasma concentrations with a favorable toxicity profile. MC regimens were developed to optimize the antitumor efficacy of agents that could impact multiple targets, including the tumor vasculature, regulatory T cells, and cancer cells. In the last two years, MC has become increasingly used in the clinical research setting due to a number of successful randomized phase II-III clinical trials in nasopharyngeal carcinoma, head and neck carcinoma, small-cell lung cancer, and epithelial ovarian cancer. The combination of MC and the antiangiogenic drug has seen its major development in the preclinical setting, where these regimes achieved high antitumor activities. Initially, the preclinical combination treatments included preferentially the anti-VEGFR-2 monoclonal antibody DC101, later the research interest focused on VEGFR-2 small tyrosine kinase inhibitors such as pazopanib, sorafenib, and sunitinib because of their oral administration route. Moreover, in the last ten years, several phase I and phase II clinical trials have combined MC and antiangiogenic drugs (e.g., bevacizumab or tyrosine kinase inhibitors, targeting VEGFR-2 and other angiogenic factor receptors), reporting mixed results (promising clinical efficacy or no clinical advantage) involving different types of cancer. In conclusion, MC has shown both preclinical and clinical activity in different types of cancer. The addition of antiangiogenic drugs into MC schedules greatly enhanced the therapeutic activity in preclinical models but these results were not completely confirmed in all the clinical trials based on the same combination. Additional phase III clinical trials will hopefully validate the promising preclinical results obtained by the combination of MC and antiangiogenic drugs.

https://doi.org/10.1016/B978-0-323-90190-1.00017-2

Abbreviations

CEP	circulating endothelial progenitor
CTX	cyclophosphamide
EGFR	epidermal growth factor receptor
EpoB	epothilone B
G4	grade 4 toxicity
HCC	hepatocellular carcinoma
HUVEC	human umbilical vascular endothelial cell
MTD	maximum tolerated dose
NSCLC	non-small cell lung cancer
OS	overall survival
PDGF	platelet derived growth factor
PFS	progression-free survival
TKI	tyrosine kinase inhibitor
TSP-1	thrombospondin-1
UFT	tegafur/uracil
VEGF	vascular endothelial growth factor
VEGFR-2	vascular endothelial growth factor receptor-2

Conflict of interest

No potential conflicts of interest were disclosed.

Introduction

Definition and multiple mechanisms of metronomic chemotherapy

Metronomic chemotherapy is the frequent, regular administration of chemotherapeutic drugs that maintain a low, prolonged and active range of plasma concentrations with a favorable toxicity profile [1]. In contrast, conventional "maximum tolerated dose" (MTD) chemotherapy provides administration of cytotoxic agents at single doses or short courses of therapy at the highest tolerated doses causing life-threatening side effects and requiring 3–4 week-long breaks between successive cycles of therapy. Metronomic chemotherapy regimens were developed to optimize the antitumor efficacy of agents that could impact multiple targets, including the tumor vasculature, regulatory T cells, and cancer cells while also reducing the overall toxicity of antineoplastic drugs [1,2]. The antiangiogenic effects of chemotherapeutic agents can be achieved by administering such drugs metronomically. Indeed, metronomic chemotherapy directly and preferentially targets endothelial cells of the growing tumor vasculature therefore can be effective against multidrug-resistant tumors [3,4]. One of the mechanisms of metronomic chemotherapy is the induction of thrombospondin-1 (TSP-1), a potent endothelial-specific inhibitor of angiogenesis [5]. Moreover, low-dose chemotherapy effectively suppresses the mobilization of circulating endothelial progenitors (CEPs), blocking neoangiogenesis [6]. Metronomic chemotherapy also plays an important role in inducing an anti-tumor immune response by impacting regulatory T cells. In addition, metronomic chemotherapy may promote the maturation of dendritic cells, maintaining the survival of memory T cells, and reducing the levels of cytokines such as interleukin-2 [7,8]. One advantage of metronomic chemotherapy, due to its low toxicity profile, is the possibility of combining it with many of the new targeted and biotechnological agents to increase antitumor effects, thus improving clinical response, often without added toxicity. In that regard, cyclophosphamide (CTX)-based metronomic chemotherapy is the most widely studied metronomic regimen as a single-agent or in combination treatments [8]. Low-dose cyclophosphamide was shown to decrease the numbers of CD4+ CD25High Foxp3+ regulatory T cells (Tregs) and to inhibit their suppressive function, resulting in the restoration of anti-tumor natural killer/T-cell activity [9,10]. Moreover, a direct effect on cancer cells has been described [11]. As an example, metronomic vinorelbine has been reported to have a direct activity in vitro on non-small cell lung cancer (NSCLC) cells if administered continuously for a prolonged period at picomolar concentrations [11]. Moreover, the simultaneous combination of vinorelbine and reversible epidermal growth factor receptor (EGFR) tyrosine-kinase inhibitors (TKIs) resulted in a strong synergistic antitumor effect, increasing the intracellular concentrations of EGFR TKIs because of the reduction of the expression of *ABCG2*, an efflux transporter of the drug. Thus, metronomic vinorelbine sensitizes resistant cells to EGFR TKIs, increasing their intracellular concentrations [11].

Recent clinical data from randomized phase II–III clinical trials involving metronomic chemotherapy

Between 2012 and 2018, many phase II-III randomized clinical trials using metronomic chemotherapy were reported. Although some trials did not show a significant advantage in terms of survival, at least four randomized phase III studies and four randomized phase II studies presented some clinical benefits [12]. Such results were obtained when the metronomic schedule was used as maintenance treatment and added to other antineoplastic agents administered at standard doses (e.g., CAIRO3 study) [13]. However, in the last two years, metronomic chemotherapy has become increasingly used in the clinical research setting due to a number of successful randomized phase II-III clinical trials. Thus, in 2020–2021, at least four randomized studies showed very interesting and significant results in advanced nasopharyngeal carcinoma, metastatic head and neck carcinoma, NSCLC, and epithelial ovarian cancer.

Among these clinical trials, one highlight was a multicenter, randomized phase III study on metronomic capecitabine administered as adjuvant therapy in locoregionally advanced nasopharyngeal carcinoma. The study enrolled 406 patients who were treated with oral metronomic capecitabine (650 mg/m^2 twice daily for 1 year) and standard therapy. Interestingly, failure-free survival at three years was significantly higher in the metronomic capecitabine group (85.3% vs. 75.7% of standard therapy, $P = .0023$). The hand-foot syndrome was the most common adverse event related to metronomic capecitabine, whereas no treatment-related deaths were reported. These results support a potential role of metronomic chemotherapy as adjuvant therapy in the treatment of nasopharyngeal carcinoma [14].

One study compared oral metronomic chemotherapy versus intravenous standard chemotherapy, and the results confirmed the clinical efficacy of the low-dose treatment in patients with recurrent, metastatic, head and neck squamous carcinoma. A randomized phase III trial with 418 patients evaluated the administration of 15 mg/m^2 methotrexate once per week plus 200 mg celecoxib twice per day in the metronomic group of patients. In the standard therapy arm, an intravenous cisplatin dose of 75 mg/m^2 once every 3 weeks was given. The median overall survival was higher in the oral metronomic chemotherapy group (7.5 vs. 6.1 months; $P = .029$), and also the high-grade adverse events (e.g., the hematological toxicities) were significantly lower in the metronomic chemotherapy arm [15].

A multicenter randomized phase II study in 167 treatment-naive patients with stage IIIB/IV NSCLC was performed, to establish the progression-free survival (PFS) without grade 4 toxicity (G4PFS) of metronomic oral vinorelbine at a dose of 50 mg × 3 versus a standard schedule of vinorelbine at 60–80 mg/m^2 weekly. The enrolled patients were unfit for platinum-based combination chemotherapy. Metronomic oral vinorelbine significantly prolonged the endpoint of the study: the median G4PFS was 4.0 months vs. 2.2 months ($P = .0068$) for the standard therapy. Based on these results, the authors concluded that metronomic vinorelbine was associated with a clear reduction in toxicity and could be considered an important option in this particular subpopulation of NSCLC patients [16].

Finally, a phase II randomized trial explored the combination of oral metronomic therapy with the antiangiogenic drug pazopanib for platinum-resistant epithelial ovarian cancer. Seventy-five patients were enrolled in two arms, one receiving oral metronomic etoposide (50 mg, day 1 to 14) and cyclophosphamide (50 mg, day 1 to 28), and the other arm received the same treatment with the addition of pazopanib (400 mg/day). The patients receiving the pazopanib-combined therapeutic strategy had a significantly higher median PFS (5.1 vs. 3.4 months; $P = .045$). The authors also evaluated the serum levels of two key angiogenic factors: vascular endothelial growth factor (VEGF) and platelet-derived growth factor (PDGF). The serum levels of both proteins decreased with metronomic chemotherapy alone or in combination with pazopanib in the course of treatment (i.e., compared to baseline levels) [17].

It should be noted that the increased activity of metronomic chemotherapy combined with another small molecule TKI targeting vascular endothelial growth factor receptor-2, fibroblast growth factor receptor and platelet-derived growth factor receptor, was not confirmed in a similar setting of patients. A phase II randomized trial of metronomic cyclophosphamide and nintedanib in an advanced ovarian, fallopian tube, or primary peritoneal cancer, demonstrated that the oral administration of daily nintedanib did not improve outcomes when added to low-dose cyclophosphamide [18].

Antiangiogenic activity of metronomic chemotherapy in preclinical and clinical setting

One classic protocol for the evaluation of metronomic dosing of a drug in vitro is by exposure to cell cultures over 144 h, with daily replacement of the growth media containing the drug. Tumor cells are compared for their response to drug exposure on human endothelial cells [human dermal microvascular endothelial cells and human umbilical vein endothelial cells (HUVECs)]. For example, in 2012 Bocci et al. [19] reported on the response in vitro of pancreatic cancer cells (Capan-1 and MIA PaCa-2) treated with ceramide analogs when compared to the response of HUVECs. Thereafter, for in vivo confirmation, MTD gemcitabine was compared against metronomic doses of the ceramide analogs by evaluating the inhibition of MIA PaCa-2 subcutaneous tumor growth in nude mice [19]. In the original study that described the 144 h in vitro protocol [4], six different drugs were evaluated. The drugs were paclitaxel, 4-hydroperoxycyclophosphamide (an in vitro active form of cyclophosphamide), the

BMS-275183 oral taxane, doxorubicin, epothilone B (EpoB), and its analog 5-methylpyridine EpoB. Subsequently, a seventh drug, BAL-9504, was also evaluated by the 144 h protocol [20]. These studies were essential to establish a protocol in vitro of 144 h exposure which led to the discovery of TSP-1 upregulation in endothelial cells exposed to metronomic chemotherapy [5].

The 144 h protocol was also used to evaluate the metronomic dosing of irinotecan, either alone or in combination with semaxinib on human endothelial cells and human colorectal cancer cells [3]. Thus, the 144 h low-dose treatment with SN-38 (the active metabolite of irinotecan) was shown to induce the expression and secretion of TSP-1 in endothelial and cancer cells. The authors then showed the efficacy of metronomic irinotecan on human colon cancers grown in vivo. The same group also investigated, in a separate study, metronomic 5-fluorouracil, oxaliplatin, and irinotecan in colorectal cancer and in HUVECs [21]. By the same in vitro 144 h methodology, metronomic dosing was also evaluated for vinorelbine on NSCLC cells [11], and ceramide analogs were evaluated on HUVECs and on human pancreatic cancer cells [19]. Overall, these studies showed that a number of metronomic regimens could be rapidly assessed in vitro. Since then, another group has used this protocol to evaluate metronomic docetaxel [22].

The 144 h in vitro exposure was originally applied to 5 cell types [4]. HUVECs, HMVEC-d, and NHDFs, and two human cancer cell lines MDA-MB-435 and its P-glycoprotein-positive multidrug-resistant variant T0.1. Evaluation of response was via thymidine incorporation assay and revealed that metronomic dosing was preferentially active on endothelial cells. Since that Bocci et al. publication [4], three of the drugs used in the study were found to be effective in vivo when administered metronomically. Overall, these studies show how the 144 h in vitro protocol can effectively identify drugs that when given metronomically may produce antiangiogenic activity, which can then be confirmed in tumor models in vivo. It should be noted that the 144 h in vitro exposure has not been employed to predict the toxicity of metronomic chemotherapy combinations. This is conceivably a consequence of the lack of adaptation of this methodology to various non-transformed cells (e.g., fibroblasts) but likely also reflects the limits of what can be assessed with in vitro approaches.

Also of note is the interplay between angiogenesis inhibitors and metronomic chemotherapy in vivo, which may also depend on whether a therapy is tested on primary tumors, or on metastatic disease. For example, a bolus plus metronomic CTX regimen was found to be effective at reducing the growth of EMT-6 breast cancer cells grown as implanted primary tumors [23]. And also effective in this primary tumor setting is the monotherapy with the anti-VEGFR2 antibody DC101. However, in the metastatic setting, where EMT-6 was grown as experimental lung metastases (rather than as a primary tumor lump) DC101 monotherapy is not effective at hindering the metastatic growth. And yet when DC101 is combined with metronomic CTX it can delay the growth of the lung metastases [23]. These results indicate that for monotherapies, and for combinations of metronomic chemotherapy with antiangiogenic drugs, the response of the tumor population to treatment may be dependent on the tumor microenvironment.

Activation of oncogenes via overexpression or mutation, most notably in the case of mutant *ras*, has been shown to increase VEGF expression in a number of tumor cell types [24]. Consequently, inhibition of angiogenesis is achievable by targeting certain oncogenes [25], leading to reduced levels of VEGF expression, and such targeting strategies can be combined with metronomic chemotherapy. One additional example of such an approach could be the targeting of Her-2 (with the antibody trastuzumab), which was evaluated in Her-2 positive human breast cancer cells grown in SCID mice [26]. In this in vivo model, the combination of trastuzumab plus metronomic CTX was found to be effective at inhibiting the primary tumors, and when drug resistance to therapy emerged months later that was accompanied by overexpression of VEGF in the drug-resistant tumors. Interestingly, the trastuzumab plus CTX combination therapy also inhibited tumor growth in the metastatic setting in this model [26]. It is noteworthy that the trastuzumab and CTX monotherapies, which were effective on primary tumors, showed no or limited effect in the metastatic setting. These results also highlight the need to test metronomic regimens on preclinical metastatic disease models least the more simplistic primary tumor models exaggerate the relative impact of some therapies. It is conceivable that the microenvironment of a secondary site is a significant factor on how a tumor responds to metronomic chemotherapy. Thus, for example, while metronomic chemotherapy can sensitize tumors to angiogenesis inhibitors in the liver [27], it remains to be demonstrated that the same response would be observed to the same tumor growing in a different site such as the lungs. One notable example of this phenomenon is the work by Cruz-Munoz and colleagues [28] who showed that effective low-dose metronomic cyclophosphamide and vinblastine chemotherapy on metastatic melanoma resulted in the eventual emergence of rare brain metastases (suggesting that the metronomic vinblastine therapy was effective at halting the growth of metastases in secondary sites other than the brain). Another example is a tegafur/uracil (UFT) plus CTX metronomic therapy combination which did not inhibit primary tumor growth of breast cancers in SCID mice yet inhibited the growth of lung metastases in the same tumor model [29,30]. In that regard, metronomic chemotherapy, and brain metastasis [23,28,31]. A number of such studies have been performed by Dr. R.S. Kerbel and colleagues [26]. Overall, these studies serve as a cautionary note that is not a simple task to define the effectiveness of a metronomic regimen as a chemosensitizer, since the resulting anti-tumor effects of metronomic chemotherapy can vary significantly depending on the organ and microenvironment in which the cancer cells have grown. Thus, the effectiveness of therapeutic combinations may have to be tailored to the different secondary sites to which tumors may metastasize.

In the clinical setting, the most straightforward measurement of the antiangiogenic activity of metronomic chemotherapy is through indirect measurements of the levels of pro- and antiangiogenic cytokines in the plasma, including assessing levels of plasma VEGF and TSP-1 [2]. In that regard, Di Desidero et al. evaluated clinical activity, and the pharmacodynamic and pharmacokinetic profiles, of oral metronomic vinorelbine plus dexamethasone in metastatic castration-resistant prostate cancer patients. Forty-one patients (92% chemotherapy-resistant) received 30 mg/day vinorelbine p.o. thrice a week plus 1 mg/day dexamethasone p.o. until disease progression. Plasma VEGF, and TSP-1, were then measured by ELISA [32]. In a broader study, the plasma levels of 14 cytokines were evaluated in gastrointestinal cancer patients treated with metronomic UFT (a 5-fluorouracil prodrug; 100 mg/twice per day p.o.) and CTX (500 mg/m^2 i.v. bolus on day 1 and then 50 mg/day p.o.) plus celecoxib (200 mg/twice a day p.o.) [33].

Pharmacogenetics studies have also tried to correlate polymorphisms of pro-angiogenic genes such as VEGF with patient response to metronomic chemotherapy [34]. Thus, forty-three patients with advanced castration-resistant prostate cancer, treated with metronomic cyclophosphamide, celecoxib, and dexamethasone were evaluated. *VEGF-A* gene SNPs (−2578A/C, −634C/G, +936C/T) were assessed; patients harboring the −634CC genotype had a median PFS of 2.2 months whereas patients with the genotypes −634CG/GG had a median PFS of 6.25 months [34].

Tumor dormancy could be considered a hallmark of an angiogenic blockade. The tumor angiogenesis theory postulates that a tumor needs its own vasculature to grow beyond a few millimeters in diameter. Effective antiangiogenic therapy should therefore maintain a tumor to such a size, which would make it clinically not relevant in the majority of cases. Since the mode of action of metronomic chemotherapy is (among others) inhibition of angiogenesis, it has been proposed that metronomic chemotherapy could induce tumor dormancy [2]. This may best be tested in the adjuvant setting, and indeed there have been reports of the efficacy of metronomic chemotherapy in these patients [35,36].

The circumstantial evidence for a role of metronomic chemotherapy in tumor dormancy includes [2]: (1) metronomic chemotherapy decreases the expression of pro-angiogenic factors (e.g., VEGF) in preclinical tumor models and in clinical trials; (2) metronomic chemotherapy induces the expression of endogenous inhibitors of angiogenesis (e.g., TSP-1) in preclinical and clinical settings; (3) metronomic chemotherapy reduces the numbers of circulating endothelial progenitors (CEPs), thereby reducing the impact of CEPs in stimulating the growth of a viable tumor rim, as can be observed preclinically after exposure to high dose chemotherapy (Fig. 1).

While tumor dormancy relies in part on induction of tumor cell cycle arrest, another relevant factor is immunosurveillance, whereby the immune system can limit the growth of a mass of tumor cells. In that regard, metronomic CTX has been reported to activate the immune system [9], and that may contribute in some cases to tumor dormancy. Another important question is whether metronomic chemotherapy can act to maintain (rather than just induce) tumor dormancy, e.g., by providing a counterbalance to mechanisms that suppress TSP-1 expression. However, in spite of the above-mentioned concepts, further studies are needed to establish the role of metronomic chemotherapy in having an impact on tumor dormancy.

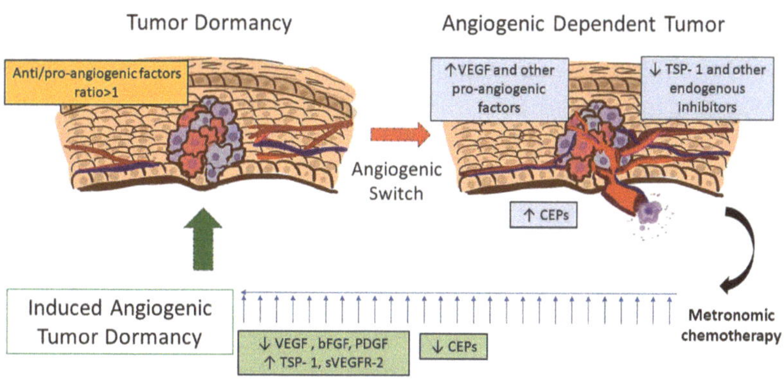

FIG. 1 The angiogenic switch causes the tumor dormancy escape. Metronomic chemotherapy, decreasing the vascular endothelial growth factor (VEGF), the basic fibroblast growth factor (bFGF), the platelet derived growth factor (PDGF), the circulating endothelial progenitors (CEPs) and upregulating thrombospondin-1 (TSP-1), the soluble VEGF receptor-2 (sVEGFR-2), reinduces the angiogenic tumor dormancy.

Combinations of metronomic chemotherapy with antiangiogenic drugs

Preclinical studies on the combination of metronomic chemotherapy and antiangiogenic drugs

The combination of metronomic chemotherapy and antiangiogenic drugs has mainly been evaluated in a preclinical setting. Combinations included predominantly the anti-VEGFR-2 monoclonal antibody DC101 [37] (the murine analog of the humanized monoclonal antibody ramucirumab [38]). Subsequent research focused on VEGFR-2 small tyrosine kinase inhibitors such as pazopanib [39], sorafenib [40], sunitinib, and semaxinib [41] partly because of their oral administration routes which were thought to facilitate the translation of preclinical results into clinical evaluation in the form of all-oral treatment with less costs for national health systems [42] and higher quality of life for oncologic patients.

Preclinical studies with anti-VEGFR-2 antibody DC101

The first combinations of metronomic chemotherapy with antiangiogenic drugs were evaluated by the group of Dr. R.S. Kerbel. They reported that orthotopic resistant human breast cancers in mice responded to metronomic chemotherapy regimens if combined with the anti-VEGFR-2 neutralizing antibody DC101 [43]. Subsequently, the same group tested the combination of metronomic cyclophosphamide, UFT, and doxorubicin with DC101 in a hepatocellular carcinoma (HCC) model [44]. The regimen including metronomic UFT plus cyclophosphamide associated with DC101 significantly improved the overall survival of mice with low toxicity [44]. The efficacy of the combination was confirmed also in the preclinical model of advanced human ovarian cancer in SCID mice [45].

Preclinical studies with VEGFR-2 tyrosine kinase inhibitors

Pazopanib

The administration of pazopanib in ovarian cancer has been explored in recent years [46], including its preclinical activity in combination with metronomic chemotherapy. Indeed, pazopanib in combination with metronomic topotecan showed significant antitumor and antiangiogenic effects in SKOV3ip1 and HeyA8 preclinical ovarian cancer models [47]. The combination therapy reduced tumor volumes by up to 85% vs. controls, increased tumor cell apoptosis 4-fold, and decreased both the tumor microvessel density and the number of pericytes in the tumor microenvironment [47]. However, as had been observed with MTD chemotherapy, this combination eventually allowed tumor acquired resistance in the same model, after more than 3months of successful therapy. The acquired resistance to the combination treatment was due to the refractoriness of ovarian cancer cells to topotecan [45]. Interestingly, this resistance was not reversed by switching to irinotecan, or to another antiangiogenic drug (e.g., DC101 or sunitinib) [45], suggesting the tumor had acquired some form of multidrug resistance. The same group investigated in the SKOV-3 human ovarian carcinoma model various low-dose metronomic chemotherapy regimens alone or in concomitant combination with pazopanib [48]. The tested chemotherapy drugs metronomically administered were oral cyclophosphamide, topotecan, irinotecan, or paclitaxel. Among them, the metronomic topotecan showed the higher antitumor activity significantly enhanced by the simultaneous administration of pazopanib [48].

The association between metronomic topotecan and pazopanib was also investigated in a triple-negative breast cancer orthotopic model [49], comparing the responses in the primary tumors and in the resulting metastases. Also, in this case, the combination of metronomic topotecan and pazopanib significantly enhanced the antitumor activity of the single treatments, prolonging the survival of mice. A marked decrease in tumor microvessel density was found as a consequence of the therapy, as well as induction of apoptosis in cancer cells. Of note, the combined regimen of metronomic topotecan with pazopanib inhibited the expression of the *ABCG2* gene in hypoxic cells, and this was associated with an increased intracellular concentration of the active form of topotecan [49]. Even more interestingly, the same combination showed very promising activity in renal cell carcinoma, a tumor type that is intrinsically chemotherapy-resistant. Moreover, the treatment of renal cell carcinoma with tyrosine kinase inhibitors (e.g., pazopanib) is also limited by both intrinsic and acquired resistance. Jedeszko and colleagues [50] used a renal cell carcinoma mouse model that can generate spontaneous metastasis, including to bone, lungs, and brain. The authors reported that the combination of metronomic topotecan plus pazopanib blocked primary tumor growth and had an antimetastatic effect that was not observed with the monotherapies. Also, in this case, the combination therapy resulted in an increased intracellular concentration of topotecan, sustaining the metastatic tumor cell dormancy (see also "Antiangiogenic activity of metronomic chemotherapy in preclinical and clinical setting" section). However, tumor progression was observed after the treatment was stopped [50].

Kumar and co-workers investigated the effect of oral metronomic topotecan and pazopanib on an SK-N-BE neuroblastoma mouse xenograft model [51]. The combination-treated animals survived for 80 days, and tumor growth was blocked for at least 50 days. As expected, the combination significantly decreased tumor microvessel density compared to the control group, with a higher pericyte coverage of the new blood vessels. Moreover, the combination treatment increased hypoxia and induced a metabolic switch toward elevated glycolysis [51]. Finally, daily oral metronomic topotecan in combination with pazopanib was tested in other pediatric solid tumor mouse models (i.e., osteosarcoma, and rhabdomyosarcoma) [52]. In vivo, the combination of topotecan plus pazopanib demonstrated a significant antitumor activity resulting in increased survival in experimental mice. The therapy produced a decrease in the levels of viable circulating endothelial cells and their progenitors, confirming the antiangiogenic activity of the combination schedule. Pharmacokinetic studies of both drugs did not reveal any drug-drug interaction [52].

Sorafenib

Sorafenib is an oral angiogenesis inhibitor that has been frequently tested in combination with oral metronomic chemotherapy to assess both anti-tumor and anti-endothelial activities. From that perspective, Murray and collaborators [53] showed the activity of the combination treatment of metronomic etoposide, paclitaxel, and temozolomide with sorafenib in in vitro experiments on microvascular endothelial cells and cancer cell lines, administering the drugs for 96 h. Sorafenib significantly augmented the anti-endothelial effect of low-dose chemotherapy and also produced an additive effect on melanoma cells [53].

The combination of sorafenib and metronomic chemotherapy was also investigated on preclinical models of HCC, a tumor for which the TKI is usually administered alone. Indeed, acquired resistance to sorafenib is a major clinical problem in the treatment of HCC. To delay such resistance, Tang and colleagues combined sorafenib with metronomic UFT, an oral 5-fluorouracil prodrug [27]. Interestingly, a metronomic regimen of daily UFT plus low-dose sorafenib (15 mg/kg per day) significantly prevented the onset of resistance, that appeared after 90 days vs. the 40 days of sorafenib monotherapy [27]. Moreover, another successful combination between metronomic fluoropyrimidines and sorafenib was tested on spontaneous HCC mouse models. The combination of metronomic S-1, an oral fluoropyrimidine [54], with sorafenib increased the antiangiogenic effects of the TKI, producing an enhanced tumor hypoxic environment as well as inducing tumor cell apoptosis [55]. Of note, recent data suggested that sorafenib maintained its antiangiogenic properties but did not block metronomic cyclophosphamide-induced innate immunity, or tumor regression [56], another important mechanism of action of metronomic chemotherapy.

Evaluation of other antiangiogenic tyrosine-kinase inhibitors

Among the antiangiogenic TKIs used in combination with metronomic chemotherapy are the following compounds: semaxinib, sunitinib, and vandetanib. Bocci and colleagues [3] showed in vitro that SN-38 (the active metabolite of irinotecan) preferentially inhibited endothelial cell proliferation when administered alone but augmented its activity synergistically if administered concomitantly with semaxinib. Moreover, these data were confirmed also in vivo where metronomic irinotecan plus semaxinib significantly inhibited tumor growth without producing any toxicity (i.e., weight loss) in the treated mice, and producing a decrease in the number of microvessels within the tumor tissues [3].

To investigate the effects of the combination of metronomic chemotherapy and sunitinib on metastatic progression, Tran Cao and co-workers used a metastatically competent model of pancreatic cancer and treated the mice with metronomic gemcitabine plus sunitinib [57]. Of note, the combination significantly prolonged the median overall survival of mice compared with controls. Moreover, the treatment with the metronomic regimen suppressed the metastatic process and this therapeutic effect was greater with the combination compared to the antiangiogenic drug alone [57].

As described above, metronomic fluoropyrimidines have been used in preclinical models of HCC alone and in combination with sorafenib. Iwamoto and collaborators [58] obtained promising results combining metronomic S-1 with the small molecule vandetanib, an EGFR and VEGFR-2 tyrosine kinase inhibitor. Metronomic S-1 significantly blocked tumor growth, and this inhibition was augmented by the combination with vandetanib. Interestingly, metronomic S-1 plus vandetanib enhanced the survival of the treated mice, decreasing the number of tumor microvessels and increasing the apoptosis in tumor cells. Notably, the combination regimen decreased the expression of VEGF in the neoplastic tissues, whereas thrombospondin-1 protein was upregulated [58].

Clinical studies on the combination of metronomic chemotherapy and antiangiogenic drugs

In the last ten years, several phase I and phase II clinical trials have combined metronomic chemotherapy and antiangiogenic drugs, in particular with bevacizumab (an anti-VEGF monoclonal antibody) or with tyrosine kinase inhibitors, targeting VEGFR-2 and other angiogenic factor receptors.

Whilst recent clinical research has focused on the combination of metronomic chemotherapy with antitumor checkpoint inhibitors, interesting data have been reported in a randomized phase III clinical trial, combining metronomic chemotherapy and pazopanib [17]. Moreover, there are numerous phase I-II clinical trials that overall delivered mixed results involving different types of cancer and different well-known antiangiogenic drugs (summarized in Table 1 the combinations including bevacizumab, in Table 2 the regimens including tyrosine kinase inhibitors).

TABLE 1 Clinical trials of the last ten years involving the combination of metronomic chemotherapy and bevacizumab.

Combination	Clinical phase	Number of patients	Type of cancer	Results	Reference
metroTMZ+ BEV+ octreotide	Phase I	15	Advanced neuroendocrine tumors	1 complete response, 8 partial responses, and 3 stable diseases	Koumarianou et al. [59]
metroCAPE + BEV+ octreotide	Phase II	45	Metastatic differentiated neuroendocrine tumors	8 partial responses	Berruti et al. [60]
metroCAPE+ metroCTX+ BEV+erlotinib	Phase II	24	HER2-negative metastatic breast cancer	Overall clinical benefit 75%, median time to progression 43 weeks	Montagna et al. [61]
weekly carboplatin and PTX+ BEV+ metroCTX	Phase II	34	Inflammatory breast cancer	10 complete responses	Palazzo et al. [62]
metroCAPE+ BEV	Phase III, randomized	558	Untreated metastatic colorectal cancer	Improved the survival until the second progression	Simken et al. [13]
metroCAPE+ metroCTX+ BEV	Phase II, randomized	232	Metastatic colorectal cancer	No significant improvement of PFS	Cremolini et al. [63]
metroCTX+ imatinib+BEV	Phase I	35	Metastatic colorectal cancer	No clinical responses	Kelley et al. [64]
metroPTX+ metroGEM+ BEV	Phase II	39	Advanced non-squamous non-small cell lung cancer	Objective response rate 56%, median PFS 8.5 months, median OS 25.5 months	Jones et al. [65]
metroCAPE+ BEV	Phase I	15	Relapsed peritoneal pseudomyxoma	20% of partial responses	Pietrantonio et al. [66]
metroTMZ+ BEV+ bortezomib	Phase I	9	Recurrent glioblastoma	Well tolerated	McCracken et al. [67]
metroTMZ+ vorinostat+ BEV	Phase I/II	9+39	Recurrent high-grade malignant gliomas	Well tolerated, no improved PFS vs. historical controls	Peters et al. [68]

metro, *metronomic;* BEV, *bevacizumab;* CAPE, *capecitabine;* CTX, *cyclophosphamide;* GEM, *gemcitabine;* PTX, *paclitaxel;* TMZ, *temozolomide.*

TABLE 2 Clinical trials of the last ten years involving the combination of metronomic chemotherapy and tyrosine kinase inhibitors.

Combination	Clinical phase	Number of patients	Type of cancer	Results	Reference
metroCTX+ SOR+ letrozole	Phase I	13	Estrogen receptor-positive breast cancer	6 complete responses	Bazzola et al. [69]
metroPTX+ SOR	Phase II	25	Mitotane-resistant adrenocortical carcinoma	Well tolerated but no clinical effects	Berruti et al. [70]
metroTMZ+ SOR	Phase II	43	Relapsed glioblastoma	5 partial responses, 18 stable diseases, median OS 7.4 months	Zustovich et al. [71]
Metro S-1+ SOR	Phase I	12	Advanced hepatocellular carcinoma	Response rate 16.7%	Suzuki et al. [55]
metroCTX+ metroMTX+ vandetanib	Phase I	20	Various solid tumors	10% partial response, 15% stable disease	Mayer et al. [72]
metroCTX+ pazopanib	Phase I	16	Platinum-resistant ovarian cancer	Median PFS 8.3 months, median OS 24.9 months	Dinkic et al. [73]
metroETO+ metroCTX+ pazopanib	Phase II, randomized	75	Platinum-resistant epithelial ovarian cancer	Significantly higher median PFS compared to metronomic chemotherapy alone (5.1 vs. 3.4 months)	Sharma et al. [17]
metroCTX+ nintedanib	Phase II, randomized	117	Advanced ovarian, fallopian tube or primary peritoneal cancer	Not improved outcomes by the combination	Hall et al. [18]

metro, *metronomic;* CTX, *cyclophosphamide;* ETO, *etoposide;* MTX, *methotrexate;* PTX, *paclitaxel;* SOR, *sorafenib;* TMZ, *temozolomide.*

Clinical trials with bevacizumab

The first antiangiogenic drug approved in oncology, bevacizumab [74], has also been tested in combination with metronomic chemotherapy. Neuroendocrine tumors are highly vascularized and express high levels of VEGF (i.e., the target of bevacizumab) together with its receptor. A phase I clinical trial enrolled 15 patients with advanced neuroendocrine tumors. The combination treatment consisted of oral temozolomide at a daily dose of 100 mg, bevacizumab 7.5 mg/m^2 every 3 weeks i.v. and 30 mg of octreotide (another antiangiogenic drug [75]) i.m. every 4 weeks. One patient had a radiological complete response, whereas the other eight patients showed a partial response, and three showed stable disease, suggesting that the combination schedule could have activity in advanced neuroendocrine tumors [59]. Another prospective phase II study tested bevacizumab (5mg/kg every 2weeks) plus octreotide (20mg monthly) and metronomic capecitabine (2000mg/daily) in 45 patients with metastatic differentiated neuroendocrine tumors. This regimen was well tolerated and active in patients, with a partial response observed in 8 patients [60].

The use of bevacizumab in combination with metronomic chemotherapy was evaluated in common cancers such as HER2-negative metastatic breast cancer. A phase II clinical trial enrolled 24 patients who were treated with metronomic oral capecitabine (500mg thrice/day), metronomic cyclophosphamide (50mg/day) plus bevacizumab (15mg/kg every 3weeks), and erlotinib (100mg/day). The overall clinical benefit of the schedule was 75% (of the total patients) and the median time to progression was 43weeks. Moreover, the researchers measured the levels of circulating endothelial progenitors at baseline, and demonstrated that a lower number of these cells correlated with improved progression-free survival [61]. Another therapeutic schedule including metronomic cyclophosphamide and bevacizumab was evaluated in a phase II study with 34 patients suffering from inflammatory breast cancer, who received weekly carboplatin and paclitaxel plus bevacizumab every 3weeks and oral metronomic cyclophosphamide for 6months. Thirty patients experienced an objective response and a pathologic complete remission was obtained in 10 patients without any marked toxicity [62].

A randomized phase III trial, named CAIRO3, recruited 558 patients with previously untreated metastatic colorectal cancer, who were treated with a maintenance therapeutic schedule of bevacizumab and low dose capecitabine. This combination significantly improved the survival until the second progression (11.7months vs. 8.5months of patients without maintenance therapy; $P < .0001$). The treatment was well tolerated [13].

Promising results were however not observed with metronomic chemotherapy in other colorectal cancer clinical trials. A phase II randomized study of maintenance treatment with bevacizumab plus metronomic chemotherapy (capecitabine 500mg/three times per day and cyclophosphamide 50mg/day) in metastatic colorectal cancer patients (i.e., MOMA trial) failed to significantly improve PFS of metastatic colorectal cancer patients [63]. Moreover, a phase I trial was conducted to determine the safety of a combined treatment including imatinib, bevacizumab, and metronomic cyclophosphamide in thirty-five patients with metastatic colorectal cancer. The combination was found to be safe and well tolerated but without any accompanying clinical response [64].

Metronomic chemotherapy of 4-week cycles of paclitaxel 80mg/m^2 and gemcitabine 300mg/m^2 weekly for three weeks, in combination with bevacizumab (10mg/kg every two weeks), in 39 patients with advanced non-squamous non-small cell lung cancer appeared to be an effective and tolerable treatment strategy. The objective response rate in this phase II clinical trial was 56% with a median PFS and overall survival (OS) of 8.5 and 25.5months, respectively. The patients with lower baseline values of angiopoietin-2 and IL-8 experienced significantly longer PFS and OS [65].

To investigate the safety of different combinations between metronomic chemotherapy and bevacizumab, various phase I clinical trials have been performed in less frequent cancer types such as peritoneal pseudomyxoma and glioblastomas. Fifteen patients with a relapsed peritoneal pseudomyxoma received the combination treatment with metronomic capecitabine (625 mg/m^2/day b.i.d.) and bevacizumab (7.5 mg/Kg three-weekly). The schedule was very well tolerated without any severe toxicity and partial responses were observed in 20% of cases [66]. Another phase I trial of dose-escalating metronomic temozolomide (25, 50, and 75 mg/m^2) plus bevacizumab and bortezomib was conducted in patients with recurrent glioblastoma. The combination treatment was well tolerated and the MTD of temozolomide was 75 mg/m^2 in combination with the antiangiogenic drug [67]. Patients affected by recurrent high-grade malignant gliomas were the subset of a phase I/II trial administering vorinostat (a histone deacetylase inhibitor [76]; 200 or 400 mg p.o.), bevacizumab (10 mg/kg iv every 2 weeks), and daily metronomic temozolomide (50 mg/m^2 p.o.). The combination therapeutic strategy was well tolerated and was found to be safe but the PFS was not statistically improved if compared to historical controls [68].

Clinical trials with tyrosine kinase inhibitors

Among tyrosine kinase inhibitors that impact signaling pathways that affect angiogenesis such as the VEGFR-2 pathway, the drug sorafenib [77] has been used in combination with metronomic chemotherapy for phase I and II clinical trials. Thirteen estrogen receptor-positive breast cancer patients received the combination of metronomic cyclophosphamide (50 mg/day), sorafenib (400 mg bid) and letrozole (a nonsteroidal aromatase inhibitor [76]; 2.5 mg/day) for 6 months. A clinical complete response was observed in 6 out of 13 patients. A significant decrease in tumor size was reported in all patients. Of note, also observed was a significant suppression of CD31 and VEGF-A expression after the treatment ($P = .01$ and $P = .007$, respectively), suggesting that the combination treatment was not only clinically active but also biologically effective as an antiangiogenic strategy [69]. Sorafenib combined with metronomic chemotherapy has been also investigated in a prospective phase II clinical study involving paclitaxel metronomically administered (60 mg/m^2 every week) and sorafenib (400 mg twice a day) as a last line of therapy in 25 patients affected by mitotane-resistant adrenocortical carcinoma. The treatment was well tolerated but the therapy failed to arrest tumor progression, leading to the interruption of the trial [70]. The combination of the same dose of sorafenib plus metronomic temozolomide (40 mg/m^2/day) was revealed to be feasible and safe in 43 patients with relapsed glioblastoma and showed a partial response in 5 patients and stable disease in 18 patients with a median overall survival of 7.4 months [71]. Recently, a prospective pilot study was performed, based on preclinical studies, to test sorafenib plus metronomic S-1 for 12 advanced HCC patients compared to historical cases. The objective response rate of the combination chemotherapy was higher than sorafenib alone (16.7% vs. 5.2%) but no differences in overall survival and time to progression were noticed [55].

Other anti-VEGFR-2 tyrosine kinase inhibitors have been evaluated in phase I clinical trials for their safety and, secondly, for clinical activity. Both vandetanib [78] and pazopanib [79] have been used to enhance the effects of metronomic chemotherapy. A phase I trial in cancer patients was performed using escalating doses of vandetanib (i.e., 100 mg, 200 mg, and 300 mg daily), metronomic cyclophosphamide (50 mg/day), and methotrexate (2.5 mg weekly). Of the 20 patients, 10% showed a partial response and 15% showed stable disease for more than 24 weeks [72]. The PACOVAR-trial was multicenter phase I clinical study enrolling 16 patients affected by platinum-resistant ovarian cancer. The trial evaluated the optimal dose, as well as the activity and tolerability of pazopanib (an escalating dose from 400 to 800 mg daily) combined with metronomic cyclophosphamide (50 mg/day). The median PFS and OS were 8.3 months and 24.9 months, respectively [73].

Conclusions

Over the last twenty years, metronomic chemotherapy has been demonstrated to have both preclinical and clinical activity in different cancer types. Recent successful randomized clinical trials have moved its use from small experimental settings to a wider clinical use and its inclusion in clinical guidelines such as the one of the Italian Association of Medical Oncology [80]. Combination studies have been a logical consequence of the preclinical and clinical development of metronomic chemotherapy to improve its therapeutic efficacy. In that regard, the addition of antiangiogenic drugs (e.g., DC101 or VEGFR-2 TKIs) into the metronomic schedules greatly enhanced the antitumor efficacy in almost all the preclinical models where they have been tested. Unfortunately, these promising results were not completely translated into clinical benefit in the clinical trials undertaken thus far. These difficulties were related to different factors: (i) the as yet controversial use of antiangiogenic drugs in the clinical setting, due to the lack of pharmacodynamic and pharmacogenetic biomarkers for the identification of responding patients; (ii) the low number of phase II and III randomized clinical trials on metronomic chemotherapy schedules which can be used as a strong platform for future combination studies; (iii) the introduction of immune checkpoint inhibitors (e.g., anti-CTLA4 strategies) into the oncology field that radically changed the perspective of combination studies, favoring the clinical research on the association between these drugs and metronomic chemotherapy, setting aside the antiangiogenic therapeutic approach. However, the possibility to have all-oral, low-cost combined therapies (many of the antiangiogenic drugs have expired patents) for neoplastic patients should be pursued because of preclinical and clinical data that in certain types of cancer are still very promising. Preclinical studies have revealed that the tumor microenvironment can be a significant factor in determining tumor response to metronomic chemotherapy combinations, particularly in the metastatic setting. Moreover, the most significant randomized clinical trial in the last year proved the efficacy of the combination of oral metronomic therapy with the antiangiogenic drug pazopanib for metastatic, platinum-resistant epithelial ovarian cancer [17]. Thus, additional phase III clinical trials will hopefully validate these results obtained by the combination of metronomic and antiangiogenic drugs and efforts should be made to characterize the biomarkers associated with the success of the therapy.

References

[1] Bocci G, Kerbel RS. Pharmacokinetics of metronomic chemotherapy: a neglected but crucial aspect. Nat Rev Clin Oncol 2016;13:659–73. https://doi.org/10.1038/NRCLINONC.2016.64.

[2] Natale G, Bocci G. Does metronomic chemotherapy induce tumor angiogenic dormancy? A review of available preclinical and clinical data. Cancer Lett 2018;432:28–37.

[3] Bocci G, Falcone A, Fioravanti A, Orlandi P, Di Paolo A, Fanelli G, Viacava P, Naccarato AG, Kerbel RS, Danesi R, Del Tacca M, Allegrini G. Antiangiogenic and anticolorectal cancer effects of metronomic irinotecan chemotherapy alone and in combination with semaxinib. Br J Cancer 2008;98:1619–29. https://doi.org/10.1038/sj.bjc.6604352.

[4] Bocci G, Nicolaou KC, Kerbel RS. Protracted low-dose effects on human endothelial cell proliferation and survival in vitro reveal a selective antiangiogenic window for various chemotherapeutic drugs. Cancer Res 2002;62(23):6938–43.

[5] Bocci G, Francia G, Man S, Lawler J, Kerbel RS. Thrombospondin 1, a mediator of the antiangiogenic effects of low-dose metronomic chemotherapy. Proc Natl Acad Sci U S A 2003;100. https://doi.org/10.1073/pnas.2135406100.

[6] Shaked Y, Bocci G, Munoz R, Man S, Ebos JML, Hicklin DJ, Bertolini F, D'Amato R, Kerbel RS. Cellular and molecular surrogate markers to monitor targeted and non-targeted antiangiogenic drug activity and determine optimal biologic dose. Curr Cancer Drug Targets 2005;5(7):551–9. https://doi.org/10.2174/156800905774574020.

[7] Orlandi P, Banchi M, Alì G, Di Desidero T, Fini E, Fontanini G, Bocci G. Active metronomic vinorelbine schedules decrease plasma interleukin-2 levels in mice with Lewis lung carcinoma. J Chemother 2020. https://doi.org/10.1080/1120009X.2020.1819069.

[8] Penel N, Adenis A, Bocci G. Cyclophosphamide-based metronomic chemotherapy: after 10 years of experience, where do we stand and where are we going? Crit Rev Oncol Hematol 2012;82. https://doi.org/10.1016/j.critrevonc.2011.04.009.

[9] Ghiringhelli F, Menard C, Puig PE, Ladoire S, Roux S, Martin F, Solary E, Le Cesne A, Zitvogel L, Chauffert B. Metronomic cyclophosphamide regimen selectively depletes CD4 +CD25+ regulatory T cells and restores T and NK effector functions in end stage cancer patients. Cancer Immunol Immunother 2007;56:641–8. https://doi.org/10.1007/s00262-006-0225-8.

[10] Lutsiak MEC, Semnani RT, De Pascalis R, Kashmiri SVS, Schlom J, Sabzevari H. Inhibition of CD4+25+ T regulatory cell function implicated in enhanced immune response by low-dose cyclophosphamide. Blood 2005;105:2862–8. https://doi.org/10.1182/blood-2004-06-2410.

[11] Orlandi P, Di Desidero T, Salvia G, Muscatello B, Francia G, Bocci G. Metronomic vinorelbine is directly active on non small cell lung cancer cells and sensitizes the $EGFR^{L858R/T790M}$ cells to reversible EGFR tyrosine kinase inhibitors. Biochem Pharmacol 2018;152. https://doi.org/10.1016/j.bcp.2018.04.011.

[12] Wichmann V, Eigeliene N, Saarenheimo J, Jekunen A. Recent clinical evidence on metronomic dosing in controlled clinical trials: a systematic literature review. Acta Oncol 2020;59:775–85. https://doi.org/10.1080/0284186X.2020.1744719.

[13] Simkens LH, van Tinteren H, May A, ten Tije AJ, Creemers GJ, Loosveld OJ, de Jongh FE, Erdkamp FL, Erjavec Z, van der Torren AM, Tol J, Braun HJ, Nieboer P, van der Hoeven JJ, Haasjes JG, Jansen RL, Wals J, Cats A, Derleyn VA, Honkoop AH, Mol L, Punt CJ, Koopman M. Maintenance treatment with capecitabine and bevacizumab in metastatic colorectal cancer (CAIRO3): a phase 3 randomised controlled trial of the Dutch Colorectal Cancer Group. Lancet 2015;385:1843–52. https://doi.org/10.1016/S0140-6736(14)62004-3.

[14] Chen Y-P, Liu X, Zhou Q, Yang K-Y, Jin F, Zhu X-D, Shi M, Hu G-Q, Hu W-H, Sun Y, Wu H-F, Wu H, Lin Q, Wang H, Tian Y, Zhang N, Wang X-C, Shen L-F, Liu Z-Z, Huang J, Luo X-L, Li L, Zang J, Mei Q, Zheng B-M, Yue D, Xu J, Wu S-G, Shi Y-X, Mao Y-P, Chen L, Li W-F, Zhou G-Q, Sun R, Guo R, Zhang Y, Xu C, Lv J-W, Guo Y, Feng H-X, Tang L-L, Xie F-Y, Sun Y, Ma J. Metronomic capecitabine as adjuvant therapy in locoregionally advanced nasopharyngeal carcinoma: a multicentre, open-label, parallel-group, randomised, controlled, phase 3 trial. Lancet 2021;398:303–13. https://doi.org/10.1016/S0140-6736(21)01123-5.

[15] Patil V, Noronha V, Dhumal SB, Joshi A, Menon N, Bhattacharjee A, Kulkarni S, Ankathi SK, Mahajan A, Sable N, Nawale K, Bhelekar A, Mukadam S, Chandrasekharan A, Das S, Vallathol D, D'Souza H, Kumar A, Agrawal A, Khaddar S, Rathnasamy N, Shenoy R, Kashyap L, Rai RK, Abraham G, Saha S, Majumdar S, Karuvandan N, Simha V, Babu V, Elamarthi P, Rajpurohit A, KAP K, Srikanth A, Ravind R, Banavali S, Prabhash K. Low-cost

oral metronomic chemotherapy versus intravenous cisplatin in patients with recurrent, metastatic, inoperable head and neck carcinoma: an open-label, parallel-group, non-inferiority, randomised, phase 3 trial. Lancet Glob Health 2020;8:e1213–22. https://doi.org/10.1016/S2214-109X(20)30275-8.

[16] Camerini A, Morabito A, Montanino A, Bernabé R, Grossi F, Ramlau R, Ciuleanu TE, Ceresoli GL, Pasello G, de Marinis F, Bosch-Barrera J, Laundreau P, Gautier S, Ta Thanh Minh C, Kowalski D. Metronomic oral vinorelbine in previously untreated advanced non-small-cell lung cancer patients unfit for platinum-based chemotherapy: results of the randomized phase II Tempo Lung trial. ESMO Open 2021;6:100051. https://doi.org/10.1016/J.ESMOOP.2021.100051.

[17] Sharma A, Singh M, Chauhan R, Malik PS, Khurana S, Mathur S, Kumar S, Sreenivas V, Kumar L. Pazopanib based oral metronomic therapy for platinum resistant/refractory epithelial ovarian cancer: a phase II, open label, randomized, controlled trial. Gynecol Oncol 2021. https://doi.org/10.1016/J.YGYNO.2021.05.025.

[18] Hall MR, Dehbi HM, Banerjee S, Lord R, Clamp A, Ledermann JA, Nicum S, Lilleywhite R, Bowen R, Michael A, Feeney A, Glasspool R, Hackshaw A, Rustin G. A phase II randomised, placebo-controlled trial of low dose (metronomic) cyclophosphamide and nintedanib (BIBF1120) in advanced ovarian, fallopian tube or primary peritoneal cancer. Gynecol Oncol 2020;159:692–8. https://doi.org/10.1016/J.YGYNO.2020.09.048.

[19] Bocci G, Fioravanti A, Orlandi P, di Desidero T, Natale G, Fanelli G, Viacava P, Naccarato AG, Francia G, Danesi R. Metronomic ceramide analogs inhibit angiogenesis in pancreatic cancer through up-regulation of caveolin-1 and thrombospondin-1 and down-regulation of cyclin D1. Neoplasia 2012;14:833–45. https://doi.org/10.1593/neo.12772.

[20] Bocci G, Danesi R, Del Tacca M, Kerbel RS. Selective anti-endothelial effects of protracted low-dose BAL-9504, a novel geranylgeranyl-transferase inhibitor. Eur J Pharmacol 2003;477. https://doi.org/10.1016/j.ejphar.2003.08.001.

[21] Fioravanti A, Canu B, Alì G, Orlandi P, Allegrini G, Di Desidero T, Emmenegger U, Fontanini G, Danesi R, Del Tacca M, Falcone A, Bocci G. Metronomic 5-fluorouracil, oxaliplatin and irinotecan in colorectal cancer. Eur J Pharmacol 2009;619:8–14. https://doi.org/10.1016/j.ejphar.2009.08.020.

[22] Wu H, Xin Y, Zhao J, Sun D, Li W, Hu Y, Wang S. Metronomic docetaxel chemotherapy inhibits angiogenesis and tumor growth in a gastric cancer model. Cancer Chemother Pharmacol 2011;68:879–87. https://doi.org/10.1007/s00280-011-1563-6.

[23] Francia G, Emmenegger U, Lee CR, Shaked Y, Folkins C, Mossoba M, Medin JA, Man S, Zhu Z, Witte L, Kerbel RS. Long-term progression and therapeutic response of visceral metastatic disease non-invasively monitored in mouse urine using β-human choriogonadotropin secreting tumor cell lines. Mol Cancer Ther 2008;7:3452–9. https://doi.org/10.1158/1535-7163.MCT-08-0200.

[24] Rak J, Mitsuhashi Y, Bayko L, Filmus J, Kerbel RS, Shirasawa S, Sasazuki T. Mutant ras oncogenes upregulate VEGF/VPF expression: implications for induction and inhibition of tumor angiogenesis. Cancer Res 1955;55:4575–80.

[25] Rak J, Filmus J, Finkenzeller G, Grugel S, Marmé D, Kerbel RS. Oncogenes as inducers of tumor angiogenesis. Cancer Metastasis Rev 1995;14:263–77. https://doi.org/10.1007/BF00690598.

[26] Kerbel RS, Guerin E, Francia G, Xu P, Lee CR, Ebos JML, Man S. Preclinical recapitulation of antiangiogenic drug clinical efficacies using models of early or late stage breast cancer metastatis. Breast 2013;22(2):S57–65. https://doi.org/10.1016/j.breast.2013.07.011.

[27] Tang TC, Man S, Xu P, Francia G, Hashimoto K, Emmenegger U, Kerbel RS. Development of a resistance-like phenotype to sorafenib by human hepatocellular carcinoma cells is reversible and can be delayed by metronomic UFT chemotherapy. Neoplasia 2010;12:928–40. https://doi.org/10.1593/neo.10804.

[28] Cruz-Munoz W, Man S, Xu P, Kerbel RS. Development of a preclinical model of spontaneous human melanoma central nervous system metastasis. Cancer Res 2008;68:4500–5. https://doi.org/10.1158/0008-5472.CAN-08-0041.

[29] Munoz R, Man S, Shaked Y, Lee CR, Wong J, Francia G, Kerbel RS. Highly efficacious nontoxic preclinical treatment for advanced metastatic breast cancer using combination oral UFT-cyclophosphamide metronomic chemotherapy. Cancer Res 2006;66:3386–91. https://doi.org/10.1158/0008-5472.CAN-05-4411.

[30] Muñoz R, Hileeto D, Cruz-Muñoz W, Wood GA, Xu P, Man S, Viloria-Petit A, Kerbel RS. Suppressive impact of metronomic chemotherapy using UFT and/or cyclophosphamide on mediators of breast cancer dissemination and invasion. PLoS One 2019;14. https://doi.org/10.1371/journal.pone.0222580.

[31] Francia G, Cruz-Munoz W, Man S, Xu P, Kerbel RS. Mouse models of advanced spontaneous metastasis for experimental therapeutics. Nat Rev Cancer 2011;11:135–41.

[32] Di Desidero T, Derosa L, Galli L, Orlandi P, Fontana A, Fioravanti A, Marconcini R, Giorgi M, Campi B, Saba A, Lucchesi S, Felipetto R, Danesi R, Francia G, Allegrini G, Falcone A, Bocci G. Clinical, pharmacodynamic and pharmacokinetic results of a prospective phase II study on oral metronomic vinorelbine and dexamethasone in castration-resistant prostate cancer patients. Investig New Drugs 2016;34. https://doi.org/10.1007/s10637-016-0385-0.
[33] Valenzuela P, Oaxaca D, Di Desidero T, Parra K, Rodriguez G, Manciu M, Allegrini G, Falcone A, Bocci G, Kirken RA, Francia G. Pharmacodynamic biomarkers in metronomic chemotherapy: multiplex cytokine measurements in gastrointestinal cancer patients. Clin Exp Med 2021;21:149–59. https://doi.org/10.1007/s10238-020-00666-9.
[34] Orlandi P, Fontana A, Fioravanti A, Di Desidero T, Galli L, Derosa L, Canu B, Marconcini R, Biasco E, Solini A, Francia G, Danesi R, Falcone A, Bocci G. VEGF-A polymorphisms predict progression-free survival among advanced castration-resistant prostate cancer patients treated with metronomic cyclophosphamide. Br J Cancer 2013;109. https://doi.org/10.1038/bjc.2013.398.
[35] Kerbel RS, Andre N. Adjuvant metronomic chemotherapy for locoregionally advanced nasopharyngeal carcinoma. Lancet 2011;398:278–9.
[36] Romero D. Benefit with adjuvant metronomic capecitabine in NPC. Nat Rev Clin Oncol 2011;18. https://doi.org/10.1038/s41571-021-00536-7.
[37] Prewett M, Huber J, Li Y, Santiago A, O'Connor W, King K, Overholser J, Hooper A, Pytowski B, Witte L, Bohlen P, Hicklin DJ. Antivascular endothelial growth factor receptor (fetal liver kinase 1) monoclonal antibody inhibits tumor angiogenesis and growth of several mouse and human tumors. Cancer Res 1999;59:5209–18.
[38] Syed YY. Ramucirumab: a review in hepatocellular carcinoma. Drugs 2020;80:315–22.
[39] Chellappan DK, Chellian J, Ng ZY, Sim YJ, Theng CW, Ling J, Wong M, Foo JH, Yang GJ, Hang LY, Nathan S, Singh Y, Gupta G. The role of pazopanib on tumour angiogenesis and in the management of cancers: a review. Biomed Pharmacother 2017;96:768–81.
[40] Fan G, Wei X, Xu X. Is the era of sorafenib over? A review of the literature. Ther Adv Med Oncol 2020;12. https://doi.org/10.1177/1758835920927602, 1758835920927602.
[41] Hartmann J, Haap M, Kopp H-G, Lipp H-P. Tyrosine kinase inhibitors – a review on pharmacology, metabolism and side effects. Curr Drug Metab 2019;10:470–81. https://doi.org/10.2174/138920009788897975.
[42] Bocci G, Tuccori M, Emmenegger U, Liguori V, Falcone A, Kerbel RS, Del Tacca M. Cyclophosphamide-methotrexate "metronomic" chemotherapy for the palliative treatment of metastatic breast cancer. A comparative pharmacoeconomic evaluation. Ann Oncol 2005;16:1243–52. https://doi.org/10.1093/annonc/mdi240.
[43] Klement G, Huang P, Mayer B, Green SK, Man S, Kerbel RS, Bohlen P, Hicklin D. Differences in therapeutic indexes of combination metronomic chemotherapy and an anti-VEGFR-2 antibody in multidrug-resistant human breast cancer xenografts. Clin Cancer Res 2002;8:221–32.
[44] Tang TC, Man S, Lee CR, Xu P, Kerbel RS. Impact of metronomic UFT/cyclophosphamide chemotherapy and antiangiogenic drug assessed in a new preclinical model of locally advanced orthotopic hepatocellular carcinoma. Neoplasia 2010;12:264–74. https://doi.org/10.1593/neo.91872.
[45] Cruz-Muñoz W, Di Desidero T, Man S, Xu P, Jaramillo ML, Hashimoto K, Collins C, Banville M, O'Connor-Mccourt MD, Kerbel RS. Analysis of acquired resistance to metronomic oral topotecan chemotherapy plus pazopanib after prolonged preclinical potent responsiveness in advanced ovarian cancer. Angiogenesis 2014;17:661–73. https://doi.org/10.1007/s10456-014-9422-9.
[46] McLachlan J, Banerjee S. Pazopanib in ovarian cancer. Expert Rev Anticancer Ther 2015;15:995–1005. https://doi.org/10.1586/14737140.2015.1081383.
[47] Merritt WM, Nick AM, Carroll AR, Lu C, Matsuo K, Dumble M, Jennings N, Zhang S, Lin YG, Spannuth WA, Kamat AA, Stone RL, Shahzad MMK, Coleman RL, Kumar R, Sood AK. Bridging the gap between cytotoxic and biologic therapy with metronomic topotecan and pazopanib in ovarian cancer. Mol Cancer Ther 2010;9:985–95. https://doi.org/10.1158/1535-7163.MCT-09-0967.
[48] Hashimoto K, Man S, Xu P, Cruz-Munoz W, Tang T, Kumar R, Kerbel RS. Potent preclinical impact of metronomic low-dose oral topotecan combined with the antiangiogenic drug pazopanib for the treatment of ovarian cancer. Mol Cancer Ther 2010;9:996–1006. https://doi.org/10.1158/1535-7163.MCT-09-0960.
[49] Di Desidero T, Xu P, Man S, Bocci G, Kerbel RS. Potent efficacy of metronomic topotecan and pazopanib combination therapy in preclinical models of primary or late stage metastatic triple-negative breast cancer. Oncotarget 2015;6:42396–410. https://doi.org/10.18632/oncotarget.6377.
[50] Jedeszko C, Paez-Ribes M, Di Desidero T, Man S, Lee CR, Xu P, Bjarnason GA, Bocci G, Kerbel RS. Postsurgical adjuvant or metastatic renal cell carcinoma therapy models reveal potent antitumor activity of metronomic oral topotecan with pazopanib. Sci Transl Med 2015;7. https://doi.org/10.1126/scitranslmed.3010722.

[51] Kumar S, Mokhtari RB, Oliveira ID, Islam S, Toledo SRC, Yeger H, Baruchel S. Tumor dynamics in response to antiangiogenic therapy with oral metronomic topotecan and pazopanib in neuroblastoma xenografts. Transl Oncol 2013;6:493–503. https://doi.org/10.1593/tlo.13286.

[52] Kumar S, Mokhtari RB, Sheikh R, Wu B, Zhang L, Xu P, Man S, Oliveira ID, Yeger H, Kerbel RS, Baruchel S. Metronomic oral topotecan with pazopanib is an active antiangiogenic regimen in mouse models of aggressive pediatric solid tumor. Clin Cancer Res 2011;17:5656–67. https://doi.org/10.1158/1078-0432.CCR-11-0078.

[53] Murray A, Little SJ, Stanley P, Maraveyas A, Cawkwell L. Sorafenib enhances the in vitro anti-endothelial effects of low dose (metronomic) chemotherapy. Oncol Rep 2010;24:1049–58. https://doi.org/10.3892/or.2010.1049.

[54] Sanford M. S-1 (Teysuno®): a review of its use in advanced gastric cancer in non-Asian populations. Drugs 2013;73:845–55.

[55] Suzuki H, Iwamoto H, Nakano M, Nakamura T, Masuda A, Sakaue T, Tanaka T, Nakano D, Kuromatsu R, Niizeki T, Okamura S, Shimose S, Shirono T, Noda Y, Kamachi N, Yano H, Kawaguchi A, Koga H, Torimura T. Efficacy and tolerability of Sorafenib plus metronomic chemotherapy S-1 for advanced hepatocellular carcinoma in preclinical and clinical assessments. Transl Oncol 2021;14:101201. https://doi.org/10.1016/j.tranon.2021.101201.

[56] Doloff JC, Chen CS, Waxman DJ. Anti-tumor innate immunity activated by intermittent metronomic cyclophosphamide treatment of 9L brain tumor xenografts is preserved by anti-angiogenic drugs that spare VEGF receptor 2. Mol Cancer 2014;13. https://doi.org/10.1186/1476-4598-13-158.

[57] Tran Cao HS, Bouvet M, Kaushal S, Keleman A, Romney E, Kim G, Fruehauf J, Imagawa DK, Hoffman RM, Katz MHG. Metronomic gemcitabine in combination with sunitinib inhibits multisite metastasis and increases survival in an orthotopic model of pancreatic cancer. Mol Cancer Ther 2010;9:2068–78. https://doi.org/10.1158/1535-7163.MCT-10-0201.

[58] Iwamoto H, Torimura T, Nakamura T, Hashimoto O, Inoue K, Kurogi J, Niizeki T, Kuwahara R, Abe M, Koga H, Yano H, Kerbel RS, Ueno T, Sata M. Metronomic S-1 chemotherapy and vandetanib: an efficacious and nontoxic treatment for hepatocellular carcinoma. Neoplasia 2011;13:187–97. https://doi.org/10.1593/neo.101186.

[59] Koumarianou A, Antoniou S, Kanakis G, Economopoulos N, Rontogianni D, Ntavatzikos A, Tsavaris N, Pectasides D, Dimitriadis G, Kaltsas G. Combination treatment with metronomic temozolomide, bevacizumab and long-acting octreotide for malignant neuroendocrine tumours. Endocr Relat Cancer 2012;19(1):L1–4. https://doi.org/10.1530/ERC-11-0287.

[60] Berruti A, Fazio N, Ferrero A, Brizzi MP, Volante M, Nobili E, Tozzi L, Bodei L, Torta M, D'Avolio A, Priola AM, Birocco N, Amoroso V, Biasco G, Papotti M, Dogliotti L. Bevacizumab plus octreotide and metronomic capecitabine in patients with metastatic well-to-moderately differentiated neuroendocrine tumors: the xelbevoct study. BMC Cancer 2014;14:184. https://doi.org/10.1186/1471-2407-14-184.

[61] Montagna E, Cancello G, Bagnardi V, Pastrello D, Dellapasqua S, Perri G, Viale G, Veronesi P, Luini IM, Calleri A, Rampinelli C, Goldhirsc A, Bertolini F, Colleoni M. Metronomic chemotherapy combined with bevacizumab and erlotinib in patients with metastatic HER2-negative breast cancer: clinical and biological activity. Clin Breast Cancer 2012;12:207–14. https://doi.org/10.1016/J.CLBC.2012.03.008.

[62] Palazzo A, Dellapasqua S, Munzone E, Bagnardi V, Mazza M, Cancello G, Ghisini R, Iorfida M, Montagna E, Goldhirsch A, Colleoni M. Phase II trial of bevacizumab plus weekly paclitaxel, carboplatin, and metronomic cyclophosphamide with or without trastuzumab and endocrine therapy as preoperative treatment of inflammatory breast cancer. Clin Breast Cancer 2018;18:328–35. https://doi.org/10.1016/j.clbc.2018.01.010.

[63] Cremolini C, Marmorino F, Bergamo F, Aprile G, Salvatore L, Masi G, Dell'Aquila E, Antoniotti C, Murgioni S, Allegrini G, Borelli B, Gemma D, Casagrande M, Granetto C, Delfanti S, Di Donato S, Schirripa M, Sensi E, Tonini G, Lonardi S, Fontanini G, Boni L, Falcone A. Phase II randomised study of maintenance treatment with bevacizumab or bevacizumab plus metronomic chemotherapy after first-line induction with FOLFOXIRI plus Bevacizumab for metastatic colorectal cancer patients: the MOMA trial. Eur J Cancer 2019;109:175–82. https://doi.org/10.1016/j.ejca.2018.12.028.

[64] Kelley R, Hwang J, Magbanua M, Watt L, Beumer J, Christner S, Baruchel S, Wu B, Fong L, Yeh B, Moore A, Ko A, Korn W, Rajpal S, Park J, Tempero M, Venook A, Bergsland E. A phase 1 trial of imatinib, bevacizumab, and metronomic cyclophosphamide in advanced colorectal cancer. Br J Cancer 2013;109:1725–34. https://doi.org/10.1038/BJC.2013.553.

[65] Jones BS, Jerome MS, Miley D, Jackson BE, DeShazo MR, Reddy VVB, Singh KP, Brown OC, Robert F. Pilot phase II study of metronomic chemotherapy in combination with bevacizumab in patients with advanced non-squamous non-small cell lung cancer. Lung Cancer 2017;106:125–30. https://doi.org/10.1016/j.lungcan.2017.02.004.

[66] Pietrantonio F, Berenato R, Maggi C, Caporale M, Milione M, Perrone F, Tamborini E, Baratti D, Kusamura S, Mariani L, Niger M, Mennitto A, Gloghini A, Bossi I, Settanni G, Busico A, Bagnoli PF, Di Bartolomeo M, Deraco M, de Braud F. GNAS mutations as prognostic biomarker in patients with relapsed peritoneal pseudomyxoma receiving metronomic capecitabine and bevacizumab: a clinical and translational study. J Transl Med 2016;14:125. https://doi.org/10.1186/s12967-016-0877-x.

[67] McCracken DJ, Celano EC, Voloschin AD, Read WL, Olson JJ. Phase I trial of dose-escalating metronomic temozolomide plus bevacizumab and bortezomib for patients with recurrent glioblastoma. J Neuro-Oncol 2016;130:193–201. https://doi.org/10.1007/s11060-016-2234-6.

[68] Peters KB, Lipp ES, Miller E, Herndon JE, McSherry F, Desjardins A, Reardon DA, Friedman HS. Phase I/II trial of vorinostat, bevacizumab, and daily temozolomide for recurrent malignant gliomas. J Neuro-Oncol 2018;137:349–56. https://doi.org/10.1007/s11060-017-2724-1.

[69] Bazzola L, Foroni C, Andreis D, Zanoni V, Cappelletti MR, Allevi G, Aguggini S, Strina C, Milani M, Venturini S, Ferrozzi F, Giardini R, Bertoni R, Turley H, Gatter K, Petronini PG, Fox SB, Harris AL, Martinotti M, Berruti A, Bottini A, Reynolds AR, Generali D. Combination of letrozole, metronomic cyclophosphamide and sorafenib is well-tolerated and shows activity in patients with primary breast cancer. Br J Cancer 2015;112:52–60. https://doi.org/10.1038/bjc.2014.563.

[70] Berruti A, Sperone P, Ferrero A, Germano A, Ardito A, Priola AM, De Francia S, Volante M, Daffara F, Generali D, Leboulleux S, Perotti P, Baudin E, Papotti M, Terzolo M. Phase II study of weekly paclitaxel and sorafenib as second/third-line therapy in patients with adrenocortical carcinoma. Eur J Endocrinol 2012;166:451–8. https://doi.org/10.1530/EJE-11-0918.

[71] Zustovich F, Landi L, Lombardi G, Porta C, Galli L, Fontana A, Amoroso D, Galli C, Andreuccetti M, Falcone A, Zagonel V. Sorafenib plus daily low-dose temozolomide for relapsed glioblastoma: a phase ii study. Anticancer Res 2013;33:3487–94. https://doi.org/10.1200/jco.2011.29.15_suppl.2080.

[72] Mayer EL, Isakoff SJ, Klement G, Downing SR, Chen WY, Hannagan K, Gelman R, Winer EP, Burstein HJ. Combination antiangiogenic therapy in advanced breast cancer: a phase 1 trial of vandetanib, a VEGFR inhibitor, and metronomic chemotherapy, with correlative platelet proteomics. Breast Cancer Res Treat 2012;136:169–78. https://doi.org/10.1007/s10549-012-2256-5.

[73] Dinkic C, Eichbaum M, Schmidt M, Grischke EM, Gebauer G, Fricke HC, Lenz F, Wallwiener M, Marme F, Schneeweiss A, Sohn C, Rom J. Pazopanib (GW786034) and cyclophosphamide in patients with platinum-resistant, recurrent, pre-treated ovarian cancer – results of the PACOVAR-trial. Gynecol Oncol 2017;146: 279–84. https://doi.org/10.1016/j.ygyno.2017.05.013.

[74] Salvatore L, Bria E, Sperduti I, Hinke A, Hegewisch-Becker S, Aparicio T, Le Malicot K, Boige V, Koeberle D, Baertschi D, Dietrich D, Tortora G, Arnold D. Bevacizumab as maintenance therapy in patients with metastatic colorectal cancer: a meta-analysis of individual patients' data from 3 phase III studies. Cancer Treat Rev 2021;97:102202. https://doi.org/10.1016/j.ctrv.2021.102202.

[75] Danesi R, Agen C, Benelli U, Di Paolo A, Nardini D, Bocci G, Basolo F, Campagni A, Del Tacca M. Inhibition of experimental angiogenesis by the somatostatin analogue octreotide acetate (SMS 201-995). Clin Cancer Res 1997;3:265–72.

[76] Eckschlager T, Plch J, Stiborova M, Hrabeta J. Histone deacetylase inhibitors as anticancer drugs. Int J Mol Sci 2017;18(7):1414. https://doi.org/10.3390/ijms18071414.

[77] Liu L, Cao Y, Chen C, Zhang X, McNabola A, Wilkie D, Wilhelm S, Lynch M, Carter C. Sorafenib blocks the RAF/MEK/ERK pathway, inhibits tumor angiogenesis, and induces tumor cell apoptosis in hepatocellular carcinoma model PLC/PRF/5. Cancer Res 2006;66:11851–8. https://doi.org/10.1158/0008-5472.can-06-1377.

[78] Vozniak JM, Jacobs JM. Vandetanib. J Adv Pract Oncol 2012;3:112–16

[79] Miyamoto S, Kakutani S, Sato Y, Hanashi A, Kinoshita Y, Ishikawa A. Drug review: pazopanib. Jpn J Clin Oncol 2018;48:503–13.

[80] AIOM. Linee guida Neoplasia della. mammella ed; 2020. https://www.aiom.it/wp-content/uploads/2020/10/2020_LG_AIOM_Neoplasie_Mammella_16022021.pdf. [Accessed 20 May 2021].

CHAPTER

9

Resistance to antiangiogenic treatments: A review

Tomas Koltai[a], *Stephan Joel Reshkin*[b], *and Rosa Angela Cardone*[b]

[a]Hospital del Centro Gallego de Buenos Aires, Buenos Aires, Argentina [b]Department of Biosciences, Biotechnology and Biopharmaceutics, University of Bari, Bari, Italy

Abstract

Angiogenesis (blood vessel formation) is essential for tissue growth in both normal development and physiology and in some diseases such as inflammation and cancer. Angiogenesis is a hallmark of cancer, however, it took many years to establish its importance. Ever since Judah Folkman's seminal publications in 1971, that clearly showed cancer angiogenesis-dependence, researchers have been investigating the mechanisms of angiogenesis and how to block them. This search blossomed with the finding of inhibitors targeting the vascular endothelial growth factor (VEGF) signaling pathways. These new molecules and monoclonal antibodies showed therapeutic efficacy in both the laboratory and human clinical settings and hopes rose. Unfortunately, the benefits did not reach all the patients and they were short-lived: sooner or later tumors resumed their growth and proliferation and became refractory to further antiangiogenic treatments. Worse, antiangiogenic treatments seemed to increase metastatic risk. The development of treatment resistance is still one of the main causes of failure in cancer therapy. Antiangiogenic treatments are no exception and a deeper knowledge of the mechanisms of resistance is necessary if we intend to delay or eliminate them.

Two different mechanisms have been identified: primary or evasive resistance and secondary or adaptive resistance.

The existence of these two mechanisms led to the non-mainstream conclusion, now shared by many authors, that there are at least two different angiogenic pathways: one is the canonical VEGF- VEGF receptor (VEGFR) axis and others, which are independent of this axis and not fully known. Primary resistance works exclusively through these independent pathways, while secondary resistance, which initially is VEGF-VEGFR-dependent, switches to the other pathways becoming non-responsive to classical antiangiogenic treatments. For the time being, the clear identification of these other pathways belongs to the realm of hypothesis. However, there is enough experimental evidence supporting their existence. We will discuss this evidence as a central issue in antiangiogenic treatment resistance. Some non-conventional pharmacologic strategies against resistance will also be considered.

https://doi.org/10.1016/B978-0-323-90190-1.00014-7

Abbreviations

ABL1	tyrosine-protein kinase ABL1
AXL1	axin-like protein 1
BA	betulinic acid
CYR61	cysteine-rich angiogenic inducer 61
Dll4	delta-like ligand 4
EC	endothelial cell
ECM	extracellular matrix
EGF	epidermal growth factor
EGFR	epidermal growth factor receptor
ET	endothelin
ETR	endothelin receptor
FABP	fatty acid binding protein
FF	fenofibrate
FGF	fibroblast growth factor
GAS6	growth arrest-specific 6
HIF-1α	hypoxia inducible factor 1 alpha
HIF-1β	hypoxia-inducible factor 1 beta
HRE	hypoxia responsive element
ICIs	immune checkpoint inhibitors
IL-1α	interleukin 1 alpha
IL-1β	interleukin 1 beta
IL-6	interleukin 6
IL-8	interleukin 8
MCP1	monocyte chemoattractant protein 1
MCT	metronomic chemotherapy
MCT1	monocarboxylate transporter 1
MDR	multidrug resistance
mGluR1	metabotropic glutamate receptor 1
MMP	matrix metalloproteases
NPL	neuropilin
PAI-1	plasminogen activator inhibitor 1
PDAC	pancreatic ductal adenocarcinoma
PDGF	platelet-derived growth factor
PK1-PK2	prokineticin 1 and 2
SDF1	stromal cell-derived factor 1
STAT3	signal transducer and activator of transcription 3
TAECs	tumor-associated endothelial cells
TNFα	tumor necrosis factor-alpha
TSP 1	thrombospondin 1
VDA	vascular disrupting agents
VE-cadherin	vascular endothelial cadherin
VEGF	vascular endothelial growth factor
VEGFR	vascular endothelial growth factor receptor
VHL	Von Hippel Lindau protein
VM	vascular mimicry

Conflict of interest

No potential conflicts of interest were disclosed.

Introduction

Oxygen and nutrients are essential elements for all complex aerobic multicellular organisms. Both are delivered to the cells via the circulatory system. The accelerated proliferation and growth of tumors render the usual circulatory system inefficient. Consequently, hypoxia develops and sets into motion a complete program of events coordinated by the transcription factor hypoxia-inducible factor 1 alpha (HIF-1α), which is released from the Von Hippel Lindau protein (VHL)) allowing HIF-1α to migrate to the nucleus where it dimerizes with hypoxia-inducible factor 1 beta (HIF-1β) and, from then on, acts as an inducer of genes harboring the hypoxia-responsive element (HRE) sequence in their promoter region. This sets in motion a genetic expression machinery that adapts the cells to unfavorable environmental conditions leading to metabolic changes that decrease oxygen consumption and stimulate the formation of new vessels [1–4]. Recently reviewed by Lee et al. [5] (Fig. 1).

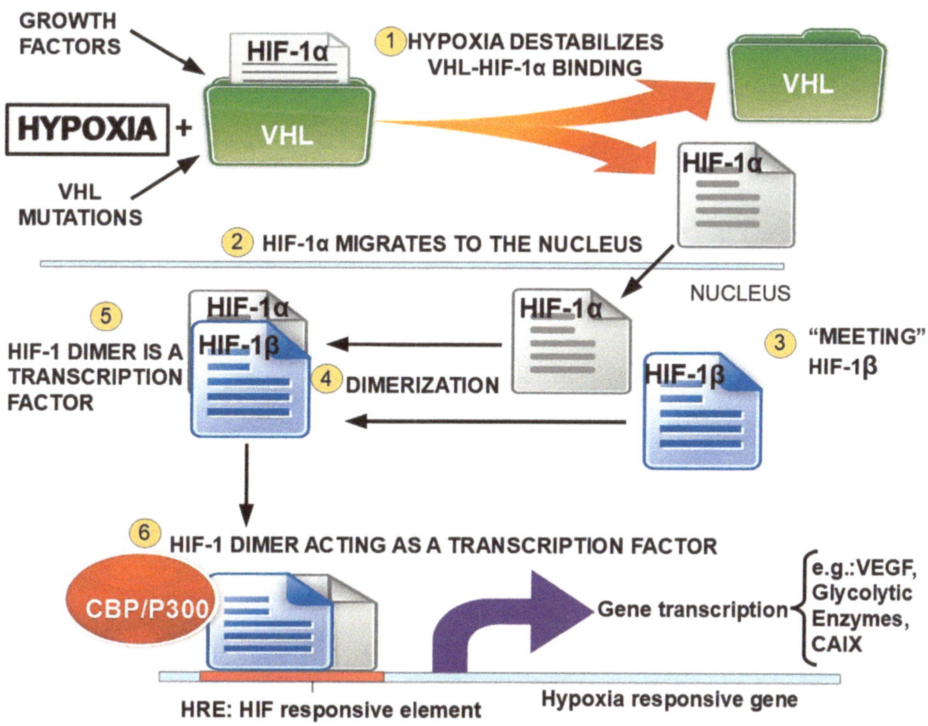

FIG. 1 Hypoxia unleashes a mechanism that leads to the transcription of hypoxia-inducible genes allowing cells to adapt to this situation. There are more than 150 genes that are HIF-1α inducible as they contain a hypoxia-responsive element (HRE) sequence in their promoter region. Many genes are involved in hypoxic adaptation including VEGF and VEGFR, nine of the ten glycolytic enzymes, and carbonic anhydrase IX (CAIX). While hypoxia is the main trigger and driver of the mechanism, other factors such as growth factors and genetic modifications (mutation, deletion) of the Von Hippel Lindau gene, can also produce HIF activation. This last situation produces a hypoxia-independent stabilization of HIF-1α allowing its migration to the nucleus where it binds the constitutional HIF-1β. The heterodimer thus formed acts as a transcription factor. *This diagram is based on references [6–22].*

Basic concepts in angiogenesis and antiangiogenesis

Angiogenesis [23] is the formation of new vessels from pre-existing capillaries [24] which requires certain specific and unspecific regulatory pathways.

The best known and principal angiogenic axis is vascular endothelial growth factor (VEGF) isoforms interacting with their receptors VEGFRs. Although imperfect and inefficient, these new vessels are essential for tumor progression.

The term angiogenesis was used for the first time in the eighteenth century by the English physician and anatomist John Hunter [25], but research on tumors was started in the 1960s by Judah Folkman. Since then, targeting tumor new vascularization has been considered one of the most promising areas of investigation. After Folkman's seminal publications in 1971 [26,27], it took thirty years for the first batch of antiangiogenic drugs to become available for clinical use. Importantly, they were successful, prolonging overall survival in many patients. However, against all expectations [28], something unpleasant happened: resistance to antiangiogenic drugs developed, seriously limiting the benefits of the treatment.

In theory, resistance was not within possibilities because antiangiogenesis acts in a completely different manner than classical chemotherapeutic drugs. Therefore, the resistance mechanisms to these chemotherapeutics were not supposed to act against antiangiogenic drugs. This is partially true. It was believed that antiangiogenic treatments represented the "resistance to the resistance" of classical cytotoxic chemotherapy [29]. However, in practice, resistance by other unexpected mechanisms was not only possible but sooner or later it appeared in almost all the patients, thus rendering the benefits of antiangiogenic therapies short-lived.

Early in their development, tumors undergo an angiogenic switch, which is the result of a change in the balance between pro-and antiangiogenic factors [30]. The main "pro" factor is represented by the VEGF-VEGFR axis, while angiostatin [31] and thrombospondin are considered by many to be the main "anti" factors [32] (Fig. 2).

Initially, the pro- and antiangiogenic systems seemed to be simple as shown in Fig. 2 in that VEGF was the pro-angiogenic protein, and thrombospondin and angiostatin were the endogenous antiangiogenic counterparts. Many pro-angiogenic and antiangiogenic factors have been identified since then. While pro- and antiangiogenic factors are balanced in health, they become strongly unbalanced in tumors where the scale tilts in favor of pro-angiogenesis.

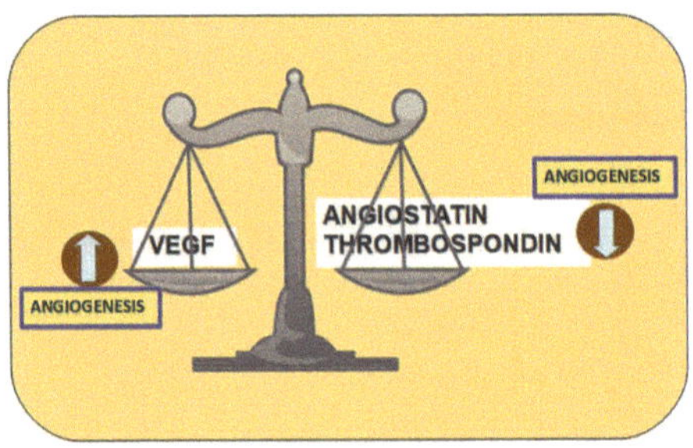

FIG. 2 The balance between the two sides of the scale determines angiogenesis or antiangiogenesis. In the tumoral angiogenic switch, VEGF is increased, while thrombospondin and angiostatin are down-regulated.

The angiogenic switch: Angiogenesis is one of the hallmarks of cancer [33,34] and the pro-angiogenic shift is an early event in a tumor life [35] and is indispensable for tumor progression and metastasis. From the beginning of the twenty-first century with the development of specific antiangiogenic treatments such as bevacizumab, sunitinib, sorafenib, and others, targeting angiogenesis has become very popular [36]. However, antiangiogenic treatments have not accomplished all that was expected. These treatments are not stand-alone therapies, and they must be used in combination with other powerful chemotherapeutic drugs. And even when used as part of a chemotherapy scheme, the results are sometimes poor over the short or medium term [37–40], while in other studies there is an improvement in overall survival and progression-free survival with antiangiogenic treatments [41–43]. Therefore, there are sound reasons to keep on using antiangiogenic treatments and to continue the search for better and less toxic drugs.

The additional cost of these treatments with, e.g., bevacizumab is around $204,000 (US dollars as of 2010) [44]. However, the benefits obtained with this high cost are limited.

Several decades ago, it was postulated that since tumors are angiogenesis-dependent, inhibiting this process would be a reasonable means to combat cancer. Since then, numerous studies have been published describing the process of tumor angiogenesis in detail and identifying new pro- and antiangiogenic factors [45–51].

Antiangiogenic treatments and classes of resistance mechanisms

Fumagillin, isolated from *Aspergillus fumigatus*, is a toxic fungal product that acts against parasitic infections in bees and has also been used in human amebiasis [52]. Fumagillin was the first compound with antiangiogenic activity discovered by Folkman and some years later a non-toxic derivative, AGM-1470, was synthesized. AGM-1470 is 50-fold more potent than Fumagillin and had two effects: it was antiangiogenic and inhibited tumor growth even though it was not cytotoxic for tumor cells at the low doses used for antiangiogenesis [53].

In 1989, recombinant α2 interferon was also successfully used for the antiangiogenic treatment of a pulmonary hemangioma in a child [54]. Clear cell renal carcinoma is considered to be an angiogenic tumor. For more than 15 years, the only available treatment for this cancer was recombinant α2 interferon [55–57], that is until the tyrosine kinase inhibitors sunitinib and sorafenib were developed. When these drugs came on the market they were first used in association with α2 interferon [58] until they fully replaced interferon.

In 1971, Baenziger et al. [59] found a glycoprotein on platelet membranes (actually it is a family of glycoproteins) that further studies revealed had powerful antiangiogenic effects: thrombospondin 1 (TSP1). It was identified as an endogenous natural antagonist of angiogenesis via its inhibition of the synthesis of VEGF [60–62] and of fibroblast growth factor (FGF) [63–65]. Angiogenesis can also be turned down by blocking VEGF's receptor: VEGFR [66–68]. TSP1, secreted by normal epithelial cells, may seem to be the ideal antiangiogenic compound due to its origin and effects: inhibiting endothelial cell migration, inducing apoptosis, and blocking growth factor stimulatory effects on the endothelial cell surface. However, its large size and additional collateral effects make it inappropriate for clinical use. From the beginning of the 1990s full-fledged research for antiangiogenic compounds developed many new molecules like monoclonal antibodies against VEGF, 2-methoxyestradiol [69], and led to the re-discovery of the antiangiogenic properties of old drugs such as thalidomide [70], and of low-dose chemotherapeutic agents that when used metronomically, were also found to prevent angiogenesis [71]. TSP1 was the model to imitate and drugs like ABT-510, a nonapeptid analog, were developed. While it was unsuccessful in clinical trials, it did reduce VEGF-C and circulating endothelial cells [72].

At the beginning of the twenty-first century different types of antiangiogenic drugs and methods have represented the proof of concept of antiangiogenesis (Fig. 3):

(1) Bevacizumab: a monoclonal antibody that binds VEGF.
(2) Aflibercept: a recombinant protein that "traps" circulating VEGF.
(3) Sunitinib and sorafenib: small molecule inhibitors of VEGF receptor tyrosine-kinases.
(4) Metronomic chemotherapy targeting the proliferation of endothelial cells.
(5) Trebananib that targets angiopoietins 1 and 2 (Fig. 3). (See Chapter 1 of the book for an updated review of the antiangiogenic drugs).

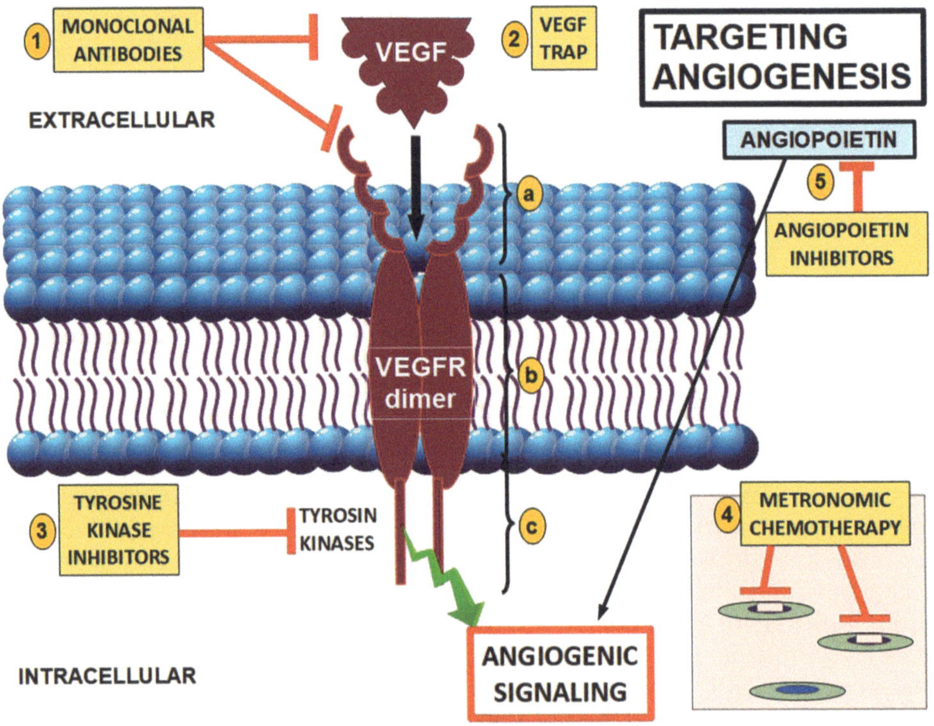

FIG. 3 Five different mechanisms of antiangiogenic therapy: (1) Monoclonal antibodies that can block VEGF or its receptor VEGFR, e.g., bevacizumab [73–76] (targets VEGF), ramucirumab [77,78] (targets VEGFR2). (2) A trapping mechanism for circulating VEGF. (3) Tyrosine kinase inhibitors that block the phosphorylation chain born in the intracellular portion of VEGFR when VEGF binds the extracellular segment, e.g., sorafenib [79], sunitinib [80], nintedanib [81], cabozantinib [82], regorafenib [83,84]. (4) Metronomic chemotherapy impeding the reproduction of endothelial cells, e.g., metronomic low dose cyclophosphamide [85], or cyclophosphamide plus methotrexate [86]. (5) Angiopoietins are natural, vascular growth factors that act through Tie-2 receptor binding. Trebananib sequesters angiopoietins I and II [87,88].

Actually, we can summarize the treatments in Fig. 3 in two basic antiangiogenic conceptual mechanisms: (i) one based on the inhibition of the various parts of the VEGF-VEGFR axis (1, 2, 3, and 5) and (ii) another impeding endothelial cell reproduction (4) or directly "destroying" endothelial cells (not shown in Fig. 3).

These mechanisms are completely different, and therefore, the cause of failure in each case is different.

Resistance to antiangiogenic treatments

Causes of antiangiogenic treatment failure

One of the reasons for poor performance was due to the fact that angiogenesis is far more complex than the mechanism shown in Fig. 2, in which it is driven by a simple imbalance between pro- and antiangiogenic factors.

While the antiangiogenic compounds have had impressive success in preclinical models, and many new agents are entering clinical trials every year, this success has proved short-lived in bedside medicine [89–91]. This lack of better clinical results is the consequence of the development of resistance to antiangiogenic treatments [92–94]. The resistance can be [95]:

(A) Primary: no response at all and supposedly due to intrinsic factors.
(B) Secondary or acquired: initial response and then resistance.

At the beginning of the "antiangiogenic era" it was widely believed that resistance would not develop [96], because the treatment was directed against endothelial cells that are not mutated, have no genetic instability, and are not supposed to over-express pro-mitotic pathways. Now we have proof that this is not so. Tumor-associated endothelial cells (TAECs) are different from their normal tissue counterparts. Genes expressed in TAECs differ from normal endothelial cells. Croix et al. identified 79 genes that were differently expressed in TAECs [97], and Lu et al. [98] found more than 400 genes differentially expressed in ovarian cancer TAECs. Akino et al. [99] also found frequent cytogenetic changes in renal cell carcinoma TAECs. Endothelial cells of glioma are resistant to apoptosis by over-expressing GRP78 (a stress chaperone protein) [100] and survivin [101]. This creates resistance to conventional chemotherapies. Therefore, TAECs can develop resistance to treatment and this is one of the reasons for the failure of the antiangiogenic approach.

However, the lack of an overwhelming success of antiangiogenesis rests on a biased concept of angiogenesis, because it is much more than the VEGF-VEGFR interaction. Some of this "much more" concept is known, but there remains a large gap of unknown factors. The "much more" participants in the angiogenic process have been called the "angiome" [102–104] in which 1233 proteins have been identified as playing a role in it. Therefore, Fig. 2 is actually a naïve over-simplification of the complex world of angiogenesis.

- **Primary resistance to antiangiogenic treatments**: Some tumors are highly resistant to hypoxia and do not depend on neovascularization. This is the case of the pancreatic ductal adenocarcinoma (PDAC) that has a large, poorly vascularized desmoplastic stroma [105,106]. Vasculogenic mimicry rather than angiogenesis is one of the features of PDAC [107], which would explain the poor results obtained with antiangiogenic treatments [108].

BOX 1

Mechanisms of resistance to antiangiogenic drugs.

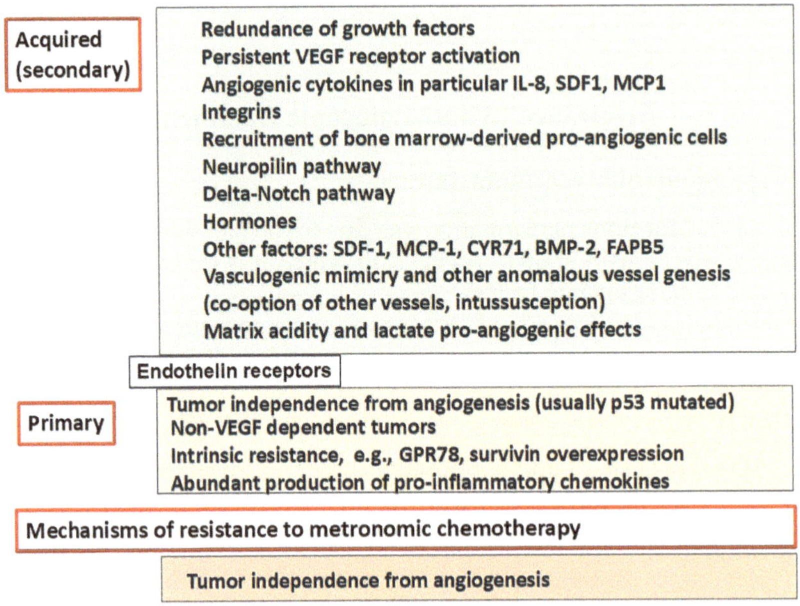

Furthermore, according to Carbone et al. [109] pancreatic cancer primary resistance to antiangiogenesis is strongly related to intensive production of pro-inflammatory chemokines from the cells and from the desmoplastic reaction of its stroma.

- **Acquired or secondary resistance to antiangiogenic treatments**: Development of resistance is a multi-origin process in which many factors combine to render antiangiogenesis little or not effective after an initial favorable response. Box 1 summarizes the main causes of resistance to antiangiogenic treatments. Endothelin receptors have a different place in the BOX because they probably participate in primary and acquired resistance to antiangiogenic treatments. It is unlikely that they have a role in metronomic chemotherapy.

Causes of antiangiogenic treatment failure with low dose metronomic chemotherapy (MCT)

Metronomic chemotherapy [110] consists of the repeated administration of anti-neoplastic drugs at very frequent (usually daily) low doses and without interruptions or with very short interruptions [111]. The main mechanism of anti-tumoral action with this scheme is anti-angiogenesis. However, there are also some other processes involved [112–115].

The low doses of chemotherapeutics used in MCT are insufficient to kill malignant cells, but they are enough to impede endothelial cell reproduction. The most frequently used drugs are cyclophosphamide, methotrexate, and capecitabine and the advantages of MCT are: oral drug administration, low toxicity with almost no adverse events, and low cost.

Two types of endothelial cells participate in angiogenesis:

(1) endothelial tip cell;
(2) endothelial stalk cell.

These two types of cells work as a team, where endothelial cells become activated and degrade their basal membrane, starting to sprout. The tip cells migrate while the stalk cells proliferate and follow the leader stalk cells [116]. It is mainly at the stalk cell level where MCT is addressed, blocking or slowing their proliferation. While Notch1 signaling regulates endothelial tip cells [117], the stalk cells are in a continuous reproductive stage and they are the real targets of MCT. The sprouting bud acquires a lumen, and develops a basal membrane, and finally recruits pericytes to form a real vessel. To achieve a functional vessel there is a necessary and permanent interaction between the VEGF-VEGFR axis and the Delta-4/Notch pathway [118]. Details are in Box 2 which is based on references [119–122].

In the experience of one of the authors (T.K.), resistance to metronomic chemotherapy in dogs harboring natural cancers, developed in all the cases after an average of 14 months of treatment (unpublished data). The causes leading to this resistance have not been fully clarified. Our initial presumption was that the mechanism of treatment failure with metronomic chemotherapy was similar to treatment failure in tumors treated with full-dose anti-mitotic drugs, but this was wrong. Thoenes et al. [123] showed that metronomic cyclophosphamide

BOX 2

Angiogenesis is a step by step process.

1) VEGF A-VGFR induces tip cell activation.

2) Tip cells increase Delta-Notch signaling and initiate migration and invasion.

3) Sprouting process starts.

4) Simultaneously stack cells proliferate in a VEGF-dependent manner.

5) Delta-Notch signaling "matures" the vessel impeding further mitosis inside the vessel, thus helping it to acquire a lumen.

6) Fusion of new vessels.

7) Expression of integrins on endothelial vessels and

8) Recruiting pericytes through Tie2-Angiopoietin (silencing Tie2?) (Silencing of Tie2 creates a pro-migratory phenotype)

therapy of prostate cancer xenografts had clear antiangiogenic effects. The treatment destroyed tumor vessels so that blood supply was seriously impaired even after the treatment failure; however, tumor growth resumed without restoring tumor vessels. Furthermore, neither multi-drug resistance (MDR) transporter activity nor new angiogenesis were found [123]. This can be interpreted as if the tumor acquired a circulation-independent ability to grow because angiogenesis did not recur after failure.

Kubish et al. [124] found that metronomic chemotherapy-induced important gene expression changes in the tumor: "*genes that encode anti-coagulation proteins were upregulated and genes that code for proteins promoting coagulation were downregulated*". However, the significance of these findings is difficult to interpret. The authors speculate that reducing coagulation may improve nutrient and oxygen supply. The only clear issue is that the mechanism of resistance is probably due to the genetic adaptation of the tumor to the lack of blood supply, without any new angiogenesis.

Causes of treatment failure with antiangiogenic therapy

This is probably a multifactorial development. Classical antiangiogenic agents target the VEGF-VEGFR axis. However, many other factors play a role in angiogenesis.

A. Growth factor redundancy

Addressing angiogenesis by down-regulating or inhibiting the VEGF-VEGFR axis as the sole treatment shows short-lived benefits because there are many other growth factor- dependent and cytokine-dependent angiogenic pathways. Among the angiogenic factors that can induce new vessel formation there are:

A.1. Other growth factors

(1) acid and basic fibroblast growth factor (FGF) [125–127];
(2) angiogenin [128];
(3) transforming growth factor (TGF) α [129] and β [130,131];
(4) epidermal growth factor (EGF) [132,133], and EGF receptor (EGFR) [134–136];
(5) platelet-derived growth factor (PDGF) [137–139];
(6) placental growth factor [140];
(7) hepatocyte growth factor/scatter factor [141,142];
(8) insulin-like growth factor [143];
(9) nerve growth factor [144];
(10) neuronal and glioma-derived stem cell factor [145];
(11) Prokineticin 1 and 2 (PK1 and PK2) [146–150].

A.2. Cytokines

(1) Tumor necrosis factor alpha (TNFα) [151,152];
(2) Interleukin-1 beta (IL-1β) [153];
(3) Interleukin-1 alpha (IL-1α) [154].

(4) Interleukin-6 (IL-6) [155–161]. The IL-6/STAT3 pathway is pro-angiogenic [162], mainly through VEGF up-regulation. However, IL-6 has also been found to be associated with decreased angiogenesis [163].

The dynamics of interleukin action are further discussed in paragraph F.

A.3. Integrins

Integrins are heterodimeric transmembrane glycoproteins that act as adhesion molecules cementing the new vessels and adhering pericytes that decrease leakage and stabilize the new structure. 24 different heterodimeric integrins have been identified in vessels [164,165], and using specific integrin antagonists seriously impairs angiogenesis [166–169]. Fig. 4 shows the

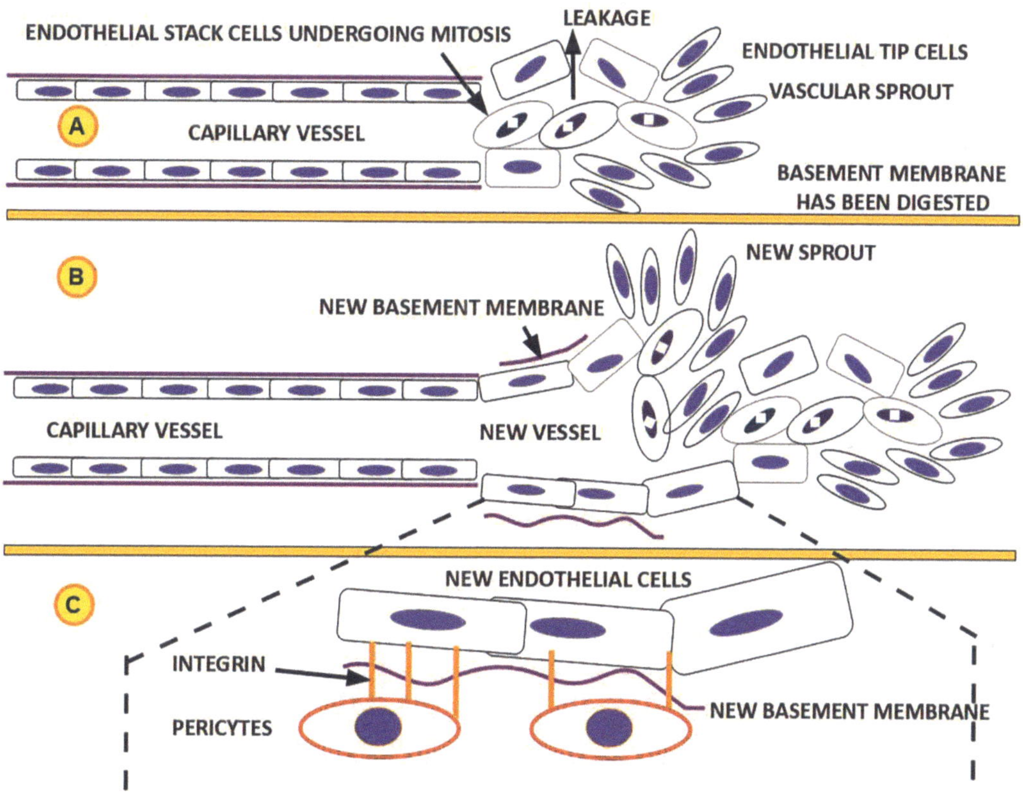

FIG. 4 Panel (A) shows the angiogenic sprouting with its two types of cells, namely stalk cells in reproduction and tip cells advancing (these cells have filopodia that follow the VEGF gradient [170]). The basement membrane has been digested. Panel (B): a portion of the new vessel has been incorporated while angiogenesis follows. Panel (C) shows the new vessel with a higher magnification: where the basement membrane has been rebuilt and integrins attach pericytes. Leakage is still present (not shown in the diagram) and the new structure is quite irregular. The new vessels are characterized by high endothelial proliferation, many sectors lack pericytes and the basement membrane with increased permeability and leakage [171].

sprouting process and the lower panel of Fig. 4C shows the important mechanical function of integrins by attracting and attaching pericytes around the newly formed vessel.

The expression of different integrins on angiogenic endothelial cells is absolutely essential for the production of new vessels [172]. For example, αv integrin is over-expressed during angiogenesis [173], and its inhibition stops the process [174]. αvβ3 integrin is over-expressed in the sprouting process and its inhibition blocks angiogenesis [175,176]. Furthermore, a β1 integrin-VEGFR2 axis has been described in PDAC cancer stem cell vasculogenic mimicry [177].

Pharmacological interruption of the VEGF-VEGFR axis only partially blocks angiogenesis because many of the other above players are still functioning.

B. VEGF Amplification

There are different types of VEGF and genetic amplification of VEGF genes have been found in some tumors.

(a) VEGF-A: gene localized at chromosome 6p12 [178] has been found to be amplified in osteosarcoma [179]. VEGF-A amplification was associated with poor outcomes in metastatic breast cancer with and without bevacizumab treatment [180]. In this study, 6.5% (21/324) of patients showed VEGF amplification. In another population study of breast cancer patients, the same authors found that 11% had VEGF gene deletion, borderline amplification in 9%, and beyond 2-fold amplification in 5% [181].

(b) VEGF-B: gene localized at 11q13

(c) VEGF-C: gene located at 4q34.

The sphingosine-1-phosphate receptor plays a role in VEGF-A/VEGFR2 signal amplification in tumors by indirectly phosphorylating tyrosine 951 of VEGFR2 [182].

The details of the relationship between VEGF-A amplification and resistance to antiangiogenic treatment have not been fully clarified; however, evidence indicates that amplification may be part of the problem. To the best of our knowledge, there are no studies on the amplification of other VEGFs.

C. VEGF Polymorphisms

VEGF polymorphisms determine the expression and activity of VEGF in non-small cell lung cancer [183,184] and renal cancer [185].

D. The neuropilin (NPL)-ABL1 pathway

Neuropilins are transmembrane glycoprotein receptors that drive neurogenesis and angiogenesis. Their function is pleiotropic and they can interact with almost all the growth factors [186–189]. NPLs have been found to be over-expressed in many tumors [190–192] and can be associated with poor prognosis [193], although this association is somewhat controversial [194]. NPLs behave as co-receptors since their short intracellular portion requires an associated receptor in order to transmit signals. Their main ligand is semaphorin 3, but various growth factors can also bind and activate them. Neuropilins in the vicinity of blood vessels function mainly as co-receptors of VEGFR, potentiating the phosphorylation activity of VEGFR, thus leading to endothelial cell migration. While this is a VEGF-dependent pathway, NPLs are also at the head of a VEGF-independent pathway. In this case, the signaling ends by activating the ABL1 protein to exert angiogenic functions [195] as shown in Fig. 5. NPLs

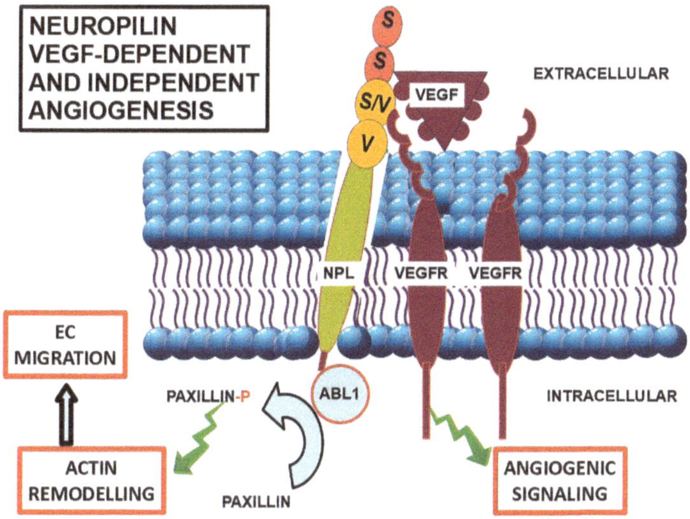

FIG. 5 Neuropilin has a dual activity in angiogenesis: one is VEGF-dependent where neuropilin stimulates VEGFR signaling; the second stimulates endothelial cell migration through a VEGF-independent mechanism by means of paxillin activation.

activated by other growth factors can initiate a signaling pathway through the ABL1 kinase, working independently of the VEGFR.

Neuropilin VEGF-dependent and independent angiogenic activity may represent the switch from non-resistance to resistance to classical antiangiogenic treatments. NPL requires the mediation of the ABL1 protein which is susceptible to blockage by imatinib. Therefore, imatinib can represent an interesting drug to be tested for gaining back antiangiogenic susceptibility. This needs experimental confirmation. In this respect, the simultaneous inhibition of NPL-1 and VEGFR2 increased radiation sensitivity of non-small cell lung cancer, while the independent inhibition of each of them, failed to do so [196].

This underlines the important corollary that NRP is a "double action" pro-angiogenic co-receptor.

E. The Delta-Notch Pathway

This is an evolutionary ancient pathway mainly associated with cell differentiation, proliferation, and signal integration during embryonic life [197]. While the Notch signaling pathway acts as a coordinator of the multiple steps of vascular development, the main function in angiogenesis is to suppress non-functional sprouts and stimulate blood vessel formation. The mechanism is based on Notch, a membrane receptor, and Delta, a one-pass transmembrane protein acting as a ligand (Fig. 6). Ligand binding releases the intracellular portion of Notch which then acts as a transcription factor for genes related to angiogenesis and other functions. Blocking the Delta-Notch pathway does not inhibit angiogenesis, but induces a functionally useless vessel, which for all practical purposes acts as an antiangiogenic proxy [200,201]. This finding confirms the coordination work carried out by the pathway. The Delta-like 4 ligand (Dll4) is the responsible agent for the differentiation of the two types of endothelial cells that participate in angiogenesis, namely tip and stalk cells in the sprouting process [117,202,203].

Hainaud et al. [204] have shown that the Delta-like 4-Notch 4-ephrin B2 pathway is essential for the production of functioning vessels. Fig. 6 summarizes this pathway.

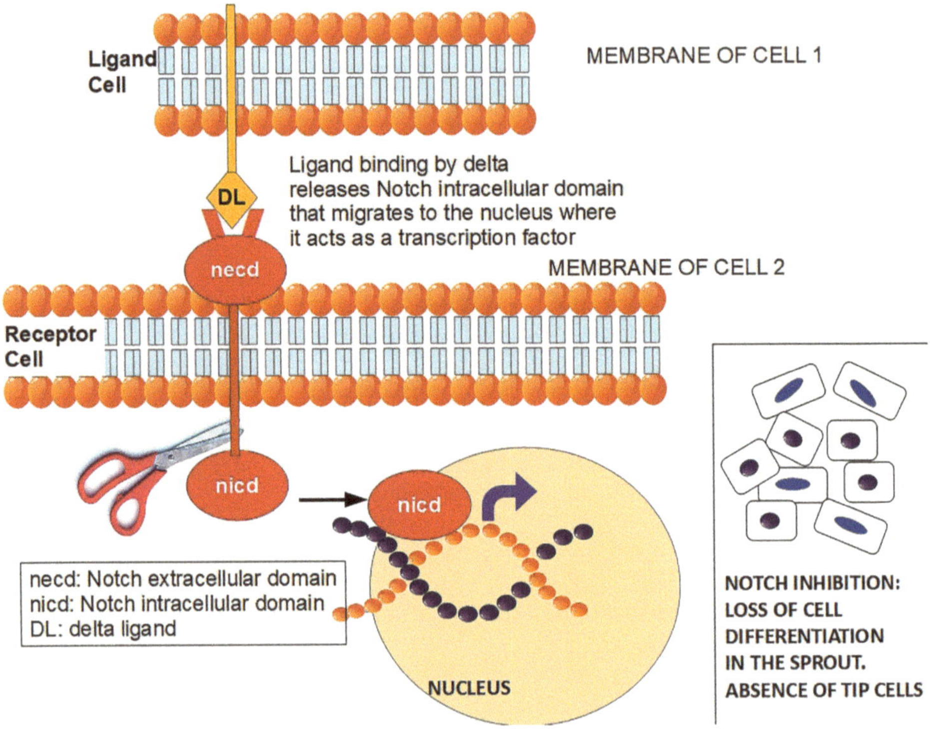

FIG. 6 Notch signaling after ligand-binding of the delta. The diagram shows the need for cell-cell interaction in order to trigger signaling. Endothelial cells express Notch1 and Notch 4 [198]. The *lower right panel* shows the lack of differentiation of the cells in the vascular sprout when Notch signaling is inhibited [199]. This renders the sprout functionally useless.

Importantly, Notch activities are both dependent and independent of VEGFs. This leaves some unanswered questions:

How independent is Notch?

Can Notch alone run the whole process of angiogenesis or does it only participate in certain steps? Evidence shows that Notch is part of a VEGF-VEGFR downstream signaling pathway [205].

As there are 4 different Notch receptors (1–4) and 5 delta ligands, does this complexity make a functional difference?

However, the most important question that we cannot yet answer, is if Notch can functionally replace the VEGF/VEGFR axis or be activated without the axis. Current knowledge seems to indicate that VEGF/VEGFR is the driver of angiogenesis and the Delta-Notch pathway is a coordinator of the process but not the driver [206]. However, there are other factors that can activate Notch signaling, such as FGF and the Wnt pathway [207]. This leaves open the question if the Notch pathway may be able to drive angiogenesis through its interaction with these alternative factors.

In conclusion, blocking the Delta-Notch pathway increases the production of useless vessels. This has been called the Delta-Notch paradox [208].

Interestingly, it has been found that Delta-Notch inhibition can result in strong anti-tumor activity after resistance to antiangiogenic treatments [209]. Research on antibodies against Delta and Notch cell surface receptors is underway.

F. Interleukin 8 (IL-8) [210,211]

myeloid angiogenic cells promote angiogenesis through IL-8 production. Extracellular IL-8 was found to transactivate VEGFR2 [212]. Transfection of IL-8 genes in gastric cancer cells promoted angiogenesis [213]. Mizukami et al. [214] showed that in spite of HIF knockdown, angiogenesis was still maintained through diminished VEGF activity in colon cancer cells, but when IL-8 was simultaneously inhibited in these cells, angiogenesis was almost completely eliminated. This experiment shows the importance of IL-8 inhibition in a tumor which is frequently treated with anti-angiogens.

In colon cancer, where antiangiogenic treatment is used frequently, IL-8 is an important player [215], therefore it would be appropriate to test IL-8 inhibition as part of the therapy. Something similar happens with melanoma [216], non small cell lung cancer [217–219], bladder [220], prostate [221,222] breast [223], thyroid [224], pancreatic [225,226], and ovarian cancer [227], nasopharyngeal carcinoma [228], gastric carcinoma [229,230], hepatocellular carcinoma [231], and glioblastoma [232,233]. Interestingly, it does not participate in small cell lung cancer [234].

Furthermore, IL-8 was found to mediate resistance to sunitinib in renal cell carcinoma, and anti-IL-8 antibody re-sensitized tumors to sunitinib [235]. Interestingly, in this case of resistance high microvessel density was found, meaning that the resistance mechanism consisted of angiogenic reinstatement and not angiogenesis-independence.

A fully humanized anti-IL-8 antibody has been synthesized (ABX-IL-8) and it proved able to inhibit angiogenesis and tumor growth in melanoma [236].

Interestingly, extracellular acidity increased VEGF and IL-8 production in tumors [237–240]. This explains the reason why a pH-centered treatment, that increases extracellular pH, showed synergy with metronomic chemo-/antiangiogenesis therapy [241].

G. Ephrins and ephrin receptors (Epr)

Although ephrin A1 [242], ephrin A2 [243], ephrin B2 [244,245], ephrin B4, and ephrin receptor (Epr) proteins are usually over-expressed during the escape from antiangiogenesis, how they contribute to this effect is still unclear. Ephrins (Eph) and their receptors seem to be promoters of endothelial cell migration and proliferation [246,247]. Some chemokines like SDF1/CXCR4 are also angiogenesis inducers [248–250]. Chemokines of the CXC family containing an ELR motif are strong angiogenesis inducers while those that lack this motif show an antagonistic action [251].

H. Insufficient angiogenesis inhibition

Laschke et al. [252] showed that inhibiting VEGF alone was insufficient to fully block angiogenesis in endometriotic lesions. It was also necessary to inhibit FGF and PDGF in order to inhibit angiogenesis and vessel maturation. Their paper confirms the importance of the other growth factors in resistance to angiogenic treatments.

I. Angiogenesis: intra- and inter-tumor heterogeneity

Different malignancies showing angiogenesis vary widely as to the degree of the process. There is considerable variation even inside the same tumor [253]. All practicing oncologists know that not all tumors are susceptible to antiangiogenic treatments. The variations in tumor endothelial cell turnover can be as low as 0.15% (percentage of endothelial cells in mitosis) in prostate cancer [254] or as high as 9.9% in colon adenocarcinoma [255]. Tumors with a low endothelial turnover are resistant to antiangiogenic treatment. It was found in breast cancer that endothelial proliferation was many-fold higher in the tumor periphery than in central areas [256]. According to the authors, there was no correlation between tumor vascularity and endothelial proliferation. Many vessels can be formed by remodeling and vessel migration rather than by endothelial proliferation. And of course, in this case, antiangiogenic treatments would have only minimal effects.

J. Vasculogenic mimicry (VM)

Vasculogenic mimicry is the formation of vessels by tumor cells without the intervention of endothelial cells [257,258]. The walls of these vessels are made up of tumor cells and a basal membrane (Fig. 7). This mechanism of vessel formation is usually found in very aggressive tumors, melanoma being the most frequent host of VM [259]. Tumors with the ability to create this type of intratumoral circulatory system express genes usually very active in embryonic life, such as vascular endothelial-cadherin (VE-cadherin). VE-cadherin, also known as CD144, is a cell-cell adhesion glycoprotein essential for cementing intercellular spaces impeding leakage [260,261]. It is specifically expressed in endothelial cells. When tumor cells express this glycoprotein it usually means that the tumor has the ability to produce VM and this is a sign of aggressive behavior [262,263].

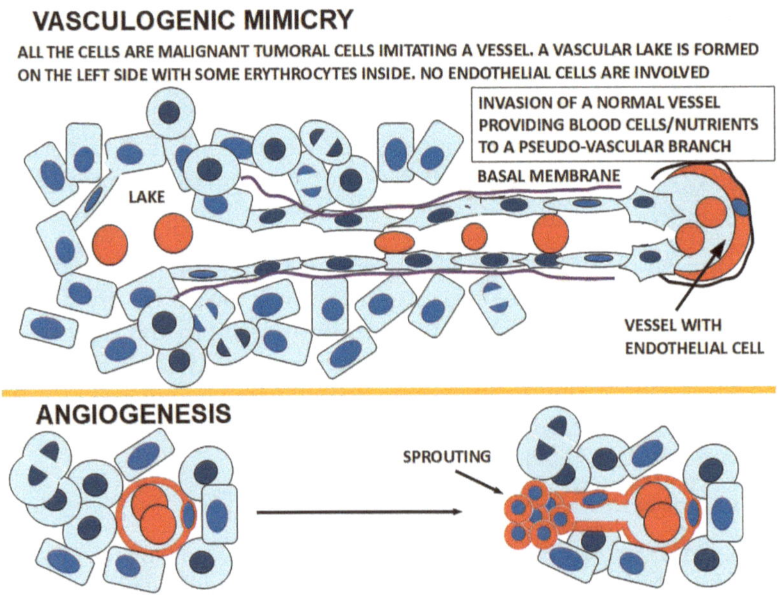

FIG. 7 Vasculogenic mimicry in the upper panel is compared to angiogenesis in the lower panel. The walls of the pseudo-vessel are formed by tumor cells "imitating" endothelial cells. There is no evidence of VEGF-VEGFR axis involvement. Angiogenesis in the lower panel shows that a sprout of endothelial cells (tip cells) grows into the tumor bulk and acquires a lumen that communicates with the capillary vessel of origin. The sprout is regulated by the VEGF-Notch pathway.

VM is not independent of VEGF stimulation. Wang et al. [264] proposed a pathway VEGFA → EphA2 → MMPs → Vascular Mimicry which reaffirms the central role of VEGF even in processes other than angiogenesis. In spite of this VEGF-VM relationship, it has been shown that VM is one of the escape mechanisms (resistance) in tumors responsive to VEGF-VEGFR [177,265]. Possibly, this could be due to the very high secretion of a variety of pro-angiogenic factors by the tumor cells with VM in both pancreatic [177] and breast cancer cells [266].

Classical antiangiogenic treatments have little or no effect on VM. There are some experimental drugs that have shown VM targeting abilities, such as CVM-1118 which is undergoing clinical trial (phase II) associated with sorafenib for hepatocarcinoma [267]. Many treatments for VM inhibition have been proposed, including:

- Gliomas: liposomes containing epirubicin and celecoxib [268].
- Triple-negative breast cancer: liposomes with vincristine and dasatinib [269].
- Melanoma: thalidomide [270].
- Breast cancer: hyaluronic acid (HA)-modified daunorubicin plus honokiol cationic liposomes [271].
- Gastric cancer: luteolin [272].
- Melanoma: Apatinib [273].
- Glioma: galunisertib [274].
- Breast cancer: 6, 6′-bis (2, 3-dimethoxybenzoyl)-a, a-D-trehalose [275].
- Hepatocellular carcinoma: doxycyclin [276].
- Pancreatic ductal adenocarcinoma (PDAC): verteporfin [277].
- Prostate cancer: kaempferol [278].
- Triple-negative breast cancer: brucine (herb extracted from seeds of *Strychnos nux-vomica*) [279].
- Hepatocellular carcinoma: curcumin [280].
- Sunitinib resistant renal carcinoma: everolimus [281].
- Osteosarcoma: zoledronic acid [282].

This is only a partial list, as there are many other known compounds with VM inhibiting abilities. Unfortunately, they have not achieved bedside utility as yet.

K. Matrix acidity and the lactate factor

(a) Matrix acidity

Tumor neovascularization, whether through angiogenesis or VM requires an essential feature of the extracellular matrix (ECM): acidity. The unaltered physiologic ECM, having a pH around 7.34 and extremely low levels of lactic acid is not the most adequate site for the generation of new vessels. Angiogenesis does not occur solely in the realm of the endothelial cell but is based upon a strong interaction between the endothelial cell and the ECM. Indeed, the remodeling of the ECM is a necessary condition to allow the creation of a new circulatory system in the tumor and this implies the activity of many degrading enzymes [283], such as metalloproteases, cathepsins, and others. All these enzymes have optimum activity at pHs much lower than that found in the normal ECM.

In this respect, a pH-centered treatment [284,285], addressed to decrease intracellular pH and increase extracellular pH, introduces a fundamental change in the ECM by creating a hostile environment for angiogenesis and VM.

Therefore, we strongly believe that antiangiogenesis and anti-vasculogenesis require the association of ECM pH modifiers in order to increase therapeutic effectiveness. We found synergy between metronomic chemotherapy and pH-centered treatments [241]. However, we do not know if this synergy is also maintained with other types of antiangiogenic treatments. On a speculative basis, we do think that this synergy would not change with other types of antiangiogenic therapies.

(b) The lactate factor

Matrix acidity in cancer is the result of two main factors: overproduction of CO_2 and more than 60-fold increase in l lactate produced by glycolysis.

Lactate alone is an important pro-angiogenic molecule [286–308]. The overwhelming evidence showing the strong relationship between the glycolytic phenotype (through lactate production) and angiogenesis, including resistance to antiangiogenic therapy, clearly calls for action.

How does lactate work in angiogenesis?

Ruan and Kazlauskas [309] found the mechanism that probably links metabolism and lactate production with the angiogenic pathways (Fig. 8).

Lactate enters the endothelial cell through the transporter activity of monocarboxylate transporter 1 (MCT1). Inside the cell, it activates the production of ligands for receptor tyrosine kinases Tie2, Axl, and VEGFR2. These ligands acting in an autocrine manner unleash the tyrosine kinase activity of these receptors which initiates the angiogenic stimulation (Fig. 8).

Therefore, to improve antiangiogenic therapy results and even to treat antiangiogenic resistance requires the correction of matrix acidity and lactate activity.

The easiest way to achieve this is through the inhibition of the MCT1 and MCT4 [310–312].

Many drugs have been proposed as MCT inhibitors [313,314]. Quercetin, at high doses, is probably the simplest and non-toxic pharmaceutical to achieve MCT inhibition [315–318], ECM lactate reduction [319], and other anti-cancer effects [320].

L. Endothelin (ET) and endothelin receptors (ETR)

Endothelins (ETs) are 21-amino acid peptides, mainly produced by endothelial cells, that have powerful vasoconstrictive abilities [321] and participate in blood pressure homeostasis via endothelin binding to its receptors (ETRs).

In addition to its physiological effects, the ET-ETR system has been found to participate actively in tumor progression and angiogenesis. ET is produced by tumors, endothelial and stromal tumor-associated cells such as fibroblasts and macrophages. ET binds ETRs which initiate signaling cascades that produce the above effects. There are different ETs and ETRs. The most frequently involved in cancer are ET-1 and ET2 and the receptor usually is the ET-B receptor (Fig. 9). Endothelin-3, a vasoactive peptide released by vascular smooth muscle cells and endothelin converting enzyme-1, a protease that activates endothelin-3 also plays a role in angiogenesis [332–334] and especially in tumor angiogenesis [335–339].

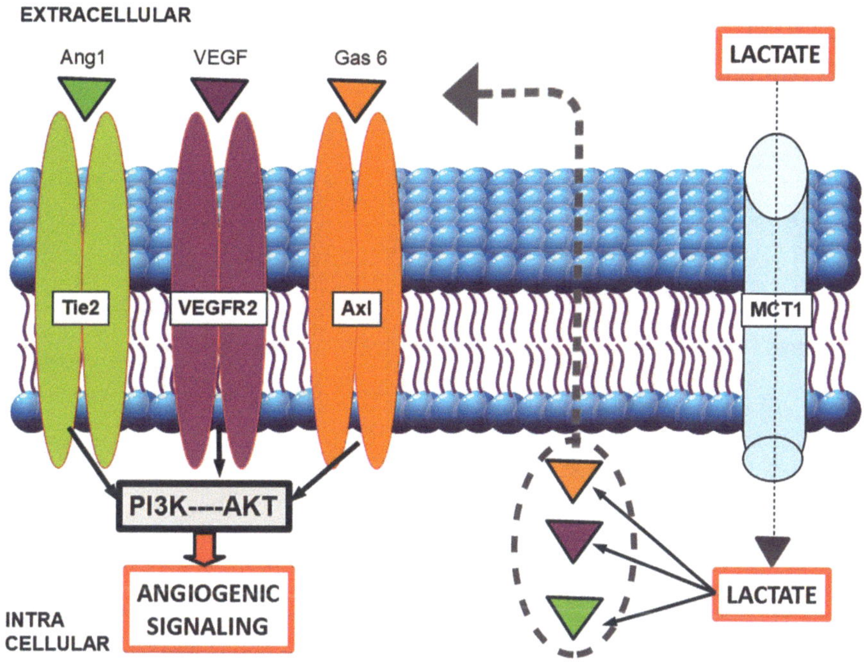

FIG. 8 Lactate pro-angiogenic mechanism. *Modified from Raun and Kazlauskas [309].*

Indeed, endothelin receptors, particularly ET-BR, play a pivotal role in tumor progression, as well as in angiogenesis and immunological escape and probably in resistance to angiogenic treatments [340]. The endothelin family and their receptors intervene in vascular tone, mitosis regulation, cell survival, angiogenesis, and immune cell regulation in a large variety of cancers [341–343]. Thus, ETRs seem to be the nexus between angiogenesis and immune escape in cancer.

Importantly, antiangiogenic treatments based on inhibiting the VEGF-VEGFR axis increase the production of endothelin, which seems to be the main cause of many of the undesired side effects of the treatment, such as hypertension, left ventricular dysfunction, cardiac ischemia, and renal function impairment, among others [344–347].

There are no doubts about the important side effects produced by endothelin in classical antiangiogenic treatments. However, we shall only focus the analysis on endothelin as a possible escape mechanism in antiangiogenic treatments.

Continuous administration of sunitinib in rats significantly increased endothelin levels in a dose-dependent manner. The ET-1 level returned to normal after sunitinib was discontinued [348]. Many side effects of sunitinib could be eliminated by pharmacologically blocking ET-1.

ETRs inhibition reduces angiogenesis and also VEFR2 expression [349]. Based on the publications discussed above, the possible resistance pathway is shown in Fig. 10.

To resolve a situation as shown in Panel 3 of Fig. 10, increased blockage of this axis would be necessary to overcome ET-ETR activity. This is not possible because of the toxicity of antiangiogenic drugs. An alternative solution would be the simultaneous inhibition of both pathways. This hypothesis is based on references [350–357].

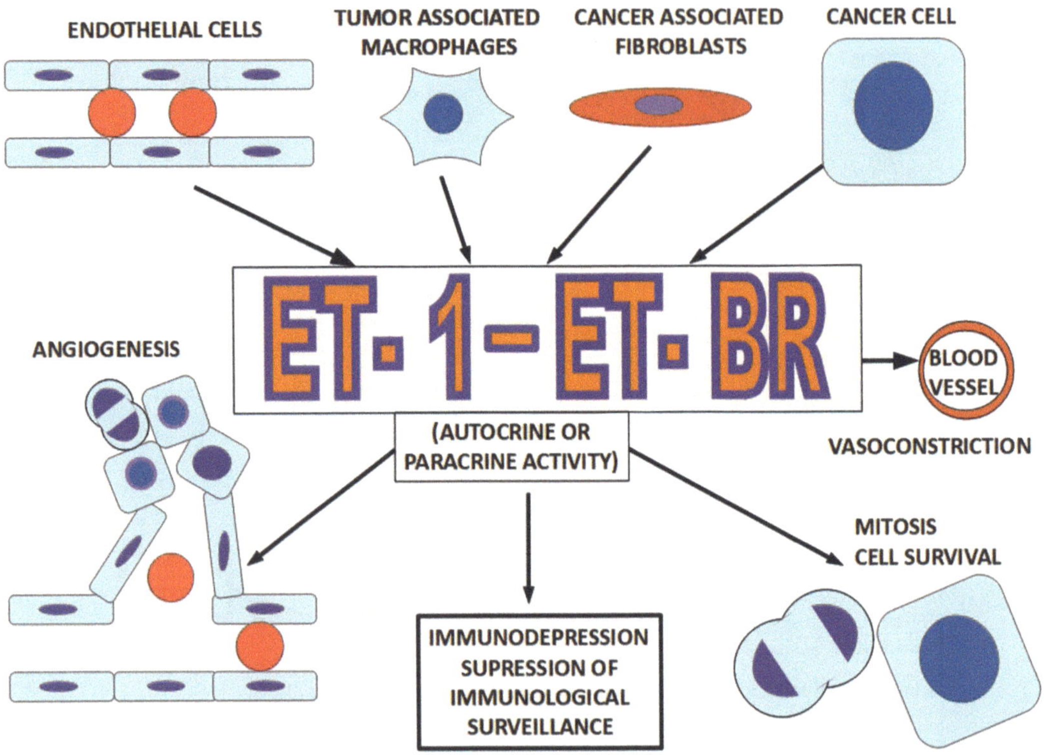

FIG. 9 Effects of the ET-1-ET-BR signaling in cancer. On the right, vasoconstriction is the main physiological function of ET-ETR. Its significant effects on cancer are shown in the lower panel. *Diagram based on references [322–331].*

M. Other important players in the angiogenesis

Cysteine-rich angiogenic inducer (61CYR61) is a heparin-binding matrix protein that promotes migration and adhesion of endothelial cells, inducing neovascularization [358]. It interacts with cell surface integrin receptors, in endothelial cells expressing αvβ3 [359]. It has an important pro-angiogenic activity [360]. Interestingly, this protein shields cancer cells from chemotherapy [361]. In inflammatory processes, it has been found that CYR61 expression is induced by IL-6 [362]. CYR61, on the other hand, induces IL-8 expression [363].

CYR61 is part of a circuit shown in Fig. 11 which is probably functional for angiogenesis even during VEGF/VEGFR inhibition and, in this way, CYR61 may work as a parallel angiogenesis pathway.

Monocyte chemoattractant protein 1 (MCP-1), also known as CCL2, and its receptor CCR2 are chemokines with angiogenic effects [380,381]. It is also a downstream protein of the CYR61 pathway. MCP1 induces the expression of VEGF-A in oral squamous cell carcinoma [382] and increases angiogenesis in melanoma [383]. The angiogenic activity of this protein seems to be carried out through VEGF-A [384].

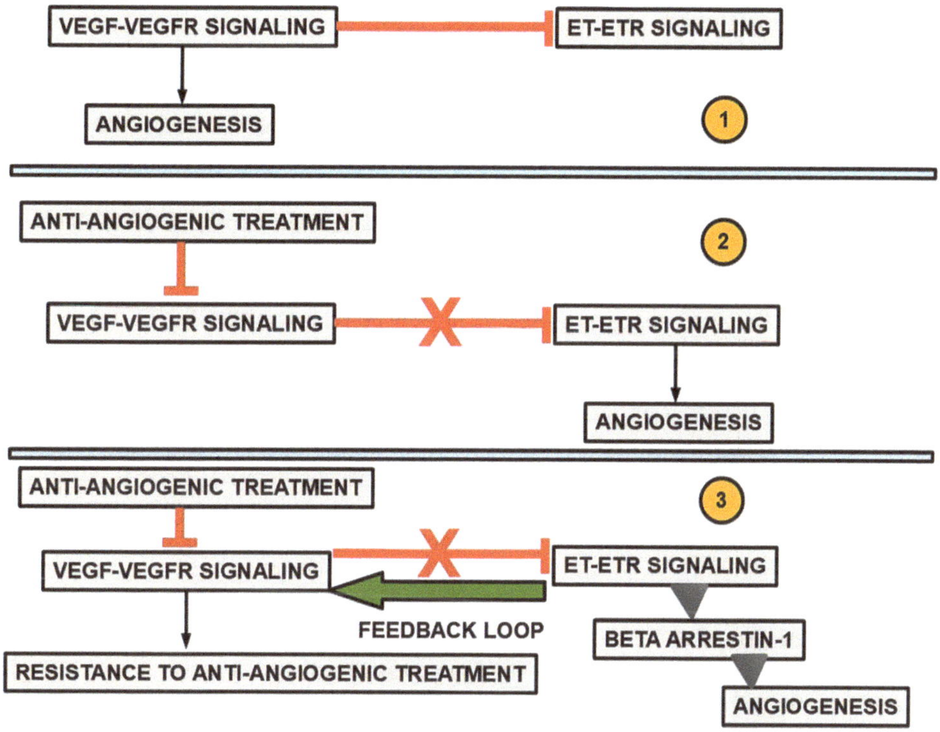

FIG. 10 A possible pathway of resistance based on ET signaling. Panel 1 shows the angiogenic mechanism found in tumors where the VEGF-VEGFR axis induces angiogenesis and at the same time down-regulates ETR. Panel 2: The antiangiogenic treatment blocks angiogenesis but triggers ET-ETR signaling that has angiogenic abilities of its own. Panel 3 shows the feedback loop in which ET-ETR signaling induces the VEGF-VEGFR axis activity. The situation of Panel 3 opens four possibilities: (1) the feedback loop overpowers VEGF-VEGFR axis inhibition; (2) ET-ETR expression and signaling reach a significant level driving angiogenesis; (3) ET-ETR additionally induces the β arrestin-1 angiogenic pathway. (4) β arrestin-1 amplifies hypoxia effects on the VEGF-VEGFR axis.

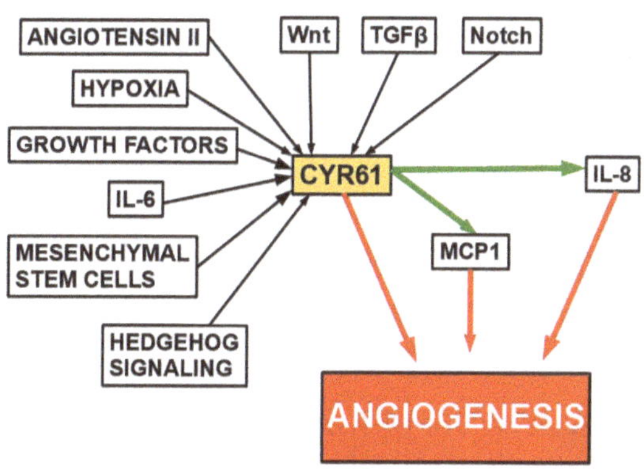

FIG. 11 CYR61 seems to be a hub of pro-angiogenic factors that may work as a parallel pathway of angiogenesis. *The diagram of CYR61 inter-connections is based on references mentioned above and [364–379].*

Other angiogenic proteins

CEA-related cell adhesion molecule-1 [385], tumor necrosis factor α [386], brain-derived neurotrophic factor [387], platelet-derived endothelial cell growth factor [388], bone morphogenetic protein 2 (BMP2) [389] are among other angiogenic proteins frequently found in tumor stroma. Endothelial cells over-expressing the fatty acid-binding proteins FABP4 and FABP5 are prone to migration, proliferation, and angiogenic response. While FABP4 is VEGF-dependent, FABP5 is VEGF-independent [390]. Metabotropic glutamate receptor-1 (mGluR1) is another player in angiogenesis [391]. Its inhibition (by riluzole) significantly inhibits the process of new vessel formation [392].

As shown above there are many angiogenic factors besides the VEGF-VEGFR axis and they interact and produce angiogenesis through this axis. However, many may independently generate new vessels. IL-8, CYR61, and endothelin-1 are good candidates for independent angiogenic agents. However, this needs experimental confirmation.

Galectin-1 [393] and galectin-3 [394] are other essential proteins for angiogenesis, [395,396]. Their down-regulation inhibits angiogenesis and decreases tumor growth [397]. Anginex [398], a designed antiangiogenic, seems to act by inhibiting galectin-1 [399] without relation to the VEGF-VEGFR axis. Endoglin, an endothelial co-receptor for TGF-β, also plays a role in resistance to antiangiogenic treatments [400].

Some hormones have also been identified as pro-angiogenic. This is the case of prolactin which is the best example of the complexity of the angiogenesis program. While the intact prolactin molecule is angiogenic, the N-terminal portion of this hormone is antiangiogenic [401]. Furthermore, when this N-terminal portion of prolactin binds plasminogen activator inhibitor-1 (PAI-1), another antiangiogenic molecule, it becomes fully antiangiogenic [402].

Clinical approach to antiangiogenic resistance

1. Vascular disrupting agents (VDA) are compounds that are able to selectively destroy tumor vessels [403]. The difference between VDAs and antiangiogenic drugs is that the first destroys existing vessels while the latter impedes the formation of new vessels [404]. Complete tumor circulatory shutdown, produced by VDAs, results in tumor necrosis/apoptosis.

VDAs address the question of what to do with tumors that already have existing vessels.

Many such compounds have been investigated, including combretastatin A4 (a tubulin poison) [405], FAA (a synthetic flavonoid), its analog DMXAA, and some proteins fused with diphtheria toxin. Most of these first-generation compounds were too toxic for the clinical setting [406,407].

The main problem of VDAs is that they kill the tumor core cells, but always leave a rim of better-irrigated cells that eventually cause resistance to treatment [408,409]. That is the reason why it has been proposed to use VDAs in association with antiangiogenic, chemo-, and/or radiotherapy.

Thiabendazole and derivatives are FDA-approved anti-helmintic, with very low toxicity [410]. Interestingly, thiabendazole and mebendazole have shown significant VDA activity [411–430]. There is also evidence showing that mebendazole acts in synergy with at least one antiangiogenic drug such as sorafenib [431]. Based on the extensive evidence accumulated with thiabendazole and its derivatives, these drugs should be further investigated as possible enhancers of antiangiogenic drugs and eventually to be used when resistance develops.

2. Fenofibrate. Looking for a complementary drug to improve and prolong the effects of antiangiogenic treatments we found, through a database search, fenofibrate (FF). This is a powerful antiangiogenic drug [432] that at the same time has the ability to act on many of the angiogenic escape factors (Fig. 12). Fenofibrate is an agonist of peroxisome proliferator-activated receptor alpha (PPARα) which is a nuclear transcription factor and is actively being used as a lipid-lowering drug. FF decreases VEGF production and increases TSP1 expression thus inhibiting angiogenesis. Added to these fundamental effects, FF also inhibits/downregulates IGF-IR [434]; inhibits the mTOR/p70S6 pathway [435]; partially inhibits the glycolytic pathway [436]; decreases TNFα and cytokines in plasma [437], and targets the TNFα/NF-kB pathway [438]; decreases the expression of HIF-1α [439,440]; inhibits 5-lipoxygenase [441]; decreases plasma platelet-activating factor (PAF) activity [442] and decreases vascular inflammatory response [443]; decreases the endothelial expression and secretion of PAI-1 [444], an indicator of poor prognosis in some cancers such as breast [445] and lung adenocarcinoma [446]; decreases EGFR activity [447,448]; although FF does not directly inhibit the VEGF-VEGFR axis, it has the ability to decrease the expression/activity of VEGF effector proteins in VEGF-dependent angiosarcomas in vitro [449]; decreases tumor intravasation [450]; increases p53 activity [451] (Fig. 12). However, there are some publications showing a pro-angiogenic effect of fenofibrate [452].

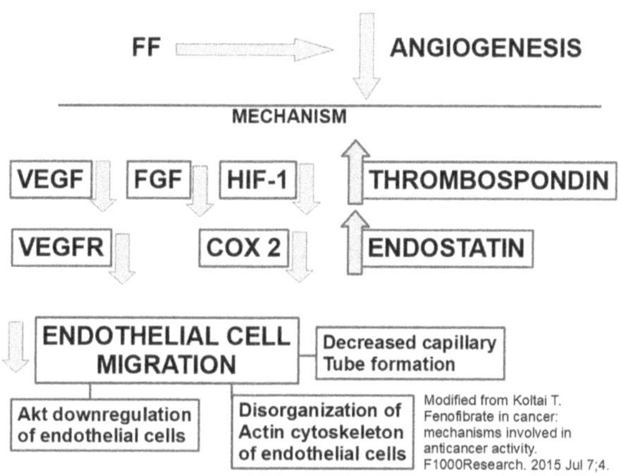

FIG. 12 Antiangiogenic mechanisms of Fenofibrate (FF) [433].

3. Betulinic acid (BA)

BA is a natural product obtained from the bark of white-birch tree. It has many anti-cancer effects [453–457]. Interestingly, BA is a compound that reduces/eliminates the activity of one of the main players in the angiogenic process, HIF-1α [458–460]. A synthetic BA derivative has been shown to block Notch signaling [461].

New pharmaceutical forms of BA conjugated with nanoparticles have shown better BA bioavailability at the tumor site [462]. This and BA lack of toxicity enhance its importance in an antiangiogenic scheme. Therefore, it deserves to be tested as a complement to antiangiogenic treatments (Fig. 13).

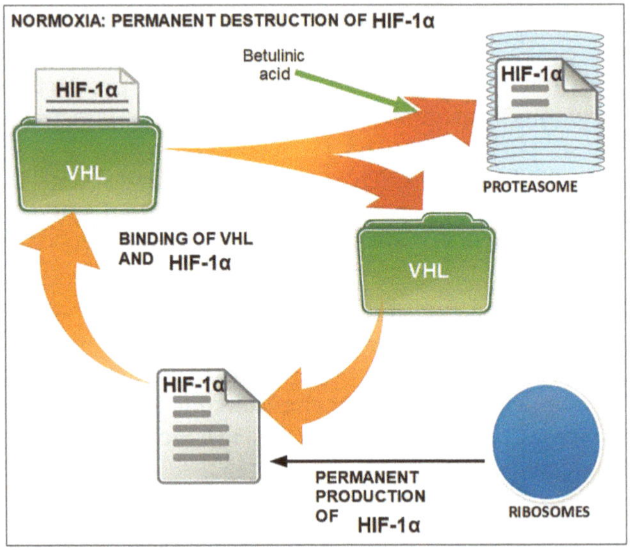

FIG. 13 When there are no hypoxia nor VHL gene modifications, HIF-1α is continuously produced and degraded at the proteasome. VHL is the protein that "carries" HIF to its final disposal. When hypoxia sets in there is no more degradation resulting in the stabilization of HIF-1α, and allowing its nuclear migration as shown in Fig. 1. Betulinic acid has the ability to increase HIF-1α proteasomal degradation. *This diagram is based on references [463–468].*

4. Cimetidine

Cimetidine, a histamine type 2 receptor antagonist, has been used for many years to reduce gastric acid production. While it has been replaced by modern proton-pump inhibitors, cimetidine has been found to inhibit angiogenesis in a VEGF-VEGFR- independent manner [469–473], making it a potentially important antiangiogenesis therapeutic option. The mechanism has not been fully elucidated.

Discussion

The large list of growth factors inducing angiogenesis explains clearly why the inhibition of VEGF or VEGFR is insufficient to fully stop angiogenesis or its recurrence. A placental growth factor is the only one that binds VEGFR, while the other growth factors bind their own receptors, but they all are able to induce angiogenesis to a greater or lesser extent. This has been called "the growth factor redundancy" [474]. This angiogenic growth factor redundancy partially explains the limited results of targeting only one of them. Furthermore, the added hypoxia produced by antiangiogenic treatments is an inducer for the added expression of other growth factors [475,476] and explains the increased rate of metastasis found after antiangiogenic treatments [477].

There are cytokines that act similarly to growth factors, with IL-8 being the most important angiogenic cytokine.

Integrins are major players in the formation of new vessels and their inhibition should not be overlooked in angiogenesis targeting.

Focalizing antiangiogenic treatments on the VEGF-VEGFR axis has been shown to be insufficient for obtaining a long-lasting remission. This is the case even when it is associated with effective chemo- and radiotherapy schemes. Undoubtedly, the success of antiangiogenic treatments has prolonged overall survival and progression-free survival in many types of cancer. However, something else is necessary for a better outcome.

We believe that resistance to treatment is probably the main problem that prevents better results.

This paper proposes some complementary drugs that may change the equation. They are all FDA-approved drugs (with the exception of betulinic acid), with no, or very low toxicity, such as thiabendazole and its derivatives, fenofibrate, cimetidine, and betulinic acid. All these drugs to be repurposed have a mechanism of action that seems to complement antiangiogenic drugs without adding toxicity. They all require clinical testing to confirm their theoretical advantages.

Some tumors, such as advanced melanoma, require also targeting vasculogenic mimicry because this mechanism of escape from antiangiogenic treatments is quite common. At present there is no effective treatment for VM, however, clinical trials with CVM-1118 may change this situation in the near future.

There is evidence that the association of inhibitors of VEGFR and/or EGFR with metronomic chemotherapy can increase the benefits obtained with antiangiogenesis [478,479]. However, this has not been incorporated into current practice.

As presented above (Section K) an important contributor to angiogenesis and resistance to antiangiogenic treatment is ECM acidity and its important load of lactate. Lactate represents the necessary link between tumor metabolism and angiogenesis and we believe it is an important target that has been neglected up to now. pH-centered treatments have the ability to decrease lactate overload and have been found to synergize with antiangiogenesis. Thus, the association of these types of treatments with antiangiogenics and to diminish resistance to antiangiogenic treatments deserves to be thoroughly explored.

Fig. 14 shows the possible relations between hypoxia and resistance to antiangiogenic treatments.

A thorough analysis of Fig. 14 leads to the conclusion that it is the extreme hypoxia produced by antiangiogenic treatments that are at the root of resistance to treatment and further progression of the tumor. Therefore, if we must identify a single cause of treatment resistance that somehow summarizes all the other issues involved, we would undoubtedly point to extreme hypoxia as the main culprit.

Intermittent antiangiogenic treatment

One mechanism to circumvent the development of extreme hypoxia as a consequence of antiangiogenic treatment would be to apply it in an intermittent fashion. This would allow short periods of lower hypoxia. There is evidence that this approach may decrease resistance

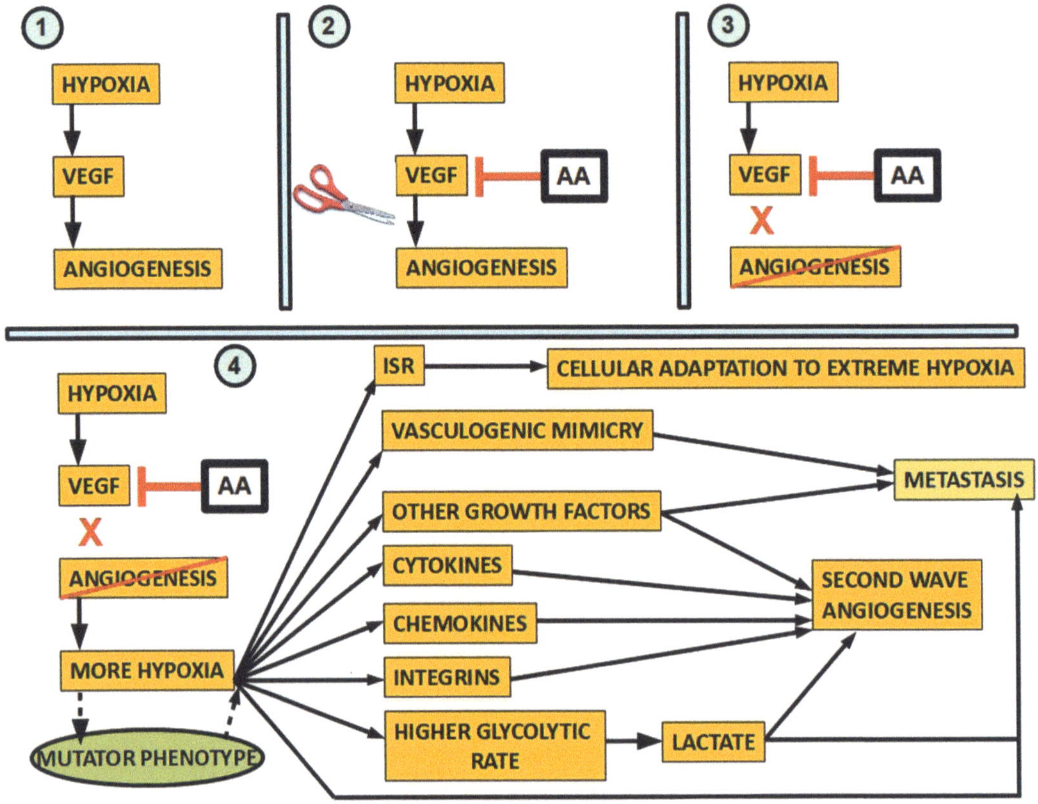

FIG. 14 Panels 1–3: A possible sequence of events leading to antiangiogenic resistance. Panel 1: classical/canonical algorithm of tumor angiogenesis. Panel 2: use of antiangiogenic drugs leading to blockage of angiogenesis (Panel 3). Panel 4: Inhibiting angiogenesis leads to extreme hypoxia that unleashes the second wave of angiogenesis, but this time produced by other pro-angiogenic proteins and pathways. *The ISR (integrated stress response) is based on references [480–499].*

and increase benefits [500,501]. However, there is not enough clinical experience to support this.

Association of antiangiogenic drugs with immunological checkpoint inhibitors

Immune checkpoint inhibitors (ICIs) block the proteins developed by the tumor (PD-1, PD-L1, and CTLA4) creating a state of immune tolerance [502], in order to evade immunological surveillance [503,504]. ICIs used as stand-alone treatments have clear anti-tumoral effects, however, the response rate is low. This situation changed dramatically when ICIs were associated with antiangiogenic drugs. The association of these two different types of drugs has shown interesting results in some tumors such as clear cell renal carcinoma [505], metastatic non-small cell lung cancer [506], hepatocarcinoma [507], among others. The exact mechanism of this "cooperation" has not been fully explained. However, there are some known aspects that deserve mention:

(1) immune checkpoint inhibition decreases hypoxia and VEGF expression;
(2) antiangiogenic treatment reprograms the tumor milieu from an immunosuppressive to an immunopermissive state [508–512];
(3) VEGF increases the expression of immune checkpoint proteins such as PD-1 [513];
(4) Lee et al. [514] proposed that the combined treatment improved vascular-immune crosstalk increasing the immune activity. According to these ideas:
   (a) angiogenic treatment improved the vascular tree and facilitated ICIs' access to the tumor;
   (b) included in this crosstalk VEGF impeded the maturation of dendritic cells and
   (c) exhaustion of cytotoxic CD8+ cells;
   (d) T regulatory cells (Tregs) produced pro-angiogenic cytokines.

Different authors have reached approximately similar conclusions [515–521].

This association is actively being used in the clinical setting and is discussed in Chapter 1 of the present book. However, as yet there is little information about resistance or the influence of the treatment on angiogenic resistance.

There are at least 50 ongoing clinical trials associating antiangiogenic drugs and ICIs.

As preliminary results in many of these studies seem to be very promising, ICIs may be an interesting group of drugs to surmount resistance to antiangiogenic treatments.

Conclusions

Present-day antiangiogenic treatments target either the VEGF-VEGFR axis (bevacizumab, sunitinib, etc.) or directly the proliferation of the endothelial cells (metronomic chemotherapy). There is evidence to support that the escape mechanism in these different treatments is also different.

With metronomic chemotherapy, angiogenesis does not resume when resistance to treatment sets in. In this case, it seems that angiogenesis-independent tumor growth develops. The mechanism has not been elucidated.

With VEGF-VEGFR targeting the mechanisms of resistance are different since there is a reinstatement of angiogenesis. Clinical resistance to anti-VEGF therapy clearly shows that if better results are expected, other angiogenic pathways need to be targeted as well.

The reinstatement of angiogenesis goes through alternate angiogenic pathways totally or partially by-passing the VEGF-VEGFR axis. Important players in the alternate pathways have been identified. These include IL-6, the Notch-pathway, ephrin-Epr signaling, growth factors redundancy, VEGF polymorphism, VEGF amplification, endothelin signaling, and the neuropilin pathway among many known and as yet unknown evasion systems. However, if we must identify one single cause for treatment resistance that somehow summarizes all the other issues involved, we would undoubtedly point to extreme hypoxia as the main culprit.

Future perspectives

With VEGF-VEGFR targeting, chemokines, integrins, and alternate pathways are an open door for escape. Therefore, targeted treatments will have to expand their range of action to include these "evaders". This means that new antiangiogenic treatments will have to attack the problem with a more integral approach also addressing chemokines, integrins, and alternative pathways. IL-8 is a significant culprit and IL-8 antibodies already exist. These antibodies have not reached bedside medicine as yet, but we hope they will soon do so as part of an integral approach. Repurposing fenofibrate and incorporating betulinic acid derivatives may improve the effectiveness and duration of classical antiangiogenic schemes. We will probably see major developments in antiangiogenic strategies associated with immune checkpoint inhibitors. Right now, this seems to be the most promising treatment since it overcomes the limitations of each of the others.

References

[1] Bunn HF, Poyton RO. Oxygen sensing and molecular adaptation to hypoxia. Physiol Rev 1996;76(3):839–85.

[2] Wenger RH. Cellular adaptation to hypoxia: O_2-sensing protein hydroxylases, hypoxia-inducible transcription factors, and O_2-regulated gene expression. FASEB J 2002;16(10):1151–62.

[3] Eales KL, Hollinshead KER, Tennant DA. Hypoxia and metabolic adaptation of cancer cells. Oncogenesis 2016; 5(1):e190.

[4] Pouyssegur J, López-Barneo J. Hypoxia in health and disease. Mol Aspects Med 2016;47:1–2.

[5] Lee P, Chandel NS, Simon MC. Cellular adaptation to hypoxia through hypoxia inducible factors and beyond. Nat Rev Mol Cell Biol 2010;21(5):268–83.

[6] Shweiki D, Itin A, Soffer D, Keshet E. Vascular endothelial growth factor induced by hypoxia may mediate hypoxia-initiated angiogenesis. Nature 1992;359(6398):843–5.

[7] Semenza GL. Regulation of mammalian O_2 homeostasis by hypoxia-inducible factor 1. Annu Rev Cell Dev Biol 1999;15(1):551–78.

[8] Semenza GL. Hypoxia-inducible factors in physiology and medicine. Cell 2012;148(3):399–408.

[9] Semenza GL. HIF-1, O2, and the 3 PHDs: how animal cells signal hypoxia to the nucleus. Cell 2001;107(1):1–3.

[10] Semenza GL. Regulation of metabolism by hypoxia-inducible factor 1. Cold Spring Harb Symp Quant Biol 2011;76:347–53.

[11] Zhong H, De Marzo AM, Laughner E, Lim M, Hilton DA, Zagzag D, Buechler P, Isaacs WB, Semenza GL, Simons JW. Overexpression of hypoxia-inducible factor 1α in common human cancers and their metastases. Cancer Res 1999;59(22):5830–5.

[12] Semenza GL, Wang GL. A nuclear factor induced by hypoxia via de novo protein synthesis binds to the human erythropoietin gene enhancer at a site required for transcriptional activation. Mol Cell Biol 1992; 12(12):5447–54.

[13] Forsythe JA, Jiang BH, Iyer NV, Agani F, Leung SW, Koos RD, Semenza GL. Activation of vascular endothelial growth factor gene transcription by hypoxia-inducible factor 1. Mol Cell Biol 1996;16(9):4604–13.

[14] Zhong H, Chiles K, Feldser D, Laughner E, Hanrahan C, Georgescu MM, Simons JW, Semenza GL. Modulation of hypoxia-inducible factor 1α expression by the epidermal growth factor/phosphatidylinositol 3-kinase/ PTEN/AKT/FRAP pathway in human prostate cancer cells: implications for tumor angiogenesis and therapeutics. Cancer Res 2000;60(6):1541–5.

[15] Gnarra JR, Tory K, Weng Y, Schmidt L, Wei MH, Li H, Latif F, Liu F, Chen F, Duh FM, et al. Mutations of the VHL tumour suppressor gene in renal carcinoma. Nat Genet 1994;7(1):85–90.

[16] Brauch H, Weirich G, Brieger J, Glavač D, Rödl H, Eichinger M, Feurer M, Weidt E, Puranakanitstha C, Neuhaus C, et al. VHL alterations in human clear cell renal cell carcinoma: association with advanced tumor stage and a novel hot spot mutation. Cancer Res 2000;60(7):1942–8.

[17] Semenza GL. Hypoxia-inducible factor 1 (HIF-1) pathway. Sci STKE 2007;2007(407). https://doi.org/10.1126/stke.4072007cm8, cm8.

[18] van Uden P, Kenneth NS, Webster R, Müller HA, Mudie S, Rocha S. Evolutionary conserved regulation of HIF-1β by NF-κB. PLoS Genet 2011;7(1), e1001285.

[19] Jiang BH, Rue E, Wang GL, Roe R, Semenza GL. Dimerization, DNA binding, and transactivation properties of hypoxia-inducible factor 1. J Biol Chem 1996;271(30):17771–8.

[20] Sokkar P, Sathis V, Ramachandran M. Computational modeling on the recognition of the HRE motif by HIF-1: molecular docking and molecular dynamics studies. J Mol Model 2012;18(5):1691–700.
[21] Kuphal S, Winklmeier A, Warnecke C, Bosserhoff AK. Constitutive HIF-1 activity in malignant melanoma. Eur J Cancer 2010;46(6):1159–69.
[22] Nagao A, Kobayashi M, Koyasu S, Chow CC, Harada H. HIF-1-dependent reprogramming of glucose metabolic pathway of cancer cells and its therapeutic significance. Int J Mol Sci 2019;20(2):238.
[23] Natale G, Bocci G, Lenzi P. Looking for the word "angiogenesis" in the history of health sciences: from ancient times to the first decades of the twentieth century. World J Surg 2017;41(6):1625–34.
[24] Griffioen AW, Molema G. Angiogenesis: potentials for pharmacologic intervention in the treatment of cancer, cardiovascular diseases, and chronic inflammation. Pharmacol Rev 2000;52(2):237–68.
[25] Hall AP. The role of angiogenesis in cancer. Comp Clin Pathol 2005;13(3):95–9.
[26] Folkman J. Tumor angiogenesis: therapeutic implications. N Engl J Med 1971;285(21):1182–6.
[27] Folkman J, Merler E, Abernathy C, Williams G. Isolation of a tumor factor responsible for angiogenesis. J Exp Med 1971;133(2):275–88.
[28] Kerbel R, Folkman J. Clinical translation of angiogenesis inhibitors. Nat Rev Cancer 2002;2(10):727–39.
[29] Kerbel RS. Inhibition of tumor angiogenesis as a strategy to circumvent acquired resistance to anti-cancer therapeutic agents. Bioessays 1991;13(1):31–6.
[30] Folkman J. New perspectives in clinical oncology from angiogenesis research. Eur J Cancer 1991;32A:2534–9.
[31] O'Reilly MS, Holmgren L, Shing Y, Chen C, Rosenthal RA, Moses M, Lane WS, Cao Y, Sage EH, Folkman J. Angiostatin: a novel angiogenesis inhibitor that mediates the suppression of metastases by a Lewis lung carcinoma. Cell 1994;79(2):315–28.
[32] Good DJ, Polverini PJ, Rastinejad F, Le Beau MM, Lemons RS, Frazier WA, Bouck NP. A tumor suppressor-dependent inhibitor of angiogenesis is immunologically and functionally indistinguishable from a fragment of thrombospondin. Proc Natl Acad Sci U S A 1990;87(17):6624–8.
[33] Hanahan D, Weinberg RA. The hallmarks of cancer. Cell 2000;100(1):57–70.
[34] Hanahan D, Weinberg RA. Hallmarks of cancer: the next generation. Cell 2011;144(5):646–74.
[35] Naumov GN, Akslen LA, Folkman J. Role of angiogenesis in human tumor dormancy: animal models of the angiogenic switch. Cell Cycle 2006;5(16):1779–87.
[36] Roukos DH, Tzakos A, Zografos G. Current concerns and challenges regarding tailored anti-angiogenic therapy in cancer. Expert Rev Anticancer Ther 2009;9(10):1413–6.
[37] Jain RK, Duda DG, Clark JW, Loeffler JS. Lessons from phase III clinical trials on anti-VEGF therapy for cancer. Nat Rev Clin Oncol 2006;3(1):24.
[38] Van Cutsem E, Vervenne WL, Bennouna J, Humblet Y, Gill S, Van Laethem JL, Sheithauer W, Shang A, Cosaert J, Moore MJ. Phase III trial of bevacizumab in combination with gemcitabine and erlotinib in patients with metastatic pancreatic cancer. J Clin Oncol 2009;27(13):2231–7.
[39] Miller K, Wang M, Gralow J, Dickler M, Cobleigh M, Perez EA, Shenkier T, Cella D, Davidson NE. Paclitaxel plus bevacizumab versus paclitaxel alone for metastatic breast cancer. N Engl J Med 2007;357(26):2666–76.
[40] Haines IE, Miklos GL. Paclitaxel plus bevacizumab for metastatic breast cancer. N Engl J Med 2008;358(15):1637.
[41] Gray R, Bhattacharya S, Bowden C, Miller K, Comis RL. Independent review of E2100: a phase III trial of bevacizumab plus paclitaxel versus paclitaxel in women with metastatic breast cancer. J Clin Oncol 2009; 27(30):4966.
[42] Cella D, Wang M, Wagner L, Miller K. Survival-adjusted health-related quality of life (HRQL) among patients with metastatic breast cancer receiving paclitaxel plus bevacizumab versus paclitaxel alone: results from Eastern Cooperative Oncology Group Study 2100 (E2100). Breast Cancer Res Treat 2011;130(3):855–61.
[43] Sandler A, Gray R, Perry MC, Brahmer J, Schiller JH, Dowlati A, Lilenbaum R, Johnson DH. Paclitaxel–carboplatin alone or with bevacizumab for non–small-cell lung cancer. Breast Cancer Res Treat 2006;355(24):2542–50.
[44] Montero AJ, Avancha K, Glück S, Lopes G. A cost-benefit analysis of bevacizumab in combination with paclitaxel in the first-line treatment of patients with metastatic breast cancer. Breast Cancer Res Treat 2012;132(2): 747–51.
[45] Griffioen AW. Anti-angiogenesis: making the tumor vulnerable to the immune system. Cancer Immunol Immunother 2008;57(10):1553–8.
[46] Griffioen AW, Molema G. Angiogenesis: potentials for pharmacologic intervention in the treatment of cancer, cardiovascular diseases, and chronic inflammation. Pharmacol Rev 2000;52(2):237–68.
[47] Hillen F, Griffioen AW. Tumour vascularization: sprouting angiogenesis and beyond. Cancer Metastasis Rev 2007;26(3):489–502.

[48] Carmeliet P, Jain RK. Principles and mechanisms of vessel normalization for cancer and other angiogenic diseases. Nat Rev Drug Discov 2011;10(6):417–27.

[49] Carmeliet P, Jain RK. Molecular mechanisms and clinical applications of angiogenesis. Nature 2011;473(7347): 298–307.

[50] Potente M, Gerhardt H, Carmeliet P. Basic and therapeutic aspects of angiogenesis. Cell 2011;146(6):873–87.

[51] Schmidt T, Carmeliet P. Angiogenesis: a target in solid tumors, also in leukemia? Hematology Am Soc Hematol Educ Program 2011;2011(1):1–8.

[52] Killough JH, Magill GB, Smith RC. The treatment of amebiasis with fumagillin. Science (Washington) 1952;115:71–2.

[53] Ingber D, Fujita T, Kishimoto S, Sudo K, Kanamaru T, Brem H, Folkman J. Synthetic analogues of fumagillin that inhibit angiogenesis and suppress tumour growth. Nature 1990;348(6301):555.

[54] White CW, Sondheimer HM, Crouch EC, Wilson H, Fan LL. Treatment of pulmonary hemangiomatosis with recombinant interferon alfa-2a. N Engl J Med 1989;320(18):1197–200.

[55] Medical Research Council Renal Cancer Collaborators. Interferon-α and survival in metastatic renal carcinoma: early results of a randomised controlled trial. Lancet 1999;353(9146):14–7.

[56] Oliver RTD, Nethersell ABW, Bottomley JM. Unexplained spontaneous regression and alpha-interferon as treatment for metastatic renal carcinoma. Br J Urol 1989;63(2):128–31.

[57] Krown SE. Interferon treatment of renal cell carcinoma. Current status and future prospects. Cancer 1987; 59(S3):647–51.

[58] Ryan CW, Goldman BH, Lara Jr PN, Mack PC, Beer TM, Tangen CM, Lemmon D, Pan CX, Drabkin HA, Crawford ED. Sorafenib with interferon alfa-2b as first-line treatment of advanced renal carcinoma: a phase II study of the Southwest Oncology Group. J Clin Oncol 2007;25(22):3296–301.

[59] Baenziger NL, Brodie GN, Majerus PW. A thrombin-sensitive protein of human platelet membranes. Proc Natl Acad Sci U S A 1971;68(1):240–3.

[60] Kaur S, Martin-Manso G, Pendrak ML, Garfield SH, Isenberg JS, Roberts DD. Thrombospondin-1 inhibits VEGF receptor-2 signaling by disrupting its association with CD47. J Biol Chem 2010;285(50):38923–32.

[61] Greenaway J, Lawler J, Moorehead R, Bornstein P, Lamarre J, Petrik J. Thrombospondin-1 inhibits VEGF levels in the ovary directly by binding and internalization via the low density lipoprotein receptor-related protein-1 (LRP-1). J Cell Physiol 2007;210(3):807–18.

[62] Chu LY, Ramakrishnan DP, Silverstein RL. Thrombospondin-1 modulates VEGF signaling via CD36 by recruiting SHP-1 to VEGFR2 complex in microvascular endothelial cells. Blood 2013;122(10):1822–32.

[63] Kanda S, Shono T, Tomasini-Johansson B, Klint P, Saito Y. Role of thrombospondin-1-derived peptide, 4N1K, in FGF-2-induced angiogenesis. Exp Cell Res 1999;252(2):262–72.

[64] Bagavandoss P, Wilks JW. Specific inhibition of endothelial cell proliferation by thrombospondin. Biochem Biophys Res Commun 1990;170(2):67–872.

[65] Colombo G, Margosio B, Ragona L, Neves M, Bonifacio S, Annis DS, Stravalaci M, Tomaselli S, Giavazzi R, Rusnati M, et al. Non-peptidic thrombospondin-1 mimics as fibroblast growth factor-2 inhibitors an integrated strategy for the development of new antiangiogenic compounds. J Biol Chem 2010;285(12):8733–42.

[66] Yakes FM, Chen J, Tan J, Yamaguchi K, Shi Y, Yu P, Qian F, Chu F, Bentzien F, Cancilla B, et al. Cabozantinib (XL184), a novel MET and VEGFR2 inhibitor, simultaneously suppresses metastasis, angiogenesis, and tumor growth. Mol Cancer Ther 2011;10(12):2298–308.

[67] Heffelfinger SC, Yan M, Gear RB, Schneider J, LaDow K, Warshawsky D. Inhibition of VEGFR2 prevents DMBA-induced mammary tumor formation. Lab Invest 2004;84(8):989.

[68] Dias S, Hattori K, Heissig B, Zhu Z, Wu Y, Witte L, Hicklin DJ, Tateno M, Bohlen P, Moore MAS, et al. Inhibition of both paracrine and autocrine VEGF/VEGFR-2 signaling pathways is essential to induce long-term remission of xenotransplanted human leukemias. Proc Natl Acad Sci U S A 2001;98(19):10857–62.

[69] Yue TL, Wang X, Louden CS, Gupta S, Pillarisetti K, Gu JL, Heart TK, Lisco PJ, Feuerstein GZ. 2-Methoxyestradiol, an endogenous estrogen metabolite, induces apoptosis in endothelial cells and inhibits angiogenesis: possible role for stress-activated protein kinase signaling pathway and Fas expression. Mol Pharmacol 1997;51(6):951–62.

[70] Masiero L, Figg WD, Kohn EC. New anti-angiogenesis agents: review of the clinical experience with carboxyamido-triazole (CAI), thalidomide, TNP-470 and interleukin 12. Angiogenesis 1997;1(1):23–35.

[71] Kerbel RS, Kamen BA. The anti-angiogenic basis of metronomic chemotherapy. Nat Rev Cancer 2004; 4(6):423.

[72] Markovic SN, Suman VJ, Rao RA, Ingle JN, Kaur JS, Erickson LA, Pitot HC, Croghan GA, McWilliams RR, Merchan J, et al. A phase II study of ABT-510 (thrombospondin-1 analog) for the treatment of metastatic melanoma. Am J Clin Oncol 2007;30(3):303–9.
[73] Hurwitz H, Fehrenbacher L, Novotny W, Cartwright T, Hainsworth J, Heim W, Berlin J, Baron A, Griffing S, Holmgren E, et al. Bevacizumab plus irinotecan, fluorouracil, and leucovorin for metastatic colorectal cancer. N Engl J Med 2004;350(23):2335–42.
[74] Ramaswamy B, Elias AD, Kelbick NT, Dodley A, Morrow M, Hauger M, Allen J, Rhoades C, Kendra K, Chen HX, et al. Phase II trial of bevacizumab in combination with weekly docetaxel in metastatic breast cancer patients. Clin Cancer Res 2006;12(10):3124–9.
[75] Yang JC, Haworth L, Sherry RM, Hwu P, Schwartzentruber DJ, Topalian SL, Steinberg SM, Chen HX, Rosenberg SA. A randomized trial of bevacizumab, an anti–vascular endothelial growth factor antibody, for metastatic renal cancer. N Engl J Med 2003;349(5):427–34.
[76] Ferrara N, Adamis AP. Ten years of anti-vascular endothelial growth factor therapy. Nat Rev Drug Discov 2016;15(6):385–403.
[77] Clarke JM, Hurwitz HI. Targeted inhibition of VEGF receptor 2: an update on ramucirumab. Expert Opin Biol Ther 2013;13(8):1187–96.
[78] Refolo MG, Lotesoriere C, Lolli IR, Messa C, D'Alessandro R. Molecular mechanisms of synergistic action of ramucirumab and paclitaxel in gastric cancers cell lines. Sci Rep 2020;10(1):1–13.
[79] Escudier B, Eisen T, Stadler WM, Szczylik C, Oudard S, Siebels M, Negrier S, Chevreau C, Solska E, Desai AA, et al. Sorafenib in advanced clear-cell renal-cell carcinoma. N Engl J Med 2007;356(2):125–34. https://doi.org/10.1056/NEJMoa060655.
[80] Motzer RJ, Hutson TE, Tomczak P, Michaelson MD, Bukowski RM, Rixe O, Oudard S, Negrier S, Szczylik C, Kim ST, et al. Sunitinib versus interferon alfa in metastatic renal-cell carcinoma. N Engl J Med 2007;356(2):115–24.
[81] Tai WT, Shiau CW, Li YS, Chang CW, Huang JW, Hsueh TT, Yu HC, Chen KF. Nintedanib (BIBF-1120) inhibits hepatocellular carcinoma growth independent of angiokinase activity. J Hepatol 2014;61(1):89–97.
[82] Roy S, Narang BK, Rastogi SK, Rawal RK. A novel multiple tyrosine-kinase targeted agent to explore the future perspectives of anti-angiogenic therapy for the treatment of multiple solid tumors: cabozantinib. Anticancer Agents Med Chem 2015;15(1):37–47. https://doi.org/10.2174/1871520614666140902153840.
[83] Strumberg D, Schultheis B. Regorafenib for cancer. Expert Opin Investig Drugs 2012;21(6):879–89.
[84] Santangelo A, Rossato M, Lombardi G, Benfatto S, Lavezzari D, De Salvo GL, Stefano I, Dechecchi MC, Prandini P, Gambari R, et al. A molecular signature associated with prolonged survival in glioblastoma patients treated with regorafenib. Neuro Oncol 2021;23(2):264–76. https://doi.org/10.1093/neuonc/noaa156.
[85] Khan KA, de Léon JLP, Benguigui M, Xu P, Chow A, Cruz-Muñoz W, Man S, Shaked Y, Kerbel S. Immunostimulatory and anti-tumor metronomic cyclophosphamide regimens assessed in primary orthotopic and metastatic murine breast cancer. NPJ Breast Cancer 2020;6(1):1–13.
[86] Lu Q, Lee K, Xu F, Xia W, Zheng Q, Hong R, Jiang K, Zhai Q, Li Y, Shi Y, et al. Metronomic chemotherapy of cyclophosphamide plus methotrexate for advanced breast cancer: real-world data analyses and experience of one center. Cancer Commun 2020;40(5):222–33.
[87] Di Francia R, Berretta M, Benincasa G, D'Avino A, Facchini S, Costagliola D, Rossi P. Pharmacogenetic-based interactions between nutraceuticals and angiogenesis inhibitors. Cells 2019;8(6):522.
[88] Monk BJ, Poveda A, Vergote I, Raspagliesi F, Fujiwara K, Bae DS, Oaknin A, Ray-Coquard I, Provencher DM, Karlan BY, et al. Final results of a phase 3 study of trebananib plus weekly paclitaxel in recurrent ovarian cancer (TRINOVA-1): long-term survival, impact of ascites, and progression-free survival-2. Gynecol Oncol 2016; 143(1):27–34.
[89] Ribatti D, Nico B. Limits of anti-angiogenic therapy. Res Clin Med 2016;1:2.
[90] Mancuso MR, Davis R, Norberg SM, O'Brien S, Sennino B, Nakahara T, Yao VJ, Inai T, Brooks T, Freimark B, et al. Rapid vascular regrowth in tumors after reversal of VEGF inhibition. J Clin Invest 2006;116(10):2610–21.
[91] Gacche RN, Meshram RJ. Angiogenic factors as potential drug target: efficacy and limitations of anti-angiogenic therapy. Biochim Biophys Acta 2014;1846(1):161–79. https://doi.org/10.1016/j.bbcan.2014.05.002.
[92] Casanovas O, Hicklin DJ, Bergers G, Hanahan D. Drug resistance by evasion of antiangiogenic targeting of VEGF signaling in late-stage pancreatic islet tumors. Cancer Cell 2005;8(4):299–309.
[93] Kindler HL, Niedzwiecki D, Hollis D, Oraefo E, Schrag D, Hurwitz H, MacLeod L, Mulcahy MF, Schilsky RF, Goldberg RM. A double-blind, placebo-controlled, randomized phase III trial of gemcitabine (G) plus

bevacizumab (B) versus gemcitabine plus placebo (P) in patients (pts) with advanced pancreatic cancer (PC): A preliminary analysis of Cancer and Leukemia Group B (CALGB). J Clin Oncol 2007;25(18-suppl):4508.
[94] Azam F, Mehta S, Harris AL. Mechanisms of resistance to antiangiogenesis therapy. Eur J Cancer 2010; 46(8):1323–32.
[95] Bergers G, Hanahan D. Modes of resistance to anti-angiogenic therapy. Nat Rev Cancer 2008;8(8):592.
[96] Kerbel RS. A cancer therapy resistant to resistance. Nature 1997;390(6658):335.
[97] Croix BS, Rago C, Velculescu V, Traverso G, Romans KE, Montgomery E, Lal A, Riggins GJ, Lengauer C, Vogelstein B, et al. Genes expressed in human tumor endothelium. Science 2000;289(5482):1197–202.
[98] Lu C, Bonome T, Li Y, Kamat AA, Han LY, Schmandt R, Coleman RL, Gershenson DM, Jaffe RB, Birrer MJ, et al. Gene alterations identified by expression profiling in tumor-associated endothelial cells from invasive ovarian carcinoma. Cancer Res 2007;67(4):1757–68.
[99] Akino T, Hida K, Hida Y, Tsuchiya K, Freedman D, Muraki C, Ohga N, Matsuda K, Akiyama K, Harabayashi T, et al. Cytogenetic abnormalities of tumor-associated endothelial cells in human malignant tumors. Am J Pathol 2009;175(6):2657–67.
[100] Virrey JJ, Dong D, Stiles C, Patterson JB, Pen L. Stress chaperone GRP78/BiP confers chemoresistance to tumor-associated endothelial cells. Mol Cancer 2008;6(8):1268–75.
[101] Virrey JJ, Guan S, Li W, Schönthal AH, Chen TC, Hofman FM. Increased survivin expression confers chemoresistance to tumor-associated endothelial cells. Am J Pathol 2008;173(2):575–85.
[102] Chu LH, Rivera CG, Popel AS, Bader JS. Constructing the angiome: a global angiogenesis protein interaction network. Physiol Genomics 2012;44(19):915–24.
[103] Chu LH, Lee E, Bader JS, Popel AS. Angiogenesis interactome and time course microarray data reveal the distinct activation patterns in endothelial cells. PLoS One 2014;9(10), e110871.
[104] Rivera CG, Chu LH, Bader JS, Popel AS. Applications of network bioinformatics to cancer angiogenesis. In: Systems biology in cancer research and drug discovery. Dordrecht: Springer; 2012. p. 229–44.
[105] Joanne LY, Rak JW, Carmeliet P, Nagy A, Kerbel RS, Coomber BL. Heterogeneous vascular dependence of tumor cell populations. Am J Pathol 2001;158(4):1325–34.
[106] Komar G, Kauhanen S, Liukko K, Seppänen M, Kajander S, Ovaska J, Nuutila P, Minn H. Decreased blood flow with increased metabolic activity: a novel sign of pancreatic tumor aggressiveness. Clin Cancer Res 2009; 15(17):5511–7.
[107] Yang J, Zhu DM, Zhou XG, Yin N, Zhang Y, Zhang ZX, Li DC, Zhou J. HIF-2α promotes the formation of vasculogenic mimicry in pancreatic cancer by regulating the binding of Twist1 to the VE-cadherin promoter. Oncotarget 2017;8(29):47801–15.
[108] Ribatti D, Annese T, Ruggieri S, Tamma R, Crivellato E. Limitations of anti-angiogenic treatment of tumors. Transl Oncol 2019;12(7):981–6.
[109] Carbone C, Moccia T, Zhu C, Paradiso G, Budillon A, Chiao PJ, Abbruzzese JL, Melise D. Anti-VEGF treatment resistant pancreatic cancers secrete proinflammatory factors that contribute to malignant progression by inducing an EMT cell phenotype. Clin Cancer Res 2011;17:5822–32.
[110] Hanahan D, Bergers G, Bergsland E. Less is more, regularly: metronomic dosing of cytotoxic drugs can target tumor angiogenesis in mice. J Clin Invest 2000;105(8):1045–7.
[111] Kerbel RS, Kamen BA. The anti-angiogenic basis of metronomic chemotherapy. Nat Rev Cancer 2004;4(6): 423–6.
[112] Browder T, Butterfield CE, Kräling BM, Shi B, Marshall B, O'Reilly MS, Folkman J. Antiangiogenic scheduling of chemotherapy improves efficacy against experimental drug-resistant cancer. Cancer Res 2000; 60(7):1878–86.
[113] Lutsiak MC, Semnani RT, De Pascalis R, Kashmiri SV, Schlom J, Sabzevari H. Inhibition of CD4+ 25+ T regulatory cell function implicated in enhanced immune response by low-dose cyclophosphamide. Blood 2005; 105(7):2862–8.
[114] Tanaka H, Matsushima H, Mizumoto N, Takashima A. Classification of chemotherapeutic agents based on their differential in vitro effects on dendritic cells. Cancer Res 2009;69(17):6978–86.
[115] Bahl A, Bakhshi S. Metronomic chemotherapy in progressive pediatric malignancies: old drugs in new package. Indian J Pediatr 2012;79(12):1617–22.
[116] Silva R, D'Amico G, Hodivala-Dilke KM, Reynolds LE. Integrins: the keys to unlocking angiogenesis. Arterioscler Thromb Vasc Biol 2008;28(10):1703–13.

[117] Hellström M, Phng LK, Hofman JJ, Wallgard E, Coultas L, Lindblom P, Alva J, Nilsson AK, Karlsson L, Gaiano N, et al. Dll4 signalling through Notch1 regulates formation of tip cells during angiogenesis. Nature 2007;445:776–80.

[118] Thurston G, Kitajewski J. VEGF and Delta-Notch: interacting signalling pathways in tumour angiogenesis. Br J Cancer 2008;99(8):1204–9.

[119] Liu ZJ, Shirakawa T, Li Y, Soma A, Oka M, Dotto GP, Fairman RM, Velazquez OC, Herlyn M. Regulation of Notch1 and Dll4 by vascular endothelial growth factor in arterial endothelial cells: implications for modulating arteriogenesis and angiogenesis. Mol Cell Biol 2003;23(1):14–25.

[120] Radtke F, Schweisguth F, Pear W. The notch 'gospel' workshop on notch signalling in development and cancer. EMBO Rep 2005;6(12):1120–5.

[121] Jain RK. Molecular regulation of vessel maturation. Nat Med 2003;9(6):685–93.

[122] Teichert M, Milde L, Holm A, Stanicek L, Gengenbacher N, Savant S, Ruckdeschel T, Hasanov Z, Srivastava K, Hu J. Pericyte-expressed Tie2 controls angiogenesis and vessel maturation. Nat Commun 2017;8(1):1–12.

[123] Thoenes L, Hoehn M, Kashirin R, Ogris M, Arnold GJ, Wagner E, Guenther M. In vivo chemoresistance of prostate cancer in metronomic cyclophosphamide therapy. J Proteomics 2010;73(7):1342–54.

[124] Kubisch R, Meissner L, Krebs S, Blum H, Günther M, Roidl A, Wagner E. A comprehensive gene expression analysis of resistance formation upon metronomic cyclophosphamide therapy. Transl Oncol 2013;6(1), 1-IN3.

[125] Folkman J, Klagsbrun M. Angiogenic factors. Science 1987;235(4787):442–7.

[126] Pepper MS, Ferrara N, Orci L, Montesano R. Potent synergism between vascular endothelial growth factor and basic fibroblast growth factor in the induction of angiogenesis in vitro. Biochem Biophys Res Commun 1992; 189(2):824–31.

[127] Gospodarowicz D, Bialecki H, Thakral TK. The angiogenic activity of the fibroblast and epidermal growth factor. Exp Eye Res 1979;28(5):501–14.

[128] Moroianu J, Riordan JF. Nuclear translocation of angiogenin in proliferating endothelial cells is essential to its angiogenic activity. Proc Natl Acad Sci U S A 1994;91(5):1677–81.

[129] Schreiber AB, Winkler ME, Derynck R. Transforming growth factor-alpha: a more potent angiogenic mediator than epidermal growth factor. Science 1986;232(4755):1250–3.

[130] Roberts AB, Sporn MB, Assoian RK, Smith JM, Roche NS, Wakefield LM, Heine UI, Liotta LA, Falanga V, Kehrl JH. Transforming growth factor type beta: rapid induction of fibrosis and angiogenesis in vivo and stimulation of collagen formation in vitro. Proc Natl Acad Sci U S A 1986;83(12):4167–71.

[131] Wiseman DM, Polverini PJ, Kamp DW, Leibovich SJ. Transforming growth factor-beta (TGFβ) is chemotactic for human monocytes and induces their expression of angiogenic activity. Biochem Biophys Res Commun 1988;157(2):793–800.

[132] Zhong H, Chiles K, Feldser D, Laughner E, Hanrahan C, Georgescu MM, Simmos JW, Semenza GL. Modulation of hypoxia-inducible factor 1α expression by the epidermal growth factor/phosphatidylinositol 3-kinase/PTEN/AKT/FRAP pathway in human prostate cancer cells: implications for tumor angiogenesis and therapeutics. Cancer Res 2000;60(6):1541–5.

[133] Petit AM, Rak J, Hung MC, Rockwell P, Goldstein N, Fendly B, Kerbel RS. Neutralizing antibodies against epidermal growth factor and ErbB-2/neu receptor tyrosine kinases down-regulate vascular endothelial growth factor production by tumor cells in vitro and in vivo: angiogenic implications for signal transduction therapy of solid tumors. Am J Pathol 1997;151(6):1523.

[134] Perrotte P, Matsumoto T, Inoue K, Kuniyasu H, Eve BY, Hicklin DJ, Radinsky R, Dinney CP. Anti-epidermal growth factor receptor antibody C225 inhibits angiogenesis in human transitional cell carcinoma growing orthotopically in nude mice. Clin Cancer Res 1999;5(2):257–64.

[135] Huang SM, Harari PM. Modulation of radiation response after epidermal growth factor receptor blockade in squamous cell carcinomas: inhibition of damage repair, cell cycle kinetics, and tumor angiogenesis. Clin Cancer Res 2000;6(6):2166–74.

[136] Viloria-Petit A, Crombet T, Jothy S, Hicklin D, Bohlen P, Schlaeppi JM, Rak J, Kerbel RS. Acquired resistance to the antitumor effect of epidermal growth factor receptor-blocking antibodies in vivo: a role for altered tumor angiogenesis. Cancer Res 2001;61(13):5090–101.

[137] Relf M, LeJeune S, Scott PA, Fox S, Smith K, Leek R, Moghaddam A, Whitehouse R, Bicknell R, Harris AL. Expression of the angiogenic factors vascular endothelial cell growth factor, acidic and basic fibroblast growth factor, tumor growth factor β-1, platelet-derived endothelial cell growth factor, placenta growth factor, and pleiotrophin in human primary breast cancer and its relation to angiogenesis. Cancer Res 1997;57(5):963–9.

[138] Risau W, Drexler H, Mironov V, Smits A, Siegbahn A, Funa K, Heldin CH. Platelet-derived growth factor is angiogenic in vivo. Growth Factors 1992;7(4):261–6.
[139] Sato N, Beitz JG, Kato J, Yamamoto M, Clark JW, Calabresi P, Raymond A, Frackelton Jr AR. Platelet-derived growth factor indirectly stimulates angiogenesis in vitro. Am J Pathol 1993;142(4):1119.
[140] Ziche M, Maglione D, Ribatti D, Morbidelli L, Lago CT, Battisti M, Paoletti I, Barra A, Tucci M, Parise G, et al. Placenta growth factor-1 is chemotactic, mitogenic, and angiogenic. Lab Invest 1997;76(4):517–31.
[141] Zhang YW, Su Y, Volpert OV, Woude GFV. Hepatocyte growth factor/scatter factor mediates angiogenesis through positive VEGF and negative thrombospondin 1 regulation. Proc Natl Acad Sci U S A 2003; 100(22):12718–23.
[142] Bussolino F, Di Renzo MF, Ziche M, Bocchietto E, Olivero M, Naldini L, Gaudino G, Tamagnone L, Coffer A, Comoglio PM. Hepatocyte growth factor is a potent angiogenic factor which stimulates endothelial cell motility and growth. J Cell Biol 1992;119(3):629–41.
[143] Grant MB, Mames RN, Fitzgerald C, Ellis EA, Aboufriekha M, Guy J. Insulin-like growth factor I acts as an angiogenic agent in rabbit cornea and retina: comparative studies with basic fibroblast growth factor. Diabetologia 1993;36(4):282–91.
[144] Calza L, Giardino L, Giuliani A, Aloe L, Levi-Montalcini R. Nerve growth factor control of neuronal expression of angiogenetic and vasoactive factors. Proc Natl Acad Sci U S A 2001;98(7):4160–5.
[145] Sun L, Hui AM, Su Q, Vortmeyer A, Kotliarov Y, Pastorino S, Passaniti A, Menon J, Walling J, Bailey R, et al. Neuronal and glioma-derived stem cell factor induces angiogenesis within the brain. Cancer Cell 2006;9(4): 287–300.
[146] Ngan ES, Tam PK. Prokineticin-signaling pathway. Int J Biochem Cell Biol 2008;40(9):1679–84.
[147] Monnier J, Samson M. Prokineticins in angiogenesis and cancer. Cancer Lett 2010;296(2):144–9.
[148] Urayama K, Guilini C, Messaddeq N, Hu K, Steenman M, Kurose H, Ert G, Nebigil CG. The prokineticin receptor-1 (GPR73) promotes cardiomyocyte survival and angiogenesis. FASEB J 2007;21(11): 2980–93.
[149] Kurebayashi H, Goi T, Shimada M, Tagai N, Naruse T, Nakazawa T, Kimura Y, Hirono Y, Yamaguchi A. Prokineticin 2 (PROK2) is an important factor for angiogenesis in colorectal cancer. Oncotarget 2015; 6(28):26242.
[150] Goi T, Nakazawa T, Hirono Y, Yamaguchi A. Prokineticin 1 expression in gastrointestinal tumors. Anticancer Res 2013;33(12):5311–5.
[151] Friedlander M, Brooks PC, Shaffer RW, Kincaid CM, Varner JA, Cheresh DA. Definition of two angiogenic pathways by distinct αv integrins. Science 1995;270(5241):1500–2.
[152] Kwon YW, Heo SC, Jeong GO, Yoon JW, Mo WM, Lee MJ, Kim JH. Tumor necrosis factor-α-activated mesenchymal stem cells promote endothelial progenitor cell homing and angiogenesis. Biochim Biophys Acta 2013;1832(12):2136–44.
[153] Margetts PJ, Kolb M, Yu L, Hoff CM, Holmes CJ, Anthony DC, Gauldie J. Inflammatory cytokines, angiogenesis, and fibrosis in the rat peritoneum. Am J Pathol 2002;160(6):2285–94.
[154] Naldini A, Carraro F. Role of inflammatory mediators in angiogenesis. Curr Drug Targets Inflamm Allergy 2005;4(1):3–8.
[155] Wei LH, Kuo ML, Chen CA, Chou CH, Lai KB, Lee CN, Hsieh CY. Interleukin-6 promotes cervical tumor growth by VEGF-dependent angiogenesis via a STAT3 pathway. Oncogene 2003;22(10):1517–27.
[156] Nilsson MB, Langley RR, Fidler IJ. Interleukin-6, secreted by human ovarian carcinoma cells, is a potent proangiogenic cytokine. Cancer Res 2005;65(23):10794–800.
[157] Cohen T, Nahari D, Cerem LW, Neufeld G, Levi BZ. Interleukin 6 induces the expression of vascular endothelial growth factor. J Biol Chem 1996;271(2):736–41.
[158] Catar R, Witowski J, Zhu N, Lücht C, Soria AD, Fernandez JU, Chen L, Jones SA, Fielding CA, Rudolf A, et al. IL-6 trans–signaling links inflammation with angiogenesis in the peritoneal membrane. J Am Soc Nephrol 2017; 28(4):1188–99.
[159] Nagasaki T, Hara M, Nakanishi H, Takahashi H, Sato M, Takeyama H. Interleukin-6 released by colon cancer-associated fibroblasts is critical for tumour angiogenesis: anti-interleukin-6 receptor antibody suppressed angiogenesis and inhibited tumour–stroma interaction. Br J Cancer 2014;110(2):469–78.
[160] Fan Y, Ye J, Shen F, Zhu Y, Yeghiazarians Y, Zhu W, Chen Y, Lawton MT, Young WL, Yang GY. Interleukin-6 stimulates circulating blood-derived endothelial progenitor cell angiogenesis in vitro. J Cereb Blood Flow Metab 2008;28(1):90–8.

[161] Huang SP, Wu MS, Shun CT, Wang HP, Lin MT, Kuo ML, Lin JT. Interleukin-6 increases vascular endothelial growth factor and angiogenesis in gastric carcinoma. J Biomed Sci 2004;11(4):517–27.

[162] Lamy S, Akla N, Ouanouki A, Lord-Dufour S, Béliveau R. Diet-derived polyphenols inhibit angiogenesis by modulating the interleukin-6/STAT3 pathway. Exp Cell Res 2012;318(13):1586–96.

[163] Hatzi E, Murphy C, Zoephel A, Rasmussen H, Morbidelli L, Ahorn H, Kunisada K, Tontsch U, Klenk M, Yamauchi-Takihara K, et al. N-myc oncogene overexpression down-regulates IL-6; evidence that IL-6 inhibits angiogenesis and suppresses neuroblastoma tumor growth. Oncogene 2002;21(22):3552–61.

[164] Albelda SM, Daise M, Levine EM, Buck CA. Identification and characterization of cell-substratum adhesion receptors on cultured human endothelial cells. J Clin Invest 1989;83(6):1992–2002.

[165] Smith LL, Cheung HK, Ling LE, Chen J, Sheppard D, Pytela R, Giachelli CM. Osteopontin N-terminal domain contains a cryptic adhesive sequence recognized by α9β1 integrin. J Biol Chem 1996;271(45):28485–91.

[166] Senger DR, Perruzzi CA, Streit M, Koteliansky VE, de Fougerolles AR, Detmar M. The α1β1 and α2β1 integrins provide critical support for vascular endothelial growth factor signaling, endothelial cell migration, and tumor angiogenesis. Am J Pathol 2002;160(1):195–204.

[167] Jin H, Su J, Garmy-Susini B, Kleeman J, Varner J. Integrin α4β1 promotes monocyte trafficking and angiogenesis in tumors. Cancer Res 2006;66(4):2146–52.

[168] Grazioli A, Alves CS, Konstantopoulos K, Yang JT. Defective blood vessel development and pericyte/pvSMC distribution in α4 integrin-deficient mouse embryos. Dev Biol 2006;293(1):165–77.

[169] Carlson TR, Feng Y, Maisonpierre PC, Mrksich M, Morla AO. Direct cell adhesion to the angiopoietins mediated by integrins. J Biol Chem 2001;276(28):26516–25.

[170] Gerhardt H, Golding M, Fruttiger M, Ruhrberg C, Lundkvist A, Abramsson A, Jeltsch M, Mitchell C, Alitalo K, Shima D, et al. VEGF guides angiogenic sprouting utilizing endothelial tip cell filopodia. J Cell Biol 2003; 161(6):1163–77.

[171] Konerding MA, Fait E, Gaumann A. 3D microvascular architecture of pre-cancerous lesions and invasive carcinomas of the colon. Br J Cancer 2001;84(10):1354–62.

[172] Weis SM, Cheresh DA. αV integrins in angiogenesis and cancer. Cold Spring Harb Perspect Med 2011;1(1): a006478. https://doi.org/10.1101/cshperspect.a006478.

[173] Brooks PC, Montgomery AM, Rosenfeld M, Reisfeld RA, Hu T, Klier G, Cheresh DA. Integrin αvβ3 antagonists promote tumor regression by inducing apoptosis of angiogenic blood vessels. Cell 1994;79(7):1157–64.

[174] Brooks PC, Strömblad S, Klemke R, Visscher D, Sarkar FH, Cheresh DA. Antiintegrin alpha v beta 3 blocks human breast cancer growth and angiogenesis in human skin. J Clin Invest 1995;96(4):1815–22.

[175] Gutheil JC, Campbell TN, Pierce PR, Watkins JD, Huse WD, Bodkin DJ, Cheresh DA. Targeted antiangiogenic therapy for cancer using Vitaxin: a humanized monoclonal antibody to the integrin αvβ3. Clin Cancer Res 2000; 6(8):3056–61.

[176] Brooks PC, Clark RA, Cheresh DA. Requirement of vascular integrin alpha v beta 3 for angiogenesis. Science 1994;264(5158):569–71.

[177] Biondani G, Zeeberg K, Greco MR, Cannone S, Dando I, Dalla Pozza E, Mastrodonato M, Forciniti S, Casavola V, Palmieri M, et al. Extracellular matrix composition modulates PDAC parenchymal and stem cell plasticity and behavior through the secretome. FEBS J 2018;85(11):2104–24.

[178] Vincenti V, Cassano C, Rocchi M, Persico MG. Assignment of the vascular endothelial growth factor gene to human chromosome 6p21.3. Circulation 1996;93(8):1493–5.

[179] Yang J, Yang D, Sun Y, Sun B, Wang G, Trent JC, Araujo DM, Chen K, Zhang W. Genetic amplification of the vascular endothelial growth factor (VEGF) pathway genes, including VEGFA, in human osteosarcoma. Cancer 2011;117(21):4925–38.

[180] Schneider BP, Gray R, Radovich M, Vance GH, Li L, Jiang G, Miller KD, Gralow JR, Dickler MN, Cobleigh MA, et al. Prognostic and predictive value of tumor VEGF gene amplification in metastatic breast cancer treated with paclitaxel with and without bevacizumab; results from ECOG 2100 trial. Clin Cancer Res 2013;19(5):1281–9.

[181] Schneider BP, Radovich M, Hancock BA, Kassem N, Vance GH, Sledge GW, Nielsen V, Müller S, Thorat MA, Badve SS. VEGFA amplification/deletion in human breast tumors. J Clin Oncol 2010;28(15_suppl), e21017.

[182] Ragunathrao VAB, Anwar M, Akhter MZ, Chavez A, Natarajan V, Lakshmikanthan S, Chrzanowska-Wodnicka M, Dudek AZ, Claesson-Welsh L, Kitajewski JK, et al. Sphingosine-1-phosphate receptor 1 activity promotes tumor growth by amplifying VEGF-VEGFR2 angiogenic signaling. Cell Rep 2019;29(11):3472–87.

[183] Koukourakis MI, Papazoglou D, Giatromanolaki A, Bougioukas G, Maltezos E, Sivridis E. VEGF gene sequence variation defines VEGF gene expression status and angiogenic activity in non-small cell lung cancer. Lung Cancer 2004;46(3):293–8.

[184] Naykoo NA, Rasool R, Shah S, Ahangar AG, Bhat IA, Qasim I, Siddiqi MA, Shah ZA. Single nucleotide polymorphisms, haplotype association and tumour expression of the vascular endothelial growth factor (VEGF) gene with lung carcinoma. Gene 2017;608:95–102.
[185] Ajaz S, Khaliq S, Abid A, Hassan AS, Hashmi A, Sultan G, Mohsin R, Mubarrak M, Naqvi SAA, Rizvi SAH, et al. Association of a single-nucleotide polymorphism in the promoter region of the VEGF gene with the risk of renal cell carcinoma. Genet Test Mol Biomarkers 2011;15(9):653–7.
[186] Kofler N, Simons M. The expanding role of neuropilin: regulation of transforming growth factor-β and platelet-derived growth factor signaling in the vasculature. Curr Opin Hematol 2016;23(3):260–7.
[187] West DC, Rees CG, Duchesne L, Patey SJ, Terry CJ, Turnbull JE, Delehedde M, Heegaard CW, Allain F, Vanpouille C, et al. Interactions of multiple heparin binding growth factors with neuropilin-1 and potentiation of the activity of fibroblast growth factor-2. J Biol Chem 2005;280(14):13457–64.
[188] Yamazaki Y, Morita T. Molecular and functional diversity of vascular endothelial growth factors. Mol Divers 2006;10(4):515–27.
[189] Migdal M, Huppertz B, Tessler S, Comforti A, Shibuya M, Reich R, Baumann H, Neufeld G. Neuropilin-1 is a placenta growth factor-2 receptor. J Biol Chem 1998;273(35):22272–8.
[190] Prud'homme GJ, Glinka Y. Neuropilins are multifunctional coreceptors involved in tumor initiation, growth, me tastasis and immunity. Oncotarget 2012;3:921–39.
[191] Hong TM, Chen YL, Wu YY, Yuan A, Chao YC, Chung YC, Wu MH, Yang SC, Pan SH, Shih JY, et al. Targeting neuropilin 1 as an antitumor strategy in lung cancer. Clin Cancer Res 2007;13(16):4759–68.
[192] Hansel DE, Wilentz RE, Yeo CJ, Schulick RD, Montgomery E, Maitra A. Expression of neuropilin-1 in high-grade dysplasia, invasive cancer, and metastases of the human gastrointestinal tract. Am J Surg Pathol 2004;28(3):347–56.
[193] Osada H, Tokunaga T, Nishi M, Hatanaka H, Abe Y, Tsugu A, Kiyima H, Yamasaki H, Ueyama Y, Nakamura M. Overexpression of the neuropilin 1 (NRP1) gene correlated with poor prognosis in human glioma. Anticancer Res 2004;24(2B):547–52.
[194] Kamiya T, Kawakami T, Abe Y, Nishi M, Onoda N, Miyazaki N, Oida Y, Yamasaki H, Ueyama Y, Nakamura M. The preserved expression of neuropilin (NRP) 1 contributes to a better prognosis in colon cancer. Oncol Rep 2006;15(2):369–73.
[195] Raimondi C, Fantin A, Lampropoulou A, Denti L, Chikh A, Ruhrberg C. Imatinib inhibits VEGF-independent angiogenesis by targeting neuropilin 1–dependent ABL1 activation in endothelial cells. J Exp Med 2014; 211(6):1167–83.
[196] Hu C, Zhu P, Xia Y, Hui K, Wang M, Jiang X. Role of the NRP-1-mediated VEGFR2-independent pathway on radiation sensitivity of non-small cell lung cancer cells. J Cancer Res Clin Oncol 2018;144(7):1329–37.
[197] Artavanis-Tsakonas S, Rand MD, Lake RJ. Notch signaling: cell fate control and signal integration in development. Science 1999;284(5415):770–6.
[198] Iso T, Hamamori Y, Kedes L. Notch signaling in vascular development. Arterioscler Thromb Vasc Biol 2003; 23(4):543–53.
[199] Siekmann AF, Lawson ND. Notch signalling limits angiogenic cell behaviour in developing zebrafish arteries. Nature 2007;445(7129):781–4.
[200] Kuhnert F, Kirshner JR, Thurston G. Dll4-Notch signaling as a therapeutic target in tumor angiogenesis. Vasc Cell 2011;3(1):1–8.
[201] Noguera-Troise I, Daly C, Papadopoulos NJ, Coetzee S, Boland P, Gale NW, Lyn HC, Yancopoulos GD, Thurston G. Blockade of Dll4 inhibits tumour growth by promoting non-productive angiogenesis. Nature 2006;444(7122):1032–7.
[202] Fernández-Chacón M, García-González I, Mühleder S, Benedito R. Role of notch in endothelial biology. Angiogenesis 2021;24(2):237–50. https://doi.org/10.1007/s10456-021-09793-7.
[203] Sainson RC, Aoto J, Nakatsu MN, Holderfield M, Conn E, Koller E, Hughes CC. Cell-autonomous notch signaling regulates endothelial cell branching and proliferation during vascular tubulogenesis. FASEB J 2005; 19(8):1027–9.
[204] Hainaud P, Contrerès JO, Villemain A, Liu LX, Plouët J, Tobelem G, Dupuy E. The role of the vascular endothelial growth factor–Delta-like 4 ligand/Notch4-ephrin B2 cascade in tumor vessel remodeling and endothelial cell functions. Cancer Res 2006;66(17):8501–10.
[205] Lobov IB, Renard RA, Papadopoulos N, Gale NW, Thurston G, Yancopoulos GD, Wiegand SJ. Delta-like ligand 4 (Dll4) is induced by VEGF as a negative regulator of angiogenic sprouting. Proc Natl Acad Sci U S A 2007; 104(9):3219–24.

[206] Bridges E, Oon CE, Harris A. Notch regulation of tumor angiogenesis. Future Oncol 2011;7(4):569–88.
[207] Thurston G, Kitajewski J. VEGF and Delta-Notch: interacting signalling pathways in tumour angiogenesis. Br J Cancer 2008;99(8):1204–9.
[208] Thurston G, Noguera-Troise I, Yancopoulos GD. The Delta paradox: DLL4 blockade leads to more tumour vessels but less tumour growth. Nat Rev Cancer 2007;7(5):327–31.
[209] Ridgway J, Zhang G, Wu Y, Stawicki S, Liang WC, Chanthery Y, Kowalski J, Watts RJ, Callahan C, Ksman I, et al. Inhibition of Dll4 signalling inhibits tumour growth by deregulating angiogenesis. Nature 2006; 444(7122):1083–7.
[210] Koch AE, Polverini PJ, Kunkel SL, Harlow LA, DiPietro LA, Elner VM, Elner SG, Strieter RM. Interleukin-8 as a macrophage-derived mediator of angiogenesis. Science 1992;258(5089):1798–801.
[211] Bonavia R, Inda MM, Vandenberg S, Cheng SY, Nagane M, Hadwiger P, Tan P, Sah DWY, Cavenee WK, Furnari FB. EGFRvIII promotes glioma angiogenesis and growth through the NF-κB, interleukin-8 pathway. Oncogene 2012;31(36):4054–66.
[212] Medina RJ, O'Neill CL, O'Doherty TM, Knott H, Guduric-Fuchs J, Gardiner TA, Stitt AW. Myeloid angiogenic cells act as alternative M2 macrophages and modulate angiogenesis through interleukin-8. Mol Med 2011; 17(9):1045–55.
[213] Kitadai Y, Takahashi Y, Haruma K, Naka K, Sumii K, Yokozaki H, Yasui W, Mukaida N, Ohmoto Y, Kajiyama G, et al. Transfection of interleukin-8 increases angiogenesis and tumorigenesis of human gastric carcinoma cells in nude mice. Br J Cancer 1999;81(4):647–53.
[214] Heidemann J, Ogawa H, Dwinell MB, Rafiee P, Maaser C, Gockel HR, Otterson MF, Ota DM, Lügering N, Domschke W, et al. Angiogenic effects of interleukin 8 (CXCL8) in human intestinal microvascular endothelial cells are mediated by CXCR2. J Biol Chem 2003;278(10):8508–15.
[215] Ning Y, Manegold PC, Hong YK, Zhang W, Pohl A, Lurje G, Lenz HJ. Interleukin-8 is associated with proliferation, migration, angiogenesis and chemosensitivity in vitro and in vivo in colon cancer cell line models. Int J Cancer 2011;128(9):2038–49.
[216] Rofstad EK, Halsør EF. Vascular endothelial growth factor, interleukin 8, platelet-derived endothelial cell growth factor, and basic fibroblast growth factor promote angiogenesis and metastasis in human melanoma xenografts. Cancer Res 2000;60(17):4932–8.
[217] Arenberg DA, Kunkel SL, Polverini PJ, Glass M, Burdick MD, Strieter RM. Inhibition of interleukin-8 reduces tumorigenesis of human non-small cell lung cancer in SCID mice. J Clin Invest 1996;97(12): 2792–802.
[218] Chen JJ, Yao PL, Yuan A, Hong TM, Shun CT, Kuo ML, Lee YC, Yang PC. Up-regulation of tumor interleukin-8 expression by infiltrating macrophages: its correlation with tumor angiogenesis and patient survival in non-small cell lung cancer. Clin Cancer Res 2003;9(2):729–37.
[219] Masuya D, Huang CL, Liu D, Kameyama K, Hayashi E, Yamauchi A, Kobayashi S, Haba R, Yokomise H. The intratumoral expression of vascular endothelial growth factor and interleukin-8 associated with angiogenesis in nonsmall cell lung carcinoma patients. Cancer 2001;92(10):2628–38.
[220] Inoue K, Slaton JW, Kim SJ, Perrotte P, Eve BY, Bar-Eli M, Radinsky R, Dinney CP. Interleukin 8 expression regulates tumorigenicity and metastasis in human bladder cancer. Cancer Res 2000;60(8):2290–9.
[221] Kim SJ, Uehara H, Karashima T, Mccarty M, Shih N, Fidler IJ. Expression of interleukin-8 correlates with angiogenesis, tumorigenicity, and metastasis of human prostate cancer cells implanted orthotopically in nude mice. Neoplasia 2001;3(1):33–42.
[222] Inoue K, Slaton JW, Eve BY, Kim SJ, Perrotte P, Balbay MD, Yano S, Bar-Eli M, Radinsky R, Pattaway CA, et al. Interleukin 8 expression regulates tumorigenicity and metastases in androgen-independent prostate cancer. Clin Cancer Res 2000;6(5):2104–19.
[223] Lin Y, Huang R, Chen L, Li S, Shi Q, Jordan C, Huang RP. Identification of interleukin-8 as estrogen receptor-regulated factor involved in breast cancer invasion and angiogenesis by protein arrays. Int J Cancer 2004; 109(4):507–15.
[224] Bauerle KT, Schweppe RE, Lund G, Kotnis G, Deep G, Agarwal R, Pozdeyev N, Wood WM, Haugen BR. Nuclear factor κB–dependent regulation of angiogenesis, and metastasis in an in vivo model of thyroid cancer is associated with secreted interleukin-8. J Clin Endocrinol Metab 2014;99(8):E1436–44.
[225] Shi Q, Abbruzzese JL, Huang S, Fidler IJ, Xiong Q, Xie K. Constitutive and inducible interleukin 8 expression by hypoxia and acidosis renders human pancreatic cancer cells more tumorigenic and metastatic. Clin Cancer Res 1999;5(11):3711–21.

[226] Trevino JG, Summy JM, Gray MJ, Nilsson MB, Lesslie DP, Baker CH, Gallick GE. Expression and activity of SRC regulate interleukin-8 expression in pancreatic adenocarcinoma cells: implications for angiogenesis. Cancer Res 2005;65(16):7214–22.

[227] Wang Y, Xu RC, Zhang XL, Niu XL, Qu Y, Li LZ, Meng XY. Interleukin-8 secretion by ovarian cancer cells increases anchorage-independent growth, proliferation, angiogenic potential, adhesion and invasion. Cytokine 2012;59(1):145–55.

[228] Yoshizaki T, Horikawa T, Qing-chun R, Wakisaka N, Takeshita H, Sheen TS, Lee SY, Sato H, Furukawa M. Induction of interleukin-8 by Epstein-Barr virus latent membrane protein-1 and its correlation to angiogenesis in nasopharyngeal carcinoma. Clin Cancer Res 2001;7(7):1946–51.

[229] Kitadai Y, Haruma K, Sumii K, Yamamoto S, Ue T, Yokozaki H, Yasui W, Ohmoto I, Kajiyama G, Fidler IJ, et al. Expression of interleukin-8 correlates with vascularity in human gastric carcinomas. Am J Pathol 1998; 152(1):93.

[230] Shi JUN, Wei PK. Interleukin-8: a potent promoter of angiogenesis in gastric cancer. Oncol Lett 2016;11(2): 1043–50.

[231] Zhu B, Lin N, Zhang M, Zhu Y, Cheng H, Chen S, Ling Y, Pan W, Xu R. Activated hepatic stellate cells promote angiogenesis via interleukin-8 in hepatocellular carcinoma. J Transl Med 2015;13(1):1–10.

[232] Desbaillets I, Diserens AC, Tribolet ND, Hamou MF, Meir EGV. Upregulation of interleukin 8 by oxygen-deprived cells in glioblastoma suggests a role in leukocyte activation, chemotaxis, and angiogenesis. J Exp Med 1997;186(8):1201–12.

[233] Ferrer FA, Pantschenko AG, Miller LJ, Anderson K, Grunnet M, McKenna PH, Kreutzer D. Angiogenesis and neuroblastomas: interleukin-8 and interleukin-8 receptor expression in human neuroblastoma. J Urol 2000; 164(3 Part 2):1016–20.

[234] Yatsunami J, Tsuruta N, Ogata K, Wakamatsu K, Takayama K, Kawasaki M, Nakanishi Y, Nobuyuki H, Hayashi SI. Interleukin-8 participates in angiogenesis in non-small cell, but not small cell carcinoma of the lung. Cancer Lett 1997;120(1):101–8.

[235] Huang D, Ding Y, Zhou M, Rini BI, Petillo D, Qian CN, Kahnoski R, Futreal PA, Furge KA, et al. Interleukin-8 mediates resistance to antiangiogenic agent sunitinib in renal cell carcinoma. Cancer Res 2010;70(3):1063–71.

[236] Huang S, Mills L, Mian B, Tellez C, McCarty M, Yang XD, Gudas JM, Bar-Eli M. Fully humanized neutralizing antibodies to interleukin-8 (ABX-IL8) inhibit angiogenesis, tumor growth, and metastasis of human melanoma. Am J Pathol 2002;161(1):125–34.

[237] Rofstad EK, Mathiesen B, Kindem K, Galappathi K. Acidic extracellular pH promotes experimental metastasis of human melanoma cells in athymic nude mice. Cancer Res 2006;66(13):6699–707.

[238] Xu L, Fidler IJ. Acidic pH-induced elevation in interleukin 8 expression by human ovarian carcinoma cells. Cancer Res 2000;60(16):4610–6.

[239] Shi Q, Le X, Wang B, Abbruzzese JL, Xiong Q, He Y, Xie K. Regulation of vascular endothelial growth factor expression by acidosis in human cancer cells. Oncogene 2001;20(28):3751–6.

[240] Nakanishi M, Morita Y, Hata K, Muragaki Y. Acidic microenvironments induce lymphangiogenesis and IL-8 production via TRPV1 activation in human lymphatic endothelial cells. Exp Cell Res 2016;345(2):180–9.

[241] Koltai T, Cardone RA, Reshkin SJ. Synergy between low dose metronomic chemotherapy and the pH-centered approach against cancer. Int J Mol Sci 2019;20(21):5438.

[242] Mosch B, Reissenweber B, Neuber C, Pietzsch J. Eph receptors and ephrin ligands: important players in angiogenesis and tumor angiogenesis. J Oncol 2010;2010:135285. https://doi.org/10.1155/2010/135285.

[243] Cheng N, Brantley DM, Chen J. The ephrins and Eph receptors in angiogenesis. Cytokine Growth Factor Rev 2002;13(1):75–85. https://doi.org/10.1016/s1359-6101(01)00031-4.

[244] Wang Y, Nakayama M, Pitulescu ME, Schmidt TS, Bochenek ML, Sakakibara A, Adams S, Davy A, Deutsch U, Lüthi U, et al. Ephrin-B2 controls VEGF-induced angiogenesis and lymphangiogenesis. Nature 2010; 465(7297):483–6.

[245] Sawamiphak S, Seidel S, Essmann CL, Wilkinson GA, Pitulescu ME, Acker T, Acker-Palmer A. Ephrin-B2 regulates VEGFR2 function in developmental and tumour angiogenesis. Nature 2010;465(7297):487–91.

[246] Steinle JJ, Meininger CJ, Forough R, Wu G, Wu MH, Granger HJ. Eph B4 receptor signaling mediates endothelial cell migration and proliferation via the phosphatidylinositol 3-kinase pathway. J Biol Chem 2002; 277(46):43830–5.

[247] Steinle JJ, Meininger CJ, Chowdhury U, Wu G, Granger HJ. Role of ephrin B2 in human retinal endothelial cell proliferation and migration. Cell Signal 2003;15(11):1011–7.

[248] Deshane J, Chen S, Caballero S, Grochot-Przeczek A, Was H, Calzi SL, Lach R, Hock TD, Chen B, Hill-Kapturczak N, et al. Stromal cell–derived factor 1 promotes angiogenesis via a heme oxygenase 1–dependent mechanism. J Exp Med 2007;204(3):605–18.
[249] Laiva AL, Raftery RM, Keogh MB, O'brien FJ. Pro-angiogenic impact of SDF-1α gene-activated collagen-based scaffolds in stem cell driven angiogenesis. Int J Pharm 2018;544(2):372–9.
[250] Yang F, Xue F, Guan J, Zhang Z, Yin J, Kang Q. Stromal-cell-derived factor (SDF) 1-alpha overexpression promotes bone regeneration by osteogenesis and angiogenesis in osteonecrosis of the femoral head. Cell Physiol Biochem 2018;46(6):2561–75.
[251] Strieter RM, Polverini PJ, Kunkel SL, Arenberg DA, Burdick MD, Kasper J, Dzuiba J, Van Damme J, Waltz A, Marriot D, et al. The functional role of the ELR motif in CXC chemokine-mediated angiogenesis. J Biol Chem 1995;270(45):27348–57.
[252] Laschke MW, Elitzsch A, Vollmar B, Vajkoczy P, Menger MD. Combined inhibition of vascular endothelial growth factor (VEGF), fibroblast growth factor and platelet-derived growth factor, but not inhibition of VEGF alone, effectively suppresses angiogenesis and vessel maturation in endometriotic lesions. Hum Reprod 2006;21(1):262–8.
[253] Eberhard A, Kahlert S, Goede V, Hemmerlein B, Plate KH, Augustin HG. Heterogeneity of angiogenesis and blood vessel maturation in human tumors: implications for antiangiogenic tumor therapies. Cancer Res 2000; 60(5):1388–93.
[254] Vartanian RK, Weidner N. Endothelial cell proliferation in prostatic carcinoma and prostatic hyperplasia: correlation with Gleason's score, microvessel density, and epithelial cell proliferation. Lab Invest 1995; 73:844–50.
[255] Vermeulen PB, Verhoeven D, Hubens G, Van Maeck E, Goovaerst G, Huyghe M, De Bruijn EA, Van Oosterom AT, Dirix LY. Microvessel density, endothelial cell proliferation and tumour cell proliferation in human colorectal adenocarcinomas. Ann Oncol 1995;6:59–64.
[256] Fox SB, Gatter KC, Bicknell R, Going JJ, Stanton P, Cooke TG, Harris AL. Relationship of endothelial cell proliferation to tumor vascularity in human breast cancer. Cancer Res 1993;53(18):4161–3.
[257] Folberg R, Hendrix MJ, Maniotis AJ. Vasculogenic mimicry and tumor angiogenesis. Am J Pathol 2000; 156(2):361–81.
[258] Maniotis AJ, Folberg R, Hess A, Seftor EA, Gardner LM, Pe'er J, Trent JM, Meltzer PS, Hendrix MJ. Vascular channel formation by human melanoma cells in vivo and in vitro: vasculogenic mimicry. Am J Pathol 1999; 155(3):739–52.
[259] Hendrix MJ, Seftor EA, Hess AR, Seftor RE. Vasculogenic mimicry and tumour-cell plasticity: lessons from melanoma. Nat Rev Cancer 2003;3(6):411–21.
[260] Corada M, Liao F, Lindgren M, Lampugnani MG, Breviario F, Frank R, Müller WA, Hicklin DJ, Bohlen P, Dejana E. Monoclonal antibodies directed to different regions of vascular endothelial cadherin extracellular domain affect adhesion and clustering of the protein and modulate endothelial permeability. Blood 2001; 97(6):1679–84.
[261] Crosby CV, Fleming PA, Argraves WS, Corada M, Zanetta L, Dejana E, Drake CJ. VE-cadherin is not required for the formation of nascent blood vessels but acts to prevent their disassembly. Blood 2005;105(7):2771–6.
[262] Hendrix MJ, Seftor EA, Meltzer PS, Gardner LM, Hess AR, Kirschmann DA, Shatteman GC, Seftor RE. Expression and functional significance of VE-cadherin in aggressive human melanoma cells: role in vasculogenic mimicry. Proc Natl Acad Sci U S A 2001;98(14):8018–23.
[263] Cao Z, Bao M, Miele L, Sarkar FH, Wang Z, Zhou Q. Tumour vasculogenic mimicry is associated with poor prognosis of human cancer patients: a systemic review and meta-analysis. Eur J Cancer 2013;49(18):3914–23.
[264] Wang JY, Sun T, Zhao XL, Zhang SW, Gu Q, Wang XH, Sun BC. Functional significance of VEGF-a in human ovarian carcinoma: role in vasculogenic mimicry. Cancer Biol Ther 2008;7(5):758–66.
[265] Schnegg CI, Yang MH, Ghosh SK, Hsu MY. Induction of vasculogenic mimicry overrides VEGF-A silencing and enriches stem-like cancer cells in melanoma. Cancer Res 2015;75(8):1682–90.
[266] Greco MR, Bon E, Rubino R, Guerra L, Bernabe-Garcia M, Cannone S, Cayuela ML, Ciaccia L, Marionneau-Lambot S, Oullier T, et al. Phosphorylation of NHERF1 S279 and S301 differentially regulates breast cancer cell phenotype and metastatic organotropism. Biochim Biophys Acta Mol Basis Dis 2019;1865(1):26–37.
[267] Clinicaltrials.gov. NCT03582618: CVM-1118 and sorafenib combination in subjects with advanced hepatocellular carcinoma. Available from https://clinicaltrials.gov/ct2/results?recrs=&cond=&term=NCT03582618&cntry=&state=&city=&dist=. [Accessed 25 February 2021].

[268] Ju RJ, Zeng F, Liu L, Mu LM, Xie HJ, Zhao Y, Yan Y, Wu JS, Hu YJ, Lu WL. Destruction of vasculogenic mimicry channels by targeting epirubicin plus celecoxib liposomes in treatment of brain glioma. Int J Nanomedicine 2016;11:1131.

[269] Zeng F, Ju RJ, Liu L, Xie HJ, Mu LM, Zhao Y, Yan Y, Hu YJ, Wu JS, Lu WL. Application of functional vincristine plus dasatinib liposomes to deletion of vasculogenic mimicry channels in triple-negative breast cancer. Oncotarget 2015;6(34):36625–42. https://doi.org/10.18632/oncotarget.5382.

[270] Zhang S, Li M, Gu Y, Liu Z, Xu S, Cui Y, Sun B. Thalidomide influences growth and vasculogenic mimicry channel formation in melanoma. J Exp Clin Cancer Res 2008;27(1):1–9.

[271] Ju RJ, Cheng L, Qiu X, Liu S, Song XL, Peng XM, Wang T, Li CQ, Li XT. Hyaluronic acid modified daunorubicin plus honokiol cationic liposomes for the treatment of breast cancer along with the elimination vasculogenic mimicry channels. J Drug Target 2018;26(9):793–805.

[272] Zang M, Hu L, Zhang B, Zhu Z, Li J, Zhu Z, Yan M, Liu B. Luteolin suppresses angiogenesis and vasculogenic mimicry formation through inhibiting Notch1-VEGF signaling in gastric cancer. Biochem Biophys Res Commun 2017;490(3):913–9.

[273] Liu ZJL, Zhou YJ, Ding RL, Xie F, Fu SZ, Wu JB, Yang LL, Wen QL. In vitro and in vivo apatinib inhibits vasculogenic mimicry in melanoma MUM-2B cells. PLoS One 2018;13(7), e0200845.

[274] Zhang C, Chen W, Zhang X, Huang B, Chen A, He Y, Wang J, Li X. Galunisertib inhibits glioma vasculogenic mimicry formation induced by astrocytes. Sci Rep 2016;6(1):1–11.

[275] Li S, Zhang Q, Zhou L, Guan Y, Chen S, Zhang Y, Han X. Inhibitory effects of compound DMBT on hypoxia-induced vasculogenic mimicry in human breast cancer. Biomed Pharmacother 2017;96:982–92.

[276] Meng J, Sun B, Zhao X, Zhang D, Zhao X, Gu Q, Dong X, Zhao N, Liu P, Liu Y. Doxycycline as an inhibitor of the epithelial-to-mesenchymal transition and vasculogenic mimicry in hepatocellular carcinoma. Mol Cancer Ther 2014;13(12):3107–22.

[277] Wei H, Wang F, Wang Y, Li T, Xiu P, Zhong J, Sun X, Li J. Verteporfin suppresses cell survival, angiogenesis and vasculogenic mimicry of pancreatic ductal adenocarcinoma via disrupting the YAP-TEAD complex. Cancer Sci 2017;108(3):478–87.

[278] Da J, Xu M, Wang Y, Li W, Lu M, Wang Z. Kaempferol promotes apoptosis while inhibiting cell proliferation via androgen-dependent pathway and suppressing vasculogenic mimicry and invasion in prostate cancer. Anal Cell Pathol (Amst) 2019;2019. https://doi.org/10.1155/2019/1907698, 1907698.

[279] Xu MR, Wei PF, Suo MZ, Hu Y, Ding W, Su L, Zhu YD, Song WJ, Tang GH, Zhang M, et al. Brucine suppresses vasculogenic mimicry in human triple-negative breast cancer cell line MDA-MB-231. Biomed Res Int 2019;2019:6543230. https://doi.org/10.1155/2019/6543230.

[280] Chiablaem K, Lirdprapamongkol K, Keeratichamroen S, Surarit R, Svasti J. Curcumin suppresses vasculogenic mimicry capacity of hepatocellular carcinoma cells through STAT3 and PI3K/AKT inhibition. Anticancer Res 2014;34(4):1857–64.

[281] Serova M, Tijeras-Raballand A, Dos Santos C, Martinet M, Neuzillet C, Lopez A, Mitchell DC, Bryan BA, Gapihan G, Janin A, et al. Everolimus affects vasculogenic mimicry in renal carcinoma resistant to sunitinib. Oncotarget 2016;7(25):38467.

[282] Fu D, He X, Yang S, Xu W, Lin T, Feng X. Zoledronic acid inhibits vasculogenic mimicry in murine osteosarcoma cell line in vitro. BMC Musculoskelet Disord 2011;12:146. https://doi.org/10.1186/1471-2474-12-146.

[283] Madsen DH, Bugge TH. The source of matrix-degrading enzymes in human cancer: problems of research reproducibility and possible solutions. J Cell Biol 2015;209(2):195–8.

[284] Koltai T, Reshkin SJ, Harguindey S, editors. An innovative approach to understanding and treating cancer: targeting ph: from etiopathogenesis to new therapeutic avenues. Academic Press. Elsevier; 2020.

[285] Cardone RA, Alfarouk KO, Elliott RL, Alqahtani SS, Ahmed S, Aljarbou AN, Greco MR, Cannone S, Reshkin SJ. The role of sodium hydrogen exchanger 1 in dysregulation of proton dynamics and reprogramming of cancer metabolism as a sequela. Int J Mol Sci 2019;20(15):3694.

[286] Jensen JA, Hunt TK, Scheuenstuhl H, Banda MJ. Effect of lactate, pyruvate, and pH on secretion of angiogenesis and mitogenesis factors by macrophages. Lab Invest 1986;54(5):574–8.

[287] Kumar VS, Viji RI, Kiran MS, Sudhakaran PR. Endothelial cell response to lactate: implication of PAR modification of VEGF. J Cell Physiol 2007;211(2):477–85.

[288] Beckert S, Farrahi F, Aslam RS, Scheuenstuhl H, Königsrainer A, Hussain MZ, Hunt TK. Lactate stimulates endothelial cell migration. Wound Repair Regen 2006;14(3):321–4.

[289] Morland C, Andersson KA, Haugen ØP, Hadzic A, Kleppa L, Gille A, Rinholm JE, Palibrk V, Diget EH, Kennedy LH, et al. Exercise induces cerebral VEGF and angiogenesis via the lactate receptor HCAR1. Nat Commun 2017;8:15557. https://doi.org/10.1038/ncomms15557.

[290] Porporato PE, Payen VL, De Saedeleer CJ, Préat V, Thissen JP, Feron O, Sonveaux P. Lactate stimulates angiogenesis and accelerates the healing of superficial and ischemic wounds in mice. Angiogenesis 2012;15(4):581–92.

[291] Constant JS, Feng JJ, Zabel DD, Yuan H, Suh DY, Scheuenstuhl H, Hunt TK, Hussain MZ. Lactate elicits vascular endothelial growth factor from macrophages: a possible alternative to hypoxia. Wound Repair Regen 2000;8(5):353–60.

[292] Végran F, Boidot R, Michiels C, Sonveaux P, Feron O. Lactate influx through the endothelial cell monocarboxylate transporter MCT1 supports an NF-κB/IL-8 pathway that drives tumor angiogenesis. Cancer Res 2011;71(7):2550–60.

[293] Polet F, Feron O. Endothelial cell metabolism and tumour angiogenesis: glucose and glutamine as essential fuels and lactate as the driving force. J Intern Med 2013;273(2):156–65.

[294] Koukourakis MI, Giatromanolaki A, Sivridis E, Bougioukas G, Didilis V, Gatter KC, Harris AL. Lactate dehydrogenase-5 (LDH-5) overexpression in non-small-cell lung cancer tissues is linked to tumour hypoxia, angiogenic factor production and poor prognosis. Br J Cancer 2003;89(5):877–85.

[295] Kes MM, Van den Bossche J, Griffioen AW, Huijbers EJ. Oncometabolites lactate and succinate drive pro-angiogenic macrophage response in tumors. Biochim Biophys Acta Rev Cancer 2020;1874(2):188427. https://doi.org/10.1016/j.bbcan.2020.188427.

[296] Allen E, Miéville P, Warren CM, Saghafinia S, Li L, Peng MW, Hanahan D. Metabolic symbiosis enables adaptive resistance to anti-angiogenic therapy that is dependent on mTOR signaling. Cell Rep 2016;15(6):1144–60.

[297] Brown TP, Ganapathy V. Lactate/GPR81 signaling and proton motive force in cancer: role in angiogenesis, immune escape, nutrition, and Warburg phenomenon. Pharmacol Ther 2020;206:107451.

[298] Dhup S, Kumar Dadhich R, Ettore Porporato P, Sonveaux P. Multiple biological activities of lactic acid in cancer: influences on tumor growth, angiogenesis and metastasis. Curr Pharm Des 2012;18(10):1319–30.

[299] Crowther M, Brown NJ, Bishop ET, Lewis CE. Microenvironmental influence on macrophage regulation of angiogenesis in wounds and malignant tumors. J Leukoc Biol 2001;70(4):478–90.

[300] Colgan SM, Mukherjee S, Major P. Hypoxia-induced lactate dehydrogenase expression and tumor angiogenesis. Clin Colorectal Cancer 2007;6(6):442–6.

[301] Lin H, Muramatsu R, Maedera N, Tsunematsu H, Hamaguchi M, Koyama Y, Kuroda M, Ono K, Sawada M, Yamashita T. Extracellular lactate dehydrogenase a release from damaged neurons drives central nervous system angiogenesis. EBioMedicine 2018;27:71–85.

[302] Albrektsson A, Bazargani F, Wieslander A, Braide M. Peritoneal dialysis fluid–induced angiogenesis in rat mesentery is increased by lactate in the presence or absence of glucose. ASAIO J 2006;52(3):276–81.

[303] Pereira-Nunes A, Afonso J, Granja S, Baltazar F. Lactate and lactate transporters as key players in the maintenance of the Warburg effect. Adv Exp Med Biol 2020;1219:51–74. https://doi.org/10.1007/978-3-030-34025-4_3.

[304] Yetkin-Arik B, Vogels IM, Nowak-Sliwinska P, Weiss A, Houtkooper RH, Van Noorden CJ, Klaasen I, Schlingemann RO. The role of glycolysis and mitochondrial respiration in the formation and functioning of endothelial tip cells during angiogenesis. Sci Rep 2019;9(1):1–14.

[305] Parra-Bonilla G, Alvarez DF, Alexeyev M, Vasauskas A, Stevens T. Lactate dehydrogenase a expression is necessary to sustain rapid angiogenesis of pulmonary microvascular endothelium. PLoS One 2013;8(9). https://doi.org/10.1371/journal.pone.0075984, e75984.

[306] Lapel M, Weston P, Strassheim D, Karoor V, Burns N, Lyubchenko T, Paucek P, Stenmark KR, Gerasimovskaya EV. Glycolysis and oxidative phosphorylation are essential for purinergic receptor-mediated angiogenic responses in vasa vasorum endothelial cells. Am J Physiol Cell Physiol 2017;312(1):C56–70.

[307] Pinheiro C, Garcia EA, Morais-Santos F, Moreira MA, Almeida FM, Jubé LF, Queiroz GS, Paula EC, Andreoli MA, Villa LL, et al. Reprogramming energy metabolism and inducing angiogenesis: co-expression of monocarboxylate transporters with VEGF family members in cervical adenocarcinomas. BMC Cancer 2015;15:835. https://doi.org/10.1186/s12885-015-1842-4.

[308] Fitzgerald G, Soro-Arnaiz I, De Bock K. The Warburg effect in endothelial cells and its potential as an anti-angiogenic target in cancer. Front Cell Dev Biol 2018;6:100. https://doi.org/10.3389/fcell.2018.00100.

[309] Ruan GX, Kazlauskas A. Lactate engages receptor tyrosine kinases Axl, Tie2, and vascular endothelial growth factor receptor 2 to activate phosphoinositide 3-kinase/Akt and promote angiogenesis. J Biol Chem 2013; 288(29):21161–72.
[310] Sonveaux P, Copetti T, De Saedeleer CJ, Végran F, Verrax J, Kennedy KM, Moon EJ, Dhup S, Danhier P, Frérart F, et al. Targeting the lactate transporter MCT1 in endothelial cells inhibits lactate-induced HIF-1 activation and tumor angiogenesis. PLoS One 2012;7(3). https://doi.org/10.1371/journal.pone.0033418, e33418.
[311] Porporato PE, Dadhich RK, Dhup S, Copetti T, Sonveaux P. Anticancer targets in the glycolytic metabolism of tumors: a comprehensive review. Front Pharmacol 2011;2:49.
[312] Puri S, Juvale K. Monocarboxylate transporter 1 and 4 inhibitors as potential therapeutics for treating solid tumours: a review with structure-activity relationship insights. Eur J Med Chem 2020;199:112393. https://doi.org/10.1016/j.ejmech.2020.112393.
[313] Ovens MJ, Davies AJ, Wilson MC, Murray CM, Halestrap AP. AR-C155858 is a potent inhibitor of monocarboxylate transporters MCT1 and MCT2 that binds to an intracellular site involving transmembrane helices 7–10. Biochem J 2010;425(3):523–30.
[314] Konishi Y, Kobayashi S, Shimizu M. Tea polyphenols inhibit the transport of dietary phenolic acids mediated by the monocarboxylic acid transporter (MCT) in intestinal Caco-2 cell monolayers. J Agric Food Chem 2003; 51(25):7296–302.
[315] Albatany M, Meakin S, Bartha R. The Monocarboxylate transporter inhibitor Quercetin induces intracellular acidification in a mouse model of glioblastoma multiforme: in-vivo detection using magnetic resonance imaging. Invest New Drugs 2019;37(4):595–601.
[316] Jones RS, Parker MD, Morris ME. Quercetin, morin, luteolin, and phloretin are dietary flavonoid inhibitors of monocarboxylate transporter 6. Mol Pharm 2017;14(9):2930–6.
[317] Shim CK, Cheon EP, Kang KW, Seo KS, Han HK. Inhibition effect of flavonoids on monocarboxylate transporter 1 (MCT1) in Caco-2 cells. J Pharm Pharmacol 2007;59(11):1515–9.
[318] Reyes-Farias M, Carrasco-Pozo C. The anti-cancer effect of quercetin: molecular implications in cancer metabolism. Int J Mol Sci 2019;20(13):3177.
[319] McKay TB, Lyon D, Sarker-Nag A, Priyadarsini S, Asara JM, Karamichos D. Quercetin attenuates lactate production and extracellular matrix secretion in keratoconus. Sci Rep 2015;5(1):1–7.
[320] Zhang W, Zhang F. Effects of quercetin on proliferation, apoptosis, adhesion and migration, and invasion of HeLa cells. Eur J Gynaecol Oncol 2009;30(1):60–4.
[321] Davenport AP, Hyndman KA, Dhaun N, Southan C, Kohan DE, Pollock JS, Pollock DM, Webb DJ, Maguire JJ. Endothelin. Pharmacol Rev 2016;68(2):357–418.
[322] Bagnato A, Rosano L. Epithelial-mesenchymal transition in ovarian cancer progression: a crucialrole forthe endothelin axis. Cells Tissues Organs 2007;185:85–94.
[323] Rybak SM, Sanovich E, Hollingshead MG, Borgel SD, Newton DL, Melillo G, Kong D, Kaur G, Sausville EA. "Vasocrine" formation of tumor cell-lined vascular spaces: implications for rational design of antiangiogenic therapies. Cancer Res 2003;63(11):2812–9.
[324] Kappes L, Amer RL, Sommerlatte S, Bashir G, Plattfaut C, Gieseler F, Gemoll T, Busch H, Altahrawi A, Al-Sbiei A, et al. Ambrisentan, an endothelin receptor type A-selective antagonist, inhibits cancer cell migration, invasion, and metastasis. Sci Rep 2020;10(1):15931. https://doi.org/10.1038/s41598-020-72960-1.
[325] Feger M, Ewendt F, Menzel M, Hocher B, Föller M. Endothelin receptor B controls the production of fibroblast growth factor 23. FASEB J 2020;34(5):6262–70.
[326] Ahn HM, Kim DG, Kim YJ. Blockade of endothelin receptor A enhances the therapeutic efficacy of gemcitabine in pancreatic cancer cells. Biochem Biophys Res Commun 2020;527(2):568–73.
[327] Tang S, Ning Q, Yang L, Mo Z, Tang S. Mechanisms of immune escape in the cancer immune cycle. Int Immunopharmacol 2020;86:106700.
[328] Lamagna C, Aurrand-Lions M, Imhof BA. Dual role of macrophages in tumor growth and angiogenesis. J Leukoc Biol 2006;80(4):705–13.
[329] Nelson JB, Udan MS, Guruli G, Pflug BR. Endothelin-1 inhibits apoptosis in prostate cancer. Neoplasia 2005; 7(7):631–7.
[330] Grant K, Loizidou M, Taylor I. Endothelin-1: a multifunctional molecule in cancer. Br J Cancer 2003;88(2):163–6.
[331] Kawanabe Y, Nauli SM. Endothelin. Cell Mol Life Sci 2011;68(2):195–203.
[332] Makita T, Sucov HM, Gariepy CE, Yanagisawa M, Ginty DD. Endothelins are vascular-derived axonal guidance cues for developing sympathetic neurons. Nature 2008;452(7188):759.

[333] Salani D, Taraboletti G, Rosano L, Di Castro V, Borsotti P, Giavazzi R, Bagnato A. Endothelin-1 induces an angiogenic phenotype in cultured endothelial cells and stimulates neovascularization in vivo. Am J Pathol 2000;157(5):1703–11.

[334] Bek EL, McMillen MA. Endothelins are angiogenic. J Cardiovasc Pharmacol 2000;36(5 Suppl 1):S135–9.

[335] Bagnato A, Spinella F. Emerging role of endothelin-1 in tumor angiogenesis. Trends Endocrinol Metab 2003; 14(1):44–50.

[336] Wülfing P, Kersting C, Tio J, Fischer RJ, Wülfing C, Poremba C, Diallo R, Böcker W, Kiesel L. Endothelin-1-, endothelin-A-, and endothelin-B-receptor expression is correlated with vascular endothelial growth factor expression and angiogenesis in breast cancer. Clin Cancer Res 2004;10(7):2393–400.

[337] Knowles J, Loizidou M, Taylor I. Endothelin-1 and angiogenesis in cancer. Curr Vasc Pharmacol 2005;3(4): 309–14.

[338] Ratna A, Das SK. Endothelin: ominous player in breast cancer. J Cancer Clin Trials 2016;1(1), e102.

[339] Zhang ZY, Chen LL, Xu W, Sigdel K, Jiang XT. Effects of silencing endothelin-1 on invasion and vascular formation in lung cancer. Oncol Lett 2017;13(6):4390–6.

[340] Kandalaft LE, Facciabene A, Buckanovich RJ, Coukos G. Endothelin B receptor, a new target in cancer immune therapy. Clin Cancer Res 2009;15(14):4521–8.

[341] Bagnato A, Rosanò L. The endothelin axis in cancer. Int J Biochem Cell Biol 2008;40(8):1443–51.

[342] Rosanò L, Spinella F, Bagnato A. Endothelin 1 in cancer: biological implications and therapeutic opportunities. Nat Rev Cancer 2013;13(9):637–51.

[343] Irani S, Salajegheh A, Smith RA, Lam AKY. A review of the profile of endothelin axis in cancer and its management. Crit Rev Oncol Hematol 2014;89(2):314–21.

[344] Kappers MH, van Esch JH, Sluiter W, Sleijfer S, Danser AJ, van den Meiracker AH. Hypertension induced by the tyrosine kinase inhibitor sunitinib is associated with increased circulating endothelin-1 levels. Hypertension 2010;56(4):675–81.

[345] Lankhorst S, Kappers MH, van Esch JH, Danser AJ, van den Meiracker AH. Hypertension during vascular endothelial growth factor inhibition: focus on nitric oxide, endothelin-1, and oxidative stress. Antioxid Redox Signal 2014;20(1):135–45. https://doi.org/10.1089/ars.2013.5244.

[346] Izzedine H, Ederhy S, Goldwasser F, Soria JC, Milano G, Cohen A, Khayat D, Spano JP. Management of hypertension in angiogenesis inhibitor-treated patients. Ann Oncol 2009;20(5):807–15.

[347] Lankhorst S, Kappers MH, Van Esch JH, Danser AJ, van den Meiracker AH. Mechanism of hypertension and proteinuria during angiogenesis inhibition: evolving role of endothelin-1. J Hypertens 2013;31(3): 444–54.

[348] Kappers MH, Smedts FM, Horn T, van Esch JH, Sleijfer S, Leijten F, Wesseling S, Strevens H, Danser J, van den Meiracker AH. The vascular endothelial growth factor receptor inhibitor sunitinib causes a preeclampsia-like syndrome with activation of the endothelin system. Hypertension 2011;58(2):295–302.

[349] Hsu SJ, Lin TY, Wang SS, Chuang CL, Lee FY, Huang HC, Hsin IF, Lee JY, Lin HC, Lee SD. Endothelin receptor blockers reduce shunting and angiogenesis in cirrhotic rats. Eur J Clin Invest 2016;46(6):572–80.

[350] Sourdon J, Lager F, Viel T, Balvay D, Moorhouse R, Bennana E, Renault G, Tharaux PL, Dhaun N, Tavitian B. Cardiac metabolic deregulation induced by the tyrosine kinase receptor inhibitor sunitinib is rescued by endothelin receptor antagonism. Theranostics 2017;7(11):2757.

[351] Rosanò L, Cianfrocca R, Tocci P, Spinella F, Di Castro V, Caprara V, Semprucci E, Ferrandina G, Natali PG, Bagnato A. Endothelin A receptor/β-arrestin signaling to the Wnt pathway renders ovarian cancer cells resistant to chemotherapy. Cancer Res 2014;74(24):7453–64.

[352] Zou L, Yang R, Chai J, Pei G. Rapid xenograft tumor progression in beta-arrestin1 transgenic mice due to enhanced tumor angiogenesis. FASEB J 2008;22(2):355–64.

[353] Tocci P, Rosanò L, Bagnato A. Targeting endothelin-1 receptor/β-arrestin-1 axis in ovarian cancer: from basic research to a therapeutic approach. Front Endocrinol 2019;10:609.

[354] Singh S, Pillai S, Chellappan S. Nicotinic acetylcholine receptor signaling in tumor growth and metastasis. J Oncol 2011;2011:456743. https://doi.org/10.1155/2011/456743.

[355] Pérez-Sayáns M, Somoza-Martín JM, Barros-Angueira F, Diz PG, Gándara Rey JM, García-García A. β-adrenergic receptors in cancer: therapeutic implications. Oncol Res 2010;19(1):45–54. https://doi.org/10.3727/096504010x12828372551867.

[356] Cianfrocca R, Tocci P, Semprucci E, Spinella F, Di Castro V, Bagnato A, Rosanò L. β-Arrestin 1 is required for endothelin-1-induced NF-κB activation in ovarian cancer cells. Life Sci 2014;118(2):179–84.

[357] Shenoy SK, Han SO, Zhao YL, Hara MR, Oliver T, Cao Y, Dewhirst MW. β-arrestin1 mediates metastatic growth of breast cancer cells by facilitating HIF-1-dependent VEGF expression. Oncogene 2012;31(3):282–92.
[358] Babic AM, Kireeva ML, Kolesnikova TV, Lau LF. CYR61, a product of a growth factor-inducible immediate early gene, promotes angiogenesis and tumor growth. Proc Natl Acad Sci U S A 1998;95(11):6355–60.
[359] Kireeva ML, Lam SCT, Lau LF. Adhesion of human umbilical vein endothelial cells to the immediate-early gene product Cyr61 is mediated through integrin αvβ3. J Biol Chem 1998;273(5):3090–6.
[360] Fataccioli V, Abergel V, Wingertsmann L, Neuville P, Spitz E, Adnot S, Calenda V, Teiger E. Stimulation of angiogenesis by Cyr61 gene: a new therapeutic candidate. Hum Gene Ther 2002;13(12):1461–70.
[361] Menéndez JA, Mehmi I, Griggs DW, Lupu R. The angiogenic factor CYR61 in breast cancer: molecular pathology and therapeutic perspectives. Endocr Relat Cancer 2003;2:141–52.
[362] Choi C, Jeong W, Ghang B, Park Y, Hyun C, Cho M, Kim J. Cyr61 synthesis is induced by interleukin-6 and promotes migration and invasion of fibroblast-like synoviocytes in rheumatoid arthritis. Arthritis Res Ther 2020;22(1):1–13.
[363] Zhu X, Xiao L, Huo R, Zhang J, Lin J, Xie J, Sun S, He Y, Sun Y, Zhou Z, et al. Cyr61 is involved in neutrophil infiltration in joints by inducing IL-8 production by fibroblast-like synoviocytes in rheumatoid arthritis. Arthritis Res Ther 2013;15(6):1–14.
[364] Estrada R, Li N, Sarojini H, An JIN, Lee MJ, Wang E. Secretome from mesenchymal stem cells induces angiogenesis via Cyr61. J Cell Physiol 2009;219(3):563–71.
[365] Kunz M, Moeller S, Koczan D, Lorenz P, Wenger RH, Glocker MO, Thiessen HJ, Gross G, Ibrahim SM. Mechanisms of hypoxic gene regulation of angiogenesis factor Cyr61 in melanoma cells. J Biol Chem 2003;278(46): 45651–60.
[366] Hilfiker A, Hilfiker-Kleiner D, Fuchs M, Kaminski K, Lichtenberg A, Rothkötter HJ, Schieffer B, Drexler H. Expression of CYR61, an angiogenic immediate early gene, in arteriosclerosis and its regulation by angiotensin II. Circulation 2002;106(2):254–60.
[367] Brigstock DR. Regulation of angiogenesis and endothelial cell function by connective tissue growth factor (CTGF) and cysteine-rich 61 (CYR61). Angiogenesis 2002;5(3):153–65.
[368] Harris LG, Pannell LK, Singh S, Samant RS, Shevde LA. Increased vascularity and spontaneous metastasis of breast cancer by hedgehog signaling mediated upregulation of cyr61. Oncogene 2012;31(28):3370–80.
[369] Meyuhas R, Pikarsky E, Tavor E, Klar A, Abramovitch R, Hochman J, Goshen Lago T, Honigman A. A key role for cyclic AMP-responsive element binding protein in hypoxia-mediated activation of the angiogenesis factor CCN1 (CYR61) in Tumor cells. Mol Cancer Res 2008;6(9):1397–409.
[370] Hirschfeld M, zur Hausen A, Bettendorf H, Jäger M, Stickeler E. Alternative splicing of Cyr61 is regulated by hypoxia and significantly changed in breast cancer. Cancer Res 2009;69(5):2082–90.
[371] Yu Y, Gao Y, Wang H, Huang L, Qin J, Guo R, Song M, Yu S, Chen J, Cul B, Gao P. The matrix protein CCN1 (CYR61) promotes proliferation, migration and tube formation of endothelial progenitor cells. Exp Cell Res 2008;314(17):3198–208.
[372] Kubota S, Takigawa M. CCN family proteins and angiogenesis: from embryo to adulthood. Angiogenesis 2007;10(1):1–11.
[373] You JJ, Yang CH, Yang CM, Chen MS. Cyr61 induces the expression of monocyte chemoattractant protein-1 via the integrin ανβ3, FAK, PI3K/Akt, and NF-κB pathways in retinal vascular endothelial cells. Cell Signal 2014; 26(1):133–40.
[374] Chintalapudi MR, Markiewicz M, Kose N, Dammai V, Champion KJ, Hoda RS, Trojanowska M, Hsu T. Cyr61/CCN1 and CTGF/CCN2 mediate the proangiogenic activity of VHL-mutant renal carcinoma cells. Carcinogenesis 2008;29(4):696–703.
[375] Si W, Kang Q, Luu HH, Park JK, Luo Q, Song WX, Jiang W, Luo X, Li X, Yin H, et al. CCN1/Cyr61 is regulated by the canonical Wnt signal and plays an important role in Wnt3A-induced osteoblast differentiation of mesenchymal stem cells. Mol Cell Biol 2006;26(8):2955–64.
[376] Hesler RA, Huang JJ, Starr MD, Treboschi VM, Bernanke AG, Nixon AB, McCall SJ, White RR, Blobe GC. TGF-β-induced stromal CYR61 promotes resistance to gemcitabine in pancreatic ductal adenocarcinoma through downregulation of the nucleoside transporters hENT1 and hCNT3. Carcinogenesis 2016;37(11):1041–51.
[377] Bartholin L, Wessner LL, Chirgwin JM, Guise TA. The human Cyr61 gene is a transcriptional target of transforming growth factor beta in cancer cells. Cancer Lett 2007;246(1–2):230–6. https://doi.org/10.1016/j.canlet.2006.02.019.

[378] Ilhan M, Kucukkose C, Efe E, Gunyuz ZE, Firatligil B, Dogan H, Ozuysal M, Yalcin-Ozuysal O. Pro-metastatic functions of Notch signaling is mediated by CYR61 in breast cells. Eur J Cell Biol 2020;99(2–3):151070.

[379] Grote K, Salguero G, Ballmaier M, Dangers M, Drexler H, Schieffer B. The angiogenic factor CCN1 promotes adhesion and migration of circulating CD34+ progenitor cells: potential role in angiogenesis and endothelial regeneration. Blood 2007;110(3):877–85.

[380] Niu J, Azfer A, Zhelyabovska O, Fatma S, Kolattukudy PE. Monocyte chemotactic protein (MCP)-1 promotes angiogenesis via a novel transcription factor, MCP-1-induced protein (MCPIP). J Biol Chem 2008;283(21): 14542–51.

[381] Salcedo R, Ponce ML, Young HA, Wasserman K, Ward JM, Kleinman HK, Murphy WJ. Human endothelial cells express CCR2 and respond to MCP-1: direct role of MCP-1 in angiogenesis and tumor progression. Blood 2000;96(1):34–40.

[382] Lien MY, Tang CH. Abstract 200: the MCP-1/CCR2 axis enhances VEGF-A-dependent angiogenesis by directly downregulating miR-29c expression in oral squamous cell carcinoma. Cancer Res 2019;79(13 Supplement): 200. https://doi.org/10.1158/1538-7445.AM2019-200.

[383] Gazzaniga S, Bravo AI, Guglielmotti A, Van Rooijen N, Maschi F, Vecchi A, Wainstok R. Targeting tumor-associated macrophages and inhibition of MCP-1 reduce angiogenesis and tumor growth in a human melanoma xenograft. J Invest Dermatol 2007;127(8):2031–41.

[384] Hong KH, Ryu J, Han KH. Monocyte chemoattractant protein-1–induced angiogenesis is mediated by vascular endothelial growth factor-A. Blood 2005;105(4):1405–7.

[385] Ergün S, Kilic N, Ziegeler G, Hansen A, Nollau P, Götze J, Wurmbach JH, Horst A, Weil J, et al. CEA-related cell adhesion molecule 1: a potent angiogenic factor and a major effector of vascular endothelial growth factor. Mol Cell 2000;5(2):311–20.

[386] Leibovich SJ, Polverini PJ, Shepard HM, Wiseman DM, Shively V, Nuseir N. Macrophage-induced angiogenesis is mediated by tumour necrosis factor-α. Nature 1987;329(6140):630.

[387] Kermani P, Hempstead B. Brain-derived neurotrophic factor: a newly described mediator of angiogenesis. Trends Cardiovasc Med 2007;17(4):140–3.

[388] Ishikawa F, Miyazono K, Hellman U, Drexler H, Wernstedt C, Hagiwara K, Usuki K, Takaku F, Risau W, Heldin CH. Identification of angiogenic activity and the cloning and expression of platelet-derived endothelial cell growth factor. Nature 1989;338(6216):557.

[389] Raida M, Clement JH, Leek RD, Ameri K, Bicknell R, Niederwieser D, Harris AL. Bone morphogenetic protein 2 (BMP-2) and induction of tumor angiogenesis. J Cancer Res Clin Oncol 2005;131(11):741–50.

[390] Yu CW, Liang X, Lipsky S, Karaaslan C, Kozakewich H, Hotamisligil GS, Bischoff J, Cataltepe S. Dual role of fatty acid-binding protein 5 on endothelial cell fate: a potential link between lipid metabolism and angiogenic responses. Angiogenesis 2016;19(1):95–106.

[391] Grabell DA, Wen Y, Semlani N, Li J, Goydos J. Pre-clinical study targeting metabotropic glutamate receptor (GRM1) angiogenesis pathway. Cancer Res 2013;73(8 Supplement):5082. https://doi.org/10.1158/1538-7445.AM2013-5082.

[392] Speyer CL, Hachem AH, Assi AA, Johnson JS, DeVries JA, Gorski DH. Metabotropic glutamate receptor-1 as a novel target for the antiangiogenic treatment of breast cancer. PLoS One 2014;9(3), e88830.

[393] Thijssen VL, Postel R, Brandwijk RJ, Dings RP, Nesmelova I, Satijn S, Verhofstad N, Nakabeppu Y, Baum LG, Bakkers J, et al. Galectin-1 is essential in tumor angiogenesis and is a target for antiangiogenesis therapy. Proc Natl Acad Sci U S A 2006;103(43):15975–80. https://doi.org/10.1073/pnas.0603883103.

[394] Nangia-Makker P, Honjo Y, Sarvis R, Akahani S, Hogan V, Pienta KJ, Raz A. Galectin-3 induces endothelial cell morphogenesis and angiogenesis. Am J Pathol 2000;156(3):899–909.

[395] Newlaczyl AU, Yu LG. Galectin-3–a jack-of-all-trades in cancer. Cancer Lett 2011;313(2):123–8.

[396] Markowska AI, Liu FT, Panjwani N. Galectin-3 is an important mediator of VEGF-and bFGF-mediated angiogenic response. J Exp Med 2010;207(9):1981–93.

[397] Thijssen VL, Barkan B, Shoji H, Aries IM, Mathieu V, Deltour L, Hackeng TM, Kiss R, Kloog Y, Poirier F, et al. Tumor cells secrete galectin-1 to enhance endothelial cell activity. Cancer Res 2010;70(15):6216–24.

[398] Griffioen AW, Van Der Schaft DW, Barendsz-Janson AF, Andrew COX, Boudier HAS, Hillen HF. Anginex, a designed peptide that inhibits angiogenesis. Biochem J 2001;354(2):233–42.

[399] Salomonsson E, Thijssen VL, Griffioen AW, Nilsson UJ, Leffler H. The anti-angiogenic peptide anginex greatly enhances galectin-1 binding affinity for glycoproteins. J Biol Chem 2011;286(16):13801–4.

[400] Duff SE, Li C, Garland JM, Kumar S. CD105 is important for angiogenesis: evidence and potential applications. FASEB J 2003;17(9):984–92.
[401] Struman I, Bentzien F, Lee H, Mainfroid V, D'Angelo G, Goffin V, Weiner RI, Martial JA. Opposing actions of intact and N-terminal fragments of the human prolactin/growth hormone family members on angiogenesis: an efficient mechanism for the regulation of angiogenesis. Proc Natl Acad Sci U S A 1999;96(4):1246–51.
[402] Bajou K, Herkenne S, Thijssen VL, D'amico S, Bouché A, Tabruyn S, Srahna M, Carabin JY, Nivelles O, Paques C, et al. PAI-1 mediates the antiangiogenic and profibrinolytic effects of 16K prolactin. Nat Med 2014;20(7):741.
[403] Denekamp J. Review article: angiogenesis, neovascular proliferation and vascular pathophysiology as targets for cancer therapy. Br J Radiol 1993;66:181–96.
[404] Siemann DW, Bibby MC, Dark GG, Dicker AP, Eskens FA, Horsman MR, Marmè D, LoRusso PM. Differentiation and definition of vascular-targeted therapies. Clin Cancer Res 2005;11(2):416–20.
[405] McGown AT, Fox BW. Structural and biochemical comparison of the anti-mitotic agents colchicine, combretastatin A4 and amphethinile. Anticancer Drug Des 1989;3(4):249–54.
[406] Thorpe PE. Vascular targeting agents as cancer therapeutics. Clin Cancer Res 2004;10(2):415–27.
[407] Subbiah IM, Lenihan DJ, Tsimberidou AM. Cardiovascular toxicity profiles of vascular-disrupting agents. Oncologist 2011;16(8):1120–30. https://doi.org/10.1634/theoncologist.2010-0432.
[408] Liang W, Ni Y, Chen F. Tumor resistance to vascular disrupting agents: mechanisms, imaging, and solutions. Oncotarget 2016;7(13):15444.
[409] Tozer GM, Prise VE, Wilson J, Cemazar M, Shan S, Dewhirst MW, Barber PR, Vojnovic B, Chaplin DJ. Mechanisms associated with tumor vascular shut-down induced by combretastatin A-4 phosphate: intravital microscopy and measurement of vascular permeability. Cancer Res 2001;61(17):6413–22.
[410] Pantziarka P, Bouche G, Meheus L, Sukhatme V, Sukhatme VP. Repurposing drugs in oncology (ReDO)—mebendazole as an anti-cancer agent. Ecancermedicalscience 2014;8:443. https://doi.org/10.3332/ecancer.2014.443.
[411] Zhang C, Zhong B, Yang S, Pan L, Yu S, Li Z, Li S, Su B, Meng X. Synthesis and biological evaluation of thiabendazole derivatives as anti-angiogenesis and vascular disrupting agents. Bioorg Med Chem 2015;23(13):3774–80.
[412] Mukhopadhyay T, Sasaki JI, Ramesh R, Roth JA. Mebendazole elicits a potent antitumor effect on human cancer cell lines both in vitro and in vivo. Clin Cancer Res 2003;8(9):2963–9.
[413] Freisleben F, Stamm H, Muschhammer J, Krispien A, Thaden V, Wellbrock J, Fiedler W. Mebendazole exerts potent anti-leukemic effects by downregulating protein levels of Hedgehog transcription factors GLI1 and GLI2. Blood 2018;132(Supplement 1):5145.
[414] Bai RY, Staedtke V, Aprhys CM, Gallia GL, Riggins GJ. Antiparasitic mebendazole shows survival benefit in 2 preclinical models of glioblastoma multiforme. Neuro Oncol 2011;13(9):974–82.
[415] Garge RK, Cha HJ, Lee C, Gollihar JD, Kachroo AH, Wallingford JB, Marcotte EM. Antifungal benzimidazoles disrupt vasculature by targeting one of nine β-tubulins. bioRxiv 2020;, 298828. bioRxiv 2020.09.15.298828 https://doi.org/10.1101/2020.09.15.298828.
[416] Guerini AE, Triggiani L, Maddalo M, Bonù ML, Frassine F, Baiguini A, Alghisi A, Tomasini D, Borghetti P, Pasinetti N, et al. Mebendazole as a candidate for drug repurposing in oncology: an extensive review of current literature. Cancers 2019;11(9):1284. https://doi.org/10.3390/cancers11091284.
[417] Sung SJ, Kim HK, Hong YK, Joe YA. Autophagy is a potential target for enhancing the anti-angiogenic effect of mebendazole in endothelial cells. Biomol Ther 2019;27(1):117.
[418] Coyne CP, Narayanan L. Mebendazole in simultaneous combination with dexamethasone-(C 21-phosphoramidate)-[anti-EGFR] generated utilizing a novel synthesis regimen: dual anti-neoplastic cytotoxicity against pulmonary adenocarcinoma (A549). J Exp Ther Oncol 2019;13(2):81–118.
[419] Torres FC, Garcia-Rubino ME, Lozano-Lopez C, Kawano D, Eifler-Lima V, von Poser G, Campos J. Imidazoles and benzimidazoles as tubulin-modulators for anti-cancer therapy. Curr Med Chem 2015;22(11):1312–23.
[420] Pourgholami MH, Khachigian LM, Fahmy RG, Badar S, Wang L, Chu SWL, Morris DL. Albendazole inhibits endothelial cell migration, tube formation, vasopermeability, VEGF receptor-2 expression and suppresses retinal neovascularization in ROP model of angiogenesis. Biochem Biophys Res Commun 2010;397(4):729–34.
[421] Nath J, Paul R, Ghosh SK, Paul J, Singha B, Debnath N. Drug repurposing and relabeling for cancer therapy: emerging benzimidazole antihelminthics with potent anticancer effects. Life Sci 2020;258:118189.
[422] Son DS, Lee ES, Adunyah SE. The antitumor potentials of benzimidazole anthelmintics as repurposing drugs. Immune Netw 2020;20(4), e29.

[423] Pourgholami MH, Szwajcer M, Chin M, Liauw W, Seef J, Galettis P, Morris DL, Links M. Phase I clinical trial to determine maximum tolerated dose of oral albendazole in patients with advanced cancer. Cancer Chemother Pharmacol 2010;65(3):597–605.

[424] Sasaki JI, Ramesh R, Chada S, Gomyo Y, Roth JA, Mukhopadhyay T. The anthelmintic drug mebendazole induces mitotic arrest and apoptosis by depolymerizing tubulin in non-small cell lung cancer cells. Mol Cancer Ther 2002;1(13):1201–9.

[425] Patil VM, Bhelekar A, Menon N, Bhattacharjee A, Simha V, Abhinav R, Abhyankar A, Sridhar E, Mahajan A, Puranik AD, et al. Reverse swing-M, phase 1 study of repurposing mebendazole in recurrent high-grade glioma. Cancer Med 2020;9(13):4676–85.

[426] Kralova V, Hanušová V, Caltová K, Špaček P, Hochmalová M, Skálová L, Rudolf E. Flubendazole and mebendazole impair migration and epithelial to mesenchymal transition in oral cell lines. Chem Biol Interact 2018;293:124–32.

[427] Bai RY, Staedtke V, Rudin CM, Bunz F, Riggins GJ. Effective treatment of diverse medulloblastoma models with mebendazole and its impact on tumor angiogenesis. Neuro Oncol 2015;17(4):545–54.

[428] Duan Q, Liu Y, Rockwell S. Fenbendazole as a potential anticancer drug. Anticancer Res 2013;33(2):355–62.

[429] Ariey-Bonnet J, Carrasco K, Le Grand M, Hoffer L, Betzi S, Feracci M, Tsvetkov P, Devred F, Collette Y, Morelli X, et al. In silico molecular target prediction unveils mebendazole as a potent MAPK14 inhibitor. Mol Oncol 2020;14(12):3083.

[430] Feng R, Li S, Lu C, Andreas C, Stolz DB, Mapara MY, Lentzsch S. Targeting the microtubular network as a new antimyeloma strategy. Mol Cancer Ther 2011;10(10):1886–96.

[431] Younis NS, Ghanim AM, Saber S. Mebendazole augments sensitivity to sorafenib by targeting MAPK and BCL-2 signalling in n-nitrosodiethylamine-induced murine hepatocellular carcinoma. Sci Rep 2019;9(1):1–16.

[432] Panigrahy D, Kaipainen A, Huang S, Butterfield CE, Barnés CM, Fannon M, Laforme AM, Chaponis DM, Folkman J, Kieran MW. PPARα agonist fenofibrate suppresses tumor growth through direct and indirect angiogenesis inhibition. Proc Natl Acad Sci U S A 2008;105(3):985–90.

[433] Koltai T. Fenofibrate in cancer: mechanisms involved in anticancer activity. F1000Research 2015;4:55. https://doi.org/10.12688/f1000research.6153.2.

[434] Drukala J, Urbanska K, Wilk A, Grabacka M, Wybieralska E, Del Valle L, Madeja Z, Reiss K. ROS accumulation and IGF-IR inhibition contribute to fenofibrate/PPARα-mediated inhibition of glioma cell motility in vitro. Mol Cancer 2010;9(1):159.

[435] Lian X, Gu J, Gao B, Li Y, Damodaran C, Wei W, Fu Y, Cai L. Fenofibrate inhibits mTOR-p70S6K signaling and simultaneously induces cell death in human prostate cancer cells. Biochem Biophys Res Commun 2018; 496(1):70–5.

[436] Han D, Wei W, Chen X, Zhang Y, Wang Y, Zhang J, Wang X, Yu T, Hu Q, Liu N, et al. NF-κB/RelA-PKM2 mediates inhibition of glycolysis by fenofibrate in glioblastoma cells. Oncotarget 2015;6(28):26119.

[437] Madej A, Okopien B, Kowalski J, Zielinski M, Wysocki J, Szygula B, Kalina Z, Herman ZS. Effects of fenofibrate on plasma cytokine concentrations in patients with atherosclerosis and hyperlipoproteinemia IIb. Int J Clin Pharmacol Ther 1998;36(6):345–9.

[438] Zak Z, Gelebart P, Lai R. Fenofibrate induces effective apoptosis in mantle cell lymphoma by inhibiting the TNFα/NF-κB signaling axis. Leukemia 2010;24(8):1476.

[439] Ge Y, Liu J, Yang X, Zhu H, Yang B, Zhao K, Wu Z, Cheng G, Wang F, Ni F, et al. Fenofibrate enhances radiosensitivity of esophageal squamous cell carcinoma by suppressing hypoxia-inducible factor-1α expression. Tumour Biol 2014;35(11):10765–71.

[440] Zhou J, Zhang S, Xue J, Avery J, Wu J, Lind SE, Ding WQ. Activation of peroxisome proliferator-activated receptor α (PPARα) suppresses hypoxia-inducible factor-1α (HIF-1α) signaling in cancer cells. J Biol Chem 2012;287(42):35161–9.

[441] Avis I, Hong SH, Martínez A, Moody T, Choi YH, Trepel J, Das R, Jett M, Mulshine JL. Five-lipoxygenase inhibitors can mediate apoptosis in human breast cancer cell lines through complex eicosanoid interactions. FASEB J 2001;15(11):2007–9.

[442] Tsimihodimos V, Kakafika A, Tambaki AP, Bairaktari E, Chapman MJ, Elisaf M, Tselepis AD. Fenofibrate induces HDL-associated PAF-AH but attenuates enzyme activity associated with apoB-containing lipoproteins. J Lipid Res 2003;44(5):927–34.

[443] Delerive P, De Bosscher K, Besnard S, Berghe WV, Peters JM, Gonzalez FJ, Fruchart JC, Tedgui A, Haegeman G, Staels B. Peroxisome proliferator-activated receptor α negatively regulates the vascular inflammatory gene response by negative cross-talk with transcription factors NF-κB and AP-1. J Biol Chem 1999;274(45):32048–54.

[444] Nilsson L, Takemura T, Eriksson P, Hamsten A. Effects of fibrate compounds on expression of plasminogen activator inhibitor-1 by cultured endothelial cells. Arterioscler Thromb Vasc Biol 1999;19(6):1577–81.
[445] Look MP, Van Putten WL, Duffy MJ, Harbeck N, Christensen IJ, Thomssen C, Kates R, Spyratos F, Fernö M, Eppenberger-Castori S, et al. Pooled analysis of prognostic impact of urokinase-type plasminogen activator and its inhibitor PAI-1 in 8377 breast cancer patients. J Natl Cancer Inst 2002;94(2):116–28.
[446] Pedersen H, Grøndahl-Hansen J, Francis D, Østerlind K, Hansen HH, Danø K, Brünner N. Urokinase and plasminogen activator inhibitor type 1 in pulmonary adenocarcinoma. Cancer Res 1994;54(1):120–3.
[447] Zhang KL, Han L, Chen LY, Shi ZD, Yang M, Ren Y, Chen LC, Zhang JX, Pu PY, Kang CS. Blockage of a miR-21/EGFR regulatory feedback loop augments anti-EGFR therapy in glioblastomas. Cancer Lett 2014; 342(1):139–49.
[448] Chang NW, Tsai MH, Lin C, Hsu HT, Chu PY, Yeh CM, Chiu CF, Yeh KT. Fenofibrate exhibits a high potential to suppress the formation of squamous cell carcinoma in an oral-specific 4-nitroquinoline 1-oxide/arecoline mouse model. Biochim Biophys Acta 2011;1812(4):558–64.
[449] Majeed Y, Upadhyay R, Alhousseiny S, Taha T, Musthak A, Shaheen Y, Jameel M, Triggel CR, Ding H. Potent and PPARα-independent anti-proliferative action of the hypolipidemic drug fenofibrate in VEGF-dependent angiosarcomas in vitro. Sci Rep 2019;9(1):1–14.
[450] Nguyen CH, Huttary N, Atanasov AG, Chatuphonprasert W, Brenner S, Fristiohady A, Hong J, Stadler S, Holzner S, Milovanovic D, et al. Fenofibrate inhibits tumour intravasation by several independent mechanisms in a 3-dimensional co-culture model. Int J Oncol 2017;50(5):1879–88.
[451] Hu D, Su C, Jiang M, Shen Y, Shi A, Zhao F, Chen R, Shen Z, Bao J, Tang W. Fenofibrate inhibited pancreatic cancer cells proliferation via activation of p53 mediated by upregulation of LncRNA MEG3. Biochem Biophys Res Commun 2016;471(2):290–5.
[452] Xin R, An D, Li Y, Fu J, Huang F, Zhu Q. Fenofibrate improves vascular endothelial function in diabetic mice. Biomed Pharmacother 2019;112:108722.
[453] Pisha E, Chai H, Lee IS, Chagwedera TE, Farnsworth NR, Cordell GA, Beecher CWW, Fong HHS, Kinghorn AD, Brown DM, et al. Discovery of betulinic acid as a selective inhibitor of human melanoma that functions by induction of apoptosis. Nat Med 1995;1(10):1046–51.
[454] Tan Y, Yu R, Pezzuto JM. Betulinic acid-induced programmed cell death in human melanoma cells involves mitogen-activated protein kinase activation. Clin Cancer Res 2003;9(7):2866–75.
[455] Zhang X, Hu J, Chen Y. Betulinic acid and the pharmacological effects of tumor suppression. Mol Med Rep 2016;14(5):4489–95.
[456] Saeed ME, Mahmoud N, Sugimoto Y, Efferth T, Abdel-Aziz H. Betulinic acid exerts cytotoxic activity against multidrug-resistant tumor cells via targeting autocrine motility factor receptor (AMFR). Front Pharmacol 2018;9:481.
[457] Zhang H, Li L, Li M, Huang X, Xie W, Xiang W, Yao P. Combination of betulinic acid and chidamide inhibits acute myeloid leukemia by suppression of the HIF1α pathway and generation of reactive oxygen species. Oncotarget 2017;8(55):94743.
[458] Shin J, Lee HJ, Jung DB, Jung JH, Lee HJ, Lee EO, Lee SG, Shim BS, Choi SH, Ko SG, et al. Suppression of STAT3 and HIF-1 alpha mediates anti-angiogenic activity of betulinic acid in hypoxic PC-3 prostate cancer cells. PLoS One 2011;6(6), e21492.
[459] Karna E, Szoka L, Palka JA. Betulinic acid inhibits the expression of hypoxia-inducible factor 1α and vascular endothelial growth factor in human endometrial adenocarcinoma cells. Mol Cell Biochem 2010;340(1):15–20.
[460] Kutkowska J, Strzadala L, Rapak A. Hypoxia increases the apoptotic response to betulinic acid and betulin in human non-small cell lung cancer cells. Chem Biol Interact 2021;333:109320.
[461] Kushwaha PP, Singh AK, Shuaib M, Prajapati KS, Vardhan PS, Gupta S, Kumar S. 3-O-(E)-p-Coumaroyl betulinic acid possess anticancer activity and inhibit Notch signaling pathway in breast cancer cells and mammosphere. Chem Biol Interact 2020;328:109200.
[462] Saneja A, Sharma L, Dubey RD, Mintoo MJ, Singh A, Kumar A, Sangwan PL, Tasaduq SA, Singh G, Mondhe DM, et al. Synthesis, characterization and augmented anticancer potential of PEG-betulinic acid conjugate. Mater Sci Eng C Mater Biol Appl 2017;73:616–26.
[463] Fandrey J, Gassmann M. Oxygen sensing and the activation of the hypoxia inducible factor 1 (HIF-1)–invited article. In: Gonzalez C, Nurse CA, Peers C, editors. Arterial chemoreceptors. Arterial chemoreceptors. Advances in experimental medicine and biology, vol. 648. Dordrecht: Springer; 2009. p. 197–206. https://doi.org/10.1007/978-90-481-2259-2_23.

[464] Salceda S, Caro J. Hypoxia-inducible factor 1α (HIF-1α) protein is rapidly degraded by the ubiquitin-proteasome system under normoxic conditions: its stabilization by hypoxia depends on redox-induced changes. J Biol Chem 1997;272(36):22642–7.

[465] Huang LE, Gu J, Schau M, Bunn HF. Regulation of hypoxia-inducible factor 1α is mediated by an O_2-dependent degradation domain via the ubiquitin-proteasome pathway. Proc Natl Acad Sci U S A 1998;95(14):7987–92.

[466] Kim HJ, Cho HS, Ban HS, Nakamura H. Suppression of HIF-1α accumulation by betulinic acid through proteasome activation in hypoxic cervical cancer. Biochem Biophys Res Commun 2020;523(3):726–32.

[467] Kumar A, Vaish M, Karuppagounder SS, Gazaryan I, Cave JW, Starkov AA, Anderson ET, Zhang S, Pinto JT, Rountree A, et al. HIF1α stability in hypoxia is not oxidant-initiated but can be tuned by redox regulation of the proteasome. bioRxiv 2020. https://doi.org/10.1101/2020.05.05.079681.

[468] Schönberger T, Fandrey J, Prost-Fingerle K. Ways into understanding HI inhibition. Cancers 2021;13(1):159.

[469] Natori T, Sata M, Nagai R, Makuuchi M. Cimetidine inhibits angiogenesis and suppresses tumor growth. Biomed Pharmacother 2005;59(1–2):56–60.

[470] Tomita K, Izumi K, Okabe S. Roxatidine-and cimetidine-induced angiogenesis inhibition suppresses growth of colon cancer implants in syngeneic mice. J Pharmacol Sci 2003;93(3):321–30.

[471] Chihara Y, Fujimoto K, Miyake M, Hiasa Y, Hirao Y. Anti-tumor effect of cimetidine via inhibiting angiogenesis factors in N-butyl-N-(4-hydroxybutyl) nitrosamine-induced mouse and rat bladder carcinogenesis. Oncol Rep 2009;22(1):23–8.

[472] Borentain P, Carmona S, Mathieu S, Jouve E, El-Battari A, Gérolami R. Inhibition of E-selectin expression on the surface of endothelial cells inhibits hepatocellular carcinoma growth by preventing tumor angiogenesis. Cancer Chemother Pharmacol 2016;77(4):847–56.

[473] Dana P, Vaeteewoottacharn K, Kariya R, Matsuda K, Wongkham S, Okada S. Repurposing cimetidine for cholangiocarcinoma: antitumor effects in vitro and in vivo. Oncol Lett 2017;13(3):1432–6.

[474] Kerbel RS. Therapeutic implications of intrinsic or induced angiogenic growth factor redundancy in tumors revealed. Cancer Cell 2005;8(4):269–71.

[475] Casanovas O, Hicklin DJ, Bergers G, Hanahan D. Drug resistance by evasion of antiangiogenic targeting of VEGF signaling in late-stage pancreatic islet tumors. Cancer Cell 2005;8(4):299–309.

[476] Ebos JM, Lee CR, Christensen JG, Mutsaers AJ, Kerbel RS. Multiple circulating proangiogenic factors induced by sunitinib malate are tumor-independent and correlate with antitumor efficacy. Proc Natl Acad Sci U S A 2007;104:17069–74.

[477] Ebos JM, Lee CR, Cruz-Munoz W, Bjarnason GA, Christensen JG, Kerbel RS. Accelerated metastasis after short-term treatment with a potent inhibitor of tumor angiogenesis. Cancer Cell 2009;15:232–9.

[478] Kamat AA, Kim TJ, Landen CN, Lu C, Han LY, Lin YG, Merritt WM, Thaker PH, Gershenson DM, Bischoff FZ, et al. Metronomic chemotherapy enhances the efficacy of antivascular therapy in ovarian cancer. Cancer Res 2007;67(1):281–8.

[479] Klement G, Huang P, Mayer B, Green SK, Man S, Bohlen P, Hicklin D, Kerbel RS. Differences in therapeutic indexes of combination metronomic chemotherapy and an anti-VEGFR-2 antibody in multidrug-resistant human breast cancer xenografts. Clin Cancer Res 2002;8(1):221–32.

[480] Jayson GC, Kerbel R, Ellis LM, Harris AL. Antiangiogenic therapy in oncology: current status and future directions. Lancet 2016;388(10043):518–29.

[481] Pan ST, Li ZL, He ZX, Qiu JX, Zhou SF. Molecular mechanisms for tumour resistance to chemotherapy. Clin Exp Pharmacol Physiol 2016;43(8):723–37.

[482] Chan N, Ali M, McCallum GP, Kumareswaran R, Koritzinsky M, Wouters BG, Bristow RG. Hypoxia provokes base excision repair changes and a repair-deficient, mutator phenotype in colorectal cancer cells. Mol Cancer Res 2014;12(10):1407–15.

[483] Ashton J, Bristow R. Bad neighbours: hypoxia and genomic instability in prostate cancer. Br J Radiol 2020; 93(1115):20200087.

[484] Fiandaca G, Delitala M, Lorenzi T. A mathematical study of the influence of hypoxia and acidity on the evolutionary dynamics of cancer. Bull Math Biol 2021;83(7):83. https://arxiv.org/abs/2009.00251.

[485] Lu X, Kang Y. Hypoxia and hypoxia-inducible factors: master regulators of metastasis. Clin Cancer Res 2010; 16(24):5928–35.

[486] Cao MQ, You AB, Cui W, Zhang S, Guo ZG, Chen L, Zhu XD, Zhang W, Zhu XL, Guo H, et al. Cross talk between oxidative stress and hypoxia via thioredoxin and HIF-2α drives metastasis of hepatocellular carcinoma. FASEB J 2020;34(4):5892–905.

[487] Wang X, Che X, Yu Y, Cheng Y, Bai M, Yang Z, Guo O, Xie X, Li D, Guo M, et al. Hypoxia-autophagy axis induces VEGFA by peritoneal mesothelial cells to promote gastric cancer peritoneal metastasis through an integrin α5-fibronectin pathway. J Exp Clin Cancer Res 2020;39(1):1–14.
[488] Li H, Peng C, Zhu C, Nie S, Qian X, Shi Z, Shi M, Liang Y, Ding X, Zhang S, et al. Hypoxia promotes the metastasis of pancreatic cancer through regulating NOX4/KDM5A-mediated histone methylation modification changes in a HIF1A-independent manner. Clin Epigenetics 2021;13(1):1–15.
[489] Donato C, Kunz L, Castro-Giner F, Paasinen-Sohns A, Strittmatter K, Szczerba BM, Scherrer R, Di Maggio N, Heusermann W, Biehlmaler O, et al. Hypoxia triggers the intravasation of clustered circulating tumor cells. Cell Rep 2020;32(10):108105.
[490] Wu H, Chu Y, Sun S, Li G, Xu S, Zhang X, Jiang Y, Gao S, Wang Q, Zhang J, et al. Hypoxia-mediated C1QBP regulates metastasis and chemoresistance in triple-negative breast cancer and modulates VCAM-1 expression via canonical Nf-Kβ signaling. Front Cell Dev Biol 2021;9:607142. https://doi.org/10.3389/fcell.2021.607142.
[491] Pajdzik K, Wilamowski M, Żurawek D, Stopa KB, Nodzyński M, Kalita A, Jura J. Anterior gradient 2 promotes tumorigenesis through upregulation of CCAAT-enhancer binding protein beta and hypoxia-inducible factor-2α and subsequent secretion of interleukin-6, interleukin-8, and vascular endothelial growth factor in the Caki-1 clear cell renal cell carcinoma cell line. IUBMB Life 2020;72(8):1807–18.
[492] Karakurum M, Shreeniwas R, Chen J, Pinsky D, Yan SD, Anderson M, Sonouchi K, Hamilton MT, Kuwabara K. Hypoxic induction of interleukin-8 gene expression in human endothelial cells. J Clin Invest 1994;93(4): 1564–70.
[493] Mizukami Y, Jo WS, Duerr EM, Gala M, Li J, Zhang X, Zimmer MA, Iliopolus O, Zukerberg LR, Kohgo Y, et al. Induction of interleukin-8 preserves the angiogenic response in HIF-1α–deficient colon cancer cells. Nat Med 2005;11(9):992–7.
[494] Shi Q, Abbruzzese JL, Huang S, Fidler IJ, Xiong Q, Xie K. Constitutive and inducible interleukin 8 expression by hypoxia and acidosis renders human pancreatic cancer cells more tumorigenic and metastatic. Clin Cancer Res 1999;5(11):3711–21.
[495] Bi M, Naczki C, Koritzinsky M, Fels D, Blais J, Hu N, Harding H, Novoa I, Varia M, Raleigh J, et al. ER stress-regulated translation increases tolerance to extreme hypoxia and promotes tumor growth. EMBO J 2005; 24(19):3470–81.
[496] Al Tameemi W, Dale TP, Al-Jumaily RMK, Forsyth NR. Hypoxia-modified cancer cell metabolism. Front Cell Dev Biol 2019;7:4.
[497] Patton MC, Zubair H, Khan MA, Singh S, Singh AP. Hypoxia alters the release and size distribution of extracellular vesicles in pancreatic cancer cells to support their adaptive survival. J Cell Biochem 2020;121(1):828–39.
[498] Muz B, de la Puente P, Azab F, Azab AK. The role of hypoxia in cancer progression, angiogenesis, metastasis, and resistance to therapy. Hypoxia 2015;3:83.
[499] Huang D, Ding Y, Zhou M, Rini BI, Petillo D, Qian CN, Kahnosky R, Futreal PA, Furke KA, Teh BT. Interleukin-8 mediates resistance to antiangiogenic agent sunitinib in renal cell carcinoma. Cancer Res 2010;70(3):1063–71.
[500] Reguera-Nuñez E, Man S, Xu P, Kerbel RS. Preclinical impact of high dose intermittent antiangiogenic tyrosine kinase inhibitor pazopanib in intrinsically resistant tumor models. Angiogenesis 2018;21(4):793–804.
[501] Wang X, Zhang L, Goldberg SN, Bhasin M, Brown V, Alsop DC, Signoretti S, Mier JW, Atkins MB, Bhatt RS. High dose intermittent sorafenib shows improved efficacy over conventional continuous dose in renal cell carcinoma. J Transl Med 2011;9(1):1–8.
[502] Kythreotou A, Siddique A, Mauri FA, Bower M, Pinato DJ. PD-L1. J Clin Pathol 2018;71(3):189–94.
[503] Yi M, Jiao D, Qin S, Chu Q, Wu K, Li A. Synergistic effect of immune checkpoint blockade and anti-angiogenesis in cancer treatment. Mol Cancer 2019;18(1):60.
[504] Yi M, Jiao D, Xu H, Liu Q, Zhao W, Han X, Wu K. Biomarkers for predicting efficacy of PD-1/PD-L1 inhibitors. Mol Cancer 2018;17(1):1–14.
[505] Georganaki M, van Hooren L, Dimberg A. Vascular targeting to increase the efficiency of immune checkpoint blockade in cancer. Front Immunol 2018;9:3081.
[506] Perdrizet K, Leighl NB. The role of angiogenesis inhibitors in the era of immune checkpoint inhibitors and targeted therapy in metastatic non-small cell lung cancer. Curr Treat Options Oncol 2019;20(3):1–14.
[507] Donisi C, Puzzoni M, Ziranu P, Lai E, Mariani S, Saba G, Impera V, Dubois M, Persano M, Migliari M, et al. Immune checkpoint inhibitors in the treatment of HCC. Front Oncol 2021;10:2808.
[508] Song Y, Fu Y, Xie Q, Zhu B, Wang J, Zhang B. Anti-angiogenic agents in combination with immune checkpoint inhibitors: a promising strategy for cancer treatment. Front Immunol 2020;11:1956.
[509] Yang J, Yan J, Liu B. Targeting VEGF/VEGFR to modulate antitumor immunity. Front Immunol 2018;9:978.

[510] Malo CS, Khadka RH, Ayasoufi K, Jin F, AbouChehade JE, Hansen MJ, Iezzi R, Pavelko KD, Johnson AJ. Immunomodulation mediated by anti-angiogenic therapy improves CD8 T cell immunity against experimental glioma. Front Oncol 2018;8:320.

[511] Long J, Hu Z, Xue H, Wang Y, Chen J, Tang F, Zhou J, Liu L, Qiu W, Zhang S, et al. Vascular endothelial growth factor (VEGF) impairs the motility and immune function of human mature dendritic cells through the VEGF receptor 2-RhoA-cofilin1 pathway. Cancer Sci 2019;110(8):2357.

[512] Voron T, Colussi O, Marcheteau E, Pernot S, Nizard M, Pointet AL, Latreche S, Bergava S, Benhamouda N, Tanchot C, et al. VEGF-A modulates expression of inhibitory checkpoints on CD8+ T cells in tumors. J Exp Med 2015;212(2):139–48.

[513] Meder L, Schuldt P, Thelen M, Schmitt A, Dietlein F, Klein S, Borchmann S, Wennhold K, Vlasic I, Oberbeck S, et al. Combined VEGF and PD-L1 blockade displays synergistic treatment effects in an autochthonous mouse model of small cell lung cancer. Cancer Res 2018;78(15):4270–81.

[514] Lee WS, Yang H, Chon HJ, Kim C. Combination of anti-angiogenic therapy and immune checkpoint blockade normalizes vascular-immune crosstalk to potentiate cancer immunity. Exp Mol Med 2020;52(9):1475–85.

[515] Fukumura D, Kloepper J, Amoozgar Z, Duda DG, Jain RK. Enhancing cancer immunotherapy using antiangiogenics: opportunities and challenges. Nat Rev Clin Oncol 2018;15(5):325.

[516] Khan KA, Kerbel RS. Improving immunotherapy outcomes with anti-angiogenic treatments and vice versa. Nat Rev Clin Oncol 2018;15(5):310.

[517] Huang Y, Kim BY, Chan CK, Hahn SM, Weissman IL, Jiang W. Improving immune–vascular crosstalk for cancer immunotherapy. Nat Rev Immunol 2018;18(3):195–203.

[518] Huang Y, Goel S, Duda DG, Fukumura D, Jain RK. Vascular normalization as an emerging strategy to enhance cancer immunotherapy. Cancer Res 2013;73(10):2943–8.

[519] Lanitis E, Irving M, Coukos G. Targeting the tumor vasculature to enhance T cell activity. Curr Opin Immunol 2015;33:55–63.

[520] Jain RK. Normalization of tumor vasculature: an emerging concept in antiangiogenic therapy. Science 2005; 307(5706):58–62.

[521] Buckanovich RJ, Facciabene A, Kim S, Benencia F, Sasaroli D, Balint K, Katsaros D, O'Brien-Jenkins A, Gimotty PA, Coukos G. Endothelin B receptor mediates the endothelial barrier to T cell homing to tumors and disables immune therapy. Nat Med 2008;14(1):28–36.

CHAPTER

10

Predictive "omic" biomarkers of drug response: Colorectal cancer as a model

Enrico Mini[a,b], Ida Landini[a,b], Antonello Di Paolo[c], Gloria Ravegnini[d], Simona Saponara[e], Maria Frosini[e], Andrea Lapucci[a,b], and Stefania Nobili[f]

[a]Department of Health Sciences, University of Florence, Florence, Italy [b]DENOTHE Excellence Center, University of Florence, Florence, Italy [c]Department of Clinical and Experimental Medicine, University of Pisa, Pisa, Italy [d]Department of Pharmacy and Biotechnology, University of Bologna, Bologna, Italy [e]Department of Life Sciences, University of Siena, Siena, Italy [f]Department of Neurosciences, Imaging and Clinical Sciences, "G. D'Annunzio" University of Chieti-Pescara, Chieti, Italy

Abstract

The introduction of biomarkers for the rational choice of drug treatment options in several cancer types has significantly improved therapeutic outcomes. At present, optimal therapeutic planning considers however a few somatic tumor gene mutations to improve drug treatment response. Other omics-based molecular biomarkers may provide a deeper knowledge of the sources of interindividual variability in cancer patient drug response. Thus, a molecular biomarker signature predictive of response linked to standard clinical, pathological, biochemical, and molecular markers would be useful for patient stratification and drug treatment customization to select personalized therapies with greater efficacy and tolerability.

Colorectal cancer represents a model of a tumor with a high impact on health, being one of the most incident and lethal cancers worldwide. Elevated death rates depend on neoplastic progression guided by several dysregulated biomolecular events, including the evolutionary metastatic process occurring in about 50% of the patients. Despite the availability of active cytotoxic and targeted drugs, including antiangiogenic agents, the efficacy of pharmacological treatment of metastatic, and even early, colorectal cancer is hampered by the occurrence of intrinsic and acquired drug resistance. This chapter describes the status of the art of known and potential biomarkers predictive of drug response and tolerability in colorectal cancer from various "omics" sciences. It also highlights promising perspectives in this field. Combined analysis by machine learning and artificial intelligence of these determinants may contribute to identifying molecular comprehensive signatures predictive of efficacy and toxicity to be validated by prospective clinical trials, for the use in precision medicine of colorectal cancer patients.

https://doi.org/10.1016/B978-0-323-90190-1.00002-0

Abbreviations

3′-UTR	3′-untranslated region
5mC	5 methylcytosine
AKT	protein kinase B
APC	APC regulator of WNT signaling pathway
APOE	apolipoprotein E
BCAR4	breast cancer anti-estrogen resistance 4
BRAF	v-Raf murine sarcoma viral oncogene homolog B
CASC19	cancer susceptibility 19 RNA
CCAT1	colon cancer-associated transcript 1
CD11B	integrin alpha M
CD73	5′ nucleotidase
CDA	cytidine deaminase
CDKN2A	cyclin-dependent kinase inhibitor 2A
cfDNA	cell-free circulating DNA
CMS	consensus molecular subtypes
CRART16	cetuximab resistance-associated RNA transcript 16 lncRNA
CRIS	cancer cell intrinsic transcriptional traits
CRISPR	clustered regularly interspaced short palindromic repeats
ctDNA	circulating tumor DNA
CTLA-4	cytotoxic T-lymphocyte antigen 4
dMMR	deficient MMR
DNMTs	DNA methyltransferases
DPYD	dihydropyrimidine dehydrogenase
DUT	deoxyuridine triphosphatase
dUTP	deoxyuridine triphosphate
dUTPase	deoxyuridine 5′-triphosphate nucleotidohydrolase
EGFR	epidermal growth factor receptor
ERBB3	v-erb-b2 avian erythroblastic leukemia viral oncogene homolog 3
ERCC1	excision repair cross-complementation group 1
ERCC2	excision repair cross-complementation group 2
ERCC5	excision repair cross-complementation group 5
EREG	epiregulin
ERK	extracellular signal-regulated kinase
ESMO	European Society for Medical Oncology
FCGRs	Fc gamma receptors
FdUTP	5-fluorodeoxyuridine triphosphate
FOLFIRI	folinic acid, 5-fluorouracil, irinotecan
FOLFOX	folinic acid, 5-fluorouracil, oxaliplatin
GC	gas-chromatography
GEO	gene expression omnibus
GST	glutathione-S-transferase
H3K27me3	trimethylation levels of lysine 27 on histone 3
HAT	histone acetyl transferases
HDACs	histone deacetylases
HDMs	histone demethylases
HER2	human epidermal growth factor receptor 2
HER3	human epidermal growth factor receptor 3
HMTs	histone methyltransferases
HSP	heat shock protein
IRS2	insulin receptor substrate 2 gene
ITGA2	integrin alpha-2

KCNQ1OT1	KCNQ1 opposite strand/antisense transcript 1
KEGG	Kyoto encyclopedia of genes and genomes
KRAS	Kirsten rat sarcoma viral oncogene homolog
LC	liquid-chromatography
LINC00941	long intergenic non-protein coding RNA 941
LINC00973	long intergenic non-protein coding RNA 973
LINE-1	long interspersed nuclear element 1
lncRNAs	long non-coding RNAs
MAPK	mitogen-activated protein kinase
MEG3	maternally expressed gene 3
MEK	mitogen-activated protein kinase kinase
MET	met proto-oncogene
miRNAs	microRNAs
MLH1	mutL homolog 1
MMR	mismatch repair
mRNA	messenger RNA
MS	mass spectrometry
MSH2	mutS homolog 2
MSH6	mutS homolog 6
MSI	microsatellite instability
MTHFR	methylenetetrahydrofolate reductase
NCCN	National Comprehensive Cancer Network
ncRNAs	non-coding RNAs
NER	nucleotide excision repair
NGS	next-generation sequencing
NMR	nuclear magnetic resonance
NRAS	neuroblastoma RAS viral oncogene homolog
NTRK	rearrangements of neurotrophic receptor tyrosine kinase
OLFM4	olfactomedin 4
PD-1	programmed cell death protein 1
PI3K	phosphoinositide 3-kinase
PI3KCA	phosphatidylinositol-4,5-bisphosphate 3-kinase catalytic subunit alpha
PMS2	PMS1 homolog 2, mismatch repair system component
PNN	pinin, desmosome associated protein
POLE1	DNA polymerase epsilon catalytic sub-unit gene
PTEN	phosphatase and tensin homolog
RNA-seq	RNA-sequencing
RT-qPCR	reverse transcription-quantitative polymerase chain reaction
SNPs	single nucleotide polymorphisms
TCGA	The Cancer Genome Atlas
TP	thymidine phosphorylase
TrkA	tropomyosin receptor kinase A
TrkB	tropomyosin receptor kinase B
TrkC	tropomyosin receptor kinase C
TYMS	thymidylate synthase
UGT1A1	uridine diphosphate glucuronosyltransferase isoform 1A1
VEGF	vascular endothelial growth factor
VEGFR	vascular endothelial growth factor receptor
wt	wild type
XPA	xeroderma pigmentosum group A protein
XPC	xeroderma pigmentosum group C protein
XPD	xeroderma pigmentosum group D protein
XPG	xeroderma pigmentosum group G protein

Conflict of interest

No potential conflicts of interest were disclosed.

Introduction

With the advancements in molecular biology of cancer initiation and progression, in genomic biotechnologies and in selective molecular targeted drug development, the identification of biomarkers has been found to play an important role in the treatment of cancer patients. Cancer biomarkers can be classified in diagnostic, prognostic, and predictive, as they may identify cancer inception or cancer progression or response to therapy and can inform the entire course of clinical cancer prevention and management.

While advanced pathobiological knowledge is nowadays linked mainly to somatic genomics of precancerous [1] and cancerous lesions [2], fueled by the rapid expansion of innovative sequencing biotechnologies, other relevant biomolecular alterations such as DNA and histone epigenome, coding and non-coding RNA, protein and metabolite modifications have acquired increasing importance for the depiction of a multi "omic", comprehensive profile of cancer aberrant molecular pathways and for consequent development of more tailored treatments.

Colorectal cancer is one of the most frequent and lethal cancers worldwide [3,4]. High death rates are related to about 50% of patients who present with metastasis at diagnosis or who develop it later, despite primary loco-regional treatment and adjuvant chemotherapy [5]. The process of metastatic progression leads to about 14% 5-years overall survival of stage IV patients [6]. Selection of optimal drug therapy of metastatic colorectal cancer among the currently available treatment options and discovery and development of new drugs active against specific colorectal cancer phenotypes, including drug-resistant ones, are thus crucial for long-term prognosis.

At present, various cytotoxic agents constitute the backbone of drug treatment of this disease (e.g., fluoropyrimidines, oxaliplatin, irinotecan), and are primarily used in combinations. Also targeted agents, both chemical and biological, have been used with substantial clinical benefit in the metastatic setting of this disease. They include anti-epidermal growth factor receptor (EGFR) monoclonal antibodies, mutated BRAF kinase inhibitors, or immune checkpoint inhibitors in the presence of specific biomarkers of oncogenic addiction or mismatch repair (MMR) deficiency [7–12]. Also, anti-vascular endothelial growth factor (receptor) (VEGF(R)) monoclonal antibodies or VEGFR tyrosine kinase inhibitors have a therapeutic role in the treatment of metastatic colorectal cancer [7]. However, their use still lacks stratification based on predictive biomarkers of drug response [13]. The complex of these drugs, used as a continuum of care, has significantly improved the overall survival of metastatic colorectal cancer patients [7].

However, therapeutic outcomes are still unsatisfactory and variable among patients receiving the same drugs [14], due to the emergence of drug-resistant populations of cancer cells, either at the time of first treatment (intrinsic resistance) or following initial objective response (acquired resistance) [15].

Genome-wide mutation analysis has provided relevant contributions to the clinical diagnosis of resistance to anti-EGFR monoclonal antibodies and of sensitivity to immune checkpoint inhibitors [16]. Depending on the tumor mutational status, targeted drugs specifically inhibit related dysregulated proteins [17]. However, a percentage of Kirsten rat sarcoma viral oncogene homolog (*KRAS*) or neuroblastoma RAS viral oncogene homolog (*NRAS*) wild-type (wt) patients is not responsive to anti-EGFR monoclonal antibodies. This could be related to additional mediators involved in metastatic colorectal cancer intrinsic resistance to these agents or to the occurrence of *RAS* and v-Raf murine sarcoma viral oncogene homolog B (*BRAF*) mutations emerging during anti-EGFR monoclonal antibody therapy [18]. Patients who progressed on biological drug-containing regimens might also become refractory to co-administered cytotoxic drugs [19]. Also, *BRAF*-mutated tumors resistant to anti-EGFR monoclonal antibodies display resistance to specific BRAF inhibitors due to compensatory mechanisms, such as increased EGFR and extracellular signal-regulated kinase (ERK) activation [20]. This condition requires complex drug regimens including inhibitors of BRAF, mitogen-activated protein kinase kinase (MEK), and EGFR [8]. Patients with left-sided wt *KRAS* metastatic colorectal cancer treated with anti-EGFR monoclonal antibodies had better progression-free survival, overall survival, and objective response rates than patients with right-sided primary tumors [21]. Thus, left- or right-sided tumors have different molecular and pathological pattern phenotypes [22].

A precise characterization of the driver gene mutations in metastatic colorectal cancer patients is therefore essential. Transcriptomic, epigenomic, proteomic, and metabolomic biomarkers may also play a fundamental role in the prediction of drug sensitivity/resistance to both cytotoxic and targeted anticancer drugs. Thus, increased knowledge in these fields may substantially improve personalized treatment in metastatic colorectal cancer patients, with increased response, survival, and quality of life. Indeed, identification and integration of these biomarkers remain at present an unmet clinical need. "Omics" methodologies may efficiently investigate molecular alterations [23]. Next-generation sequencing (NGS) has greatly contributed to identifying many different mutations, and the genetic landscape of cancer has become much more complex than expected. Through the years, the study of DNA mutations improved knowledge not only of colorectal cancer initiation and progression but also of drug sensitivity and resistance mechanisms [24]. These advances have allowed the development of alternative active drug treatments for metastatic colorectal cancer cases harboring somatic oncogenic mutations responsible for cancer progression and drug resistance, which have recently become actionable by specific inhibitors (e.g., mutations of human epidermal growth factor receptor 2—*HER2*, DNA polymerase epsilon catalytic sub-unit gene—*POLE1*, rearrangements of neurotrophic receptor tyrosine kinases—*NTRKs*, etc) [25,26]. The NGS technology is at present widely used not only in research but also in clinical oncology practice applied both to tumor tissue and circulating tumor DNA (ctDNA) or tumor cells [27–29]. In particular, the minimally invasive liquid biopsy approach to colorectal cancer molecular diagnostics based on ctDNA has the potential to change clinical practice by monitoring drug responses and tracking clonal dynamics related to drug treatments [30].

Gene expression analysis is able to detect tumor drug resistance mechanisms in preclinical studies [31–34] and translational clinical trials [13,35,36], although it has so far failed to identify validated biomarkers of clinical value [37]. Recently, RNA-sequencing (RNA-seq) applied to colorectal cancer has allowed accurate analysis of whole-genome expression profiles and identified prognostic consensus molecular subtypes [38]. Preliminary studies performed to verify if these classifiers may also predict drug response [39] did not provide conclusive results. Epigenomic determinants such as methylated DNA and non-coding RNAs may play a key role in cancer initiation and progression and represent an important class of molecules eligible as cancer biomarkers [40], especially when detected by a minimally invasive procedure (liquid biopsy), for instance from circulating tumor-derived plasma exosomes. Metastatic colorectal cancer-associated regulatory RNAs have been suggested as predictive biomarkers of drug response or resistance [41]. Also, soluble proteins involved in the pharmacodynamic effects of drugs used in metastatic colorectal cancer treatment may provide information on disease biology and insights into therapeutic outcomes [42]. For instance, key regulators of inflammation, angiogenesis, and extracellular matrix remodeling may influence the efficacy of antiangiogenic drugs [43,44]. Metabolomic analysis on sera from colorectal cancer patients has also the potential for identifying novel metabolites associated with therapeutic outcomes [45,46].

Overall, the clinical identification and validation of biomarkers predictive of drug response in colorectal cancer have been mainly performed in retrospective studies that can be influenced by multiple biases [47]. Therefore, prospective studies with well-defined protocols to identify and validate biomarkers with the aim of their standard use in clinical practice to address different treatment options would be desirable [48,49] in order to offer a personalized treatment to metastatic colorectal cancer patients.

An additional aspect that affects drug response in colorectal cancer patients is related to the toxicity commonly associated with an anticancer drug treatment that may lead to severe morbidity and consequently to treatment delay as well as to treatment discontinuation with a direct impact on efficacy. Heritable allelic variants (i.e., germline) that influence drug pharmacokinetics and pharmacodynamics may be responsible for interindividual variability of adverse events [50] and have now been identified for various drugs. The main example of pharmacogenetic clinical testing is represented by the pre-emptive detection of four major deleterious variants of the dihydropyrimidine dehydrogenase (*DPYD*) gene that affects DPD enzymatic activity and are responsible for grade 3–4 and even fatal toxicity in colorectal cancer patients treated with a fluoropyrimidine-based therapy [51].

Germline gene variants may also be associated with the activity and efficacy of anticancer drugs [52]. Further complexity is linked to the correlations between their toxicity and clinical outcome due to their narrow therapeutic window, especially when cytotoxic agents are used [53]. Dose reductions based on observed toxicity following chemotherapy or aimed at preventing toxicity following pre-emptive deleterious gene variant screening would in principle endanger therapeutic efficacy.

This chapter describes the status of the art of "omic" biomarkers predictive of response in colorectal cancer and highlights the most promising perspectives in this field.

Somatic genomics

Over the years, colorectal cancer development and progression have been widely studied and characterized and it is now well established that these processes involve the multistep acquisition of genetic alterations [54]. In particular, the activation of the Wnt signaling pathway [55,56], the inhibition of the p53 pathway, and TGF-β signaling deregulation represent the fundamental steps in colorectal cancer progression.

Together with these events, additional oncogenes concur in colorectal cancer evolutionary processes. Among the most common ones, mutations of the *RAS* and *BRAF* genes occur in 55%–60% of cases [57]. Also, about 30% of colorectal cancer cases have somatic gain-of-function events on phosphatidylinositol-4,5-bisphosphate 3-kinase catalytic subunit alpha (*PI3KCA*) gene [58–60]. Alternative genetic alterations may hit phosphatase and tensin homolog (*PTEN*) (in 20%–40% of patients), an inhibitor of phosphoinositide 3-kinase (PI3K) signaling, promoting its loss of expression, or amplification of the insulin receptor substrate 2 gene (*IRS2*). As previously mentioned, the recent research advances have allowed us to identify specific biomarkers able to drive the clinical response. The selection of the right targeted treatment for a specific metastatic colorectal cancer patient depends on the underlying gene mutations which drive the disease.

At present, standard molecular tumor profiling consists of pathologic testing of tumor tissue to assess for microsatellite instability (MSI) or MMR deficiency and for variants in *KRAS*, *NRAS*, and *BRAF* genes [23,48]. So far, the research advances have in fact led to the identification of a few somatic alterations which can promote or hinder the therapeutic effects in colorectal cancer. However, the increasing application of NGS technologies on tumor tissue samples and ctDNA or circulating tumor cells has opened up the way to the identification of rare but relevant hot-spot driver gene mutations with high coverage and accuracy in the clinical context and to the discovery and development of further selective treatments [61].

EGFR-targeted therapy

Cetuximab and panitumumab, two anti-EGFR monoclonal antibodies, target the EGFR receptor and inhibit the mitogen-activated protein kinase (MAPK) signaling pathway which is dysregulated in the malignant transformation process of tumor cells. Initially, only mutations in the *KRAS* gene, occurring in about 40% of metastatic colorectal cancer patients [62], later also mutations in *NRAS* gene, occurring in only 3%–5% of these cases [63], have been associated with tumor resistance to both cetuximab and panitumumab. Indeed, it is well recognized that *KRAS* or *NRAS* mutations occurring in their respective exons 2, 3, and 4 bypass the EGFR inhibition. Also, mutations in the *BRAF* gene occurring in about 8%–10% of cases are major determinants of resistance to anti-EGFR monoclonal antibodies [64,65]. In this context, several studies have shown that cetuximab and panitumumab obtain objective responses and prolong the overall survival in *RAS* and *BRAF* wt colorectal cancer patients only [63,66]. The mentioned tumor mutations promote primary and acquired drug resistance and the patients must be tested at a somatic level before treatment [7].

In addition to the RAS-RAF-MAPK pathway, it has been reported that the oncogenic activation of the intracellular signaling pathway of PI3K-PTEN-protein kinase B (AKT), downstream of EGFR, is a possible mechanism for promoting resistance to anti-EGFR monoclonal antibody and the mutations on the exon 20 of *PI3KCA* and *PTEN* loss have been correlated with a worse outcome in term of progression-free survival and overall survival [23,64,65,67].

HER2-targeted therapy

Several trials suggest that the *HER2* gene amplification could be a predictive biomarker of resistance to anti-EGFR monoclonal antibody in patients with *RAS/RAF* wt metastatic colorectal cancer.

Overall, HER2 overexpression/amplification is reported in 1.3% to 6.3% of colorectal cancer, with a higher percentage of *HER2* amplification among *RAS* and *BRAF* wt colorectal cancers [68]. In this context, HER2 could serve as a potential target to inhibit tumor growth.

Based on the current data, *HER2* amplification seems to predict a lack of response to anti-EGFR monoclonal antibody therapy, and dual HER2 blockade with specific monoclonal antibodies (trastuzumab and pertuzumab) or the combination of a monoclonal antibody (trastuzumab) with a tyrosine kinase inhibitor (lapatinib) promotes durable tumor responses in about 30%–40% of patients with *HER2/RAS* wt metastatic colorectal cancer refractory to standard therapy. Also, an anti-HER2 monoclonal antibody-drug immunoconjugate (trastuzumab deruxtecan) has shown activity in the clinic. However, these data should be confirmed in further clinical trials.

Tyrosine kinase-targeted therapy

The *BRAF*V600E mutation represents a clinically actionable oncogenic mutation for combined drug therapy with a BRAF tyrosine kinase inhibitor (encorafenib) and an upstream signaling pathway inhibitor, cetuximab [9] or panitumumab in metastatic colorectal cancer pretreated patients. Other rarer *BRAF* mutations may also be targets for selective tyrosine kinase inhibitors. For patients with *KRAS* or *NRAS* mutations, effective targeted treatments are still lacking. The remarkable single-agent activity of AMG-510, a direct a *KRAS* G12C inhibitor, in mutant lung cancer has not been confirmed in tyrosine kinase G12C mutated colorectal cancer [69].

Also, amplification of the met proto-oncogene (*MET*) drives acquired resistance to anti-EGFR therapy in colorectal cancer [70]. It occurs rarely in *de novo* metastatic colorectal cancer (1%–2% of cases) but can be identified in a significant subset of patients that are refractory to anti-EGFR antibodies (23% of cases) [71]. This suggests the use of MET inhibitors in patients displaying resistance as a result of *MET* amplification. A clinical study is currently underway in anti-EGFR antibody therapy-resistant metastatic colorectal cancer and *MET* amplification, to verify if a highly selective, potent MET tyrosine kinase inhibitor, tepotinib, plus an anti-EGFR agent, cetuximab, could achieve disease control by targeting MET pathway activation and maintaining EGFR-pathway inhibition [72]. A similar strategy has been successful in *EGFR*-mutant, *MET*-amplified non-small cell lung cancer with acquired EGFR tyrosine kinase inhibitor resistance [73].

Regorafenib is a multi-target tyrosine kinase inhibitor with prevalent VEGFR1-3 inhibitory effects that has obtained significant survival benefits in patients with previously treated metastatic colorectal cancer in randomized placebo-controlled phase 3 trials (CORRECT and CONCUR) [74,75]. The role of somatic alterations in genes involved in its mechanism of action is at present still unknown and translational studies are needed. In the CORRECT trial, patients with metastatic colorectal cancer progression after receiving all standard therapies underwent a retrospective exploratory analysis on the clinical activity of regorafenib in biomarker subgroups of the study population [76]. Regorafenib seemed to be consistently associated with a clinical benefit, in a range of patient subgroups, irrespective of the mutational status of *KRAS* and *PI3KCA*. This is consistent with the antiangiogenic mechanism of action of regorafenib.

Microsatellite instability/mismatch repair-targeted therapy

Genetic defects inducing genome instability are classical pathological events that lead to the initiation and progression of a substantial proportion of colorectal cancers. One genome instability pathway that has received extensive evaluation both as a pathogenic event and as a target for cancer therapy is the MSI pathway. MSI occurs in about 10%–15% of sporadic colorectal cancers and is triggered by a deficient DNA MMR system (dMMR), resulting from somatic or germline mutations in MMR genes (mutL homolog 1—*MLH1*, mutS homolog 2—*MSH2*, mutS homolog 6—*MSH6*, PMS1 homolog 2, mismatch repair system component—*PMS2*) or from the epigenetic silencing of *MLH1*. The inactivation of one of these MMR genes results in an increased rate of somatic gene mutations. The sporadic cases are frequently associated with the *BRAF* V600E mutation. MSI/MMR deficiency is detected in approximately 10%–15% of localized colorectal cancer and 4%–5% of metastatic colorectal cancer and predisposes patients to respond to specific treatments. In particular, patients whose tumors are MSI-high and MMR deficient benefit from the treatment with inhibitors of the immune checkpoint component programmed cell death protein 1 (PD-1), pembrolizumab, or nivolumab [10,11]. More recently, ipilimumab, a monoclonal antibody that blocks cytotoxic T-lymphocyte antigen 4 (CTLA-4), has been approved for use in combination with nivolumab for the treatment of dMMR or MSI-high colorectal cancer patients previously treated with chemotherapy, based on benefit observed in objective response (primary endpoint) [12]. *POLE1* mutant metastatic colorectal cancer patients are also sensitive to pembrolizumab [77].

NTRK fusion-targeted therapy

Recently, it has been observed that among the MSI-high metastatic colorectal cancer cases there is an enrichment of *NTRK* gene fusions (about 4% versus 0.5%–2% in MS stable metastatic colorectal cancers). The *NTRK1*, *NTRK2*, and *NTRK3* genes, encoding for tropomyosin receptor kinase A (TrkA), tropomyosin receptor kinase B (TrkB), and tropomyosin receptor kinase C (TrkC), respectively, have been recently identified as important partners in fusion events in multiple cancers and effective selective therapies are now available for these targets [78]. In particular, the tyrosine kinase inhibitors larotrectinib and entrectinib have been approved by the FDA for the treatment of refractory solid tumors harboring *NTRK* gene fusions, independently by the tumor type thus including colorectal cancer [79]. The main somatic oncogene alterations recognized as specific drivers in colorectal cancer and as actionable targets for selective drug treatments are summarized in Table 1, with particular reference to their therapeutic levels of evidence.

Germline pharmacogenetics/genomics

The relatively poor tolerability to chemical drugs and the variable response to various therapeutic regimens fueled the study of the impact and the influence of germline gene variants on the activity and toxicity of antineoplastic drugs. In this context, the search for predictive pharmacogenetic and genomic markers in colorectal cancer has addressed the discovery and

TABLE 1 Summary of somatic alterations predictive of response or resistance to drugs in the treatment of colorectal cancer.

Gene	Alterations	Cancer types	Drugs	Therapeutic level of evidence[1]
BRAF	V600E	Colorectal Cancer	Encorafenib + Cetuximab	1
BRAF	V600E	Colorectal Cancer	Encorafenib + Panitumumab	2
BRAF	G464, G469A, G469R, G469V	All Solid Tumors	PLX8394	4
BRAF	K601	All Solid Tumors	PLX8394	4
BRAF	L597	All Solid Tumors	PLX8394	4
KRAS	Wild-type	Colorectal Cancer	Cetuximab	1
KRAS	Wild-type	Colorectal Cancer	Panitumumab	1
KRAS	Oncogenic Mutations	All Solid Tumors	Cobimetinib, Trametinib, Binimetinib	4
KRAS	Oncogenic Mutations	Colorectal Cancer	Cetuximab, Panitumumab	R1
NRAS	Wild-type	Colorectal Cancer	Panitumumab	1
NRAS	Oncogenic Mutations	Colorectal Cancer	Panitumumab, Cetuximab	R1
Other Biomarkers	Microsatellite Instability-High	All Solid Tumors	Pembrolizumab	1
Other Biomarkers	Microsatellite Instability-High	Colorectal Cancer	Ipilimumab + Nivolumab	1
Other Biomarkers	Microsatellite Instability-High	Colorectal Cancer	Nivolumab	1
Other Biomarkers	Tumor Mutational Burden-High	All Solid Tumors	Pembrolizumab	1
NTRK1	Fusions	All Solid Tumors	Entrectinib	1
NTRK1	Fusions	All Solid Tumors	Larotrectinib	1
NTRK1	G595R	All Solid Tumors	Larotrectinib	R1
NTRK1	G595R	All Solid Tumors	Entrectinib	R2
NTRK2	Fusions	All Solid Tumors	Entrectinib	1
NTRK2	Fusions	All Solid Tumors	Larotrectinib	1
NTRK3	Fusions	All Solid Tumors	Entrectinib	1
NTRK3	F617L	All Solid Tumors	Larotrectinib	R1
NTRK3	G623R	All Solid Tumors	Larotrectinib	R1
NTRK3	G696A	All Solid Tumors	Larotrectinib	R1
NTRK3	Fusions	All Solid Tumors	Larotrectinib	1
ERBB2	Amplification	Colorectal Cancer	Lapatinib + Trastuzumab	2
ERBB2	Amplification	Colorectal Cancer	Trastuzumab + Pertuzumab	2
ERBB2	Amplification	Colorectal Cancer	Trastuzumab Deruxtecan	2
FGFR1	Oncogenic Mutations	All Solid Tumors	Debio1347, Erdafitinib, AZD4547, Infigratinib	4
FGFR2	Oncogenic Mutations	All Solid Tumors	AZD4547, Infigratinib, Debio1347, Erdafitinib	4

TABLE 1 Summary of somatic alterations predictive of response or resistance to drugs in the treatment of colorectal cancer—cont'd

FGFR3	Oncogenic Mutations	All Solid Tumors	AZD4547, Debio1347, Erdafitinib, Infigratinib	4
CDK12	Truncating Mutations	All Solid Tumors	Nivolumab, Cemiplimab, Pembrolizumab	4
CDKN2A	Oncogenic Mutations	All Solid Tumors	Abemaciclib, Ribociclib, Palbociclib	4
MET	Fusions	All Solid Tumors	Crizotinib	4
MTOR	Oncogenic Mutations	All Solid Tumors	Everolimus, Temsirolimus	4
NF1	Oncogenic Mutations	All Solid Tumors	Trametinib, Cobimetinib	4
PI3K	Oncogenic Mutations	All Solid Tumors	AZD8186, GSK2636771	4

Note: Data are derived from a search for colorectal cancer available gene alterations – drug couples from the publicly available precision oncology knowledge database OncoKB [26].

[1] *Therapeutic level of evidence reported in the table are derived from the OncoKB Therapeutic Level of Evidence V2 classification (12/20/2019), reporting the following definitions in decreasing order of evidence, from standard care to investigational, to hypothetical (1–4) and from standard care resistance to investigational resistance (R1–R2).*

1, FDA-recognized biomarker predictive of response to an FDA-approved drug in this indication.

2, Standard care biomarker recommended by the NCCN or other professional guidelines predictive of response to an FDA-approved drug in this indication.

3A, Compelling clinical evidence supports the biomarker as being predictive of response to a drug in this indication.

3B, Standard care or investigational biomarker predictive of response to an FDA-approved or investigational drug in another indication.

4, Compelling biological evidence supports the biomarker as being predictive of response to a drug.

R1, Standard care biomarker predictive of resistance to an FDA-approved drug in this indication.

R2, Compelling clinical evidence supports the biomarker as being predictive.

validation of heritable allelic variants (i.e., germline) that may influence the pharmacokinetics and pharmacodynamics of these drugs. Notably, this approach tends to exclude the analysis of messenger RNA (mRNA) levels of target genes, because some differences exist when comparing healthy and neoplastic tissues. However, through the analysis of biochemical pathways, candidate genes encoding for target proteins were investigated, taking advantage from the collection of healthy tissues (i.e., nucleated blood cells or epithelial cells from buccal swabs).

When considering the classes of drugs prescribed for colorectal cancer therapy, we may divide them into chemical agents, either cytotoxic or targeted, and biological agents. Historically, the first-class includes the antimetabolite fluoropyrimidines (5-fluorouracil, its oral prodrug capecitabine and, more recently, a preconstituted formulation of trifluridine and tipiracil), the topoisomerase I inhibitor irinotecan, and the platinum derivative oxaliplatin. Along the time, these agents were combined with monoclonal antibodies targeting both EGFR, namely cetuximab and panitumumab, and VEGF or VEGFR (bevacizumab, aflibercept, and ramucirumab) to treat metastatic disease. Tyrosine kinase inhibitors of VEGFR (regorafenib) and mutated BRAF (encorafenib) have also been used.

Chemical drugs

The platinum derivative oxaliplatin acts through DNA damage by forming adducts mainly with guanine and adenine residues within the helices. Several cellular mechanisms counteract this damage, by identifying the damaged helices (xeroderma pigmentosum proteins XPD, XPC, and XPA) and removing the adducts (xeroderma pigmentosum group G—XPG, and excision repair cross-complementation group 1—ERCC1) or through the inactivation of platinum compounds (glutathione-S-transferase, GST). Therefore, the variable activity of these mechanisms of defense against oxaliplatin may result or not in the expected clinical benefit.

The investigation of GST gene isoforms (i.e., *GSTP1*, *GSTT1*, and *GSTM1*) did prove that single nucleotide polymorphisms (SNPs) could be associated with modified enzyme activity, albeit the confirmation was not conclusive. For instance, the *GSTP1* variant Ile105Val was associated with improved survival in colorectal cancer patients [80–85], but other studies did not confirm this result [86,87]. Among the several confounding factors, the relatively low number of patients, the different regimens and the stage of the disease could contribute to the variability of this scenario. In addition, colorectal cancer patients harboring one copy of the *GSTM1* gene had significantly better survival than individuals with two copies [83].

Interesting findings were obtained about the relationship between nucleotide excision repair (NER) system polymorphisms and the survival of colorectal cancer patients treated with oxaliplatin. Indeed, variant alleles in *ERCC1* (rs11615C>T) and excision repair cross-complementation group 2 (*ERCC2*) (rs13181T>G) were associated with reduced response and poor progression-free survival and overall survival in Asian and Caucasian patients, respectively [88]. Similar results were obtained in the following meta-analysis, being the *ERCC1* 118T allele associated with an unfavorable prognosis in Asian patients but not in Caucasian individuals [89]. However, subsequent studies found a significant association between the T allele of the *ERCC1* C118T SNP and longer survival especially in Asian patients [85,90,91]. Recent literature showed that the association between *ERCC1* SNPs and efficacy was even feebler in the adjuvant setting [92,93]. Further SNPs belonging to the excision repair cross-complementation group 5 (*ERCC5)* (A−763G and A+25G) were correlated with response to oxaliplatin [94]. Interestingly, the *ERCC5* promoter SNPs were associated with an increased gene expression [95].

Fluoropyrimidines have several candidate genes in their pharmacodynamic and metabolic pathways that could be actionable as predictive biomarkers. SNPs in methylenetetrahydrofolate reductase (*MTHFR*) and thymidylate synthase (*TYMS*) genes have been some of the most important targets for pharmacogenetic analyses. Together with variant alleles of the *DPYD* gene, they concur with the overall tolerability [96] and efficacy of fluoropyrimidines. For example, the *MTHFR* c.677C>T genotypes (eventually combined with c.1298A>C genotypes) predicted overall survival in patients treated with capecitabine or 5-fluorouracil, regardless of ethnic origins and of the regimens investigated (i.e., adjuvant or first-line chemotherapy) [97–101], despite this relationship was weaker in a more recent study [102].

The allelic variants within the promoter and the 3′-untranslated region (3′-UTR) of the *TYMS* gene have received much attention [103,104], being predictive of overall survival in colorectal cancer patients [102]. Indeed, both 28-bp tandem repeats in the promoter region and 6-bp variants in the 3′UTR are associated with survival after fluoropyrimidine-based chemotherapy [105–107].

DPYD allelic variants are of utmost importance to stratify patients according to the risk of severe toxicities from fluoropyrimidines [108]. The pre-emptive detection of four major deleterious variants (*DPYD*2A, DPYD*13, DPYD*_c.2846C > T, and *DPYD*-HapB3) may prevent grade 3–4 toxicity in about 3%–5% of cases [109,110] and represents an effective measure to contain high toxicity costs associated with fluoropyrimidine treatment of unscreened carriers of *DPYD* deleterious variants [111]. The incidence and importance of further variants associated with fluoropyrimidines-induced toxicity are not yet clear [112,113]. NGS has been used to identify various new casual SNPs in the *DPYD* exome that may be responsible for the high fluoropyrimidine toxicity observed [114].

Further research for the prediction of severe or life-threatening toxicity of fluoropyrimidines is warranted since these drugs are still widely used in the treatment of colorectal cancer and of other epithelial tumors, and about 30% of treated patients develop high-grade adverse events which are not predicted by current pre-emptive pharmacogenetic *DPYD* screening [110]. Also, up to 30%–50% of patients who suffer from ≥G3 toxicity have decreased DPD enzyme activity. Besides additional still unknown, possibly rare, deleterious variants detectable by extensive NGS analysis of the gene exome, other mechanisms of DPD deficiency may be in place, including *DPYD* promoter methylation [115] and *DPYD* inhibitory miRNAs [116].

In the remaining cases, in which no deficient DPD activity is involved, research should be addressed to interindividual variation of activating pyrimidine pathway genes and to an overall analysis of all the 5-fluorouracil nucleoside and nucleotide anabolic metabolites in peripheral blood mononuclear cells [117].

However, it has also been suggested that *DPYD* genotypes may contribute to the efficacy of fluoropyrimidines. Interestingly, patients homozygous for the 496G allele or the 85C allele had shorter disease-free survival than wt homozygous individuals [118]. Another clinical trial found a significant association between the *DPYD*5* allele and a higher overall response rate [119].

Capecitabine has a peculiar metabolic pathway of activation that includes both cytidine deaminase (CDA) and thymidine phosphorylase (TP), meaning that the presence of specific germline genotypes may influence patients' survival. For example, the *CDA* 79CC genotype (rs2072671) was associated with shorter progression-free survival [120].

Finally, irinotecan has a complex metabolism with different biochemical pathways leading to both active (SN-38) and inactive metabolites [121]. The interest in irinotecan pharmacogenetics is mainly based on the prevention of severe treatment-associated toxicities, being the uridine diphosphate glucuronosiltransferase isoform 1A1 (*UGT1A1*) the principal target for these analyses [122]. This strategy has led to the definition of dose reductions in the presence of low enzymatic activities (i.e., *UGT1A1**28) [123], although it has been observed that no dose-limiting toxicities were observed in *UGT1A1*1/*28* patients treated at 310 mg/m^2, a dose level substantially higher than that considered as the maximum tolerated one in the FOLFIRI regimen (i.e., 180 mg/m^2) [124]. Therefore, these standardized approaches suggested the possibility that *UGT1A1* genotyping could be also predictive of treatment outcome, as confirmed by Liu et al. [119] who found a higher response rate in patients carrying the UGT1A1*28 polymorphic allele with respect to UGT1A1 wild-type individuals. It is likely that the observational design of another clinical trial contributed to discordant findings obtained in a cohort of metastatic colorectal cancer Chinese patients [125] As a matter of fact, the low-activity *UGT1A1*28* variant was associated with both a worse progression-free survival and overall survival.

Biological drugs

As anticipated above, the use of monoclonal antibodies cetuximab, panitumumab, and bevacizumab has increased over the years aimed at improving response rate and disease control.

Former pharmacogenetic studies identified an *EGFR* polymorphism, the number of CA repeats in the intron 1, as a predictive biomarker of survival in patients treated with cetuximab plus 5-fluorouracil, but discordant results were observed according to the cut-off to discriminate between short (<17 or <20 repeats) and long alleles (≥17 or ≥20 repeats) [126–128]. However, recent acquisitions demonstrate that the activity of cetuximab may depend on the immune system, the microenvironment, and inflammation [129]. For example, the polymorphisms in receptors for the constant fraction of immunoglobulin genes (*FCGRs*) were evaluated in cetuximab-treated patients, but these variants were actionable only in particular subgroups of individuals [130,131]. Notably, a meta-analysis did not find significant results after evaluating several polymorphisms in different genes (included those cited above) as predictive biomarkers of anti-EGFR monoclonal antibodies [132], hence suggesting a lack of replication of former results.

In the case of bevacizumab, many studies have been principally focused on *VEGF-A* variants, resulting in a significant association between promoter polymorphisms (i.e., −1154G>A, −1498C>T) and survival [133,134].

Transcriptomics

Besides somatic mutations, the expression of several genes at the mRNA level may contribute to tumor drug sensitivity/resistance. Thus, transcriptomics represents a useful "omic" approach to identify additional predictive and actionable tumor biomarkers [135]. Qualitative and quantitative differences in tumor microenvironment components (e.g., stroma, vasculature, immune cells) may also be detected by transcriptomic analysis and provide further useful information on drug sensitivity/resistance characteristics [136].

Initially, mRNA gene expression analysis of single or limited numbers of candidate genes predictive of sensitivity/resistance to drugs with clinical activity against colorectal cancer according to information from classical biochemical and molecular studies in preclinical models [137–141] have been carried out in tumor explants. This approach led to the identification of various determinants of 5-fluorouracil clinical sensitivity/resistance (e.g., high/low gene expression levels of the target enzyme thymidylate synthase—*TYMS* [142–146], of the nucleoside cleavage enzyme/angiogenic factor *TP* [147], and of the rate-limiting enzyme for 5-fluorouracil metabolism *DPYD* [148]). A subsequent study showed that the use of more than one of these independent determinants of response to 5-fluorouracil permitted the identification of a high percentage of responding colorectal cancer patients [35]. However, a prospective study of tailored chemotherapy for metastatic colorectal cancer with either 5-fluorouracil and folinic acid or oxaliplatin and irinotecan based on mRNA gene expression of *TYMS* and *DPYD* did not confirm the high response rates reported in previous retrospective studies [36]. Thus, no single gene expression biomarkers of clinical value have been so far validated [37].

Determinants of drug sensitivity/resistance in the context of pyrimidine metabolism and sites of action for fluoropyrimidines, of repair pathways of platinum-DNA adducts for oxaliplatin, of DNA replication enzymes, and cellular responses of camptothecin-topoisomerase I-DNA complex formation for irinotecan and of functionally and pharmacologically addicted to kinase genes have been investigated by gene expression profiling in preclinical models [31–34].

Significant variations in mRNA expression levels between paired tumor and non-tumor samples were observed in patients with stage III colon cancer between 5-fluorouracil pathway relevant candidate genes [149]. Two distinct groups of patients based on their gene expression levels were detected using unsupervised hierarchical clustering. No correlation of pathway-defined groups with clinical outcome was however performed in this study. More recently, reverse transcription quantitative polymerase chain reaction (RT-qPCR) analysis of a relevant number of genes involved in metabolism and mechanism of action of 5-fluorouracil in tumor samples from patients with stage II or stage III colorectal cancer who underwent 5-fluorouracil adjuvant chemotherapy, identified a significant relationship between high mRNA expression levels of deoxyuridine triphosphatase (*DUT*) and shorter disease-free survival [150]. *DUT* encodes for the enzyme deoxyuridine 5′-triphosphate nucleotidohydrolase (dUTPase), a key regulator of deoxyuridine triphosphate (dUTP) pools, that normally prevents incorporation of dUTP and of the 5-fluorouracil nucleotide 5-fluorodeoxyuridine triphosphate (FdUTP) into DNA [151]. dUTPase is a clinically relevant target of biochemical modulation by TAS-114, a potent dual inhibitor of dUTPase/DPD that improves the therapeutic efficacy of fluoropyrimidine-based chemotherapy [152]. TAS-114 is at present investigated in various phase I and II clinical trials in solid tumors [153–155].

Later, prognostic gene expression profiles analyzed by microarray and/or RT-qPCR have been obtained with a supervised approach for early-stage (stage II and/or stage III) colorectal cancer. Some gene expression classifiers have been validated and are currently commercialized in the USA as prognosticators of stage II colorectal cancer outcome (e.g., Oncotype DX Colon, an RT-qPCR- based assay including 7 candidate genes and 5 reference genes; ColoPrint, a whole-genome microarray-based array including 18 genes; GeneFX Colon, a microarray-based assay including a 634 probe set (482 genes) signature). These multigene expression signatures include a limited number of genes with little overlap between the different platforms and, being devoid of a predictive role of response to adjuvant chemotherapy in colorectal cancer, are not recommended for clinical use (either for prognosis and response prediction) by National Comprehensive Cancer Network (NCCN) Guidelines [7] and European Society for Medical Oncology (ESMO) Guidelines [156].

More recently, by applying unsupervised clustering methods to transcriptomic data from all colorectal cancer stages, new molecular classifications, able to divide colorectal cancer into subtypes according to prognosis have been defined. The best known are the CMS, "consensus molecular subtypes" [38], and the CRIS, "cancer cell intrinsic transcriptional traits" [157] classifiers. The CMS classifier, including 4 prognostic subtypes (i.e., MSI-enriched CMS1 subtype: good prognosis; canonical CMS2 and metabolic CMS3 subtypes: intermediate prognosis; mesenchymal CMS4 subtype: poor prognosis), is the result of a network-based approach aimed at evaluating the association among six previous colorectal cancer classification systems [158–163]. The CMS classifier is based on any stage colorectal cancer samples. The CRIS classifier is instead based on patient-derived xenografts of liver metastasis from colorectal cancer patients. The latter mainly defines the biology of the colorectal cancer epithelial component. However, also these unsupervised signatures are currently not recommended in clinical practice for the prediction of patient prognosis [7].

The availability of RNA-seq technology, allowing rapid and accurate analysis of whole-genome expression profiles has been thereafter used to study the potential predictive role of drug response of the above-mentioned classifiers, either in the early and in the advanced disease. The predictive role of CMS and CRIS classifications was evaluated on a transcriptomic basis in about 150 samples from stage II-III colorectal cancer patients by Allen et al. [164]. These results, integrated with further case series available in GEO datasets (i.e., GSE GSE39582 and GSE14333), showed a predominance of the CMS2 subtype that, redistributed among the CRIS subgroups, mainly fell back into the CRIS-C subgroup. In particular, patients whose tumors belonged to the CMS2 subtype and/or the CRIS-C subgroup were those that mostly benefited from adjuvant chemotherapy. The additional evaluation of CD8 expression was able to identify high-risk CMS2/CRIS-C stage II patients susceptible to clinical benefit after adjuvant fluorouracil-based chemotherapy, thus suggesting CD8 as a useful marker in clinical practice.

Interestingly, in a mixed colorectal population (any colorectal cancer stages), the integration of the CMS classifier with the composition of the tumor microenvironment was evaluated in order to suggest potential immunotherapeutic approaches suitable for each subtype [136]. The authors showed that colorectal cancer molecular subtypes and microenvironmental signatures were highly correlated. In particular, CMS1 was characterized by the overexpression of genes specific to cytotoxic lymphocytes whereas CMS4 expressed markers of lymphocytes and of cells of monocytic origin. CMS4 also displayed an angiogenic, inflammatory, and immunosuppressive signature. CMS2 and CMS3 subtypes with intermediate prognosis exhibited low immune and inflammatory signatures [136].

The predictive role of the available colorectal cancer classifiers in response to pharmacological treatment has also been studied mainly in metastatic colorectal cancer patients. Through the last years, a number of studies, based on retrospective cohorts as well as on cohorts from randomized controlled prospective clinical trials, reported potential relationships between the CMS classifier and response to metastatic colorectal cancer standard pharmacological treatments [39,165–167]. Irinotecan-based chemotherapy was shown to be significantly superior to oxaliplatin-based chemotherapy in terms of progression-free survival and overall survival in the CMS4 subtype. Among patients receiving chemotherapy treatment including an anti-EGFR antibody, CMS1 and CMS2 subtypes showed significantly shorter and longer progression-free survival, respectively, compared with the other subtypes. CMS2 and CMS3 tumors were found to preferentially benefit from the addition of bevacizumab to first-line fluoropyrimidine-based chemotherapy compared with the other CMS subtypes [166]. Two studies compared patients treated with bevacizumab or cetuximab according to the CMS subtypes [39,167]. The study of Lenz et al. [39] showed that among patients whose tumor belonged to the CMS1 subtype, those treated with bevacizumab had a significantly longer overall survival compared with those treated with cetuximab, whereas the opposite behavior was reported for the CMS2 subtype [39]. The study of Stintzing et al. [167] showed that within the RAS wt population, the CMS4 subtype benefitted from a prolonged overall survival if treated with FOLFIRI plus cetuximab compared with FOLFIRI plus bevacizumab. With regard to CMS3, overall survival showed a trend in favor of the cetuximab arm, although it was comparable in CMS1 and CMS2 subtypes, independently from the targeted therapy.

Based on an RNA-seq transcriptomic approach, we recently identified two biomarkers predictive of fluoropyrimidine resistance (i.e., pinin, desmosome associated protein - *PNN*, and *KCNQ1OT1*) [168] in stage III colorectal cancer patients. *PNN* codifies the 140-kDa phosphoprotein pinin, whereas KCNQ1OT1 is a long noncoding RNA (KCNQ1 opposite strand/antisense transcript 1). For both of them, findings related to their potential role in predicting prognosis and/or colorectal cancer drug resistance, are increasing [169–171]. *PNN* and *KCNQ1OT1* were identified in two extreme prognosis cohorts of stage III colorectal cancer patients treated with fluorouracil-based adjuvant chemotherapy and validated in independent GEO datasets (i.e., GSE14333, GSE39582, GSE103479) [168]. A subsequent validation study performed in an independent cohort of stage II-III colorectal cancer patients confirmed the role of *PNN* and *KCNQ1OT1* as predictive biomarkers of fluoropyrimidine resistance [172]. The predictive role of both genes was confirmed by the lack of correlation observed when untreated patients from Gene Expression Omnibus (GEO) datasets (i.e., GSE14333, GSE39582, GSE103479) were analyzed according to *PNN* and *KCNQ1OT1* gene expression levels [168,172].

An interesting study investigated the expression of 14 candidate EGFR axis-related genes in metastatic colorectal cancer patients treated with FOLFOX or FOLFIRI +/− cetuximab within the CALGB 80203 trial [173]. Although high tumor mRNA expression levels of *HER2* and epiregulin (*EREG*) were prognostic markers associated with longer progression-free survival independently from the type of treatment, human epidermal growth factor receptor 3 (*HER3*) and *CD73* expression levels emerged as potential predictive markers of benefit from cetuximab. In *KRAS* wt tumors, low *HER3* expression levels were associated with longer overall survival from cetuximab treatment. High 5′ nucleotidase (*CD73*) expression was associated with longer progression-free survival from cetuximab treatment in patients with *KRAS*-wt and *KRAS*-mutant tumors.

Notably, all the above reported correlative studies were retrospective and were performed only on available tumor tissue samples from the entire clinical study populations.

Epigenomics

DNA methylation and histone deacetylation

Colorectal cancer may be considered as the sum of genetic and epigenetic alterations of healthy colonic mucosa, and those alterations can lead to the acquisition of the drug-resistant phenotype [174]. For example, the presence of super-enhancers has been associated with the acquisition of a drug-resistant phenotype by cancer cells [175], opening a broader scenario for the screening, treatment, and outcome monitoring of neoplastic diseases.

DNA methyltransferases (DNMTs) cause DNA methylation at the C5 position of cytosine residues (5mC) thus resulting in the modulation of DNA transcription. In particular, the cytosine methylation occurs at CpG dinucleotides (which are generally methylated) or at CpG islands, which are CpG rich sequences, generally unmethylated in humans, and often present in the gene promoter. Abnormal methylation patterns especially at CpG islands (i.e., hypermethylation) may alter normal gene transcription, resulting in silencing of tumor suppressor genes as cyclin-dependent kinase inhibitor 2A (*CDKN2A*), *MLH1*, and APC regulator of WNT signaling pathway (*APC*). On the contrary, hypomethylation may directly activate proto-oncogenes, at the promoter regions, distant regulatory regions (super-enhancers), and antisense promoters (i.e., long interspersed nuclear elements 1—LINE-1). Therefore, alterations in DNA methylation (genome-wide hypomethylation and hypermethylation) are hallmarks of colorectal cancer during its growth and development [176].

Several studies have been performed to evaluate the capability of methylation markers to predict treatment outcomes [177]. For example, in Egyptian colorectal cancer patients, the global DNA methylation did correlate with the overexpression of DNMT 3A and 3B isoforms responsible for the *de novo* methylation [178]. More interestingly, fluoropyrimidines significantly reduced DNA methylation and DNMTs expression over time, while patients with 5mC levels >0.55 (which was the median basal level of 5mC) had a significantly higher event-free survival. Notably, the level of gene methylation as a predictive biomarker can be measured in cell-free circulating DNA (cfDNA) isolated from blood samples, as it occurs for diagnostic and prognostic methylated markers [179]. Indeed, the early decrease in methylated HPP1 cfDNA below the detection threshold was significantly associated with overall survival in a large cohort of colorectal cancer patients treated with fluoropyrimidine, oxaliplatin, and bevacizumab [180]. Analogously, the time-dependent reduction of methylation markers was predictive of histological response of colorectal cancer liver metastases to neoadjuvant chemotherapy [181]. Notably, the changes in methylation of 5 cancer-specific loci (*EYA4, GRIA4, ITGA4, MAP3K14-AS1, MSC*) were significantly correlated with objective tumor response and progression-free survival for most currently available drug treatment regimens in a set of metastatic colorectal cancer patients treated with chemotherapy or targeted therapy [182]. These results suggest that detection of DNA methylation in cfDNA by liquid biopsy may contribute to the prediction of tumor response in cases where patient-specific mutations cannot be detected by monitoring tumor burden dynamics, although some issues remain [183].

Another class of epigenetic markers is represented by the histone modifications of eukaryotic cells that modulate the DNA-histone interaction thus influencing gene expression profiles and ultimately patients' survival [184]. There are several histone modifications, being the methylation and acetylation the two processes more often involved in the regulation of gene transcription. As a matter of fact, acetylation (by histone acetyltransferases, HAT) causes the loss of electrostatic interaction between histones and DNA, so that genes can be transcribed. On the contrary, hypoacetylation due to the histone deacetylases (HDACs) increases the nucleosome formation and stabilization, and, while being in its condensed status, the chromatin cannot be transcribed. Histone methylation and demethylation, respectively catalyzed by histone methyltransferases (HMTs) and histone demethylases (HDMs), can alter the mRNA levels of thousands of genes. The different patterns of post-translational modifications of histones can differentiate between normal and neoplastic tissues [185]. Moreover, the highest trimethylation levels of lysine 27 on histone 3 (H3K27me3) were associated with the highest cytotoxic effect of oxaliplatin and with the longer metastases-free survival of colorectal cancer patients [186]. The liquid biopsies can offer the possibility to consider circulating nucleosome and their modifications as biomarkers for colorectal cancer [187]. In a former study, serum levels of nucleosomes predicted the probability of response in colorectal cancer patients who received selective internal radiation therapy [188]. Improved analytical techniques may increase the capability to discern tumors from healthy circulating nucleosomes [189].

Non-coding RNA

According to an estimate from the latest GENCODE v. 22 [190], only 2% of the human genome codes for proteins while 75%–90% of it is transcribed as non-coding RNAs (ncRNAs) [191]. The ncRNAs comprise various groups of RNA molecules that play an important role in several cellular processes, mainly in the regulation of genetic information from DNA to protein. Mutation or imbalance in the ncRNA repertoire is involved in the development of human diseases including cancer. Advances in the understanding of cancer molecular biology have demonstrated the role of ncRNAs in colorectal cancer progression and drug resistance [192]. NcRNAs are classified into two sub-families based on their length: short ncRNAs which are less than 200 nucleotides in length and long ncRNAs (lncRNAs) which have a length greater than 200 nucleotides [193].

MicroRNAs

Among short-length RNAs, microRNAs (miRNAs) represent a family of small ncRNAs of about 19 to 25 nucleotides in length which play an important role in regulating the expression of their target genes at the post-transcriptional level. They mainly act through binding to the 3′ untranslated region (3′-UTR) of target mRNAs by a seed sequence of 2–8 nucleotides and inhibiting the protein expression by degrading the mRNA or blocking the translational machinery with relevant implications in cancer [194,195]. In this way, miRNAs exert a crucial role in many biological and oncogenic processes such as cell proliferation, differentiation, growth and development, metabolism, and cell death [196]. It has been shown that miRNAs may act as tumor promoters (oncomiRNAs) or tumor suppressors (anti-oncomiRNAs) [197]. Associations between miRNAs deregulation and clinical response in colorectal cancer have been described in the literature at both tissue and circulating levels [198].

miRNAs in tissue and clinical response. Many efforts have been done to identify specific miRNA signatures in colorectal cancer tissues whose expression levels might predict response to drug treatment and thus promote personalization of patients' management and improve drug efficacy. However, little concordance among studies has been reported and clinical translation of data obtained in colorectal cancer tissues has been challenging, with poor results. Expression of miRNAs in tumor tissues may be dependent on the tumor microenvironment diverse composition and its detection requires a more precise and specific spatial analysis of the tissue samples. Given a large amount of data, Table 2 summarizes the main findings with regard to the potential relationship between the key treatments in colorectal cancer and miRNAs in tumor tissues.

Circulating miRNAs and clinical response. More recently, research focused on liquid biopsy. In particular, miRNA-based biomarkers seem to be particularly exploitable due to their high stability within body fluids. Indeed, circulating miRNAs are protected from enzymatic

TABLE 2 Tissue miRNAs associated with drug treatment in colorectal cancer.

miRNA	Better response	Reference
Capecitabine, oxaliplatin		
miR-126	High expression	Hansen et al. [199]
miR-506	High expression	Zhou et al. [200]
miR-625-3p	Low expression	Rasmussen et al. [201]
miR-4299	Low expression	Hu et al. [202]
miR-196b	High expression	Hu et al. [202]
Capecitabine, oxaliplatin, bevacizumab		
miR-143	Low expression	Mekenkamp et al. [203]
miR-200b	High expression	Mekenkamp et al. [203]
miR-455-5p	Low expression	Boisen et al. [204]
miR-664-3p	High expression	Boisen et al. [204]
miR-92b-3p, miR-3156-5p, miR-10a-5p, miR-125a-5p	High expression	Kiss et al. [205]
5-FU		
miR-21	Low expression	Caramés et al. [206] Schetter et al. [207]
miR-148a	High expression	Takahashi et al. [208]
miR-150	High expression	Ma et al. [209]
5-FU + Radiotherapy		
miR-21-5p	High expression	Lopes-Ramos et al. [210]
miR-215, miR-190b, miR29b-2	High expression	Svoboda et al. [211]
Let-7e, miR-196, miR-450a, miR-450b-5p, miR-99a	High expression	Boisen et al. [204]
5-FU + oxaliplatin		
miR-320e	Low expression	Perez-Carbonell et al. [212]
miR-506	High expression	Zhou et al. [200]
Anti-EGFR MoAbs		
miR-31-3p	Low expression	Mosakhani et al. [213]
miR-31-5p	Low expression	Igarashi et al. [214]
miR-181a	High expression	Pichler et al. [215]
miR-592	High expression	Mosakhani et al. [213]
miR-7	High expression	Suto et al. [216]
miR-146b-3p	Low expression	Ragusa et al. [217]
miR-486-5p	Low expression	Ragusa et al. [217]

degradation thanks to specific RNA binding proteins or lipoproteins or storage in exosomes or vesicles. Given the limited invasiveness of liquid biopsy, circulating miRNAs are potential good biomarkers to monitor clinical efficacy over time through serial blood samples. In this context, several miRNAs have been described. Providing the reader with an exhaustive list of miRNAs suggested as potential predictive biomarkers of cancer progression/drug response is out of the scope of this chapter and only the main examples are discussed thereafter. More detailed information is reported in Jung et al. [218].

Kjersem et al. [219] showed that the plasmatic level of a set of three miRNAs can predict the response to 5-fluorouracil/oxaliplatin in metastatic colorectal cancer patients. Specifically, measuring the expression profile of miR-106a, miR-130b, and miR-484 before and after four cycles of chemotherapy, the authors observed an upregulation in non-responder patients. On the same wave, Zhang et al. [220], comparing oxaliplatin responsive and resistant patients reported that a panel of 5 miRNAs (miR-20a, miR-130, miR-145, miR-216, and miR-372) was able to discriminate oxaliplatin-chemoresistant colorectal cancer patients from the chemosensitive ones. Besides these miRNA combinations, some research groups have highlighted the role of single circulating miRNAs in mediating pharmacological treatments. For example, two independent teams showed that high levels of serum miR-155 and miR-345 were associated with a lower response to 5-fluorouracil/leucovorin/cetuximab and irinotecan/cetuximab treatments, respectively, in metastatic colorectal cancer [221,222].

Another study by Chen et al. [223] showed that serum miR-19a is significantly downregulated in first-line FOLFOX sensitive metastatic colorectal cancer patients. Similarly, an independent report showed that exosomal miR-208b was significantly decreased in FOLFOX responsive patients compared with non-responder patients [224] and it could be used as a non-invasive marker. Jin et al. [41] reported that miR-21-5p, miR-1246, miR-1229-5p, miR-135b, miR-425, and miR-96-5p are up-regulated in serum from patients resistant to oxaliplatin/5-fluorouracil and in exosomes from culture media of resistant colorectal cancer cells. Han et al. [225] described a panel of 6 plasma exosomal miRNAs (miR-100, miR-92a, miR-16, miR-30e, miR-144-5p, and let-7i) able to significantly stratify oxaliplatin-based chemotherapy-resistant and sensitive patients.

However, as highlighted by the above-reported data, there is no overlap among the miRNAs potentially predictive of drug response in colorectal cancer. Thus, it appears quite difficult to identify a unique set of miRNAs able to predict clinical response. For this reason, Madurantakam Royam et al. [226] pooled 39 studies in a meta-analysis analyzing both tissue and circulating miRNAs. The study included 2822 colorectal cancer patients treated with single or combination therapies, for a total of 14 anticancer drugs. Based on that, they were able to identify 28 miRNAs associated with drug sensitivity, 20 with drug resistance, and one with differential expression and radiosensitivity. The work allowed to show that multiple miRNAs could act as clinical predictors of drug resistance and sensitivity in colorectal cancer, however further large-scale prospective studies are needed to confirm these findings.

Profiling of 762 different miRNAs using RT-qPCR arrays was performed to identify serum miRNAs associated with response to regorafenib treatment in metastatic colorectal cancer patients [227]. In the identification step, two small-size patient cohorts with different treatment outcomes (i.e., complete response + partial response + stable disease vs. progression of the disease), were analyzed. Hierarchical clustering of serum miRNAs indicated different expression profiles between favorable and non-favorable responder patients, suggesting that serum miRNAs could be a helpful tool for patient stratification. A panel of 12 miRNAs was then selected for validation in a larger patient cohort. High levels of miR-185-5p in serum were associated with a favorable response to treatment.

Indeed, although a large amount of data has been reported in the literature, there is still no clinical application of miRNAs as predictive biomarkers of drug response. In particular, the development of a "liquid biomarker", which can be easily isolated from any body fluids, represents a great opportunity for drug response monitoring. Despite the promising expectation, the research is still in its embryonic phase and thus a global effort should be considered mandatory to translate the use of "liquid miRNAs" into the clinic.

Long non-coding RNAs

LncRNAs are non-translational RNAs >200 nucleotides in length that are also involved in many crucial biological processes and may execute different molecular functions such as signals, decoys/sponges, guides, or scaffolds [228–230]. Several lncRNAa possess multiple characteristics that in combination are critical to their biological function. LncRNAs, serving as pivotal regulators and acting as oncogenes or tumor suppressors, may play important roles in many different types of cancers and have the ability to affect tumor development, metastasis, and drug resistance [231,232]. LncRNAs in colorectal cancer are involved in processes such as cellular proliferation, apoptosis, development of metastasis, cell cycle regulation, epithelial-mesenchymal transition, maintenance of cancer cell stemness, and development of drug-resistant phenotypes, ultimately influencing patient prognosis [233]. In a recent study, about 200 differentially expressed lncRNAs were identified in tumor tissues as compared to adjacent normal mucosa of colorectal cancer patients by RNA sequencing from The Cancer Genome Atlas (TCGA) database [234]. Validation in an independent set of normal colon and patient-derived colon cancer cell lines identified a novel lncRNA, lincDUSP, as a potential candidate oncogene. Knockdown of lincDUSP in patient-derived colon cancer cell lines resulted in decreased cell proliferation and clonogenicity, and increased susceptibility to apoptosis [234]. Studies on the potential contribution of lncRNAs to the onset of drug resistance in colorectal cancer have been conducted to identify potential predictive biomarkers of drug treatment response [192]. Such studies, mainly carried out in colorectal cancer cell lines and xenografts in mouse models but more recently also in clinical tumor samples, have identified lncRNAs associated with drug resistance. In particular, by means of a transcriptomic approach, five lncRNAs were identified [235]. The long intergenic non-protein coding RNA 973 (LINC00973) and RNA 941 (LINC00941), the cancer susceptibility 19 RNA (CASC19), the colon cancer-associated transcript 1 (CCAT1), and the breast cancer anti-estrogen resistance 4 (BCAR4) were in fact differentially expressed after treatments with 5-fluorouracil, oxaliplatin, and irinotecan both *in vivo* and *in vitro*. It was shown that LINC00973 was the most strongly and consistently upregulated lncRNA in colorectal cancer cells treated with all three chemotherapeutic drugs [235]. This demonstrated that deregulation of lncRNAs may contribute to the development of drug resistance in colorectal cancer models.

It was later reported that high expression of lncRNA CCAT1 is associated with decreased apoptosis and with 5-fluorouracil resistance in colon cancer cell lines [236]. Also, expression levels of CCAT1 in colorectal tumor tissues were higher than those in non-cancer tissues [236]. Correlation between CCAT1 high expression in human colon tumor samples and clinical resistance to 5-fuorouracil remains, however, to be demonstrated.

Another lncRNA, the 1-stranded/opposite antisense transcript KCNQ1 (KCNQ1OT1), is involved in the resistance to methotrexate in colorectal cancer cells by the regulation of the miR-760/PPP1R1B axis by the cAMP signaling pathway [237]. Its expression is also associated with oxaliplatin resistance in colorectal cancer cell lines. This resistance-promoting activity was mediated by sponging miR-34a and consequent deregulation of the expression of the Atg4B gene by enhancing protective autophagy [170]. Also, expression levels of KCNQ1OT1 were significantly increased in tumor tissues compared with adjacent normal tissues. In tissue samples of stage I and II patients the expression levels of KCNQ1OT1 were lower than those in stage III-IV patients. High expression of KCNQ1OT1 was associated with poor overall survival of colon cancer patients suggesting that these patients might be less sensitive to chemotherapy. However, no data on this correlation were reported [170].

In more recent studies, KCNQ1OT1 was shown to be upregulated in colorectal cancer tissues and significantly associated with poor prognosis in stage II and III patients treated with adjuvant fluoropyrimidine chemotherapy, but not in untreated patients from the same cohorts, suggesting a specific predictive role of clinical outcome [168,172].

A novel cetuximab resistance-associated RNA transcript 16 lncRNA (CRART16) upregulated in Caco-2 colorectal cancer cell line has also been identified [238]. The overexpression of CRART16 was associated with the reversal of the inhibitory effects of cetuximab on cell viability and to the reduction of cetuximab-induced apoptosis. It was demonstrated that CRART16 acts as a sponge for miR-371a-5p to regulate the v-erb-b2 avian erythroblastic leukemia viral oncogene homolog 3 (*ERBB3* or *HER3*) gene expression. The Kyoto Encyclopedia of Genes and Genomes (KEGG) pathway analysis reported that CRART16 upregulation resulted mainly in the enrichment of differentially expressed mRNAs in the MAPK signaling pathway. Thus, CRART16 may contribute to cetuximab resistance through the miR-371a-5p/ERBB3/MAPK axis pathway. Furthermore, CRART16 was associated with the contribution of the acquisition of stemness properties of colorectal cancer cells [238].

Studies have, however, shown that lncRNAs may also revert drug resistance. An elevated level of the lncRNA maternally expressed gene 3 (MEG3) may in fact revert at least in part resistance to oxaliplatin in resistant colorectal cancer cells *in vitro*, while low expression levels of MEG3 increased oxaliplatin-induced apoptosis in parental sensitive cells. In the same study, it was also found that the proportion of patients that responded to oxaliplatin treatment was significantly higher in the high MEG3-expressing group than in the low MEG3-expressing group [239].

By applying unsupervised consensus clustering to lncRNAs expression profiles, five molecular subtypes of colorectal cancer were identified by Chen et al. [240] with distinct clinical outcomes in patients with stage II-III from GEO datasets. This classifier may provide a tool for clinical prognosis evaluation, pending further validation steps. Also, the predictive value of this classifier in influencing treatment decisions remains to be demonstrated, since no attempts were made to analyze gene expression data according to adjuvant chemotherapy received by these patients [240].

While the above data suggest that lncRNAs can serve as biomarkers for the prediction of colorectal cancer response to drug therapy, further translational studies are needed to confirm these preliminary data. Targeting lncRNAs may also represent a novel approach for selective treatment of colorectal cancer by the development of RNA-based therapeutics, such as small interfering RNAs, antisense oligonucleotides, clustered regularly interspaced short palindromic repeats (CRISPR) interference, for specific disruption of important signaling pathways or reversal of tumor drug resistance [241,242].

Proteomics

Proteomics consists of a wide range of technologies used for large-scale identification, measurement, characterization, and analysis of proteins. Translational proteomics research helps to define protein expression profiles that reflect phenotypic changes and contributes to clinical application and utility [243]. The majority of biomarker studies use quantitative mass spectrometry-based technologies for the identification and profiling of disease-associated or disease-specific protein markers.

In the last years, the sensitivity of proteomics has significantly improved due to better sample preparation, advances in current technologies, and the introduction of new ultrasensitive tools. This led to an intensive search in expression proteomics (i.e., the down- and upregulation of protein levels) as well as in functional proteomics (i.e., discovering biological functions of unknown proteins and the definition of molecular mechanisms) [244].

The identification of specific protein in colorectal cancer may contribute to the discovery of additional useful biomarkers for predicting response to drug treatments, guiding the decision-making process, and novel therapeutic targeting [245,246].

In patients with newly diagnosed stage IV colorectal cancer, Martin et al. [246] identified 68 differentially expressed proteins in serum and tissue samples between responders and nonresponders to chemotherapy and bevacizumab. Among these, apolipoprotein E (APOE), angiotensinogen, and vitamin D-binding protein were proven to be correlated with better survival outcomes [246]. Quantitative proteomic analysis of the plasma of patients with metastatic colorectal cancer was also used to assess the response to cetuximab [247], showing that phosphorylated-EGFR is linked with sensitivity to the drug.

More recently, microsatellite instability has emerged as a predictor of sensitivity to immunotherapy-based treatments, including patients with metastatic colorectal cancer [248]. In this regard, some studies have demonstrated promising prognostic biomarkers. The overexpression of heat shock protein (HSP) 110 or HSP 70 and protein ß2-microglobulin, for example, has been suggested to be associated with poor overall survival and disease-free survival [249,250]. Finally, an immuno-proteomic strategy performed on matched sets of tissues from primary colon cancer, liver metastases, and adjacent benign tissues was used to discover potential cancer markers. It was found that the tumor tissue autoantigens olfactomedin 4 (OLFM4), integrin alpha M (CD11b), and integrin alpha-2 (ITGA2) were significantly overexpressed in both primary colon tumors and liver metastases, suggesting that these may serve as promising markers for developing differential diagnostics and immunotherapies [251].

Regarding the potential prognostic markers, some of them have been also identified using proteomic-based approaches [252–256] along with an attempt to correlate them to progression-free survival [257].

In conclusion, despite numerous proteomic studies, no biomarkers identified after the discovery and/or validation steps are currently investigated in clinical studies [258]. Efforts to discover protein biomarkers predictive of treatment response must, nevertheless, continue in order to improve the quality of life of these patients, whilst offering them the best treatments and therefore, a better chance of survival.

Metabolomics

Metabolomics is a field of omics science that comprehensively characterizes the metabolome, the complete mixture of metabolites, or small molecule chemicals, such as carbohydrates, lipids, amino acids, nucleic acids, alcohols, or organic acids present in cells, tissues, body fluids [259]. Both liquid- or gas-chromatography (LC or GC) coupled to mass spectrometry (MS) and nuclear magnetic resonance (NMR) spectroscopy in conjunction with advanced computational methods are used to describe an individual's metabolic profile, resulting from genetics, epigenetics, environmental and the gut microbiota interplay, in health and disease [260]. As dysregulated metabolism is one of the hallmarks of cancer, metabolomics has the potential to provide the functional phenotype of colorectal cancer and identify novel colorectal cancer diagnostic, prognostic and predictive biomarkers in patient's serum/plasma, urine, feces, and tissue.

The majority of studies have attempted to use metabolomics for colorectal cancer early diagnosis, staging, and prognosis [261–267] exploiting metabolic perturbations correlated with its initiation and progression (the Warburg effect and glutaminolysis, for example).

The most commonly identified metabolites for colorectal cancer were carbohydrates (lactate, pyruvate, etc.), amino acids (glutamine, glutamate, alanine, etc.), and fatty acids (propionate, butyrate, etc.) [268].

Nevertheless, none of these biomarkers has reached routine clinical practice yet, due to heterogeneity in study populations, biospecimen, analytic platforms, and data analyses, as well as the extent of internal and external validation [269].

Pharmacometabolomics is a field of metabolomics that pursues the achievement of tailored therapeutic regimens [260], using metabolic profiling to predict the efficacy and adverse reactions of cancer treatment [270]. It will complement pharmacogenomics by capturing environmental and microbiota-level influences on response to drug therapy [271].

So far a limited number of pharmacometabolic phenotyping studies have been conducted on metastatic colorectal cancer patients.

Backshall et al. [272], among 54 patients with locally advanced or metastatic inoperable colorectal cancer, identified those who had a high susceptibility to capecitabine toxicity (diarrhea, hand-foot syndrome, and fatigue), based on ^{1}H NMR pre-treatment serum metabolic profiles. Higher levels of low-density lipoprotein–derived lipids, including polyunsaturated fatty acids and choline phospholipids, reasonably due to alteration in hepatic metabolism caused by the presence of circulating proinflammatory cytokines, are predictive of higher grade toxicity over the treatment period. The authors proposed that lipids may affect the protein binding of capecitabine itself and its metabolite, strongly influencing the drug distribution and/or clearance [272].

In a retrospective study on 62 patients with stage III and IV colorectal cancer, Dykstra et al. [273] developed a ^{1}H NMR urine metabolomics-based biomarker panel that was moderately accurate in predicting treatment delay, mainly due to reactions to chemotherapy and other medical condition, but less accurate in predicting cancer recurrence and chemotherapy dose reduction.

In a phase II, non-randomized, open-label, single-center trial, Iemoto et al. [274] used GC/MS and LC/MS to identify serum biomarkers for predicting the response to chemotherapy with either cetuximab plus irinotecan or cetuximab alone in 31 patients with surgically unresectable metastatic colorectal cancer, who have already received ineffective irinotecan, oxaliplatin or fluoropyrimidine-based therapy. Among 111 metabolites, octanoic acid and 1,5-anhydro-D-glucitol were identified as biomarker candidates and, in the quantitative analysis, the serum concentrations of both metabolites differed significantly between the partial response and non-partial response groups. However, only octanoic acid was significantly associated with the time to progression, and patients with lower serum octanoic acid levels exhibited better prognoses compared with patients with higher serum levels.

Although the findings reviewed here have been inconsistent and lacking in clinical applicability, they demonstrate that pharmacometabolomics could provide new opportunities to improve treatment outcomes in metastatic colorectal cancer, elucidating individual variability in drug response.

Discussion and conclusions

Colorectal cancer represents a model of a tumor with a high impact on health, being one of the most incident and lethal cancers [4]. Elevated death rates depend on neoplastic progression events guided by several dysregulated biomolecular processes [275] including the evolutionary metastatic process occurring in about half of the patients: about 20% of cases already at diagnosis (stage IV) and about 25% of cases thereafter, following primary treatment, often including adjuvant chemotherapy (stage I–III) [5]. Despite the availability of active cytotoxic and targeted agents, including antiangiogenic drugs, the efficacy of pharmacological treatment of patients with metastatic colorectal cancer is hampered by the occurrence of intrinsic and acquired resistance. Somatic genetic biomarkers used nowadays in the clinic for therapeutic patient management are undoubtedly useful and included in the clinical recommendations for predicting efficacy. Also, germline pharmacogenetic biomarkers are clinically used for pre-emptive stratification of colorectal cancer patients aimed at predicting toxicity and adjusting dosing for some cytotoxic drugs. However, a molecular comprehensive signature of new multi "omic" biomarkers predictive of drug outcome together with clinical, pathological, hemato-chemical, and standard molecular determinants would be of great usefulness for personalizing drug treatment of colorectal cancer patients, especially in, but not limited to, the metastatic setting [276,277]. Further research in the areas addressed by this chapter may also provide further insight into the knowledge of the biological evolution of colorectal cancer and of the development of metastasis and drug-resistant tumor cell phenotypes [278]. Combined analysis by machine learning and artificial intelligence will contribute to identifying these signatures [279,280]. Prospective confirmatory clinical trials will also be crucial for the validation of predictive biomolecular classifiers [48].

Acknowledgment

This manuscript has been partially supported by the H2020 - EU Research and Innovation Programme – REVERT, grant agreement No. 848098 (to E.M.)

References

[1] Kato S, Lippman SM, Flaherty KT, Kurzrock R. The conundrum of genetic "drivers" in benign conditions. J Natl Cancer Inst 2016;108, djw036.
[2] Vogelstein B, Papadopoulos N, Velculescu VE, Zhou S, Diaz Jr LA, Kinzler KW. Cancer genome landscapes. Science 2013;339:1546–58.
[3] Sung H, Ferlay J, Siegel RL, Laversanne M, Soerjomataram I, Jemal A, Bray F. Global Cancer Statistics 2020: GLOBOCAN estimates of incidence and mortality worldwide for 36 cancers in 185 countries. CA Cancer J Clin. 2021;71:209–49.
[4] Dekker E, Tanis PJ, Vleugels JLA, Kasi PM, Wallace MB. Colorectal cancer. Lancet 2019;394:1467–80.
[5] Biller LH, Schrag D. Diagnosis and treatment of metastatic colorectal cancer: a review. JAMA 2021;325:669–85.
[6] National Cancer Institute. Surveillance, epidemiology, end end results program, https://seer.cancer.gov/statfacts/.
[7] NCCN. Guideline clinical practice guidelines in oncology—colon cancer version 2.2021, https://www.nccn.org/professionals/physician_gls/pdf/colon.pdf.
[8] Kopetz S, Grothey A, Yaeger R, Van Cutsem E, Desai J, Yoshino T, Wasan H, Ciardiello F, Loupakis F, Hong YS, Steeghs N, Guren TK, Arkenau HT, Garcia-Alfonso P, Pfeiffer P, Orlov S, Lonardi S, Elez E, Kim TW, Schellens JHM, Guo C, Krishnan A, Dekervel J, Morris V, Calvo Ferrandiz A, Tarpgaard LS, Braun M, Gollerkeri A, Keir C, Maharry K, Pickard M, Christy-Bittel J, Anderson L, Sandor V, Tabernero J. Encorafenib, binimetinib, and cetuximab in BRAF V600E-mutated colorectal cancer. N Engl J Med 2019;381:1632–43.
[9] Tabernero J, Grothey A, Van Cutsem E, Yaeger R, Wasan H, Yoshino T, Desai J, Ciardiello F, Loupakis F, Hong YS, Steeghs N, Guren TK, Arkenau HT, Garcia-Alfonso P, Elez E, Gollerkeri A, Maharry K, Christy-Bittel J, Kopetz S. Encorafenib plus cetuximab as a new standard of care for previously treated BRAF V600E-mutant metastatic colorectal cancer: updated survival results and subgroup analyses from the BEACON study. J Clin Oncol 2021;39:273–84.
[10] Le DT, Durham JN, Smith KN, Wang H, Bartlett BR, Aulakh LK, Lu S, Kemberling H, Wilt C, Luber BS, Wong F, Azad NS, Rucki AA, Laheru D, Donehower R, Zaheer A, Fisher GA, Crocenzi TS, Lee JJ, Greten TF, Duffy AG, Ciombor KK, Eyring AD, Lam BH, Joe A, Kang SP, Holdhoff M, Danilova L, Cope L, Meyer C, Zhou S, Goldberg RM, Armstrong DK, Bever KM, Fader AN, Taube J, Housseau F, Spetzler D, Xiao N, Pardoll DM, Papadopoulos N, Kinzler KW, Eshleman JR, Vogelstein B, Anders RA, Diaz Jr LA. Mismatch repair deficiency predicts response of solid tumors to PD-1 blockade. Science 2017;357:409–13.
[11] Overman MJ, McDermott R, Leach JL, Lonardi S, Lenz HJ, Morse MA, Desai J, Hill A, Axelson M, Moss RA, Goldberg MV, Cao ZA, Ledeine JM, Maglinte GA, Kopetz S, André T. Nivolumab in patients with metastatic DNA mismatch repair-deficient or microsatellite instability-high colorectal cancer (CheckMate 142): an open-label, multicentre, phase 2 study. Lancet Oncol 2017;18:1182–91.
[12] Overman MJ, Lonardi S, Wong KYM, Lenz HJ, Gelsomino F, Aglietta M, Morse MA, Van Cutsem E, McDermott R, Hill A, Sawyer MB, Hendlisz A, Neyns B, Svrcek M, Moss RA, Ledeine JM, Cao ZA, Kamble S, Kopetz S, André T. Durable clinical benefit with nivolumab plus Ipilimumab in DNA mismatch repair-deficient/microsatellite instability-high metastatic colorectal cancer. J Clin Oncol 2018;36:773–9.
[13] Haibe Y, Kreidieh M, El Hajj H, Khalifeh I, Mukherji D, Temraz S, Shamseddine A. Resistance mechanisms to anti-angiogenic therapies in cancer. Front Oncol 2020;10:221.
[14] Van Cutsem E, Cervantes A, Adam R, Sobrero A, Van Krieken JH, Aderka D, Aranda Aguilar E, Bardelli A, Benson A, Bodoky G, Ciardiello F, D'Hoore A, Diaz-Rubio E, Douillard JY, Ducreux M, Falcone A, Grothey A, Gruenberger T, Haustermans K, Heinemann V, Hoff P, Köhne CH, Labianca R, Laurent-Puig P,

Ma B, Maughan T, Muro K, Normanno N, Österlund P, Oyen WJ, Papamichael D, Pentheroudakis G, Pfeiffer P, Price TJ, Punt C, Ricke J, Roth A, Salazar R, Scheithauer W, Schmoll HJ, Tabernero J, Taïeb J, Tejpar S, Wasan H, Yoshino T, Zaanan A, Arnold D. ESMO consensus guidelines for the management of patients with metastatic colorectal cancer. Ann Oncol 2016;27:1386–422.
[15] Mullard A. Stemming the tide of drug resistance in cancer. Nat Rev Drug Discov 2020;19:221–3.
[16] Cancer Genome Atlas Network. Comprehensive molecular characterization of human colon and rectal cancer. Nature 2012;487:330–7.
[17] Kopetz S. New therapies and insights into the changing landscape of colorectal cancer. Nat Rev Gastroenterol Hepatol 2019;16:79–80.
[18] Diaz Jr LA, Williams RT, Wu J, Kinde I, Hecht JR, Berlin J, Allen B, Bozic I, Reiter JG, Nowak MA, Kinzler KW, Oliner KS, Vogelstein B. The molecular evolution of acquired resistance to targeted EGFR blockade in colorectal cancers. Nature 2012;486:537–40.
[19] Ciardiello F, Normanno N, Martinelli E, Troiani T, Pisconti S, Cardone C, Nappi A, Bordonaro AR, Rachiglio M, Lambiase M, Latiano TP, Modoni G, Cordio S, Giuliani F, Biglietto M, Montesarchio V, Barone C, Tonini G, Cinieri S, Febbraro A, Rizzi D, De Vita F, Orditura M, Colucci G, Maiello E, CAPRI-GOIM Investigators. Cetuximab continuation after first progression in metastatic colorectal cancer (CAPRI-GOIM): a randomized phase II trial of FOLFOX plus cetuximab versus FOLFOX. Ann Oncol 2016;27:1055–61.
[20] Prahallad A, Sun C, Huang S, Di Nicolantonio F, Salazar R, Zecchin D, Beijersbergen RL, Bardelli A, Bernards R. Unresponsiveness of colon cancer to BRAF(V600E) inhibition through feedback activation of EGFR. Nature 2012;483:100–3.
[21] Arnold D, Lueza B, Douillard JY, Peeters M, Lenz HJ, Venook A, Heinemann V, Van Cutsem E, Pignon JP, Tabernero J, Cervantes A, Ciardiello F. Prognostic and predictive value of primary tumour side in patients with RAS wild-type metastatic colorectal cancer treated with chemotherapy and EGFR directed antibodies in six randomized trials. Ann Oncol 2017;28:1713–29.
[22] Lee MS, Menter DG, Kopetz S. Right versus left colon cancer biology: integrating the consensus molecular subtypes. J Natl Compr Canc Netw 2017;15:411–9.
[23] Sveen A, Kopetz S, Lothe RA. Biomarker-guided therapy for colorectal cancer: strength in complexity. Nat Rev Clin Oncol 2020;17:11–32.
[24] ICGC/TCGA Pan-Cancer Analysis of Whole Genomes Consortium. Pan-cancer analysis of whole genomes. Nature 2020;578:82–93.
[25] Chakravarty D, Gao J, Phillips SM, Kundra R, Zhang H, Wang J, Rudolph JE, Yaeger R, Soumerai T, Nissan MH, Chang MT, Chandarlapaty S, Traina TA, Paik PK, Ho AL, Hantash FM, Grupe A, Baxi SS, Callahan MK, Snyder A, Chi P, Danila D, Gounder M, Harding JJ, Hellmann MD, Iyer G, Janjigian Y, Kaley T, Levine DA, Lowery M, Omuro A, Postow MA, Rathkopf D, Shoushtari AN, Shukla N, Voss M, Paraiso E, Zehir A, Berger MF, Taylor BS, Saltz LB, Riely GJ, Ladanyi M, Hyman DM, Baselga J, Sabbatini P, Solit DB, Schultz N. OncoKB: a precision oncology knowledge base. JCO Precis Oncol 2017;2017, PO.17.00011.
[26] Anon. Precision oncology knowledge base, https://www.oncokb.org/.
[27] Wakai T, Prasoon P, Hirose Y, Shimada Y, Ichikawa H, Nagahashi M. Next-generation sequencing-based clinical sequencing: toward precision medicine in solid tumors. Int J Clin Oncol 2019;24:115–22.
[28] Tarazona N, Gimeno-Valiente F, Gambardella V, Zuñiga S, Rentero-Garrido P, Huerta M, Roselló S, Martinez-Ciarpaglini C, Carbonell-Asins JA, Carrasco F, Ferrer-Martínez A, Bruixola G, Fleitas T, Martín J, Tébar-Martínez R, Moro D, Castillo J, Espí A, Roda D, Cervantes A. Targeted next-generation sequencing of circulating-tumor DNA for tracking minimal residual disease in localized colon cancer. Ann Oncol 2019;30:1804–12.
[29] Galarza Fortuna GM, Dvir K. Circulating tumor DNA: where are we now? A mini review of the literature. World J Clin Oncol 2020;11:723–31.
[30] Dasari A, Morris VK, Allegra CJ, Atreya C, Benson 3rd AB, Boland P, Chung K, Copur MS, Corcoran RB, Deming DA, Dwyer A, Diehn M, Eng C, George TJ, Gollub MJ, Goodwin RA, Hamilton SR, Hechtman JF, Hochster H, Hong TS, Innocenti F, Iqbal A, Jacobs SA, Kennecke HF, Lee JJ, Lieu CH, Lenz HJ, Lindwasser OW, Montagut C, Odisio B, Ou FS, Porter L, Raghav K, Schrag D, Scott AJ, Shi Q, Strickler JH, Venook A, Yaeger R, Yothers G, You YN, Zell JA, Kopetz S. ctDNA applications and integration in colorectal cancer: an NCI colon and rectal-anal task forces whitepaper. Nat Rev Clin Oncol 2020;17:757–70.
[31] Allen WL, Coyle VM, Jithesh PV, Proutski I, Stevenson L, Fenning C, Longley DB, Wilson RH, Gordon M, Lenz HJ, Johnston PG. Clinical determinants of response to irinotecan-based therapy derived from cell line models. Clin Cancer Res 2008;14:6647–55.

[32] Boyer J, Allen WL, McLean EG, Wilson PM, McCulla A, Moore S, Longley DB, Caldas C, Johnston PG. Pharmacogenomic identification of novel determinants of response to chemotherapy in colon cancer. Cancer Res 2006;66:2765–77.
[33] Mariadason JM, Arango D, Shi Q, Wilson AJ, Corner GA, Nicholas C, Aranes MJ, Lesser M, Schwartz EL, Augenlicht LH. Gene expression profiling-based prediction of response of colon carcinoma cells to 5-fluorouracil and camptothecin. Cancer Res 2003;63:8791–812.
[34] Medico E, Russo M, Picco G, Cancelliere C, Valtorta E, Corti G, Buscarino M, Isella C, Lamba S, Martinoglio B, Veronese S, Siena S, Sartore-Bianchi A, Beccuti M, Mottolese M, Linnebacher M, Cordero F, Di Nicolantonio F, Bardelli A. The molecular landscape of colorectal cancer cell lines unveils clinically actionable kinase targets. Nat Commun 2015;6:7002.
[35] Salonga D, Danenberg KD, Johnson M, Metzger R, Groshen S, Tsao-Wei DD, Lenz HJ, Leichman CG, Leichman L, Diasio RB, Danenberg PV. Colorectal tumors responding to 5-fluorouracil have low gene expression levels of dihydropyrimidine dehydrogenase, thymidylate synthase, and thymidine phosphorylase. Clin Cancer Res 2000;6:1322–7.
[36] Smorenburg CH, Peters GJ, van Groeningen CJ, Noordhuis P, Smid K, van Riel AM, Dercksen W, Pinedo HM, Giaccone G. Phase II study of tailored chemotherapy for advanced colorectal cancer with either 5-fluouracil and leucovorin or oxaliplatin and irinotecan based on the expression of thymidylate synthase and dihydropyrimidine dehydrogenase. Ann Oncol 2006;17:35–42.
[37] O'Dwyer PJ, Eckhardt SG, Haller DG, Tepper J, Ahnen D, Hamilton S, Benson 3rd AB, Rothenberg M, Petrelli N, Lenz HJ, Diasio R, DuBois R, Sargent D, Sloan J, Johnson CD, Comis RL, O'Connell MJ. Gastrointestinal scientific leadership council of the coalition of cancer cooperative groups. Priorities in colorectal cancer research: recommendations from the gastrointestinal scientific leadership council of the coalition of cancer cooperative groups. J Clin Oncol 2007;25:2313–21.
[38] Guinney J, Dienstmann R, Wang X, de Reyniès A, Schlicker A, Soneson C, Marisa L, Roepman P, Nyamundanda G, Angelino P, Bot BM, Morris JS, Simon IM, Gerster S, Fessler E, De Sousa E, Melo F, Missiaglia E, Ramay H, Barras D, Homicsko K, Maru D, Manyam GC, Broom B, Boige V, Perez-Villamil B, Laderas T, Salazar R, Gray JW, Hanahan D, Tabernero J, Bernards R, Friend SH, Laurent-Puig P, Medema JP, Sadanandam A, Wessels L, Delorenzi M, Kopetz S, Vermeulen L, Tejpar S. The consensus molecular subtypes of colorectal cancer. Nat Med 2015;21:1350–6.
[39] Lenz HJ, Ou FS, Venook AP, Hochster HS, Niedzwiecki D, Goldberg RM, Mayer RJ, Bertagnolli MM, Blanke CD, Zemla T, Qu X, Wirapati P, Tejpar S, Innocenti F, Kabbarah O. Impact of consensus molecular subtype on survival in patients with metastatic colorectal cancer: results from CALGB/SWOG 80405 (Alliance). J Clin Oncol 2019;37:1876–85.
[40] Avgeris M, Panoutsopoulou K, Papadimitriou MA, Scorilas A. Circulating exosomal miRNAs: clinical significance in human cancers. Expert Rev Mol Diagn 2019;19:979–95.
[41] Jin G, Liu Y, Zhang J, Bian Z, Yao S, Fei B, Zhou L, Yin Y, Huang Z. A panel of serum exosomal microRNAs as predictive markers for chemoresistance in advanced colorectal cancer. Cancer Chemother Pharmacol 2019;84:315–25.
[42] Van Gool A, Corrales F, Čolović M, Krstić D, Oliver-Martos B, Martínez-Cáceres E, Jakasa I, Gajski G, Brun V, Kyriacou K, Burzynska-Pedziwiatr I, Wozniak LA, Nierkens S, Pascual García C, Katrlik J, Bojic-Trbojevic Z, Vacek J, Llorente A, Antohe F, Suica V, Suarez G, t'Kindt R, Martin P, Penque D, Martins IL, Bodoki E, Iacob BC, Aydindogan E, Timur S, Allinson J, Sutton C, Luider T, Wittfooth S, Sammar M. Analytical techniques for multiplex analysis of protein biomarkers. Expert Rev Proteomics 2020;17:257–73.
[43] Varkaris A, Katsiampoura A, Davis JS, Shah N, Lam M, Frias RL, Ivan C, Shimizu M, Morris J, Menter D, Overman M, Tran H, Heymach J, Chun YS, Vauthey JN, Calin G, Kopetz S. Circulating inflammation signature predicts overall survival and relapse-free survival in metastatic colorectal cancer. Br J Cancer 2019;120:340–5.
[44] Socovich AM, Naba A. The cancer matrisome: from comprehensive characterization to biomarker discovery. Semin Cell Dev Biol 2019;89:157–66.
[45] Bertini I, Cacciatore S, Jensen BV, Schou JV, Johansen JS, Kruhøffer M, Luchinat C, Nielsen DL, Turano P. Metabolomic NMR fingerprinting to identify and predict survival of patients with metastatic colorectal cancer. Cancer Res 2012;72:356–64.
[46] Zhang Y, Du Y, Song Z, Liu S, Li W, Wang D, Suo J. Profiling of serum metabolites in advanced colon cancer using liquid chromatography-mass spectrometry. Oncol Lett 2020;19:4002–10.
[47] Goossens N, Nakagawa S, Sun X, Hoshida Y. Cancer biomarker discovery and validation. Transl Cancer Res 2015;4:256–69.

[48] Patil H, Saxena SG, Barrow CJ, Kanwar JR, Kapat A, Kanwar RK. Chasing the personalized medicine dream through biomarker validation in colorectal cancer. Drug Discov Today 2017;22:111–9.

[49] Peluso G, Incollingo P, Calogero A, Tammaro V, Rupealta N, Chiacchio G, Sandoval Sotelo ML, Minieri G, Pisani A, Riccio E, Sabbatini M, Bracale UM, Dodaro CA, Carlomagno N. Current tissue molecular markers in colorectal cancer: a literature review. Biomed Res Int 2017;2017:2605628.

[50] Church D, Kerr R, Domingo E, Rosmarin D, Palles C, Maskell K, Tomlinson I, Kerr D. 'Toxgnostics': an unmet need in cancer medicine. Nat Rev Cancer 2014;14:440–5.

[51] Gray R, Barnwell J, McConkey C, Hills RK, Williams NS, Kerr DJ. Quasar Collaborative Group Adjuvant chemotherapy versus observation in patients with colorectal cancer: a randomised study. Lancet 2007;370:2020–9.

[52] Mini E, Nobili S. "Pharmacogenetics of cancer"—cancer drug resistance special issue. Cancer Drug Resist 2020;3:225–31.

[53] Soveri LM, Hermunen K, de Gramont A, Poussa T, Quinaux E, Bono P, André T, Sterlund OE. Association of adverse events and survival in colorectal cancer patients treated with adjuvant 5-fluorouracil and leucovorin: is efficacy an impact of toxicity? Eur J Cancer 2014;50:2966–74.

[54] Markowitz SD, Bertagnolli MM. Molecular origins of cancer: molecular basis of colorectal cancer. N Engl J Med 2009;361:2449–60.

[55] Novellasdemunt L, Antas P, Li VS. Targeting Wnt signaling in colorectal cancer. A review in the theme: cell signaling: proteins, pathways and mechanisms. Am J Physiol Cell Physiol 2015;309:C511–21.

[56] Clevers H, Nusse R. Wnt/β-catenin signaling and disease. Cell 2012;149(6):1192–205.

[57] Nandan MO, Yang VW. An update on the biology of RAS/RAF mutations in colorectal cancer. Curr Colorectal Cancer Rep 2011;7:113–20.

[58] Papadatos-Pastos D, Rabbie R, Ross P, Sarker D. The role of the PI3K pathway in colorectal cancer. Crit Rev Oncol Hematol 2015;94:18–30.

[59] Laplante M, Sabatini DM. mTOR signaling in growth control and disease. Cell 2012;149(2):274–93.

[60] Koveitypour Z, Panahi F, Vakilian M, Peymani M, Seyed Forootan F, Nasr Esfahani MH, Ghaedi K. Signaling pathways involved in colorectal cancer progression. Cell Biosci 2019;9:97.

[61] Chang MT, Asthana S, Gao SP, Lee BH, Chapman JS, Kandoth C, Gao J, Socci ND, Solit DB, Olshen AB, Schultz N, Taylor BS. Identifying recurrent mutations in cancer reveals widespread lineage diversity and mutational specificity. Nat Biotechnol 2016;34:155–63.

[62] Khambata-Ford S, Garrett CR, Meropol NJ, Basik M, Harbison CT, Wu S, Wong TW, Huang X, Takimoto CH, Godwin AK, Tan BR, Krishnamurthi SS, Burris 3rd HA, Poplin EA, Hidalgo M, Baselga J, Clark EA, Mauro DJ. Expression of epiregulin and amphiregulin and K-ras mutation status predict disease control in metastatic colorectal cancer patients treated with cetuximab. J Clin Oncol 2007;25:3230–7.

[63] Douillard JY, Oliner KS, Siena S, Tabernero J, Burkes R, Barugel M, Humblet Y, Bodoky G, Cunningham D, Jassem J, Rivera F, Kocákova I, Ruff P, Błasińska-Morawiec M, Šmakal M, Canon JL, Rother M, Williams R, Rong A, Wiezorek J, Sidhu R, Patterson SD. Panitumumab-FOLFOX4 treatment and RAS mutations in colorectal cancer. N Engl J Med 2013;369:1023–34.

[64] De Roock W, Claes B, Bernasconi D, De Schutter J, Biesmans B, Fountzilas G, Kalogeras KT, Kotoula V, Papamichael D, Laurent-Puig P, Penault-Llorca F, Rougier P, Vincenzi B, Santini D, Tonini G, Cappuzzo F, Frattini M, Molinari F, Saletti P, De Dosso S, Martini M, Bardelli A, Siena S, Sartore-Bianchi A, Tabernero J, Macarulla T, Di Fiore F, Gangloff AO, Ciardiello F, Pfeiffer P, Qvortrup C, Hansen TP, Van Cutsem E, Piessevaux H, Lambrechts D, Delorenzi M, Tejpar S. Effects of KRAS, BRAF, NRAS, and PIK3CA mutations on the efficacy of cetuximab plus chemotherapy in chemotherapy-refractory metastatic colorectal cancer: a retrospective consortium analysis. Lancet Oncol 2010;11:753–62.

[65] De Roock W, De Vriendt V, Normanno N, Ciardiello F, Tejpar S. KRAS, BRAF, PIK3CA, and PTEN mutations: implications for targeted therapies in metastatic colorectal cancer. Lancet Oncol 2011;12:594–603.

[66] Clarke SJ, Karapetis CS, Gibbs P, Pavlakis N, Desai J, Michael M, Tebbutt NC, Price TJ, Tabernero J. Overview of biomarkers in metastatic colorectal cancer: tumour, blood and patient-related factors. Crit Rev Oncol Hematol 2013;85:121–35.

[67] Dienstmann R, Vermeulen L, Guinney J, Kopetz S, Tejpar S, Tabernero J. Consensus molecular subtypes and the evolution of precision medicine in colorectal cancer. Nat Rev Cancer 2017;17:79–92.

[68] De Cuyper A, Van Den Eynde M, Machiels JP. HER2 as a predictive biomarker and treatment target in colorectal cancer. Clin Colorectal Cancer 2020;19:65–72.

[69] Dienstmann R, Connor K, Byrne AT, COLOSSUS Consortium. Precision therapy in RAS mutant colorectal cancer. Gastroenterology 2020;158:806–11.

[70] Bardelli A, Corso S, Bertotti A, Hobor S, Valtorta E, Siravegna G, Sartore-Bianchi A, Scala E, Cassingena A, Zecchin D, Apicella M, Migliardi G, Galimi F, Lauricella C, Zanon C, Perera T, Veronese S, Corti G, Amatu A, Gambacorta M, Diaz Jr LA, Sausen M, Velculescu VE, Comoglio P, Trusolino L, Di Nicolantonio F, Giordano S, Siena S. Amplification of the MET receptor drives resistance to anti-EGFR therapies in colorectal cancer. Cancer Discov 2013;3:658–73.

[71] Raghav K, Morris V, Tang C, Morelli P, Amin HM, Chen K, Manyam GC, Broom B, Overman MJ, Shaw K, Meric-Bernstam F, Maru D, Menter D, Ellis LM, Eng C, Hong D, Kopetz S. MET amplification in metastatic colorectal cancer: an acquired response to EGFR inhibition, not a de novo phenomenon. Oncotarget 2016;7:54627–31.

[72] Tabernero J, Van Cutsem E, Prausová J, Isambert N, Neuzillet C, Siena S, Pietrantonio F, Falk S, Wainberg ZA, Raghav KPS, Campbell N, Liu E, Esser R, Salim S, Beier F, Adrian S, Elez E, Bekaii-Saab TS. Tepotinib plus cetuximab in patients (pts) with RAS/BRAF wild-type left-sided metastatic colorectal cancer (mCRC) and acquired resistance to anti-EGFR antibody therapy due to MET amplification (METamp). J Clin Oncol 2021; 39 (3 suppl). Abstract #TPS149.

[73] Wu YL, Cheng Y, Zhou J, Lu S, Zhang Y, Zhao J, Kim DW, Soo RA, Kim SW, Pan H, Chen YM, Chian CF, Liu X, Tan DSW, Bruns R, Straub J, Johne A, Scheele J, Park K, Yang JC, INSIGHT Investigators. Tepotinib plus gefitinib in patients with EGFR-mutant non-small-cell lung cancer with MET overexpression or MET amplification and acquired resistance to previous EGFR inhibitor (INSIGHT study): an open-label, phase 1b/2, multicentre, randomised trial. Lancet Respir Med 2020;8:1132–43.

[74] Grothey A, Van Cutsem E, Sobrero A, Siena S, Falcone A, Ychou M, Humblet Y, Bouché O, Mineur L, Barone C, Adenis A, Tabernero J, Yoshino T, Lenz HJ, Goldberg RM, Sargent DJ, Cihon F, Cupit L, Wagner A, Laurent D, CORRECT Study Group. Regorafenib monotherapy for previously treated metastatic colorectal cancer (CORRECT): an international, multicentre, randomised, placebo-controlled, phase 3 trial. Lancet 2013; 381:303–12.

[75] Li J, Qin S, Xu R, Yau TC, Ma B, Pan H, Xu J, Bai Y, Chi Y, Wang L, Yeh KH, Bi F, Cheng Y, Le AT, Lin JK, Liu T, Ma D, Kappeler C, Kalmus J, Kim TW, CONCUR Investigators. Regorafenib plus best supportive care versus placebo plus best supportive care in Asian patients with previously treated metastatic colorectal cancer (CONCUR): a randomised, double-blind, placebo-controlled, phase 3 trial. Lancet Oncol 2015;16:619–29.

[76] Tabernero J, Lenz HJ, Siena S, Sobrero A, Falcone A, Ychou M, Humblet Y, Bouché O, Mineur L, Barone C, Adenis A, Yoshino T, Goldberg RM, Sargent DJ, Wagner A, Laurent D, Teufel M, Jeffers M, Grothey A, Van Cutsem E. Analysis of circulating DNA and protein biomarkers to predict the clinical activity of regorafenib and assess prognosis in patients with metastatic colorectal cancer: a retrospective, exploratory analysis of the CORRECT trial. Lancet Oncol 2015;16:937–48.

[77] Gong J, Wang C, Lee PP, Chu P, Fakih M. Response to PD-1 blockade in microsatellite stable metastatic colorectal cancer harboring a POLE mutation. J Natl Compr Canc Netw 2017;15:142–7.

[78] Kheder ES, Hong DS. Emerging targeted therapy for tumors with *NTRK* fusion proteins. Clin Cancer Res 2018;24:5807–14.

[79] Cohen R, Pudlarz T, Delattre JF, Colle R, André T. Molecular targets for the treatment of metastatic colorectal cancer. Cancers (Basel) 2020;12:2350.

[80] Stoehlmacher J, Park DJ, Zhang W, Groshen S, Tsao-Wei DD, Yu MC, Lenz HJ. Association between glutathione S-transferase P1, T1, and M1 genetic polymorphism and survival of patients with metastatic colorectal cancer. J Natl Cancer Inst 2002;94:936–42.

[81] Stoehlmacher J, Park DJ, Zhang W, Yang D, Groshen S, Zahedy S, Lenz HJ. A multivariate analysis of genomic polymorphisms: prediction of clinical outcome to 5-FU/oxaliplatin combination chemotherapy in refractory colorectal cancer. Br J Cancer 2004;91:344–54.

[82] Jun L, Haiping Z, Beibei Y. Genetic polymorphisms of GSTP1 related to response to 5-FU-oxaliplatin-based chemotherapy and clinical outcome in advanced colorectal cancer patients. Swiss Med Wkly 2009;139:724–8.

[83] Funke S, Timofeeva M, Risch A, Hoffmeister M, Stegmaier C, Seiler CM, Brenner H, Chang-Claude J. Genetic polymorphisms in GST genes and survival of colorectal cancer patients treated with chemotherapy. Pharmacogenomics 2010;11:33–41.

[84] Hong J, Han SW, Ham HS, Kim TY, Choi IS, Kim BS, Oh DY, Im SA, Kang GH, Bang YJ, Kim TY. Phase II study of biweekly S-1 and oxaliplatin combination chemotherapy in metastatic colorectal cancer and pharmacogenetic analysis. Cancer Chemother Pharmacol 2011;67:1323–31.

[85] Li HY, Ge X, Huang GM, Li KY, Zhao JQ, Yu XM, Bi WS, Wang YL. GSTP1, ERCC1 and ERCC2 polymorphisms, expression and clinical outcome of oxaliplatin-based adjuvant chemotherapy in colorectal cancer in Chinese population. Asian Pac J Cancer Prev 2012;13:3465–9.

[86] Le Morvan V, Smith D, Laurand A, Brouste V, Bellott R, Soubeyran I, Mathoulin-Pelissier S, Robert J. Determination of ERCC2 Lys751Gln and GSTP1 Ile105Val gene polymorphisms in colorectal cancer patients: relationships with treatment outcome. Pharmacogenomics 2007;8:1693–703.

[87] Kweekel DM, Gelderblom H, Antonini NF, Van der Straaten T, Nortier JW, Punt CJ, Guchelaar HJ. Glutathione-S-transferase pi (GSTP1) codon 105 polymorphism is not associated with oxaliplatin efficacy or toxicity in advanced colorectal cancer patients. Eur J Cancer 2009;45:572–8.

[88] Yin M, Yan J, Martinez-Balibrea E, Graziano F, Lenz HJ, Kim HJ, Robert J, Im SA, Wang WS, Etienne-Grimaldi MC, Wei Q. ERCC1 and ERCC2 polymorphisms predict clinical outcomes of oxaliplatin-based chemotherapies in gastric and colorectal cancer: a systemic review and meta-analysis. Clin Cancer Res 2011;17:1632–40.

[89] Qian YY, Liu XY, Wu Q, Song X, Chen XF, Liu YQ, Pei D, Shen LZ, Shu YQ. The ERCC1 C118T polymorphism predicts clinical outcomes of colorectal cancer patients receiving oxaliplatin-based chemotherapy: a meta-analysis based on 22 studies. Asian Pac J Cancer Prev 2014;15:8383–90.

[90] Lu X, Xiao S, Jin C, van der Straaten T, Li X. ERCC1 and XPD/ERCC2 polymorphisms' predictive value of oxaliplatin-based chemotherapies in advanced colorectal cancer has an ethnic discrepancy: a meta-analysis.-J Clin Lab Anal 2012;26:10–5.

[91] Shahnam A, Ridha Z, Wiese MD, Kichenadasse G, Sorich MJ. Pharmacogenetic and ethnicity influence on oxaliplatin therapy for colorectal cancer: a meta-analysis. Pharmacogenomics 2016;17:1725–32.

[92] Formica V, Doldo E, Antonetti FR, Nardecchia A, Ferroni P, Riondino S, Morelli C, Arkenau HT, Guadagni F, Orlandi A, Roselli M. Biological and predictive role of ERCC1 polymorphisms in cancer. Crit Rev Oncol Hematol 2017;111:133–43.

[93] Duran G, Cruz R, Simoes AR, Barros F, Giráldez JM, Bernárdez B, Anido U, Candamio S, López-López R, Carracedo Á, Lamas MJ. Efficacy and toxicity of adjuvant chemotherapy on colorectal cancer patients: how much influence from the genetics? J Chemother 2020;32:310–22.

[94] Chen J, Xie F, Chen K, Wang D, Jiang H, Li J, Pan F, Chen S, Zhang Y, Ruan Z, Huang H, Zou L, Liang H. ERCC5 promoter polymorphisms at −763 and +25 predict the response to oxaliplatin-based chemotherapy in patients with advanced colorectal cancer. Cancer Biol Ther 2009;8:1424–30.

[95] Chen J, Luo X, Xie G, Chen K, Jiang H, Pan F, Li J, Ruan Z, Pang X, Liang H. Functional analysis of SNPs in the ERCC5 promoter in advanced colorectal cancer patients treated with oxaliplatin-based chemotherapy. Medicine (Baltimore) 2016;95, e3652.

[96] Loganayagam A, Arenas Hernandez M, Corrigan A, Fairbanks L, Lewis CM, Harper P, Maisey N, Ross P, Sanderson JD, Marinaki AM. Pharmacogenetic variants in the DPYD, TYMS, CDA and MTHFR genes are clinically significant predictors of fluoropyrimidine toxicity. Br J Cancer 2013;108:2505–15.

[97] Sharma R, Hoskins JM, Rivory LP, Zucknick M, London R, Liddle C, Clarke SJ. Thymidylate synthase and methylenetetrahydrofolate reductase gene polymorphisms and toxicity to capecitabine in advanced colorectal cancer patients. Clin Cancer Res 2008;14:817–25.

[98] Zhu L, Wang F, Hu F, Wang Y, Li D, Dong X, Cui B, Zhao Y. Association between MTHFR polymorphisms and overall survival of colorectal cancer patients in Northeast China. Med Oncol 2013;30:467.

[99] Cecchin E, Perrone G, Nobili S, Polesel J, De Mattia E, Zanusso C, Petreni P, Lonardi S, Pella N, D'Andrea M, Errante D, Rizzolio F, Mazzei T, Landini I, Mini E, Toffoli G. MTHFR-1298 A>C (rs1801131) is a predictor of survival in two cohorts of stage II/III colorectal cancer patients treated with adjuvant fluoropyrimidine chemotherapy with or without oxaliplatin. Pharmacogenomics J 2015;15:219–25.

[100] Yeh CC, Lai CY, Chang SN, Hsieh LL, Tang R, Sung FC, Lin YK. Polymorphisms of MTHFR C677T and A1298C associated with survival in patients with colorectal cancer treated with 5-fluorouracil-based chemotherapy. Int J Clin Oncol 2017;22:484–93.

[101] Kim KR, Yoon JH, Shim HJ, Hwang JE, Bae WK, Chung IJ, Kim HN, Shin MH, Cho SH. Role of depth of response and MTHFR genotype as predictors of fluorouracil rechallenge therapy for refractory metastatic colorectal cancer. Oncol Lett 2017;14:2491–8.

[102] Ose J, Botma A, Balavarca Y, Buck K, Scherer D, Habermann N, Beyerle J, Pfütze K, Seibold P, Kap EJ, Benner A, Jansen L, Butterbach K, Hoffmeister M, Brenner H, Ulrich A, Schneider M, Chang-Claude J, Burwinkel B, Ulrich CM. Pathway analysis of genetic variants in folate-mediated one-carbon metabolism-related genes and survival in a prospectively followed cohort of colorectal cancer patients. Cancer Med 2018;7:2797–807.

[103] Di Paolo A, Chu E. The role of thymidylate synthase as a molecular biomarker. Clin Cancer Res 2004;10:411–2.
[104] Marsh S. Thymidylate synthase pharmacogenetics. Invest New Drugs 2005;23:533–7.
[105] Páez D, Paré L, Altés A, Sancho-Poch FJ, Petriz L, Garriga J, Monill JM, Salazar J, del Rio E, Barnadas A, Marcuello E, Baiget M. Thymidylate synthase germline polymorphisms in rectal cancer patients treated with neoadjuvant chemoradiotherapy based on 5-fluorouracil. J Cancer Res Clin Oncol 2010;136:1681–9.
[106] Páez D, Salazar J, Paré L, Pertriz L, Targarona E, del Rio E, Barnadas A, Marcuello E, Baiget M. Pharmacogenetic study in rectal cancer patients treated with preoperative chemoradiotherapy: polymorphisms in thymidylate synthase, epidermal growth factor receptor, GSTP1, and DNA repair genes. Int J Radiat Oncol Biol Phys 2011;81:1319–27.
[107] Joerger M, Huitema AD, Boot H, Cats A, Doodeman VD, Smits PH, Vainchtein L, Rosing H, Meijerman I, Zueger M, Meulendijks D, Cerny TD, Beijnen JH, Schellens JH. Germline TYMS genotype is highly predictive in patients with metastatic gastrointestinal malignancies receiving capecitabine-based chemotherapy. Cancer Chemother Pharmacol 2015;75:763–72.
[108] Abdullah-Koolmees H, van Keulen AM, Nijenhuis M, Deneer VHM. Pharmacogenetics guidelines: overview and comparison of the DPWG, CPIC, CPNDS, and RNPGx guidelines. Front Pharmacol 2021;11:595219.
[109] Anon. CPIC® guideline for fluoropyrimidines and DPYD, https://cpicpgx.org/guidelines/guideline-for-fluoropyrimidines-and-dpyd/.
[110] Meulendijks D, Cats A, Beijnen JH, Schellens JH. Improving safety of fluoropyrimidine chemotherapy by individualizing treatment based on dihydropyrimidine dehydrogenase activity—ready for clinical practice? Cancer Treat Rev 2016;50:23–34.
[111] Toffoli G, Innocenti F, Polesel J, De Mattia E, Sartor F, Dalle Fratte C, Ecca F, Dreussi E, Palazzari E, Guardascione M, Buonadonna A, Foltran L, Garziera M, Bignucolo A, Nobili S, Mini E, Favaretto A, Berretta M, D'Andrea M, De Paoli A, Roncato R, Cecchin E. The genotype for DPYD risk variants in patients with colorectal cancer and the related toxicity management costs in clinical practice. Clin Pharmacol Ther 2019;105:994–1002.
[112] Toffoli G, Giodini L, Buonadonna A, Berretta M, De Paoli A, Scalone S, Miolo G, Mini E, Nobili S, Lonardi S, Pella N, Lo Re G, Montico M, Roncato R, Dreussi E, Gagno S, Cecchin E. Clinical validity of a DPYD-based pharmacogenetic test to predict severe toxicity to fluoropyrimidines. Int J Cancer 2015;137:2971–80.
[113] Boige V, Vincent M, Alexandre P, Tejpar S, Landolfi S, Le Malicot K, Greil R, Cuyle PJ, Yilmaz M, Faroux R, Matzdorff A, Salazar R, Lepage C, Taieb J, Laurent-Puig P. DPYD genotyping to predict adverse events following treatment with fluorouracil-based adjuvant chemotherapy in patients with stage III colon cancer: a secondary analysis of the PETACC-8 randomized clinical trial. JAMA Oncol 2016;2:655–62.
[114] Etienne-Grimaldi MC, Boyer JC, Beroud C, Mbatchi L, van Kuilenburg A, Bobin-Dubigeon C, Thomas F, Chatelut E, Merlin JL, Pinguet F, Ferrand C, Meijer J, Evrard A, Llorca L, Romieu G, Follana P, Bachelot T, Chaigneau L, Pivot X, Dieras V, Largillier R, Mousseau M, Goncalves A, Roché H, Bonneterre J, Servant V, Dohollou N, Château Y, Chamorey E, Desvignes JP, Salgado D, Ferrero JM, Milano G. New advances in DPYD genotype and risk of severe toxicity under capecitabine. PLoS One 2017;12, e0175998.
[115] Ezzeldin HH, Lee AM, Mattison LK, Diasio RB. Methylation of the DPYD promoter: an alternative mechanism for dihydropyrimidine dehydrogenase deficiency in cancer patients. Clin Cancer Res 2005;11:8699–705.
[116] Amstutz U, Offer SM, Sistonen J, Joerger M, Diasio RB, Largiadèr CR. Polymorphisms in MIR27A associated with early-onset toxicity in fluoropyrimidine-based chemotherapy. Clin Cancer Res 2015;21:2038–44.
[117] Machon C, Catez F, Venezia ND, Vanhalle F, Guyot L, Vincent A, Garcia M, Roy B, Diaz JJ, Guitton J. Study of intracellular anabolism of 5-fluorouracil and incorporation in nucleic acids based on an LC-HRMS method. J Pharm Anal 2021;11:77–87.
[118] Yousef AM, Zawiah M, Al-Yacoub S, Kadi T, Tantawi DA, Al-Ramadhani H. The association of polymorphisms in folate-metabolizing genes with response to adjuvant chemotherapy of colorectal cancer. Cancer Chemother Pharmacol 2018;82:237–43.
[119] Liu D, Li J, Gao J, Li Y, Yang R, Shen L. Examination of multiple UGT1A and DPYD polymorphisms has limited ability to predict the toxicity and efficacy of metastatic colorectal cancer treated with irinotecan-based chemotherapy: a retrospective analysis. BMC Cancer 2017;17:437.
[120] Liu D, Li X, Li X, Zhang M, Zhang J, Hou D, Tong Z, Dong M. CDA and MTHFR polymorphisms are associated with clinical outcomes in gastroenteric cancer patients treated with capecitabine-based chemotherapy. Cancer Chemother Pharmacol 2019;83:939–49.
[121] Di Paolo A, Bocci G, Polillo M, Del Re M, Di Desidero T, Lastella M, Danesi R. Pharmacokinetic and pharmacogenetic predictive markers of irinotecan activity and toxicity. Curr Drug Metab 2011;12:932–43.

[122] Lévesque E, Bélanger AS, Harvey M, Couture F, Jonker D, Innocenti F, Cecchin E, Toffoli G, Guillemette C. Refining the UGT1A haplotype associated with irinotecan-induced hematological toxicity in metastatic colorectal cancer patients treated with 5-fluorouracil/irinotecan-based regimens. J Pharmacol Exp Ther 2013;345:95–101.
[123] Fujii H, Yamada Y, Watanabe D, Matsuhashi N, Takahashi T, Yoshida K, Suzuki A. Dose adjustment of irinotecan based on UGT1A1 polymorphisms in patients with colorectal cancer. Cancer Chemother Pharmacol 2019;83:123–9.
[124] Toffoli G, Cecchin E, Gasparini G, D'Andrea M, Azzarello G, Basso U, Mini E, Pessa S, De Mattia E, Lo Re G, Buonadonna A, Nobili S, De Paoli P, Innocenti F. Genotype-driven phase I study of irinotecan administered in combination with fluorouracil/leucovorin in patients with metastatic colorectal cancer. J Clin Oncol 2010;28:866–71.
[125] Yu Q, Zhang T, Xie C, Qiu H, Liu B, Huang L, Peng P, Feng J, Chen J, Zang A, Yuan X. UGT1A polymorphisms associated with worse outcome in colorectal cancer patients treated with irinotecan-based chemotherapy. Cancer Chemother Pharmacol 2018;82:87–98.
[126] Zhang W, Stoehlmacher J, Park DJ, Yang D, Borchard E, Gil J, Tsao-Wei DD, Yun J, Gordon M, Press OA, Rhodes K, Groshen S, Lenz HJ. Gene polymorphisms of epidermal growth factor receptor and its downstream effector, interleukin-8, predict oxaliplatin efficacy in patients with advanced colorectal cancer. Clin Colorectal Cancer 2005;5:124–31.
[127] Graziano F, Ruzzo A, Loupakis F, Canestrari E, Santini D, Catalano V, Bisonni R, Torresi U, Floriani I, Schiavon G, Andreoni F, Maltese P, Rulli E, Humar B, Falcone A, Giustini L, Tonini G, Fontana A, Masi G, Magnani M. Pharmacogenetic profiling for cetuximab plus irinotecan therapy in patients with refractory advanced colorectal cancer. J Clin Oncol 2008;26:1427–34.
[128] Lai CY, Sung FC, Hsieh LL, Tang R, Chiou HY, Wu FY, Yeh CC. Associations between genetic polymorphisms of epidermal growth factor receptor (EGFR) and survival of colorectal cancer (CRC) patients treated with 5-fluorouracil-based chemotherapy. Ann Surg Oncol 2013;20(Suppl. 3):S599–606.
[129] De Mattia E, Bignucolo A, Toffoli G, Cecchin E. Genetic markers of the host to predict the efficacy of colorectal cancer targeted therapy. Curr Med Chem 2020;27:4249–73.
[130] Kjersem JB, Skovlund E, Ikdahl T, Guren T, Kersten C, Dalsgaard AM, Yilmaz MK, Fokstuen T, Tveit KM, Kure EH. FCGR2A and FCGR3A polymorphisms and clinical outcome in metastatic colorectal cancer patients treated with first-line 5-fluorouracil/folinic acid and oxaliplatin +/− cetuximab. BMC Cancer 2014;14:340.
[131] Geva R, Vecchione L, Kalogeras KT, Jensen BV, Lenz HJ, Yoshino T, Paez D, Montagut C, Souglakos J, Cappuzzo F, Cervantes A, Frattini M, Fountzilas G, Johansen JS, Høgdall EV, Zhang W, Yang D, Yamazaki K, Nishina T, Papamichael D, Vincenzi B, Macarulla T, Loupakis F, De Schutter J, Spindler KL, Pfeiffer P, Ciardiello F, Piessevaux H, Tejpar S. FCGR polymorphisms and cetuximab efficacy in chemorefractory metastatic colorectal cancer: an international consortium study. Gut 2015;64:921–8.
[132] Morgen EK, Lenz HJ, Jonker DJ, Tu D, Milano G, Graziano F, Zalcberg J, Karapetis CS, Dobrovic A, O'Callaghan CJ, Liu G. Germline polymorphisms as biomarkers of tumor response in colorectal cancer patients treated with anti- EGFR monoclonal antibodies: a systematic review and meta-analysis. Pharmacogenomics J 2017;17:535–42.
[133] Koutras AK, Antonacopoulou AG, Eleftheraki AG, Dimitrakopoulos FI, Koumarianou A, Varthalitis I, Fostira F, Sgouros J, Briasoulis E, Bournakis E, Bafaloukos D, Bompolaki I, Galani E, Kalogeras KT, Pectasides D, Fountzilas G, Kalofonos HP. Vascular endothelial growth factor polymorphisms and clinical outcome in colorectal cancer patients treated with irinotecan-based chemotherapy and bevacizumab. Pharmacogenomics J 2012;12:468–75.
[134] Loupakis F, Cremolini C, Yang D, Salvatore L, Zhang W, Wakatsuki T, Bohanes P, Schirripa M, Benhaim L, Lonardi S, Antoniotti C, Aprile G, Graziano F, Ruzzo A, Lucchesi S, Ronzoni M, De Vita F, Tonini G, Falcone A, Lenz HJ. Prospective validation of candidate SNPs of VEGF/VEGFR pathway in metastatic colorectal cancer patients treated with first-line FOLFIRI plus bevacizumab. PLoS One 2013;8, e66774.
[135] Malone ER, Oliva M, Sabatini PJB, Stockley TL, Siu LL. Molecular profiling for precision cancer therapies. Genome Med 2020;12:8.
[136] Becht E, de Reyniès A, Giraldo NA, Pilati C, Buttard B, Lacroix L, Selves J, Sautès-Fridman C, Laurent-Puig P, Fridman WH. Immune and stromal classification of colorectal cancer is associated with molecular subtypes and relevant for precision immunotherapy. Clin Cancer Res 2016;22:4057–66.

[137] Aschele C, Sobrero A, Faderan MA, Bertino JR. Novel mechanism(s) of resistance to 5-fluorouracil in human colon cancer (HCT-8) sublines following exposure to two different clinically relevant dose schedules. Cancer Res 1992;52:1855–64.

[138] Wang FS, Aschele C, Sobrero A, Chang YM, Bertino JR. Decreased folylpolyglutamate synthetase expression: a novel mechanism of fluorouracil resistance. Cancer Res 1993;53:3677–80.

[139] Tempestini A, Caciagli B, Morganti M, Witort E, Nobili S, Papucci L, Schiavone N, Donnini M, Landini I, Lapucci A, Perna F, Lulli M, Mazzei T, Sobrero A, Mini E, Capaccioli S. Molecular characterization of established human colon carcinoma cell lines (HCT-8) made resistant to 5-fluorouracil by different selection schedules. Oncol Res 2005;16:143–56.

[140] Brabec V, Kasparkova J. Molecular aspects of resistance to antitumor platinum drugs. Drug Resist Updat 2002;5:147–61.

[141] Xu Y, Villalona-Calero MA. Irinotecan: mechanisms of tumor resistance and novel strategies for modulating its activity. Ann Oncol 2002;13:1841–51.

[142] Johnston PG, Fisher ER, Rockette HE, Fisher B, Wolmark N, Drake JC, Chabner BA, Allegra CJ. The role of thymidylate synthase expression in prognosis and outcome of adjuvant chemotherapy in patients with rectal cancer. J Clin Oncol 1994;12:2640–7.

[143] Leichman CG, Lenz HJ, Leichman L, Danenberg K, Baranda J, Groshen S, Boswell W, Metzger R, Tan M, Danenberg PV. Quantitation of intratumoral thymidylate synthase expression predicts for disseminated colorectal cancer response and resistance to protracted-infusion fluorouracil and weekly leucovorin. J Clin Oncol 1997;15:3223–9.

[144] Popat S, Matakidou A, Houlston RS. Thymidylate synthase expression and prognosis in colorectal cancer: a systematic review and meta-analysis. J Clin Oncol 2004;22:529–36.

[145] Morganti M, Ciantelli M, Giglioni B, Putignano AL, Nobili S, Papi L, Landini I, Napoli C, Valanzano R, Cianchi F, Boddi V, Tonelli F, Cortesini C, Mazzei T, Genuardi M, Mini E. Relationships between promoter polymorphisms in the thymidylate synthase gene and mRNA levels in colorectal cancers. Eur J Cancer 2005;41:2176–83.

[146] Vignoli M, Nobili S, Napoli C, Putignano AL, Morganti M, Papi L, Valanzano R, Cianchi F, Tonelli F, Mazzei T, Mini E, Genuardi M. Thymidylate synthase expression and genotype have no major impact on the clinical outcome of colorectal cancer patients treated with 5-fluorouracil. Pharmacol Res 2011;64:242–8.

[147] Metzger R, Danenberg K, Leichman CG, Salonga D, Schwartz EL, Wadler S, Lenz HJ, Groshen S, Leichman L, Danenberg PV. High basal level gene expression of thymidine phosphorylase (platelet-derived endothelial cell growth factor) in colorectal tumors is associated with nonresponse to 5-fluorouracil. Clin Cancer Res 1998;4:2371–6.

[148] Beck A, Etienne MC, Chéradame S, Fischel JL, Formento P, Renée N, Milano G. A role for dihydropyrimidine dehydrogenase and thymidylate synthase in tumour sensitivity to fluorouracil. Eur J Cancer 1994;30A:1517–22.

[149] Kidd EA, Yu J, Li X, Shannon WD, Watson MA, McLeod HL. Variance in the expression of 5-fluorouracil pathway genes in colorectal cancer. Clin Cancer Res 2005;11:2612–9.

[150] Nobili S, Napoli C, Landini I, Morganti M, Cianchi F, Valanzano R, Tonelli F, Cortesini C, Mazzei T, Mini E. Identification of potential pharmacogenomic markers of clinical efficacy of 5-fluorouracil in colorectal cancer. Int J Cancer 2011;128:1935–45.

[151] Peters GJ. Novel developments in the use of antimetabolites. Nucleosides Nucleotides Nucleic Acids 2014;33:358–74.

[152] Yano W, Yokogawa T, Wakasa T, Yamamura K, Fujioka A, Yoshisue K, Matsushima E, Miyahara S, Miyakoshi H, Taguchi J, Chong KT, Takao Y, Fukuoka M, Matsuo K. TAS-114, a first-in-class dual dUTPase/DPD inhibitor, demonstrates potential to improve therapeutic efficacy of fluoropyrimidine-based chemotherapy. Mol Cancer Ther 2018;17:1683–93.

[153] Doi T, Yoh K, Shitara K, Takahashi H, Ueno M, Kobayashi S, Morimoto M, Okusaka T, Ueno H, Morizane C, Okano N, Nagashima F, Furuse J. First-in-human phase 1 study of novel dUTPase inhibitor TAS-114 in combination with S-1 in Japanese patients with advanced solid tumors. Invest New Drugs 2019;37:507–18.

[154] Yamamoto N, Hayashi H, Planchard D, Morán T, Gregorc V, Dowell J, Sakai H, Yoh K, Nishio M, Cortot AB, Benhadji KA, Soni N, Huang J, Makris L, Cedres S. A randomized, phase 2 study of deoxyuridine triphosphatase inhibitor, TAS-114, in combination with S-1 versus S-1 alone in patients with advanced non-small-cell lung cancer. Invest New Drugs 2020;38:1588–97.

[155] Kawazoe A, Takahari D, Keisho C, Nakamura Y, Ikeno T, Wakabayashi M, Nomura S, Tamura H, Fukutani M, Hirano N, Saito Y, Kambe M, Sato A, Shitara K. A multicenter phase II study of TAS-114 in combination with S-1 in patients with pretreated advanced gastric cancer (EPOC1604). Gastric Cancer 2021;24:190–6.

[156] Argilés G, Tabernero J, Labianca R, Hochhauser D, Salazar R, Iveson T, Laurent-Puig P, Quirke P, Yoshino T, Taieb J, Martinelli E, Arnold D, ESMO Guidelines Committee. Localised colon cancer: ESMO clinical practice guidelines for diagnosis, treatment and follow-up. Ann Oncol 2020;31:1291–305.
[157] Isella C, Brundu F, Bellomo SE, Galimi F, Zanella E, Porporato R, Petti C, Fiori A, Orzan F, Senetta R, Boccaccio C, Ficarra E, Marchionni L, Trusolino L, Medico E, Bertotti A. Selective analysis of cancer-cell intrinsic transcriptional traits defines novel clinically relevant subtypes of colorectal cancer. Nat Commun 2017;8:15107.
[158] Budinska E, Popovici V, Tejpar S, D'Ario G, Lapique N, Sikora KO, Di Narzo AF, Yan P, Hodgson JG, Weinrich S, Bosman F, Roth A, Delorenzi M. Gene expression patterns unveil a new level of molecular heterogeneity in colorectal cancer. J Pathol 2013;231:63–76.
[159] Marisa L, de Reyniès A, Duval A, Selves J, Gaub MP, Vescovo L, Etienne-Grimaldi MC, Schiappa R, Guenot D, Ayadi M, Kirzin S, Chazal M, Fléjou JF, Benchimol D, Berger A, Lagarde A, Pencreach E, Piard F, Elias D, Parc Y, Olschwang S, Milano G, Laurent-Puig P, Boige V. Gene expression classification of colon cancer into molecular subtypes: characterization, validation, and prognostic value. PLoS Med 2013;10, e1001453.
[160] Roepman P, Schlicker A, Tabernero J, Majewski I, Tian S, Moreno V, Snel MH, Chresta CM, Rosenberg R, Nitsche U, Macarulla T, Capella G, Salazar R, Orphanides G, Wessels LF, Bernards R, Simon IM. Colorectal cancer intrinsic subtypes predict chemotherapy benefit, deficient mismatch repair and epithelial- to-mesenchymal transition. Int J Cancer 2014;134:552–62.
[161] De Sousa EMF, Wang X, Jansen M, Fessler E, Trinh A, de Rooij LP, de Jong JH, de Boer OJ, van Leersum R, Bijlsma MF, Rodermond H, van der Heijden M, van Noesel CJ, Tuynman JB, Dekker E, Markowetz F, Medema JP, Vermeulen L. Poor- prognosis colon cancer is defined by a molecularly distinct subtype and develops from serrated precursor lesions. Nat Med 2013;19:614–8.
[162] Sadanandam A, Lyssiotis CA, Homicsko K, Collisson EA, Gibb WJ, Wullschleger S, Ostos LC, Lannon WA, Grotzinger C, Del Rio M, Lhermitte B, Olshen AB, Wiedenmann B, Cantley LC, Gray JW, Hanahan D. A colorectal cancer classification system that associates cellular phenotype and responses to therapy. Nat Med 2013;19:619–25.
[163] Schlicker A, Beran G, Chresta CM, McWalter G, Pritchard A, Weston S, Runswick S, Davenport S, Heathcote K, Castro DA, Orphanides G, French T, Wessels LF. Subtypes of primary colorectal tumors correlate with response to targeted treatment in colorectal cell lines. BMC Med Genomics 2012;5:66.
[164] Allen WL, Dunne PD, McDade S, Scanlon E, Loughrey M, Coleman H, McCann C, McLaughlin K, Nemeth Z, Syed N, Jithesh P, Arthur K, Wilson R, Coyle V, McArt D, Murray GI, Samuel L, Nuciforo P, Jimenez J, Argiles G, Dienstmann R, Tabernero J, Messerini L, Nobili S, Mini E, Sheahan K, Ryan E, Johnston PG, Van Schaeybroeck S, Lawler M, Longley DB. Transcriptional subtyping and CD8 immunohistochemistry identifies poor prognosis stage II/III colorectal cancer patients who benefit from adjuvant chemotherapy. JCO Precis Oncol 2018;2018, PO.17.00241.
[165] Okita A, Takahashi S, Ouchi K, Inoue M, Watanabe M, Endo M, Honda H, Yamada Y, Ishioka C. Consensus molecular subtypes classification of colorectal cancer as a predictive factor for chemotherapeutic efficacy against metastatic colorectal cancer. Oncotarget 2018;9:18698–711.
[166] Mooi JK, Wirapati P, Asher R, Lee CK, Savas P, Price TJ, Townsend A, Hardingham J, Buchanan D, Williams D, Tejpar S, Mariadason JM, Tebbutt NC. The prognostic impact of consensus molecular subtypes (CMS) and its predictive effects for bevacizumab benefit in metastatic colorectal cancer: molecular analysis of the AGITG MAX clinical trial. Ann Oncol 2018;29:2240–6.
[167] Stintzing S, Wirapati P, Lenz HJ, Neureiter D, Fischer von Weikersthal L, Decker T, Kiani A, Kaiser F, Al-Batran S, Heintges T, Lerchenmüller C, Kahl C, Seipelt G, Kullmann F, Moehler M, Scheithauer W, Held S, Modest DP, Jung A, Kirchner T, Aderka D, Tejpar S, Heinemann V. Consensus molecular subgroups (CMS) of colorectal cancer (CRC) and first-line efficacy of FOLFIRI plus cetuximab or bevacizumab in the FIRE3 (AIO KRK-0306) trial. Ann Oncol 2019;30:1796–803.
[168] Mini E, Lapucci A, Perrone G, D'Aurizio R, Napoli C, Brugia M, Landini I, Tassi R, Picariello L, Simi L, Mancini I, Messerini L, Magi A, Pinzani P, Mazzei T, Tonelli F, Nobili S. RNA sequencing reveals PNN and KCNQ1OT1 as predictive biomarkers of clinical outcome in stage III colorectal cancer patients treated with adjuvant chemotherapy. Int J Cancer 2019;145:2580–93.
[169] Wei Z, Ma W, Qi X, Zhu X, Wang Y, Xu Z, Luo J, Wang D, Guo W, Li X, Xin S, Yu J, Li G. Pinin facilitated proliferation and metastasis of colorectal cancer through activating EGFR/ERK signaling pathway. Oncotarget 2016;7:29429–39.

[170] Li Y, Li C, Li D, Yang L, Jin J, Zhang B. lncRNA KCNQ1OT1 enhances the chemoresistance of oxaliplatin in colon cancer by targeting the miR-34a/ATG4B pathway. Onco Targets Ther 2019;12:2649–60.

[171] Lin ZB, Long P, Zhao Z, Zhang YR, Chu XD, Zhao XX, Ding H, Huan SW, Pan YL, Pan JH. Long noncoding RNA KCNQ1OT1 is a prognostic biomarker and mediates CD8+ T cell exhaustion by regulating CD155 expression in colorectal cancer. Int J Biol Sci 2021;17:1757–68.

[172] Lapucci A, Perrone G, Di Paolo A, Napoli C, Landini I, Roviello G, Calosi L, Naccarato AG, Falcone A, Bani D, Mini E, Nobili S. *PNN* and *KCNQ1OT1* can predict the efficacy of adjuvant fluoropyrimidine-based chemotherapy in colorectal cancer patients. Oncol Res 2021;16(28):631–44.

[173] Cushman SM, Jiang C, Hatch AJ, Shterev I, Sibley AB, Niedzwiecki D, Venook AP, Owzar K, Hurwitz HI, Nixon AB. Gene expression markers of efficacy and resistance to cetuximab treatment in metastatic colorectal cancer: results from CALGB 80203 (Alliance). Clin Cancer Res 2015;21:1078–86.

[174] Crea F, Nobili S, Paolicchi E, Perrone G, Napoli C, Landini I, Danesi R, Mini E. Epigenetics and chemoresistance in colorectal cancer: an opportunity for treatment tailoring and novel therapeutic strategies. Drug Resist Updat 2011;14:280–96.

[175] Li GH, Qu Q, Qi TT, Teng XQ, Zhu HH, Wang JJ, Lu Q, Qu J. Super-enhancers: a new frontier for epigenetic modifiers in cancer chemoresistance. J Exp Clin Cancer Res 2021;40:174.

[176] Weisenberger DJ, Liang G, Lenz HJ. DNA methylation aberrancies delineate clinically distinct subsets of colorectal cancer and provide novel targets for epigenetic therapies. Oncogene 2018;37:566–77.

[177] Barchitta M, Maugeri A, Li Destri G, Basile G, Agodi A. Epigenetic biomarkers in colorectal cancer patients receiving adjuvant or neoadjuvant therapy: a systematic review of epidemiological studies. Int J Mol Sci 2019;20:3842.

[178] Fouad MA, Salem SE, Hussein MM, Zekri ARN, Hafez HF, El Desouky ED, Shouman SA. Impact of global DNA methylation in treatment outcome of colorectal cancer patients. Front Pharmacol 2018;9:1173.

[179] Ma Z, Williams M, Cheng YY, Leung WK. Roles of methylated DNA biomarkers in patients with colorectal cancer. Dis Markers 2019;2019:2673543.

[180] Herbst A, Vdovin N, Gacesa S, Ofner A, Philipp A, Nagel D, Holdt LM, Op den Winkel M, Heinemann V, Stieber P, Graeven U, Reinacher-Schick A, Arnold D, Ricard I, Mansmann U, Hegewisch-Becker S, Kolligs FT. Methylated free- circulating HPP1 DNA is an early response marker in patients with metastatic colorectal cancer. Int J Cancer 2017;140:2134–44.

[181] Bhangu JS, Beer A, Mittlböck M, Tamandl D, Pulverer W, Schönthaler S, Taghizadeh H, Stremitzer S, Kaczirek K, Gruenberger T, Gnant M, Bergmann M, Mannhalter C, Weinhäusel A, Oehler R, Bachleitner-Hofmann T. Circulating free methylated tumor DNA markers for sensitive assessment of tumor burden and early response monitoring in patients receiving systemic chemotherapy for colorectal cancer liver metastasis. Ann Surg 2018;268:894–902.

[182] Barault L, Amatu A, Siravegna G, Ponzetti A, Moran S, Cassingena A, Mussolin B, Falcomatà C, Binder AM, Cristiano C, Oddo D, Guarrera S, Cancelliere C, Bustreo S, Bencardino K, Maden S, Vanzati A, Zavattari P, Matullo G, Truini M, Grady WM, Racca P, Michels KB, Siena S, Esteller M, Bardelli A, Sartore-Bianchi A, Di Nicolantonio F. Discovery of methylated circulating DNA biomarkers for comprehensive non-invasive monitoring of treatment response in metastatic colorectal cancer. Gut 2018;67:1995–2005.

[183] Reece M, Saluja H, Hollington P, Karapetis CS, Vatandoust S, Young GP, Symonds EL. The use of circulating tumor DNA to monitor and predict response to treatment in colorectal cancer. Front Genet 2019;10:1118.

[184] Qin J, Wen B, Liang Y, Yu W, Li H. Histone modifications and their role in colorectal cancer (review). Pathol Oncol Res 2020;26:2023–33.

[185] Karczmarski J, Rubel T, Paziewska A, Mikula M, Bujko M, Kober P, Dadlez M, Ostrowski J. Histone H3 lysine 27 acetylation is altered in colon cancer. Clin Proteomics 2014;11:24.

[186] Wang Q, Chen X, Jiang Y, Liu S, Liu H, Sun X, Zhang H, Liu Z, Tao Y, Li C, Hu Y, Liu D, Ye D, Liu Y, Wang M, Zhang X. Elevating H3K27me3 level sensitizes colorectal cancer to oxaliplatin. J Mol Cell Biol 2020;12:125–37.

[187] McAnena P, Brown JA, Kerin MJ. Circulating nucleosomes and nucleosome modifications as biomarkers in cancer. Cancers (Basel) 2017;9:5.

[188] Fahmueller YN, Nagel D, Hoffmann RT, Tatsch K, Jakobs T, Stieber P, Holdenrieder S. Predictive and prognostic value of circulating nucleosomes and serum biomarkers in patients with metastasized colorectal cancer undergoing selective internal radiation therapy. BMC Cancer 2012;12:5.

[189] Van den Ackerveken P, Lobbens A, Turatsinze JV, Solis-Mezarino V, Völker-Albert M, Imhof A, Herzog M. A novel proteomics approach to epigenetic profiling of circulating nucleosomes. Sci Rep 2021;11:7256.

[190] GENECODE, https://www.gencodegenes.org/.
[191] Harrow J, Frankish A, Gonzalez JM, Tapanari E, Diekhans M, Kokocinski F, Aken BL, Barrell D, Zadissa A, Searle S, Barnes I, Bignell A, Boychenko V, Hunt T, Kay M, Mukherjee G, Rajan J, Despacio-Reyes G, Saunders G, Steward C, Harte R, Lin M, Howald C, Tanzer A, Derrien T, Chrast J, Walters N, Balasubramanian S, Pei B, Tress M, Rodriguez JM, Ezkurdia I, van Baren J, Brent M, Haussler D, Kellis M, V-alencia A, Reymond A, Gerstein M, Guigó R, Hubbard TJ. GENCODE: the reference human genome annotation for The ENCODE Project. Genome Res 2012;22:1760–74.
[192] Wei L, Wang X, Lv L, Zheng Y, Zhang N, Yang M. The emerging role of noncoding RNAs in colorectal cancer chemoresistance. Cell Oncol (Dordr) 2019;42:757–68.
[193] Mattick JS. Non-coding RNAs: the architects of eukaryotic complexity. EMBO Rep 2001;2:986–91.
[194] Drusco A, Croce CM. MicroRNAs and cancer: a long story for short RNAs. Adv Cancer Res 2017;135:1–24.
[195] Romano G, Veneziano D, Acunzo M, Croce CM. Small non-coding RNA and cancer. Carcinogenesis 2017;38:485–91.
[196] Babashah S. In: Sadegh, editor. MicroRNAs: key regulators of oncogenesis. Switzerland: Springer International Publishing; 2014.
[197] Babashah S, Soleimani M. The oncogenic and tumour suppressive roles of microRNAs in cancer and apoptosis. Eur J Cancer 2011;47:1127–37.
[198] Vaghari-Tabari M, Majidinia M, Moein S, Qujeq D, Asemi Z, Alemi F, Mohamadzadeh R, Targhazeh N, Safa A, Yousefi B. MicroRNAs and colorectal cancer chemoresistance: new solution for old problem. Life Sci 2020;259:118255.
[199] Hansen TF, Carlsen AL, Heegaard NH, Sørensen FB, Jakobsen A. Changes in circulating microRNA-126 during treatment with chemotherapy and bevacizumab predicts treatment response in patients with metastatic colorectal cancer. Br J Cancer 2015;112:624–9.
[200] Zhou H, Lin C, Zhang Y, Zhang X, Zhang C, Zhang P, Xie X, Ren Z. miR-506 enhances the sensitivity of human colorectal cancer cells to oxaliplatin by suppressing MDR1/P-gp expression. Cell Prolif 2017;50, e12341.
[201] Rasmussen MH, Jensen NF, Tarpgaard LS, Qvortrup C, Rømer MU, Stenvang J, Hansen TP, Christensen LL, Lindebjerg J, Hansen F, Jensen BV, Hansen TF, Pfeiffer P, Brünner N, Ørntoft TF, Andersen CL. High expression of microRNA-625-3p is associated with poor response to first-line oxaliplatin based treatment of metastatic colorectal cancer. Mol Oncol 2013;7:637–46.
[202] Hu J, Xu Y, Cai S. Specific microRNAs as novel biomarkers for combination chemotherapy resistance detection of colon adenocarcinoma. Eur J Med Res 2015;20:95.
[203] Mekenkamp LJ, Tol J, Dijkstra JR, de Krijger I, Vink-Börger ME, van Vliet S, Teerenstra S, Kamping E, Verwiel E, Koopman M, Meijer GA, van Krieken JH, Kuiper R, Punt CJ, Nagtegaal ID. Beyond KRAS mutation status: influence of KRAS copy number status and microRNAs on clinical outcome to cetuximab in metastatic colorectal cancer patients. BMC Cancer 2012;12:292.
[204] Boisen MK, Dehlendorff C, Linnemann D, Nielsen BS, Larsen JS, Osterlind K, Nielsen SE, Tarpgaard LS, Qvortrup C, Pfeiffer P, Holländer NH, Keldsen N, Hansen TF, Jensen BB, Høgdall EV, Jensen BV, Johansen JS. Tissue microRNAs as predictors of outcome in patients with metastatic colorectal cancer treated with first line capecitabine and oxaliplatin with or without bevacizumab. PLoS One 2014;9, e109430.
[205] Kiss I, Mlčochová J, Součková K, Fabian P, Poprach A, Halamkova J, Svoboda M, Vyzula R, Slaby O. MicroRNAs as outcome predictors in patients with metastatic colorectal cancer treated with bevacizumab in combination with FOLFOX. Oncol Lett 2017;14:743–50.
[206] Caramés C, Cristóbal I, Moreno V, del Puerto L, Moreno I, Rodriguez M, Marín JP, Correa AV, Hernández R, Zenzola V, Hernández T, León A, Martín JI, Sánchez-Fayos P, García-Olmo D, Rojo F, Goel A, Fernandez-Aceñero MJ, García-Foncillas J. MicroRNA-21 predicts response to preoperative chemoradiotherapy in locally advanced rectal cancer. Int J Colorectal Dis 2015;30:899–906.
[207] Schetter AJ, Leung SY, Sohn JJ, Zanetti KA, Bowman ED, Yanaihara N, Yuen ST, Chan TL, Kwong DL, Au GK, Liu CG, Calin GA, Croce CM, Harris CC. MicroRNA expression profiles associated with prognosis and therapeutic outcome in colon adenocarcinoma. JAMA 2008;299:425–36.
[208] Takahashi M, Cuatrecasas M, Balaguer F, Hur K, Toiyama Y, Castells A, Boland CR, Goel A. The clinical significance of MiR-148a as a predictive biomarker in patients with advanced colorectal cancer. PLoS One 2012; 7, e46684.
[209] Ma Y, Zhang P, Wang F, Zhang H, Yang J, Peng J, Liu W, Qin H. miR-150 as a potential biomarker associated with prognosis and therapeutic outcome in colorectal cancer. Gut 2012;61:1447–53.

[210] Lopes-Ramos CM, Habr-Gama A, Quevedo Bde S, Felício NM, Bettoni F, Koyama FC, Asprino PF, Galante PA, Gama-Rodrigues J, Camargo AA, Perez RO, Parmigiani RB. Overexpression of miR-21-5p as a predictive marker for complete tumor regression to neoadjuvant chemoradiotherapy in rectal cancer patients. BMC Med Genomics 2014;7:68.

[211] Svoboda M, Sana J, Fabian P, Kocakova I, Gombosova J, Nekvindova J, Radova L, Vyzula R, Slaby O. MicroRNA expression profile associated with response to neoadjuvant chemoradiotherapy in locally advanced rectal cancer patients. Radiat Oncol 2012;7:195.

[212] Perez-Carbonell L, Sinicrope FA, Alberts SR, Oberg AL, Balaguer F, Castells A, Boland CR, Goel A. MiR-320e is a novel prognostic biomarker in colorectal cancer. Br J Cancer 2015;113:83–90.

[213] Mosakhani N, Lahti L, Borze I, Karjalainen-Lindsberg ML, Sundström J, Ristamäki R, Osterlund P, Knuutila S, Sarhadi VK. MicroRNA profiling predicts survival in anti-EGFR treated chemorefractory metastatic colorectal cancer patients with wild-type KRAS and BRAF. Cancer Genet 2012;205:545–51.

[214] Igarashi H, Kurihara H, Mitsuhashi K, Ito M, Okuda H, Kanno S, Naito T, Yoshii S, Takahashi H, Kusumi T, Hasegawa T, Sukawa Y, Adachi Y, Okita K, Hirata K, Imamura Y, Baba Y, Imai K, Suzuki H, Yamamoto H, Nosho K, Shinomura Y. Association of MicroRNA-31-5p with clinical efficacy of anti-EGFR therapy in patients with metastatic colorectal cancer. Ann Surg Oncol 2015;22:2640–8.

[215] Pichler M, Winter E, Ress AL, Bauernhofer T, Gerger A, Kiesslich T, Lax S, Samonigg H, Hoefler G. miR-181a is associated with poor clinical outcome in patients with colorectal cancer treated with EGFR inhibitor. J Clin Pathol 2014;67:198–203.

[216] Suto T, Yokobori T, Yajima R, Morita H, Fujii T, Yamaguchi S, Altan B, Tsutsumi S, Asao T, Kuwano H. MicroRNA-7 expression in colorectal cancer is associated with poor prognosis and regulates cetuximab sensitivity via EGFR regulation. Carcinogenesis 2015;36:338–45.

[217] Ragusa M, Majorana A, Statello L, Maugeri M, Salito L, Barbagallo D, Guglielmino MR, Duro LR, Angelica R, Caltabiano R, Biondi A, Di Vita M, Privitera G, Scalia M, Cappellani A, Vasquez E, Lanzafame S, Basile F, Di Pietro C, Purrello M. Specific alterations of microRNA transcriptome and global network structure in colorectal carcinoma after cetuximab treatment. Mol Cancer Ther 2010;9:3396–409.

[218] Jung E, Choi J, Kim JS, Han TS. MicroRNA-based therapeutics for drug-resistant colorectal cancer. Pharmaceuticals (Basel) 2021;14:136.

[219] Kjersem JB, Ikdahl T, Lingjaerde OC, Guren T, Tveit KM, Kure EH. Plasma microRNAs predicting clinical outcome in metastatic colorectal cancer patients receiving first-line oxaliplatin-based treatment. Mol Oncol 2014;8:59–67.

[220] Zhang J, Zhang K, Bi M, Jiao X, Zhang D, Dong Q. Circulating microRNA expressions in colorectal cancer as predictors of response to chemotherapy. Anticancer Drugs 2014;25:346–52.

[221] Chen J, Wang W, Zhang Y, Chen Y, Hu T. Predicting distant metastasis and chemoresistance using plasma miRNAs. Med Oncol 2014;31:799.

[222] Schou JV, Rossi S, Jensen BV, Nielsen DL, Pfeiffer P, Høgdall E, Yilmaz M, Tejpar S, Delorenzi M, Kruhøffer M, Johansen JS. miR-345 in metastatic colorectal cancer: a non-invasive biomarker for clinical outcome in non-KRAS mutant patients treated with 3rd line cetuximab and irinotecan. PLoS One 2014;9, e99886.

[223] Chen Q, Xia HW, Ge XJ, Zhang YC, Tang QL, Bi F. Serum miR-19a predicts resistance to FOLFOX chemotherapy in advanced colorectal cancer cases. Asian Pac J Cancer Prev 2013;14:7421–6.

[224] Ning T, Li J, He Y, Zhang H, Wang X, Deng T, Liu R, Li H, Bai M, Fan Q, Zhu K, Ying G, Ba Y. Exosomal miR-208b related with oxaliplatin resistance promotes Treg expansion in colorectal cancer. Mol Ther 2021;29:2723–36.

[225] Han J, Sun W, Liu R, Zhou Z, Zhang H, Chen X, Ba Y. Plasma exosomal mirna expression profile as oxaliplatin-based chemoresistant biomarkers in colorectal adenocarcinoma. Front Oncol 2020;10:1495.

[226] Madurantakam Royam M, Kumarasamy C, Baxi S, Gupta A, Ramesh N, Kodiveri Muthukaliannan G, Jayaraj R. Current evidence on miRNAs as potential theranostic markers for detecting chemoresistance in colorectal cancer: a systematic review and meta-analysis of preclinical and clinical studies. Mol Diagn Ther 2019;23:65–82.

[227] Conde E, Earl J, Crespo-Toro L, Blanco-Agudo C, Ramos-Muñoz E, Rodríguez-Serrano EM, Martínez Ávila JC, Salinas-Muñoz L, Serrano-Huertas S, Ferreiro R, Rodriguez-Garrote M, Sainz Jr B, Massuti B, Alfonso PG, Benavides M, Aranda E, García-Bermejo ML, Carrato A. Biomarkers associated with regorafenib first-line treatment benefits in metastatic colorectal cancer patients: REFRAME molecular study. Cancers (Basel) 2021;13:1710.

[228] Wang KC, Chang HY. Molecular mechanisms of long noncoding RNAs. Mol Cell 2011;43:904–14.
[229] Paraskevopoulou MD, Hatzigeorgiou AG. Analyzing MiRNA-LncRNA interactions. Methods Mol Biol 2016;1402:271–86.
[230] Bergmann JH, Spector DL. Long non-coding RNAs: modulators of nuclear structure and function. Curr Opin Cell Biol 2014;26:10–8.
[231] Zhang Y, Liao G, Bai J, Zhang X, Xu L, Deng C, Yan M, Xie A, Luo T, Long Z, Xiao Y, Li X. Identifying cancer driver lncRNAs bridged by functional effectors through integrating multi-omics data in human cancers. Mol Ther Nucleic Acids 2019;17:362–73.
[232] Guzel E, Okyay TM, Yalcinkaya B, Karacaoglu S, Gocmen M, Akcakuyu MH. Tumor suppressor and oncogenic role of long non-coding RNAs in cancer. North Clin Istanb 2019;7:81–6.
[233] Chen S, Shen X. Long noncoding RNAs: functions and mechanisms in colon cancer. Mol Cancer 2020;19:167.
[234] Forrest ME, Saiakhova A, Beard L, Buchner DA, Scacheri PC, LaFramboise T, Markowitz S, Khalil AM. Colon cancer-upregulated long non-coding RNA lincDUSP regulates cell cycle genes and potentiates resistance to apoptosis. Sci Rep 2018;8:7324.
[235] Zinovieva OL, Grineva EN, Prokofjeva MM, Karpov DS, Zheltukhin AO, Krasnov GS, Snezhkina AV, Kudryavtseva AV, Chumakov PM, Mashkova TD, Prassolov VS, Lisitsyn NA. Expression of long non-coding RNA LINC00973 is consistently increased upon treatment of colon cancer cells with different chemotherapeutic drugs. Biochimie 2018;151:67–72.
[236] Yang C, Pan Y, Deng SP. Downregulation of lncRNA CCAT1 enhances 5-fluorouracil sensitivity in human colon cancer cells. BMC Mol Cell Biol 2019;20:9.
[237] Xian D, Zhao Y. LncRNA KCNQ1OT1 enhanced the methotrexate resistance of colorectal cancer cells by regulating miR-760/PPP1R1B via the cAMP signalling pathway. J Cell Mol Med 2019;23:3808–23.
[238] Zhang X, Wen L, Chen S, Zhang J, Ma Y, Hu J, Yue T, Wang J, Zhu J, Bu D, Wang X. The novel long noncoding RNA CRART16 confers cetuximab resistance in colorectal cancer cells by enhancing ERBB3 expression via miR-371a-5p. Cancer Cell Int 2020;20:68.
[239] Qi FF, Yang Y, Zhang H, Chen H. Long non-coding RNAs: key regulators in oxaliplatin resistance of colorectal cancer. Biomed Pharmacother 2020;128:110329.
[240] Chen H, Xu J, Hong J, Tang R, Zhang X, Fang JY. Long noncoding RNA profiles identify five distinct molecular subtypes of colorectal cancer with clinical relevance. Mol Oncol 2014;8:1393–403.
[241] Schwarzmueller L, Bril O, Vermeulen L, Léveillé N. Emerging role and therapeutic potential of lncRNAs in colorectal cancer. Cancers (Basel) 2020;12:3843.
[242] Xiong XD, Ren X, Cai MY, Yang JW, Liu X, Yang JM. Long non-coding RNAs: an emerging powerhouse in the battle between life and death of tumor cells. Drug Resist Updat 2016;26:28–42.
[243] Ma H, Chen G, Guo M. Mass spectrometry based translational proteomics for biomarker discovery and application in colorectal cancer. Proteomics Clin Appl 2016;10:503–15.
[244] Abyadeh M, Meyfour A, Gupta V, Zabet Moghaddam M, Fitzhenry MJ, Shahbazian S, Hosseini Salekdeh G, Mirzaei M. Recent advances of functional proteomics in gastrointestinal cancers – a path towards the identification of candidate diagnostic, prognostic, and therapeutic molecular biomarkers. Int J Mol Sci 2020;21:8532.
[245] Fayazfar S, Zali H, Arefi Oskouie A, Asadzadeh Aghdaei H, Rezaei Tavirani M, Nazemalhosseini Mojarad E. Early diagnosis of colorectal cancer via plasma proteomic analysis of CRC and advanced adenomatous polyp. Gastroenterol Hepatol Bed Bench 2019;12:328–39.
[246] Martin P, Noonan S, Mullen MP, Scaife C, Tosetto M, Nolan B, Wynne K, Hyland J, Sheahan K, Elia G, O'Donoghue D, Fennelly D, O'Sullivan J. Predicting response to vascular endothelial growth factor inhibitor and chemotherapy in metastatic colorectal cancer. BMC Cancer 2014;14:887.
[247] Katsila T, Juliachs M, Gregori J, Macarulla T, Villarreal L, Bardelli A, Torrance C, Elez E, Tabernero J, Villanueva J. Circulating pEGFR is a candidate response biomarker of cetuximab therapy in colorectal cancer. Clin Cancer Res 2014 Dec;20:6346–56.
[248] Battaglin F, Naseem M, Lenz HJ, Salem ME. Microsatellite instability in colorectal cancer: overview of its clinical significance and novel perspectives. Clin Adv Hematol Oncol 2018;16:735–45.
[249] Gao G, Liu S, Yao Z, Zhan Y, Chen W, Liu Y. The prognostic significance of Hsp70 in patients with colorectal cancer patients: a PRISMA-compliant meta-analysis. Biomed Res Int 2021;2021:5526327.
[250] Oh HJ, Kim JH, Lee TH, Park HE, Bae JM, Lee HS, Kang GH. Dominant high expression of wild-type HSP110 defines a poor prognostic subgroup of colorectal carcinomas with microsatellite instability: a whole-section immunohistochemical analysis. APMIS 2017;125:1076–83.

[251] Yang Q, Bavi P, Wang JY, Roehrl MH. Immuno-proteomic discovery of tumor tissue autoantigens identifies olfactomedin 4, CD11b, and integrin alpha-2 as markers of colorectal cancer with liver metastases. J Proteomics 2017;168:53–65.

[252] Snoeren N, Emmink BL, Koerkamp MJ, van Hooff SR, Goos JA, van Houdt WJ, de Wit M, Prins AM, Piersma SR, Pham TV, Belt EJ, Bril H, Stockmann HB, Meijer GA, van Hillegersberg R, Holstege FC, Jimenez CR, Fijneman RJ, Kranenburg OW, Rinkes IH. Maspin is a marker for early recurrence in primary stage III and IV colorectal cancer. Br J Cancer 2013;109:1636–47.

[253] Surinova S, Radová L, Choi M, Srovnal J, Brenner H, Vitek O, Hajdúch M, Aebersold R. Non-invasive prognostic protein biomarker signatures associated with colorectal cancer. EMBO Mol Med 2015;7:1153–65.

[254] Marfà S, Marti J, Reyes A, Casals G, Fernández-Varo G, Carvajal S, García-Valdecasas JC, Fuster J, Jiménez W. Metastatic tissue proteomic profiling predicts 5-year outcomes in patients with colorectal liver metastases. Transl Oncol 2016;9:445–52.

[255] Kirana C, Peng L, Miller R, Keating JP, Glenn C, Shi H, Jordan TW, Maddern GJ, Stubbs RS. Combination of laser microdissection, 2D-DIGE and MALDI-TOF MS to identify protein biomarkers to predict colorectal cancer spread. Clin Proteomics 2019;16:3.

[256] Lenz HJ, Argiles G, Yoshino T, Tejpar S, Ciardiello F, Braunger J, Salnikov AV, Gabrielyan O, Schmid R, Höfler J, Kitzing T, Van Cutsem E. Association of consensus molecular subtypes and molecular markers with clinical outcomes in patients with metastatic colorectal cancer: biomarker analyses from LUME-Colon 1. Clin Colorectal Cancer 2021;20:84–95.

[257] Clarke SJ, Burge M, Feeney K, Gibbs P, Jones K, Marx G, Molloy MP, Price T, Reece WHH, Segelov E, Tebbutt NC. The prognostic role of inflammatory markers in patients with metastatic colorectal cancer treated with bevacizumab: a translational study [ASCENT]. PLoS One 2020;15, e0229900.

[258] Chauvin A, Boisvert FM. Clinical proteomics in colorectal cancer, a promising tool for improving personalised medicine. Proteomes 2018;6:49.

[259] Wishart DS. Emerging applications of metabolomics in drug discovery and precision medicine. Nat Rev Drug Discov 2016;15:473–84.

[260] Mussap M, Loddo C, Fanni C, Fanos V. Metabolomics in pharmacology—a delve into the novel field of pharmacometabolomics. Expert Rev Clin Pharmacol 2020;13:115–34.

[261] Jiménez B, Mirnezami R, Kinross J, Cloarec O, Keun HC, Holmes E, Goldin RD, Ziprin P, Darzi A, Nicholson JK. 1H HR-MAS NMR spectroscopy of tumor-induced local metabolic "field-effects" enables colorectal cancer staging and prognostication. J Proteome Res 2013;12:959–68.

[262] Cross AJ, Moore SC, Boca S, Huang WY, Xiong X, Stolzenberg-Solomon R, Sinha R, Sampson JN. A prospective study of serum metabolites and colorectal cancer risk. Cancer 2014;120:3049–57.

[263] Mirnezami R, Jiménez B, Li JV, Kinross JM, Veselkov K, Goldin RD, Holmes E, Nicholson JK, Darzi A. Rapid diagnosis and staging of colorectal cancer via high-resolution magic angle spinning nuclear magnetic resonance (HR-MAS NMR) spectroscopy of intact tissue biopsies. Ann Surg 2014;259:1138–49.

[264] Qiu Y, Cai G, Zhou B, Li D, Zhao A, Xie G, Li H, Cai S, Xie D, Huang C, Ge W, Zhou Z, Xu LX, Jia W, Zheng S, Yen Y, Jia W. A distinct metabolic signature of human colorectal cancer with prognostic potential. Clin Cancer Res 2014;20:2136–46.

[265] Farshidfar F, Weljie AM, Kopciuk KA, Hilsden R, McGregor SE, Buie WD, MacLean A, Vogel HJ, Bathe OF. A validated metabolomic signature for colorectal cancer: exploration of the clinical value of metabolomics. Br J Cancer 2016;115:848–57.

[266] Martín-Blázquez A, Díaz C, González-Flores E, Franco-Rivas D, Jiménez-Luna C, Melguizo C, Prados J, Genilloud O, Vicente F, Caba O, Pérez Del Palacio J. Untargeted LC-HRMS-based metabolomics to identify novel biomarkers of metastatic colorectal cancer. Sci Rep 2019;9:20198.

[267] Long Z, Zhou J, Xie K, Wu Z, Yin H, Daria V, Tian J, Zhang N, Li L, Zhao Y, Wang F, Wang M, Cui Y. Metabolomic markers of colorectal tumor with different clinicopathological features. Front Oncol 2020; 10:981.

[268] Cai Y, Rattray NJW, Zhang Q, Mironova V, Santos-Neto A, Muca E, Vollmar AKR, Hsu KS, Rattray Z, Cross JR, Zhang Y, Paty PB, Khan SA, Johnson CH. Tumor tissue-specific biomarkers of colorectal cancer by anatomic location and stage. Metabolites 2020;10:257.

[269] Erben V, Bhardwaj M, Schrotz-King P, Brenner H. Metabolomics biomarkers for detection of colorectal neoplasms: a systematic review. Cancers (Basel) 2018;10:246.

[270] Beger RD, Dunn W, Schmidt MA, Gross SS, Kirwan JA, Cascante M, Brennan L, Wishart DS, Oresic M, Hankemeier T, Broadhurst DI, Lane AN, Suhre K, Kastenmüller G, Sumner SJ, Thiele I, Fiehn O, Kaddurah-Daouk R, for "Precision Medicine and Pharmacometabolomics Task Group"-Metabolomics Society Initiative.

Metabolomics enables precision medicine: "a white paper, community perspective". Metabolomics 2016; 12:149.
[271] Kaddurah-Daouk R, Weinshilboum RM, Pharmacometabolomics Research Network. Pharmacometabolomics: implications for clinical pharmacology and systems pharmacology. Clin Pharmacol Ther 2014;95:154–67.
[272] Backshall A, Sharma R, Clarke SJ, Keun HC. Pharmacometabonomic profiling as a predictor of toxicity in patients with inoperable colorectal cancer treated with capecitabine. Clin Cancer Res 2011;17:3019–28.
[273] Dykstra MA, Switzer N, Eisner R, Tso V, Foshaug R, Ismond K, Fedorak R, Wang H. Urine metabolomics as a predictor of patient tolerance and response to adjuvant chemotherapy in colorectal cancer. Mol Clin Oncol 2017;7:767–70.
[274] Iemoto T, Nishiumi S, Kobayashi T, Fujigaki S, Hamaguchi T, Kato K, Shoji H, Matsumura Y, Honda K, Yoshida M. Serum level of octanoic acid predicts the efficacy of chemotherapy for colorectal cancer. Oncol Lett 2019;17:831–42.
[275] Sottoriva A, Kang H, Ma Z, Graham TA, Salomon MP, Zhao J, Marjoram P, Siegmund K, Press MF, Shibata D, Curtis C. A big Bang model of human colorectal tumor growth. Nat Genet 2015;47:209–16.
[276] Taieb J, Jung A, Sartore-Bianchi A, Peeters M, Seligmann J, Zaanan A, Burdon P, Montagut C, Laurent-Puig P. The evolving biomarker landscape for treatment selection in metastatic colorectal cancer. Drugs 2019;79:1375–94.
[277] Sinicrope FA, Chakrabarti S, Laurent-Puig P, Huebner L, Smyrk TC, Tabernero J, Mini E, Goldberg RM, Zaanan A, Folprecht G, Van Laethem JL, Le Malicot K, Shi Q, Alberts SR, Taieb J. Prognostic variables in low and high risk stage III colon cancers treated in two adjuvant chemotherapy trials. Eur J Cancer 2021;144:101–12.
[278] Allgayer H, Leupold JH, Patil N. Defining the "metastasome": perspectives from the genome and molecular landscape in colorectal cancer for metastasis evolution and clinical consequences. Semin Cancer Biol 2020;60:1–13.
[279] Tong D, Tian Y, Zhou T, Ye Q, Li J, Ding K, Li J. Improving prediction performance of colon cancer prognosis based on the integration of clinical and multi-omics data. BMC Med Inform Decis Mak 2020;20:22.
[280] Herrmann M, Probst P, Hornung R, Jurinovic V, Boulesteix AL. Large-scale benchmark study of survival prediction methods using multi-omics data. Brief Bioinform 2021;22, bbaa167.

CHAPTER

11

General conclusions and future perspectives

Lucia Morbidelli

Department of Life Sciences, University of Siena, Siena, Italy

Abstract

The normalization of the tumor ecosystem, consisting in many different cell types and components, is an important new tool for cancer treatment. This chapter discusses the potential targets within the tumor microenvironment (TME) with pharmacological perspective on tumor vascularization and optimization of cancer treatments: hypoxic, acidic and oxidative environment responsible for angiogenic switch and chaotic neovascularization; switch of normal ECs (NECs) to tumor ECs (TECs) in term of cell behavior, metabolism and signaling; cancer associated fibroblasts (CAFs) mainly responsible for angiogenic factor release and processing of the extracellular matrix (ECM); macrophage recruitment in the TME to become tumor associated macrophages (TAMs) and differentiation toward M2 phenotype; normalization of the tumor vasculature; activation of the antitumoral activity by immune cells; miRNAs and epigenetic determinants. The combination of therapies targeting different stromal components, together with traditional antitumor agents, represents the key element to definitely impair cancer progression. Ongoing studies in the field which focus on evaluating the TME with an integrative approach bear the potential to significantly control tumor angiogenesis and broaden the spectrum of current anticancer treatments.

Abbreviations

ALDH	aldehyde dehydrogenase
CA	carbonic anhydrase
CAF	cancer associated fibroblast
CRC	colerectal cancer
CSC	cancer stem cell
CXCL12	CXC chemokine ligand 12
CXCR4	CXC chemokine receptor type 4
DMNT	DNA methyltransferase
EC	endothelial cell
ECM	extracellular matrix
EMT	epithetial-mesenchymal transition
eNOS	endothelial nitric oxide synthase
ERK	extracellular signal-related kinase

https://doi.org/10.1016/B978-0-323-90190-1.00004-4

HDAC	histone deacetylase
HIF-1α	hypoxia inducible factor-1α
Hsp	heat shock protein
MAPK	mitogen-activated protein kinase
MMP	matrix metalloproteases
NEC	normal endothelial cell
NO	nitric oxide
NSCLC	non-small cell lung cancer
PDGF	platelet derived growth factor
PDGFR	platelet derived growth factor receptor
PI-3K	phosphatidyl inositol-3 kinase
RCC	renal cell carcinoma
ROS	reactive oxygen species
TAM	tumor associated macrophage
TEC	tumor endothelial cell
TGF	transforming growth factor
TK	tyrosine kinase
TKI	tyrosine kinase inhibitor
TME	tumor microenvironment
TNF	tumor necrosis factor
VEGF	vascular endothelial growth factor
VEGFR	vascular endothelial growth factor receptor
vHL	von-Hippel Lindau

Conflict of interest

No potential conflicts of interest were disclosed.

General statements on tumor angiogenesis

Endothelial cells (ECs) receive multiple stimuli from the tumor microenvironment (TME) that eventually induce them to progress along all the stages of angiogenesis, until new vessels are formed [1]. Angiogenic signals that promote EC survival, proliferation, migration, and finally differentiation are the result of a complex network, involving cell-cell and cell-extracellular matrix (ECM) interactions and the action of soluble mediators on their receptors [2,3]. ECs express a large number of receptors that make them responsive to several growth factors and cytokines involved in the promotion of sprouting angiogenesis, but the most important and specific for its action is vascular endothelial growth factor receptor (VEGFR) [4]. In particular, the binding of vascular endothelial growth factor-A (VEGF-A), secreted by cells in the hypoxic environment during tumor growth, to VEGFR-2 is the major way by which EC migration and proliferation are promoted [5]. Specifically, like other tyrosine kinase (TK) receptors, VEGFR-2 undergoes ligand (VEGF-A) induced dimerization and oligomerization, which activate its intrinsic tyrosine kinase activity, resulting in auto- and transphosphorylation on specific tyrosine residues in the cytoplasmic domain. The ignition of the Ras/Raf/MAPK pathway mainly regulates the progression of the cell cycle and proliferation. Another signaling pathway downstream to TK receptor activation and involved in cell survival and inhibition of apoptosis is PI-3K/Akt/protein kinase B [6,7]. These pathways lead to the downstream activation of intracellular proteins which induce (i) proliferation, sprouting, and tube formation of ECs; (ii) vasodilation and increased permeability of newly formed vessels, linked to VE-cadherin internalization and tight junctions loosening [8]. Efficient angiogenesis in tissues is crucially dependent on the proper interactions among different types of cells (ECs, pericytes, stromal cells), cells, and basement membrane, and is the result of the balance between pro- and antiangiogenic factors [9]. In tumors the balance is toward the proangiogenic stimuli, producing an aberrant and disorganized neovascular growth. The resulting circulation is chaotic and impaired due to the lack of pericytes and integral basement membrane [10,11].

Based on these findings antiangiogenic therapies have been developed in the form of monoclonal antibodies against VEGF or VEGFR, and small molecules inhibiting the kinase activity of the receptors and cytosolic pathways [11]. Bevacizumab (Avastin) is a humanized monoclonal antibody able to neutralize soluble VEGF-A, approved for clinical use in 2004. Its therapeutic use alone or in combination is detailed in Chapter 2. Ramucirumab, a human antibody that binds VEGFR-2 preventing VEGF binding, has been approved for cancers of the gastrointestinal system (see Chapter 3 of the present book). Tyrosine kinase inhibitors (TKIs) are orally available small molecules that can be used to hamper angiogenesis via VEGFRs. Axitinib is an example. This drug is highly selective for VEGFR and is approved by FDA for targeting advanced renal cell carcinoma (RCC) [12]. Other TKIs developed to target VEGFR are sorafenib, sunitinib, and pazopanib [13]. Their mechanism of action is multitarget acting on different kinases both in endothelial and tumor cells. They suppress the TK activity of VEGFR-1, 2 and 3, and inhibits c-Kit and platelet-derived growth factor receptor (PDGFR) [14]. Depending on the drug, additional signaling targets are inhibited.

In various neoplasms, antiangiogenic therapy may benefit from combination therapy with conventional, target therapy, and hormone therapy to strengthen its efficacy and to reduce drug resistance and disease recurrence. Specific examples are reported in this book for breast and lung cancers and various other neoplasms in Chapter 2, gastrointestinal tumors and colorectal cancer (CRC) in Chapter 3, renal and urogenital cancer in Chapters 4 and 5, melanoma in Chapter 6, hematological tumors in Chapter 7. The state of the art of metronomic therapy in combination with antiangiogenic drugs is reported in Chapter 8, while Chapter 9 reports all the strategies adopted by tumors to resist antiangiogenic strategies.

The higher success of multitarget drugs with respect to drugs acting on single signaling leads to consider that tumor angiogenesis cannot be only the result of the hypoxia-driven angiogenic switch, but more complex and intricated signaling pathways are active within the TME. Nowadays, the majority of expectations for successful therapies are from the TME and cell communications within it [15]. Cancer cells communicate with ECs and induce angiogenesis to fulfill the energetic requirements of proliferating tumor cells. Cancer cells have a wide and complex variety of means of communicating with ECs [1,16]. Based on the information presented in this book, we argue that it is fundamental to better understand the pathways of angiogenesis activated within the tumor, but also to have a more comprehensive knowledge of how cancer and stromal cells trigger these pathways. The investigation of the role of cell-to-cell communication between cancer and ECs in different types of cancers and in the different stages of tumor progression will enable the development of better therapies. Finally, we also support the idea that cancer cells might change their type of cell-to-cell communication targeting ECs to overcome antiangiogenic therapy.

Potential targets within the TME with a pharmacological perspective on tumor vascularization and optimization of cancer treatments are:

- Hypoxic, acidic, and oxidative environment responsible for the angiogenic switch and chaotic neovascularization;
- Transformation of normal ECs (NECs) to tumor ECs (TECs) in terms of cell behavior, metabolism, and signaling;
- Cancer-associated fibroblasts (CAFs) mainly responsible for angiogenic factor release and processing of the ECM;
- Macrophage recruitment in the TME to become tumor-associated macrophages (TAMs) and differentiation toward M2 phenotype;
- Normalization of the tumor vasculature;
- Activation of the antitumoral activity of the immune cells;
- miRNAs and epigenetic determinants.

All these aspects are closely entangled and sometimes difficult to be considered separately (Fig. 1). Following there are some reports, experimental or early clinical, of the potential strategies designed for the therapeutic readdressing of the TME with positive outcomes on tumor vascularization and tissue perfusion.

Strategies acting on hypoxic, acidic and oxidative environment

Hypoxia-mediated angiogenesis is directly correlated to hypoxia-inducible factor-1α (HIF-1α) stabilization, nuclear translocation, and transcriptional activity of enzymes and factors able to adapt cell metabolism. Since the many factors involved, this pathway can be pharmacologically targetable. HIF inhibition has emerged as an attractive therapeutic option against cancer, thus disrupting the pathological angiogenic switch prompted by tumor cells.

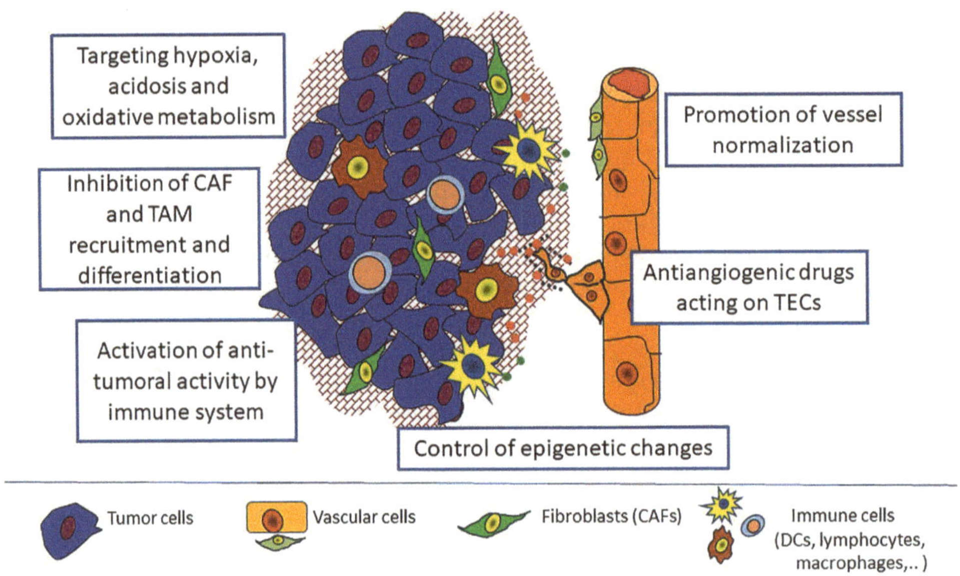

FIG. 1 Schematic representation of the novel strategies against the components of the tumor microenvironment under development with the aim of inhibiting tumor angiogenesis or promote vessel normalization.

Mechanistically, it can involve two major approaches: (i) direct HIF inhibition to interrupt HIF stability and function, and (ii) indirect HIF inhibitors that can block upstream or downstream transcriptional activity [17].

Acriflavine, a direct HIF inhibitor able to prevent HIF heterodimerization, has been shown to inhibit tumor growth and vascularization [18], particularly in glioblastoma where the poorer clinical outcome is associated with hypoxia-induced gene overexpression. Biodegradable polymers for local acriflavine delivery through the blood-brain barrier demonstrated their efficacy in experimental brain tumors [19]. Echinomycin is another direct HIF inhibitor able to hamper the binding of HIF-1α to the hypoxia-responsive element sequence in the VEGF promoter [20]. The nano-delivery of echinomycin induces autophagy-mediated death in pancreatic cancer in vivo [21], wherein syndecan-1-encapsulated echinomycin, a custom-designed tumor-specific delivery method, resulted in significantly greater survival [21].

Indirect HIF inhibition can be obtained by interfering with heat shock proteins (particularly Hsp90 and Hsp70) activity. These Hsp chaperones bind to HIF-1α and block von-Hippel Lindau (vHL)-independent proteasome degradation, allowing proper conformation of HIF-1α heterodimers for their subsequent interaction with downstream proteins [22]. The Hsp70 inhibitor IDF-11774, approved for clinical trials phase I study by the Korean FDA, has shown a significant reduction of VEGF expression in vivo, together with angiogenesis disturbance [23]. More recently, AT-533, a novel Hsp90 inhibitor, has been demonstrated to directly impair HIF-1α/VEGF/VEGFR2-mediated angiogenesis in vitro and in breast carcinoma xenografts in vivo [24]. Thus, Hsp inhibitors, severely disrupting tumor-associated-angiogenesis, show potential benefits for targeted cancer therapy. In addition, a very promising antiangiogenic and antitumor drug, 2-Methoxyestradiol (2ME2, Panzem®) NanoCrystal® Dispersion (NCD®) can downregulate HIF-1α expression by depolarizing cytoplasmatic microtubules that ultimately block HIF-1α translation [17,25]. A combination of 2ME2 and bevacizumab has also led to significant tumor reduction in patients with prostate cancer [26].

Other molecules have been described to interact with the HIF pathway to disrupt tumor-associated angiogenesis. For example, the antimicrobial agent salinomycin has shown anticancer potential [27] and inhibition of angiogenesis in a murine breast carcinoma model [28]. Another study has revealed that SN-38, an active irinotecan metabolite, can suppress HIF-1α expression induced by radiotherapy, with a reduction in VEGF expression in CRC cells [29]. In addition, HIF-2α inhibitors are under development, especially to treat RCC, as recently reviewed [30]. Although HIF inhibition seems to be advantageous in cancer therapy, it is important to note that, to date, there are no clinically approved HIF inhibitors. The physiological significance of the HIF pathway, accompanied by unspecific delivery systems, may alert the cautious use of HIF inhibitors.

Besides hypoxia, other biochemical features of TME are pH alteration and change in glucose and lactate secretion [31]. To maintain an intracellular pH (pHi) compatible with cell survival, different transporters extrude the excess lactate and protons in the extracellular milieu. Subsequently, the pHi increases with respect to normal cells, while the extracellular pH (pHe) decreases. This "reversed pH gradient" in cancer cells is considered an emerging hallmark of cancer [32], promoting tumor growth and metastasis via various mechanisms. Among them, there is pH-dependent degradation of the ECM via activation of matrix metalloproteases (MMPs) and cathepsins, and modulation of integrin-mediated cell-matrix adhesion [33]. Furthermore, alkaline pHi can promote cancer cell migration by cytoskeletal rearrangements [34]. Finally, acidification of pHe can influence the invasive and metastatic potential of tumors through the induction of epithelial-mesenchymal transition (EMT) [35,36]. Several proteins are involved in the pH regulation including monocarboxylate transporters 1 and 4 (MCT1 and MCT4), Na+/H+ exchangers (NHE1), Na+/HCO3- co-transporters, and carbonic anhydrase IX (CA-IX) [33].

CA-IX belongs to the CA family, ubiquitous zinc enzymes, encoded by eight distinct, evolutionary unrelated gene families. CAs catalyze the reversible hydration of CO_2 to HCO_3^- and H+ playing an essential role in pHe and pHi regulation, respiration, metabolism, and other processes that use CO_2 and HCO_3^- [37].

While CA-IX is normally expressed in gut epithelial cells [38], it results upregulated in many solid tumors including breast and lung [39–41]. Its expression is under the control of HIF-1α, and its membrane localization is elevated in chronically hypoxic tumor cells. Accumulating evidence suggests that CA-IX expression correlates with poor prognosis [40,42] because it is induced by hypoxia, functionally linked to acidosis, tumor invasiveness, and drug resistance [43].

The CA-IX inhibitor SLC-0111 is a ureido benzene-sulfonamide, currently in phase Ib/II clinical trial for the treatment of solid tumors associated with a hypoxic micro-environment [44–46]. The aim of the trial is to evaluate SLC-0111 efficacy in combination with gemcitabine in metastatic pancreatic cancer patients whose tumors are CA-IX positive (ClinicalTrials.gov Identifier: NCT03450018).

We have recently contributed to characterize SLC-0111 activity on the breast (MDA-MB-231) and lung (A549) cancer cells [47]. The CA-IX inhibitor inhibited cell proliferation, migration, and invasion by reverting the mesenchymal phenotype toward an epithelial one and by impairing MMP-2 activity. The antitumor activity was elicited via activation of the apoptosis pathway. Upregulation of p53 was observed, which in turn regulated the activation of caspase-3. Inhibition of proteolytic activity was accompanied by upregulation of the endogenous tissue inhibitor TIMP-2 [47], strengthening CA-IX role in TME control [48].

Interestingly, in triple-negative breast cancer xenograft, the administration of SLC-0111 significantly reduced overall metastatic burden, whereas exposure to sunitinib increased hypoxia and CA-IX expression in primary tumors, and failed to inhibit metastasis. The administration of SLC-0111 significantly decreased primary tumor vascular density and permeability, and reduced metastasis to the lung and liver. In addition, combining SLC-0111 and sunitinib significantly reduced both primary tumor growth and sunitinib-induced metastasis to the lung. These findings support the concept that targeting both angiogenesis and hypoxia effectors can be a promising novel strategy for the effective treatment of patients with triple-negative breast cancer [49].

A pro-oxidant-rich TME is tightly associated with angiogenesis promotion. Indeed, high levels of reactive oxygen species (ROS) upregulate HIF-1α and VEGF. Thus, several therapeutic strategies with antioxidant activity have been developed to target ROS-induced angiogenesis [50]. However, since the efficacy of some chemotherapeutics and radiotherapy is mediated by ROS production, the combination of conventional therapies with ROS scavengers can be questionable in the clinic.

Nevertheless, simvastatin, a statin with antioxidant properties, has been shown to reduce hypoxia-induced endothelial leakage and decrease ROS-induced HIF-1α and VEGF expression, attenuating VEGF-derived tumor vessel hyperpermeability and improving cisplatin and cyclophosphamide efficacies, with a protective effect against doxorubicin-mediated cardiotoxicity [51–54].

In this context, recently in our laboratory, we have been interested in the contribution of some isoenzymes of the aldehyde dehydrogenase (ALDH) family, such as ALDH1A1, in cancer progression [55]. The latter is a cytosolic enzyme involved in the detoxification of cells from toxic reactive aldehydes, including acetaldehyde deriving from ethanol metabolism, and retinaldehyde which is then converted into retinoic acid. High ALDH1A1 activity is closely related to the stemness phenotype of several tumors, as well as to the pro-angiogenic switch of NECs into TECs, possibly contributing to cancer progression and diffusion in the body [56]. We have documented that the overexpression of ALDH1A1 in cancer stem cells is associated with its ability to control the angiogenic phenotype of TECs through the release of angiogenic factors [57]. Pharmacological inhibition of ALDH1A1 activity or downregulation of its expression significantly reduces angiogenic function in ECs co-cultured with breast cancer cells, as well the tumor angiogenesis in a breast cancer cell xenograft model [57]. These effects were due to the activation of HIF-1α and VEGF signaling pathways by ALDH1A1 in tumor cells. However, since no drug targeting ALDH1A1 is currently available, these data remain in the preclinical phase.

Other examples of antioxidants are nutraceuticals, foodstuff (such as fortified food or dietary supplements, many of which rich in polyphenols) which provide health benefits in addition to their nutritional value, including anti-tumor and antiangiogenic effects [58]. The use of nutraceuticals with antioxidant and anti-inflammatory activities could be a therapeutic/preventive option to blunt aberrant angiogenesis. An example is baicalein, a phenolic flavonoid that employs its antioxidant properties to block cancer progression, vascularization, and metastasis [59]. Similarly, OptiBerry, an anthocyanins-rich berry extract, inhibits H_2O_2 and tumor necrosis factor α (TNF-α) induced VEGF expression and diminishes the ability of TECs to form hemangioma, demonstrating antiangiogenic, antioxidant, and anticancer properties [60].

We have contributed to demonstrating that hydroxytyrosol, a product present in extra-virgin olive oil, blocks microsomal prostaglandin-E synthase-1 and HIF-1α-dependent VEGF expression, thus reducing tumor angiogenesis [61], and promotes epidermal growth factor receptor degradation in colon cancer cells, thus sensitizing them to anticancer drugs [62,63]. Oleuropein, the most abundant phenolic compound in olives, inhibits the progression of melanoma in mice exposed high-fat-diet-induced obesity, by reducing the expression of angiogenesis, lymphangiogenesis, and hypoxia markers. In addition, oleuropein directly inhibited in vitro tube formation of vascular and lymphatic ECs [64]. Based on these findings, even if there is no approved drug with these activities, it is important to emphasize the beneficial effect of the Mediterranean diet on cancer prevention, due to the presence of many diet-derived protective bioactives. From a nutraceutical point of view, a plethora of natural compounds acting on HIF and metabolic pathways is under preclinical evaluation with potential clinical exploitations.

Vascular normalizing strategies

There are several strategies to promote the normalization of tumor vasculature besides the hypoxia control seen above [65]: targeting of EC glycolysis [66], nitric oxide delivery [67], M2-to-M1 polarization of TAMs [68], microRNA targeting (i.e., miRNA-200) [69], radiotherapy [70], application of agents with vascular normalizing activity, such as chloroquine (an antimalaria drug) [71] and VEGF inhibitors.

NECs transformation in TECs is accompanied by an important metabolic reprogramming [72]. EC metabolism is a key regulator of angiogenesis that has been increasingly in the spotlight during the past 10 years and now can be of particular value for novel therapy development given that most metabolic enzymes of the glycolytic cascade are druggable. Lowering (not eliminating) glycolysis in TECs can normalize vascular abnormalities, reduce the chance of metastasis and improve the efficacy of chemotherapy without causing systemic toxicity [66].

Nitric oxide (NO) controlled delivery represents a novel strategy for vessel normalization. Sung et al. [67] reported an NO delivery system with a prolonged half-life and sustained release to regulate angiogenesis and drug delivery. The authors developed NanoNO, a nanocarrier formulation that effectively targeted NO to hepatocellular carcinoma. NanoNO improved the delivery and effectiveness of chemotherapies and also tumor necrosis factor-related, apoptosis-inducing, ligand (TRAIL)-induced therapy in both primary and metastatic tumors [67].

Careful dosage of anti-VEGF drugs is able to restore normal levels of angiogenic signals in different types of tumors, provoking decreased permeability by recruiting pericytes and tightening cell-cell junctions [73]. Benefits of vascular normalization have been observed in different types of tumors. The combination of bevacizumab, together with chemotherapy, produced a positive outcome in a subset of breast cancer patients [74]. Moreover, it is recommended that the antiangiogenic agent be administered in intermittent cycles to induce vessel normalization and proper delivery of a chemotherapeutic agent. Furthermore, combined inhibition of VEGFR and angiopoietin-2 improves survival of mouse glioblastoma tumor models, by increasing vessel normalization and reprogramming TAMs [75]. Another example of the benefits of vessel normalization includes the use of trebananib, a fusion protein that inhibits angiogenesis by blocking angiopoietin 1 and 2 binding to the Tie-2 receptor. Trebananib received in 2013 the orphan designation by FDA and EMA for the treatment of ovarian cancer (https://www.ema.europa.eu/en/medicines/human/orphan-designations/eu3131207). Recently, the combination of trebananib and chemotherapy demonstrated benefits in progression-free survival in patients with cancers of the reproductive system [76].

Another opportunity is targeting perivascular cells. Pericytes in the TME are partly derived from cancer stem cells (CSCs) and contribute to immunosuppression and chemotherapy resistance [11]. CSCs can also be attracted toward these niches through the CXC chemokine ligand 12 (CXCL12)/CXC chemokine receptor type 4 (CXCR4) signaling axis [77]. Here, CSCs can undergo pericyte differentiation, which is mainly promoted under the influence of transforming growth factor-beta (TGF-β) [78]. Inhibiting this pleiotropic immunosuppressive cytokine may lead to suppression of cancer growth and progression [79,80].

Since the large overlap in the expression of biomarkers between pericytes and SMCs, a close cooperation among the two cell types in the formation of new blood vessels is suggested. Oct4 is considered a marker of CSCs. Interestingly, Oct4 knockout in perivascular cells (SMCs and pericytes) has been found to suppress angiogenesis [81]. All these data support the crucial role of CSCs in the promotion of angiogenesis in perivascular niches and that these cells can be pharmacologically targeted to normalize tumor vascularization.

Finally, to get vessel normalization there has been the interest in repurposing drugs already active in other diseases, such as metformin and statins. The favorable safety profile, the low cost, and immediate clinical availability support the repositioning of old drugs for new clinical purposes [82].

Thalidomide is a good example, as the teratogenic effect of thalidomide is due to the prevention of the physiological embryo angiogenesis [83,84]. Thalidomide demonstrated antiangiogenic effects in mouse models of neuroblastoma and breast cancer [85,86]. Thalidomide also increases the remodeling and stabilization of the abnormal tumor vasculature in models of solid and hematologic tumors [87,88].

Effects similar to the above described for simvastatin have been reported for the most prescribed antidiabetic drug metformin, which is able to influence angiogenesis by epigenetic mechanisms [89]. Metformin has shown its ability to inhibit tumor angiogenesis in metastatic breast cancer models. Wang et al. [90] demonstrated that metformin significantly reduced microvessel density, leakage, and hypoxia, and increased vascular mural cell coverage and perfusion with consequently reduced lung metastasis and improved chemosensitivity. Further results of genetic screening and in vivo experiments showed that the downregulation of platelet-derived growth factor B (PDGF-B) greatly contributed to the metformin-induced vessel normalization [90].

These data widen the portfolio of available antiangiogenic and antitumor drugs, documenting the protective effects of therapies universally prescribed for the control of chronic metabolic disorders.

Control of lymphocyte trafficking into the TME

Clinical data clearly document that anti-VEGF drugs used alone may result in limited clinical benefits and/or failure in remodulation of TME. The higher success is coming from the combination of antiangiogenic drugs with conventional chemotherapy, target therapy, and immune therapy (see Chapter 2 of this book). Again, the best results derive from multi-target TK inhibitors with respect to anti-VEGF/VEGFR drugs as bevacizumab. It is now clear that vascular targeting/antiangiogenic agents in combination with immune checkpoint blockade may promote tumor immunity by inducing the formation of high-endothelial venules, able to facilitate lymphocyte trafficking. Interestingly, pharmacological blockade of the angiogenic axis restores adhesion molecule expression on ECs and induces vessel normalization that promotes T cell migration within the tumor, thus restoring immune cell attack against the tumor [91]. In various types of cancer, immune checkpoint inhibitors (ICIs) synergize with antiangiogenic therapy by allowing the cytotoxic potential of effector T cells [10,92–94]. The combination however suffers from negative drawbacks as failure or severe side effects, thus prompting the research toward other opportunities.

An increased understanding of the crosstalk between tumor cells, endothelial cells, and immune cells during immune checkpoint blockade therapy may lead to new combinatorial treatment regimens that enhance the abundance of activated T-cells in tumor tissue. This can ultimately increase the number of patients that safely respond to ICIs.

Strategies acting on miRNA and epigenetic determinants

miRNAs play a pivotal role in the regulation of gene expression at both the transcriptional and post-transcriptional levels. Characterized by small dimensions, miRNAs regulate lots of signaling implied in apoptosis, cell proliferation, differentiation, and migration [95]. miRNA-mediated gene silencing affects the regulation of endothelial function, metabolic and homeostasis conditions in cancer by targeting tumor angiogenesis pathways, like VEGF and its TK receptor expression downstream of HIF-1α [96]. Therefore, miRNAs controlling the angiogenic mechanisms are collectively grouped as angiomiRs [95].

Long noncoding RNAs (lncRNAs) were considered for a long time as cancer biomarkers. Instead, they play an active role in cancer progression and angiogenesis [97]. Indeed, the deregulation of lncRNA is involved in developing the hallmarks of cancer and specifically is implied in the regulation of the TME and the modulation of tumor angiogenesis due to the interaction with miRNA and other oncogenic pathways [98]. The direct and indirect correlations between lncRNA and tumor angiogenesis arise from a dysregulated transcription of lncRNA under hypoxic conditions and, in turn, lncRNA can regulate the HIF-related pathway [99]. LncRNAs indirectly increase VEGF levels due to the activation of STAT3, or the Wnt/β-catenin signaling axis in cancer cells. Interestingly, lncRNAs regulate angiogenesis by modulating the transcription factor NF-κB and operating a divergent effect on angiogenesis: a proangiogenic function in tumor cells and antiangiogenic role in ECs [100]. Furthermore, lncRNAs are directly involved in the regulation of enzymatic activity by binding to MMPs and other upstream or downstream proteins in different signaling pathways including HIF, VEGF, and TGF-β. Hence, lncRNAs can also directly affect tumor angiogenesis [99].

As shown at the preclinical level, the inhibition of miRNAs can contribute to the development of novel antiangiogenic therapies. At the clinical level, MesomiR-1 was the first-in-human Phase I study about the application of TargomiRs as 2nd- or 3rd-line treatment in patients with recurrent malignant pleural mesothelioma and non-small cell lung cancer (NSCLC) (ClinicalTrials.gov Identifier: NCT02369198), obtaining promising results [101]. In the complex, the identification of miRNAs as onco-miRNAs or oncosuppressor miRNA with a critical role in angiogenesis has raised new hopes for the treatment of tumors. They may be candidates even for specific targeting as antagonists or mimics [102]. The potential of miRNA-based therapy is under preclinical and clinical evaluations, however, their development as drugs is still far from the clinical application because there are several criticisms to be carefully considered, as tissue-specific delivery and adequate cellular uptake. In addition to the critical issues concerning the pharmacokinetics and pharmacodynamics of miRNA-drugs, it should be stressed that there is still much to discover in the field of miRNAs. A clear profile of cancer-specific epigenetic imprinting, as well as the patient-specific pattern of epigenetic modification, is needed. The sequencing of miRNome and the mapping of miRNA/mRNA interactions are far from being complete due to the recognized challenge of computational prediction of mRNA/miRNA interactions. In most studies, methods of high throughput analysis based on Next Generation Sequencing (NGS) technology and computational analysis are applied. However, these experimental approaches to define the role of small non-coding RNAs in the biological and pathological processes are not enough. The morphology-based approach, such as in situ Dual ISH-IHC Technology, should walk together with the most modern computational and molecular technologies for a more precise characterization of the biological events in a more realistic molecular and cellular context [103,104].

Among the causes which determine the transformation of NECs in TECs, there are epigenetic determinants [105], which represent a promising chance to target vascularized tumors. The imprinting given by the hypoxic conditions and the changes in TME to activate NECs involve epigenetic modifications that, in turn, continue to support tumor progression and aggressiveness and result in resistance to many of the antiangiogenic therapies used today [106]. Epigenetic events, such as DNA hypermethylation and histone acetylation, gene silencing, or reactivation, trigger the main phenotypic modifications of TECs [107]. DNA methylation is due to the addition of methyl groups to CpG islands by DNA methyltransferase (DNMTs) [108,109]. In particular, poor prognosis has been demonstrated to be associated with hypermethylation of tumor-suppressor genes and hypomethylation of tumor-promoting genes. Interestingly, key genes regulating angiogenesis are affected by the selective demethylation and hypermethylation of CpG islands in the regulatory sequences.

Histone molecular modification occurs on N-terminal tails extending from the compact nucleosomal cores, controlling histone interaction and gene regulation [110]. Generally, acetylation of histones, increasing DNA accessibility, is correlated with the promotion of transcription, whereas histone deacetylation is associated with chromatin condensation and transcriptional repression [111]. DNA methylation and post-transcriptional modifications on the histone code cooperate to give pro-angiogenic features to the ECs and sustain TEC functions in tumors [112–115].

Even if they do not involve ECs, the epigenetic modifications decrease the efficacy of antiangiogenic therapy in multiple ways. As reported by Maiti et al. [116], histone deacetylation is involved also in the vascular mimicry process associated with resistance to antiangiogenic therapy in triple-negative breast cancer. Therefore, the inhibition of vascular mimicry epigenetically restores the expression of antiangiogenic genes and downregulates VEGF-A transcription as well as the EMT-related genes [116,117]. Therefore, the use of epigenetic drugs may be an effective and novel strategy to re-express tumor suppressor and angiogenesis inhibitor genes in order to control vascular mimicry-dependent cancers.

Rather than the aberrant DNA mutation, the reversibility of epigenetic imprinting makes the epigenetic mechanisms a valid target for antiangiogenic therapy development [109,118]. On the other hand, the same antiangiogenic therapy could improve epigenetically driven adaptation of TECs and tumor cells, resulting in drug resistance. In fact, the aberrant expression of receptors targeted by antiangiogenic drugs stimulates the genetic adaptation of tumor cells mainly through epigenetic modifications. Significantly, EZH2, a specific histone methyltransferase, drives drug resistance to sunitinib in kidney cancers [119].

Following the primary preclinical data, therapeutic approaches on reactivation of epigenetically silenced genes were considered to restore the angiostatic properties of cells and to balance phenotypical features. To regain the activity of angiogenesis-regulator genes, many inhibitors of DNMT and histone deacetylase (HDAC), namely epi-drugs, were characterized for their indirect angiostatic effects in in vivo experiments. HDAC down-regulates pro-angiogenic genes such as VEGF and endothelial nitric oxide synthase (eNOS), leading to an impairment of cancer angiogenesis, cellular stress response pathways, cell mobility, and metastatic process [120].

Many HDAC inhibitors are in a different stage of clinical development. Trichostatin A, a specific HDAC inhibitor, was proposed as an antiangiogenic therapeutic agent in cancer for its ability to up-regulate the expression of p53 and vHL genes, down-regulate HIF-1α, and VEGF, and therefore inhibiting hypoxia-induced angiogenesis [121]. Similarly, dacinostat (NVP-LAQ824) inhibits angiogenesis and down-regulates the expression of eNOS [122]. To date, six epi-drugs (HDAC and DNMT inhibitors) have been approved by FDA for treatment of different tumors: 5-azacytidine (azacytidine or AZA Celgene as Vidaza), vorinostat (SAHA, Merck), belinostat (BELEODAQ, Spectrum Pharmaceuticals, Inc.), romidepsin (Istodax, Celgene), tucidinostat (formerly known as chidamide) and panobinostat (Farydak, Novartis) [123]. The antiangiogenic activity of both HDAC and DNMT inhibitors in cancer treatment, due to the modulation of the tumor cells or by affecting TECs, provides a rationale for combining the two types of therapies. For instance, a phase I trial of panobinostat plus bevacizumab in patients with advanced solid tumors was concluded (ClinicalTrials.gov Identifier: NCT01055795), while panobinostat alone was proposed against advanced NSCLC and head-neck tumors (NCT00738751). For detail see Ciesielski et al. [105].

Final considerations

Undoubtedly, the normalization of the tumor ecosystem, consisting of many different cell types and components, is an important new tool for cancer treatment. Based on this observation, the combination of different therapies targeting different stromal components, together with traditional antitumor agents, could hold the key to impair cancer progression. Ongoing studies in the field which focus on studying the TME with an integrative approach bear the potential to significantly control tumor angiogenesis and broaden the spectrum of current anticancer treatments. There are also expectations from phase III clinical trials with antiangiogenic drugs combined with metronomic therapy, where the antiangiogenics enhance the therapeutic activity of the metronomic schedule (see Chapter 8 of this book).

The concept of EC heterogeneity within and between (diseased) organs summits at the level of the individual patient and pertains to attempting personalized medicine. It is highly likely that the proportional contribution of specific EC subpopulations within the aberrant vasculature of the same tumor type differs between patients. It remains to be determined to which extent this can be explored as a prognostic marker for therapy outcome. Along the same lines, the newly identified molecular players in alternative vascularization modes such as vessel cooption and vascular mimicry could be of potential value as biomarkers for the development of novel antiangiogenic treatments or the proper choice of drugs and drug combination.

The readdressing of TME is an important challenge in the development of experimental cancer models. In order to develop a reliable tool for personalized cancer therapy and drug development, it is essential to preserve the key characteristics of the original tumor. Recent advances on three dimensional (3D) platforms through the setup of complex tumorspheres and the use of lab-on-chip and microfluidic devices [124] have provided an enormous opportunity to better stimulate the function and biology of the single components of TME and to bridge the translational gap between preclinical and clinical settings [125].

From a clinical perspective, it continues to be fundamental to define circulatory biomarkers for the choice of the proper therapy and to monitor therapy efficacy and disease progression (see Chapter 10 of this book).

Further research in the areas is needed to provide insight into the knowledge of the biological evolution of specific cancers and of the development of metastasis and drug-resistant tumor cell phenotypes [126]. Combined analysis by machine learning and artificial intelligence will contribute to identifying these signatures in a personalized view [127,128]. Prospective confirmatory clinical trials will also be crucial for the validation of predictive biomolecular classifiers [129].

Summarizing, tumor stroma, acting as a co-director for the development of vascularized growing mass, will give the rationale driving the definition of new antitumor therapies with final antiangiogenic activity. It is expected that the development of drugs and drug formulations (as nanomaterials) able to modify the TME, will be designed with the potential aim to further sensitize the tumor toward conventional and innovative antitumor agents given in combination.

References

[1] Zuazo-Gaztelu I, Casanovas O. Unraveling the role of angiogenesis in cancer ecosystems. Front Oncol 2018;8:248. https://doi.org/10.3389/fonc.2018.00248.

[2] Lamalice L, Le Boeuf F, Huo J. Endothelial cell migration during angiogenesis. Circ Res 2007;100(6):782–94. https://doi.org/10.1161/01.RES.0000259593.07661.1e.

[3] Muñoz-Chápuli R, Quesada AR, Medina MA. Angiogenesis and signal transduction in endothelial cells. Cell Mol Life Sci 2004;61(17):2224–43.

[4] Hofer E, Schweighofer B. Signal transduction induced in endothelial cells by growth factor receptors involved in angiogenesis. Thromb Haemost 2007;97(3):355–63.

[5] Olsson AK, Dimberg A, Kreuger J, Claesson-Welsh L. VEGF receptor signalling – in control of vascular function. Nat Rev Mol Cell Biol 2006;7(5):359–71.

[6] Gerber HP, McMurtrey A, Kowalski J, Yan M, Keyt BA, Dixit V, Ferrara N. Vascular endothelial growth factor regulates endothelial cell survival through the phosphatidylinositol 3′- kinase/Akt signal transduction pathway requirement for Flk-1/KDR activation. J Biol Chem 1998;273:30336–43. https://doi.org/10.1074/jbc.273.46.30336.

[7] Hicklin DJ, Ellis LM. Role of the vascular endothelial growth factor pathway in tumor growth and angiogenesis. J Clin Oncol 2005;23:1011–27. https://doi.org/10.1200/JCO.2005.06.081.

[8] Gavard J, Gutkind JS. VEGF controls endothelial-cell permeability by promoting the beta-arrestin-dependent endocytosis of VE-cadherin. Nat Cell Biol 2006;8(11):1223–34. https://doi.org/10.1038/ncb1486.

[9] Chung AS, Lee J, Ferrara N. Targeting the tumour vasculature: insights from physiological angiogenesis. Nat Rev Cancer 2010;10:505. https://doi.org/10.1038/nrc2868.

[10] Eelen G, Treps L, Li X, Carmeliet P. Basic and therapeutic aspects of angiogenesis updated. Circ Res 2020;127(2):310–29. https://doi.org/10.1161/CIRCRESAHA.120.316851.

[11] Majidpoor J, Mortezaee K. Angiogenesis as a hallmark of solid tumors – clinical perspectives. Cell Oncol (Dordr) 2021;44(4):715–37. https://doi.org/10.1007/s13402-021-00602-3.

[12] Motzer RJ, Penkov K, Haanen J, Rini B, Albiges L, Campbell MT, Venugopal B, Kollmannsberger C, Negrier S, Uemura M, Lee JL, Vasiliev A, Miller Jr WH, Gurney H, Schmidinger M, Larkin J, Atkins MB, Bedke J, Alekseev B, Wang J, Mariani M, Robbins PB, Chudnovsky A, Fowst C, Hariharan S, Huang B, di Pietro A, Choueiri TK. Avelumab plus Axitinib versus Sunitinib for advanced renal-cell carcinoma. N Engl J Med 2019;380(12):1103–15. https://doi.org/10.1056/NEJMoa1816047.

[13] Porta C, Cosmai L, Leibovich BC, Powles T, Gallieni M, Bex A. The adjuvant treatment of kidney cancer: a multidisciplinary outlook. Nat Rev Nephrol 2019;15(7):423–33. https://doi.org/10.1038/s41581-019-0131-x.

[14] Duska LR, Petroni GR, Varhegyi N, Brown J, Jelovac D, Moore KN, McGuire WP, Darus C, Barroilhet LM, Secord AA. A randomized phase II evaluation of weekly gemcitabine plus pazopanib versus weekly gemcitabine alone in the treatment of persistent or recurrent epithelial ovarian, fallopian tube or primary peritoneal carcinoma. Gynecol Oncol 2020;157(3):585–92. https://doi.org/10.1016/j.ygyno.2019.10.014.

[15] Baghban R, Roshangar L, Jahanban-Esfahlan R, Seidi K, Ebrahimi-Kalan A, Jaymand M, Kolahian S, Javaheri T, Zare P. Tumor microenvironment complexity and therapeutic implications at a glance. Cell Commun Signal 2020;18(1):59. https://doi.org/10.1186/s12964-020-0530-4.

[16] Lopes-Bastos BM, Jiang WG, Cai J. Tumour-endothelial cell communications: important and indispensable mediators of tumour angiogenesis. Anticancer Res 2016;36(3):1119–26.

[17] Fallah J, Rini BI. HIF inhibitors: status of current clinical development. Curr Oncol Rep 2019;21:6. https://doi.org/10.1007/s11912-019-0752-z.

[18] Lee K, Zhang H, Qian DZ, Rey S, Liu JO, Semenza GL. Acriflavine inhibits HIF-1 dimerization, tumor growth, and vascularization. Proc Natl Acad Sci U S A 2009;106:17910–5. https://doi.org/10.1073/pnas.0909353106.

[19] Mangraviti A, Raghavan T, Volpin F, Skuli N, Gullotti D, Zhou J, Asnaghi L, Sankey E, Liu A, Wang Y, Lee DH, Gorelick N, Serra R, Peters M, Schriefer D, Delaspre F, Rodriguez FJ, Eberhart CG, Brem H, Olivi A, Tyler B. HIF-1α-targeting acriflavine provides long term survival and radiological tumor response in brain cancer therapy. Sci Rep 2017;7(1):14978. https://doi.org/10.1038/s41598-017-14990-w.

[20] Kong D, Park EJ, Stephen AG, Calvani M, Cardellina JH, Monks A, Fisher RJ, Shoemaker RH, Melillo G. Echinomycin, a small-molecule inhibitor of hypoxia-inducible factor-1 DNA-binding activity. Cancer Res 2005;65:9047–55. https://doi.org/10.1158/0008-5472.CAN-05-1235.

[21] Thomas A, Samykutty A, Gomez-Gutierrez JG, Yin W, Egger ME, McNally M, Chuong P, MacCuaig WM, Albeituni S, Zeiderman M, Li M, Edil BH, Grizzle WE, McMasters KM, McNally LR. Actively targeted nanodelivery of echinomycin induces autophagy-mediated death in chemoresistant pancreatic cancer in vivo. Cancers (Basel) 2020;12(8):2279. https://doi.org/10.3390/cancers12082279.

[22] Hur E, Kim HH, Choi SM, Kim JH, Yim S, Kwon HJ, Choi Y, Kim DK, Lee MO, Park H. Reduction of hypoxia-induced transcription through the repression of hypoxia-inducible factor-1alpha/aryl hydrocarbon receptor nuclear translocator DNA binding by the 90-kDa heat-shock protein inhibitor radicicol. Mol Pharmacol 2002;62:975–82. https://doi.org/10.1124/mol.62.5.975.

[23] Ban HS, Kim BK, Lee H, Kim HM, Harmalkar D, Nam M, Park SK, Lee K, Park JT, Kim I, Lee K, Hwang GS, Won M. The novel hypoxia-inducible factor-1α inhibitor IDF-11774 regulates cancer metabolism, thereby suppressing tumor growth. Cell Death Dis 2017;8(6). https://doi.org/10.1038/cddis.2017.235, e2843.

[24] Zhang PC, Liu X, Li MM, Ma YY, Sun HT, Tian XY, Wang Y, Liu M, Fu LS, Wang YF, Chen HY, Liu Z. AT-533, a novel Hsp90 inhibitor, inhibits breast cancer growth and HIF-1α/VEGF/VEGFR-2-mediated angiogenesis in vitro and in vivo. Biochem Pharmacol 2020;172:113771. https://doi.org/10.1016/j.bcp.2019.113771.

[25] Mabjeesh NJ, Escuin D, LaVallee TM, Pribluda VS, Swartz GM, Johnson MS, Willard MT, Zhong H, Simons JW, Giannakakou P. 2ME2 inhibits tumor growth and angiogenesis by disrupting microtubules and dysregulating HIF. Cancer Cell 2003;3:363–75. https://doi.org/10.1016/s1535-6108(03)00077-1.

[26] Kulke MH, Chan JA, Meyerhardt JA, Zhu AX, Abrams TA, Blaszkowsky LS, Regan E, Sidor C, Fuchs CS. A prospective phase II study of 2-methoxyestradiol administered in combination with bevacizumab in patients with metastatic carcinoid tumors. Cancer Chemother Pharmacol 2011;68:293–300. https://doi.org/10.1007/s00280-010-1478-7.

[27] Zhou S, Wang F, Wong ET, Fonkem E, Hsieh TC, Wu JM, Wu E. Salinomycin: a novel anti-cancer agent with known anti-coccidial activities. Curr Med Chem 2013;20:4095–101. https://doi.org/10.2174/15672050113109990199.

[28] Dewangan J, Srivastava S, Mishra S, Divakar A, Kumar S, Rath SK. Salinomycin inhibits breast cancer progression via targeting HIF-1alpha/VEGF mediated tumor angiogenesis in vitro and in vivo. Biochem Pharmacol 2019;164:326–35. https://doi.org/10.1016/j.bcp.2019.04.026.

[29] Okuno T, Kawai K, Hata K, Murono K, Emoto S, Kaneko M, Sasaki K, Nishikawa T, Tanaka T, Nozawa H. SN-38 Acts as a radiosensitizer for colorectal cancer by inhibiting the radiation-induced up-regulation of HIF-1alpha. Anticancer Res 2018;38:3323–31. https://doi.org/10.21873/anticanres.12598.

[30] Choueiri TK, Kaelin Jr WG. Targeting the HIF2-VEGF axis in renal cell carcinoma. Nat Med 2020;26:1519–30. https://doi.org/10.1038/s41591-020-1093-z.

[31] Boedtkjer E, Pedersen SF. The acidic tumor microenvironment as a driver of cancer. Annu Rev Physiol 2020;82:103–26.

[32] Sharma M, Astekar M, Soi S, Manjunatha BS, Shetty DC, Radhakrishnan R. pH gradient reversal: an emerging hallmark of cancers. Recent Pat Anticancer Drug Discov 2015;10:244–58.

[33] Neri D, Supuran CT. Interfering with pH regulation in tumours as a therapeutic strategy. Nat Rev Drug Discov 2011;10:767–77.

[34] Stock C, Schwab A. Protons make tumor cells move like clockwork. Pflugers Arch 2009;458:981–92.

[35] Suzuki A, Maeda T, Baba Y, Shimamura K, Kato Y. Acidic extracellular pH promotes epithelial mesenchymal transition in Lewis lung carcinoma model. Cancer Cell Int 2014;14:129.

[36] Riemann A, Rauschner M, Gießelmann M, Reime S, Haupt V, Thews O. Extracellular acidosis modulates the expression of epithelial-mesenchymal transition (EMT) markers and adhesion of epithelial and tumor cells. Neoplasia 2019;21:450–8.

[37] Supuran CT. Carbonic anhydrases: novel therapeutic applications for inhibitors and activators. Nat Rev Drug Discov 2008;7:168–81.

[38] Saarnio J, Parkkila S, Parkkila AK, Waheed A, Casey MC, Zhou XY, Pastoreková S, Pastorek J, Karttunen T, Haukipuro K, Kairaluoma MI, Sly WS. Immunohistochemistry of carbonic anhydrase isozyme IX (MN/CA IX) in human gut reveals polarized expression in the epithelial cells with the highest proliferative capacity. J Histochem Cytochem 1998;46(4):497–504. https://doi.org/10.1177/002215549804600409.

[39] Ilie M, Mazure NM, Hofman V, Ammadi RE, Ortholan C, Bonnetaud C, Havet K, Venissac N, Mograbi B, Mouroux J, Pouysségur J, Hofman P. High levels of carbonic anhydrase IX in tumour tissue and plasma are biomarkers of poor prognostic in patients with non-small cell lung cancer. Br J Cancer 2010;102(11): 1627–35. https://doi.org/10.1038/sj.bjc.6605690.

[40] Forker L, Gaunt P, Sioletic S, Shenjere P, Potter R, Roberts D, Irlam J, Valentine H, Hughes D, Hughes A, Billingham L, Grimer R, Seddon B, Choudhury A, Robinson M, West CML. The hypoxia marker CAIX is prognostic in the UK phase III VorteX-biobank cohort: an important resource for translational research in soft tissue sarcoma. Br J Cancer 2018;118(5):698–704. https://doi.org/10.1038/bjc.2017.430.

[41] Lou Y, McDonald PC, Oloumi A, Chia S, Ostlund C, Ahmadi A, Kyle A, Auf dem Keller U, Leung S, Huntsman D, Clarke B, Sutherland BW, Waterhouse D, Bally M, Roskelley C, Overall CM, Minchinton A, Pacchiano F, Carta F, Scozzafava A, Touisni N, Winum JY, Supuran CT, Dedhar S. Targeting tumor hypoxia: suppression of breast tumor growth and metastasis by novel carbonic anhydrase IX inhibitors. Cancer Res 2011;71(9):3364–76. https://doi.org/10.1158/0008-5472.CAN-10-4261 [Erratum in: Cancer Res 2011;71 (13):4733. Erratum in: Cancer Res 2011;71(12):4325].

[42] van Kuijk SJ, Yaromina A, Houben R, Niemans R, Lambin P, Dubois LJ. Prognostic significance of carbonic anhydrase IX expression in cancer patients: a meta-analysis. Front Oncol 2016;6:69.

[43] Pastorekova S, Gillies RJ. The role of carbonic anhydrase IX in cancer development: links to hypoxia, acidosis, and beyond. Cancer Metastasis Rev 2019;38:65–77.

[44] Pacchiano F, Carta F, McDonald PC, Lou Y, Vullo D, Scozzafava A, Dedhar S, Supuran CT. Ureido-substituted benzenesulfonamides potently inhibit carbonic anhydrase IX and show antimetastatic activity in a model of breast cancer metastasis. J Med Chem 2011;54:1896–902.

[45] Angeli A, Tanini D, Viglianisi C, Panzella L, Capperucci A, Menichetti S, Supuran CT. Evaluation of selenide, diselenide and selenoheterocycle derivatives as carbonic anhydrase I, II, IV, VII and IX inhibitors. Bioorg Med Chem 2017;25:2518–23.

[46] McDonald PC, Chia S, Bedard PL, Chu Q, Tang L, Singh M, Zhang Z, Supuran CT, Renouf DJ, Dedhar S. A phase 1 study of SLC-0111, a novel inhibitor of carbonic anhydrase IX, in patients with advanced solid tumors. Am J Clin Oncol 2020;43(7):484–90. https://doi.org/10.1097/COC.0000000000000691.

[47] Ciccone V, Filippelli A, Angeli A, Supuran CT, Morbidelli L. Pharmacological inhibition of CA-IX impairs tumor cell proliferation, migration and invasiveness. Int J Mol Sci 2020;21(8):2983. https://doi.org/10.3390/ijms21082983.

[48] Angeli A, Carta F, Nocentini A, Winum JY, Zalubovskis R, Akdemir A, Onnis V, Eldehna WM, Capasso C, Simone G, Monti SM, Carradori S, Donald WA, Dedhar S, Supuran CT. Carbonic anhydrase inhibitors targeting metabolism and tumor microenvironment. Metabolites 2020;10(10):412. https://doi.org/10.3390/metabo10100412.

[49] Hedlund EE, McDonald PC, Nemirovsky O, Awrey S, Jensen LDE, Dedhar S. Harnessing induced essentiality: targeting carbonic anhydrase IX and angiogenesis reduces lung metastasis of triple negative breast cancer xenografts. Cancers (Basel) 2019;11(7):1002. https://doi.org/10.3390/cancers11071002.

[50] Yang J. The role of reactive oxygen species in angiogenesis and preventing tissue injury after brain ischemia. Microvasc Res 2019;123:62–7. https://doi.org/10.1016/j.mvr.2018.12.005.
[51] Kureishi Y, Luo Z, Shiojima I, Bialik A, Fulton D, Lefer DJ, Sessa WC, Walsh K. The HMG-CoA reductase inhibitor simvastatin activates the protein kinase Akt and promotes angiogenesis in normocholesterolemic animals. Nat Med 2000;6:1004–10.
[52] Bitto A, Minutoli L, Altavilla D, Polito F, Fiumara T, Marini H, Galeano M, Calo M, Lo Cascio P, Bonaiuto M, Migliorato A, Caputi AP, Squadrito F. Simvastatin enhances VEGF production and ameliorates impaired wound healing in experimental diabetes. Pharmacol Res 2008;57:159–69.
[53] Chen Y, Zhang S, Peng G, Yu J, Liu T, Meng R, Li Z, Zhao Y, Wu G. Endothelial NO synthase and reactive oxygen species mediated effect of simvastatin on vessel structure and function: pleiotropic and dose-dependent effect on tumor vascular stabilization. Int J Oncol 2013;42(4):1325–36. https://doi.org/10.3892/ijo.2013.1833.
[54] Wilkinson EL, Sidaway JE, Cross MJ. Statin regulated ERK5 stimulates tight junction formation and reduces permeability in human cardiac endothelial cells. J Cell Physiol 2018;233(1):186–200. https://doi.org/10.1002/jcp.26064.
[55] Vassalli G. Aldehyde dehydrogenases: not just markers, but functional regulators of stem cells. Stem Cells Int 2019;2019. https://doi.org/10.1155/2019/3904645, 3904645.
[56] Ohmura-Kakutani H, Akiyama K, Maishi N, Ohga N, Hida Y, Kawamoto T, Iida J, Shindoh M, Tsuchiya K, Shinohara N, Hida K. Identification of tumor endothelial cells with high aldehyde dehydrogenase activity and a highly angiogenic phenotype. PLoS ONE 2014;9, e113910.
[57] Ciccone V, Terzuoli E, Donnini S, Giachetti A, Morbidelli L, Ziche M. Stemness marker ALDH1A1 promotes tumor angiogenesis via retinoic acid/HIF-1α/VEGF signalling in MCF-7 breast cancer cells. J Exp Clin Cancer Res 2018;37(1):311. https://doi.org/10.1186/s13046-018-0975-0 [Erratum in: J Exp Clin Cancer Res. 2019;38 (1):45].
[58] Morbidelli L, Terzuoli E, Donnini S. Use of nutraceuticals in angiogenesis-dependent disorders. Molecules 2018;23(10):2676. https://doi.org/10.3390/molecules23102676.
[59] Park C, Choi EO, Kim GY, Hwang HJ, Kim BW, Yoo YH, Park HT, Choi YH. Protective effect of baicalein on oxidative stress-induced DNA damage and apoptosis in RT4-D6P2T Schwann cells. Int J Med Sci 2019;16(1):8–16. https://doi.org/10.7150/ijms.29692.
[60] Bagchi D, Sen CK, Bagchi M, Atalay M. Anti-angiogenic, antioxidant, and anti-carcinogenic properties of a novel anthocyanin-rich berry extract formula. Biochemistry 2004;69(1):75–80. https://doi.org/10.1023/b:biry.0000016355.19999.93.
[61] Terzuoli E, Donnini S, Giachetti A, Iñiguez MA, Fresno M, Melillo G, Ziche M. Inhibition of hypoxia inducible factor-1alpha by dihydroxyphenylethanol, a product from olive oil, blocks microsomal prostaglandin-E synthase-1/vascular endothelial growth factor expression and reduces tumor angiogenesis. Clin Cancer Res 2010;16(16):4207–16. https://doi.org/10.1158/1078-0432.CCR-10-0156.
[62] Terzuoli E, Giachetti A, Ziche M, Donnini S. Hydroxytyrosol, a product from olive oil, reduces colon cancer growth by enhancing epidermal growth factor receptor degradation. Mol Nutr Food Res 2016;60:519–29. https://doi.org/10.1002/mnfr.201500498.
[63] Terzuoli E, Nannelli G, Frosini M, Giachetti A, Ziche M, Donnini S. Inhibition of cell cycle progression by the hydroxytyrosol-cetuximab combination yields enhanced chemotherapeutic efficacy in colon cancer cells. Oncotarget 2017;8:83207–24. https://doi.org/10.18632/oncotarget.20544.
[64] Song H, Lim DY, Jung JI, Cho HJ, Park SY, Kwon GT, Kang YH, Lee KW, Choi MS, Park JHY. Dietary oleuropein inhibits tumor angiogenesis and lymphangiogenesis in the B16F10 melanoma allograft model: a mechanism for the suppression of high-fat diet-induced solid tumor growth and lymph node metastasis. Oncotarget 2017;8:32027–42. https://doi.org/10.18632/oncotarget.16757.
[65] Collet G, Skrzypek K, Grillon C, Matejuk A, El Hafni-Rahbi B, Lamerant-Fayel N, Kieda C. Hypoxia control to normalize pathologic angiogenesis: potential role for endothelial precursor cells and miRNAs regulation. Vasc Pharmacol 2012;56(5–6):252–61. https://doi.org/10.1016/j.vph.2012.03.001.
[66] Falkenberg KD, Rohlenova K, Luo Y, Carmeliet P. The metabolic engine of endothelial cells. Nat Metab 2019;1 (10):937–46. https://doi.org/10.1038/s42255-019-0117-9.
[67] Sung YC, Jin PR, Chu LA, Hsu FF, Wang MR, Chang CC, Chiou SJ, Qiu JT, Gao DY, Lin CC, Chen YS, Hsu YC, Wang J, Wang FN, Yu PL, Chiang AS, Wu AY, Ko JJ, Lai CP, Lu TT, Chen Y. Delivery of nitric oxide with a

nanocarrier promotes tumour vessel normalization and potentiates anti-cancer therapies. Nat Nanotechnol 2019;14(12):1160–9. https://doi.org/10.1038/s41565-019-0570-3.

[68] Rolny C, Mazzone M, Tugues S, Laoui D, Johansson I, Coulon C, Squadrito ML, Segura I, Li X, Knevels E, Costa S, Vinckier S, Dresselaer T, Åkerud P, De Mol M, Salomäki H, Phillipson M, Wyns S, Larsson E, Buysschaert I, Botling J, Himmelreich U, Van Ginderachter JA, De Palma M, Dewerchin M, Claesson-Welsh L, Carmeliet P. HRG inhibits tumor growth and metastasis by inducing macrophage polarization and vessel normalization through downregulation of PlGF. Cancer Cell 2011;19(1):31–44. https://doi.org/10.1016/j.ccr.2010.11.009.

[69] Pecot CV, Rupaimoole R, Yang D, Akbani R, Ivan C, Lu C, Wu S, Han HD, Shah MY, Rodriguez-Aguayo C, Bottsford-Miller J, Liu Y, Kim SB, Unruh A, Gonzalez-Villasana V, Huang L, Zand B, Moreno-Smith M, Mangala LS, Taylor M, Dalton HJ, Sehgal V, Wen Y, Kang Y, Baggerly KA, Lee JS, Ram PT, Ravoori MK, Kundra V, Zhang X, Ali-Fehmi R, Gonzalez-Angulo AM, Massion PP, Calin GA, Lopez-Berestein G, Zhang W, Sood AK. Tumour angiogenesis regulation by the miR-200 family. Nat Commun 2013;4:2427. https://doi.org/10.1038/ncomms3427.

[70] Herrera FG, Irving M, Kandalaft LE, Coukos G. Rational combinations of immunotherapy with radiotherapy in ovarian cancer. Lancet Oncol 2019;20(8):e417–33. https://doi.org/10.1016/S1470-2045(19)30401-2 [Erratum in: Lancet Oncol 2019;20(9):e468].

[71] Maes H, Kuchnio A, Peric A, Moens S, Nys K, De Bock K, Quaegebeur A, Schoors S, Georgiadou M, Wouters J, Vinckier S, Vankelecom H, Garmyn M, Vion AC, Radtke F, Boulanger C, Gerhardt H, Dejana E, Dewerchin M, Ghesquière B, Annaert W, Agostinis P, Carmeliet P. Tumor vessel normalization by chloroquine independent of autophagy. Cancer Cell 2014;26(2):190–206. https://doi.org/10.1016/j.ccr.2014.06.025.

[72] Li X, Sun X, Carmeliet P. Hallmarks of endothelial cell metabolism in health and disease. Cell Metab 2019;30:414–33. https://doi.org/10.1016/j.cmet.2019.08.011.

[73] Viallard C, Larrivée B. Tumor angiogenesis and vascular normalization: alternative therapeutic targets. Angiogenesis 2017;20:409–26. https://doi.org/10.1007/s10456-017-9562-9.

[74] Tolaney SM, Boucher Y, Duda DG, Martin JD, Seano G, Ancukiewicz M, Barry WT, Goel S, Lahdenrata J, Isakoff SJ, Yeh ED, Jain SR, Golshan M, Brock J, Snuderl M, Winer EP, Krop IE, Jain RK. Role of vascular density and normalization in response to neoadjuvant bevacizumab and chemotherapy in breast cancer patients. Proc Natl Acad Sci U S A 2015;112:14325–30. https://doi.org/10.1073/pnas.1518808112.

[75] Kloepper J, Riedemann L, Amoozgar Z, Seano G, Susek K, Yu V, Dalvie N, Amelung RL, Datta M, Song JW, Askoxylakis V, Taylor JW, Lu-Emerson C, Batista A, Kirkpatrick ND, Jung K, Snuderl M, Muzikansky A, Stubenrauch KG, Krieter O, Wakimoto H, Xu L, Munn LL, Duda DG, Fukumura D, Batchelor TT, Jain RK. Ang-2/VEGF bispecific antibody reprograms macrophages and resident microglia to anti-tumor phenotype and prolongs glioblastoma survival. Proc Natl Acad Sci U S A 2016;113:4476–81. https://doi.org/10.1073/pnas.1525360113.

[76] Monk BJ, Poveda A, Vergote I, Raspagliesi F, Fujiwara K, Bae DS, Oaknin A, Ray-Coquard I, Provencher DM, Karlan BY, Lhommé C, Richardson G, Rincón DG, Coleman RL, Marth C, Brize A, Fabbro M, Redondo A, Bamias A, Ma H, Vogl FD, Bach BA, Oza AM. Final results of a phase 3 study of trebananib plus weekly paclitaxel in recurrent ovarian cancer (TRINOVA-1): long-term survival, impact of ascites, and progression-free survival-2. Gynecol Oncol 2016;143:27–34. https://doi.org/10.1016/j.ygyno.2016.07.112.

[77] Mortezaee K. CXCL12/CXCR4 axis in the microenvironment of solid tumors: a critical mediator of metastasis. Life Sci 2020;249:117534.

[78] Cheng L, Huang Z, Zhou W, Wu Q, Donnola S, Liu JK, Fang X, Sloan AE, Mao Y, Lathia JD, Min W, McLendon RE, Rich JN, Bao S. Glioblastoma stem cells generate vascular pericytes to support vessel function and tumor growth. Cell 2013;153(1):139–52. https://doi.org/10.1016/j.cell.2013.02.021.

[79] Mortezaee K, Najafi M. Immune system in cancer radiotherapy: resistance mechanisms and therapy perspectives. Crit Rev Oncol Hematol 2021;57:103180. https://doi.org/10.1016/j.critrevonc.2020.103180.

[80] Najafi M, Mortezaee K, Majidpoor J. Stromal reprogramming: a target for tumor therapy. Life Sci 2019;239:117049. https://doi.org/10.1016/j.lfs.2019.117049.

[81] Hess DL, Kelly-Goss MR, Cherepanova OA, Nguyen AT, Baylis RA, Tkachenko S, Annex BH, Peirce SM, Owens GK. Perivascular cell-specific knockout of the stem cell pluripotency gene Oct4 inhibits angiogenesis. Nat Commun 2019;10(1):967. https://doi.org/10.1038/s41467-019-08811-z.

[82] Pantziarka P, Verbaanderd C, Sukhatme V, Rica Capistrano I, Crispino S, Gyawali B, Rooman I, Van Nuffel AM, Meheus L, Sukhatme VP, Bouche G. ReDO_DB: the repurposing drugs in oncology database. Ecancermedicalscience 2018;12:886. https://doi.org/10.3332/ecancer.2018.886.
[83] Dredge K, Marriott JB, Macdonald CD, Man HW, Chen R, Muller GW, Stirling D, Dalgleish AG. Novel thalidomide analogues display anti-angiogenic activity independently of immunomodulatory effects. Br J Cancer 2002;87(10):1166–72. https://doi.org/10.1038/sj.bjc.6600607.
[84] Dredge K, Horsfall R, Robinson SP, Zhang LH, Lu L, Tang Y, Shirley MA, Muller G, Schafer P, Stirling D, Dalgleish AG, Bartlett JB. Orally administered lenalidomide (CC-5013) is anti-angiogenic in vivo and inhibits endothelial cell migration and Akt phosphorylation in vitro. Microvasc Res 2005;69(1–2):56–63. https://doi.org/10.1016/j.mvr.2005.01.002.
[85] Kaicker S, McCrudden KW, Beck L, New T, Huang J, Frischer JS, Serur A, Kadenhe-Chiweshe A, Yokoi A, Kandel JJ, Yamashiro DJ. Thalidomide is anti-angiogenic in a xenograft model of neuroblastoma. Int J Oncol 2003;23(6):1651–5.
[86] Maria de Souza C, Fonseca de Carvalho L, da Silva Vieira T, Cândida Araújo E Silva A, Teresa Paz Lopes M, Alves Neves Diniz Ferreira M, Passos Andrade S, Dantas Cassali G. Thalidomide attenuates mammary cancer associated-inflammation, angiogenesis and tumor growth in mice. Biomed Pharmacother 2012;66(7):491–8. https://doi.org/10.1016/j.biopha.2012.04.005.
[87] Du W, Hattori Y, Hashiguchi A, Kondoh K, Hozumi N, Ikeda Y, Sakamoto M, Hata J, Yamada T. Tumor angiogenesis in the bone marrow of multiple myeloma patients and its alteration by thalidomide treatment. Pathol Int 2004;54(5):285–94. https://doi.org/10.1111/j.1440-1827.2004.01622.x.
[88] Shen Y, Li S, Wang X, Wang M, Tian Q, Yang J, Wang J, Wang B, Liu P, Yang J. Tumor vasculature remolding by thalidomide increases delivery and efficacy of cisplatin. J Exp Clin Cancer Res 2019;38(1):427. https://doi.org/10.1186/s13046-019-1366-x.
[89] Wang G, Lin F, Wan Q, Wu J, Luo M. Mechanisms of action of metformin and its regulatory effect on microRNAs related to angiogenesis. Pharmacol Res 2021;164:105390. https://doi.org/10.1016/j.phrs.2020.105390.
[90] Wang JC, Li GY, Wang B, Han SX, Sun X, Jiang YN, Shen YW, Zhou C, Feng J, Lu SY, Liu JL, Wang MD, Liu PJ. Metformin inhibits metastatic breast cancer progression and improves chemosensitivity by inducing vessel normalization via PDGF-B downregulation. J Exp Clin Cancer Res 2019;38(1):235. https://doi.org/10.1186/s13046-019-1211-2.
[91] Daum S, Hagen H, Naismith E, Wolf D, Pircher A. The role of anti-angiogenesis in the treatment landscape of non-small cell lung cancer – new combinational approaches and strategies of neovessel inhibition. Front Cell Dev Biol 2021;8:610903. https://doi.org/10.3389/fcell.2020.610903.
[92] Lee MS, Ryoo BY, Hsu CH, Numata K, Stein S, Verret W, Hack SP, Spahn J, Liu B, Abdullah H, Wang Y, He AR, Lee KH. GO30140 investigators. Atezolizumab with or without bevacizumab in unresectable hepatocellular carcinoma (GO30140): an open-label, multicentre, phase 1b study. Lancet Oncol 2020;21(6):808–20. https://doi.org/10.1016/S1470-2045(20)30156-X [Erratum in: Lancet Oncol 2020;21(7):e341].
[93] Hack SP, Spahn J, Chen M, Cheng AL, Kaseb A, Kudo M, Lee HC, Yopp A, Chow P, Qin S. IMbrave 050: a phase III trial of atezolizumab plus bevacizumab in high-risk hepatocellular carcinoma after curative resection or ablation. Future Oncol 2020;16(15):975–89. https://doi.org/10.2217/fon-2020-0162.
[94] Choueiri TK, Motzer RJ, Rini BI, Haanen J, Campbell MT, Venugopal B, Kollmannsberger C, Gravis-Mescam G, Uemura M, Lee JL, Grimm MO, Gurney H, Schmidinger M, Larkin J, Atkins MB, Pal SK, Wang J, Mariani M, Krishnaswami S, Cislo P, Chudnovsky A, Fowst C, Huang B, di Pietro A, Albiges L. Updated efficacy results from the JAVELIN Renal 101 trial: first-line avelumab plus axitinib versus sunitinib in patients with advanced renal cell carcinoma. Ann Oncol 2020;31(8):1030–9. https://doi.org/10.1016/j.annonc.2020.04.010.
[95] Tiwari A, Mukherjee B, Dixit M. MicroRNA key to angiogenesis regulation: MiRNA biology and therapy. Curr Cancer Drug Targets 2018;18(3):266–77. https://doi.org/10.2174/1568009617666170630142725.
[96] Annese T, Tamma R, De Giorgis M, Ribatti D. microRNAs biogenesis, functions and role in tumor angiogenesis. Front Oncol 2021;10:581007. https://doi.org/10.3389/fonc.2020.581007.
[97] Ling H, Vincent K, Pichler M, Fodde R, Berindan-Neagoe I, Slack FJ, Calin GA. Junk DNA and the long non-coding RNA twist in cancer genetics. Oncogene 2015;34(39):5003–11. https://doi.org/10.1038/onc.2014.456.
[98] Guzel E, Okyay TM, Yalcinkaya B, Karacaoglu S, Gocmen M, Akcakuyu MH. Tumor suppressor and oncogenic role of long non-coding RNAs in cancer. North Clin Istanb 2019;7(1):81–6. https://doi.org/10.14744/nci.2019.46873.

[99] Teppan J, Barth DA, Prinz F, Jonas K, Pichler M, Klec C. Involvement of long non-coding RNAs (lncRNAs) in tumor angiogenesis. Non-Coding RNA 2020;6(4):42. https://doi.org/10.3390/ncrna6040042.

[100] Tabruyn SP, Griffioen AW. NF-kappa B: a new player in angiostatic therapy. Angiogenesis 2008;11(1):101–6. https://doi.org/10.1007/s10456-008-9094-4.

[101] Reid G, Pel ME, Kirschner MB, Cheng YY, Mugridge N, Weiss J, Williams M, Wright C, Edelman JJ, Vallely MP, McCaughan BC, Klebe S, Brahmbhatt H, MacDiarmid JA, van Zandwijk N. Restoring expression of miR-16: a novel approach to therapy for malignant pleural mesothelioma. Ann Oncol 2013;24(12):3128–35. https://doi.org/10.1093/annonc/mdt412.

[102] Garzon R, Marcucci G, Croce CM. Targeting microRNAs in cancer: rationale, strategies and challenges. Nat Rev Drug Discov 2010;9:775–89. https://doi.org/10.1038/nrd3179.

[103] Parilla M, Kadri S, Patil SA, Fitzpatrick C, Ritterhouse L, Segal J, Collins J, Pytel P. Integrating a large next-generation sequencing panel into the clinical diagnosis of gliomas provides a comprehensive platform for classification from FFPE tissue or smear preparations. J Neuropathol Exp Neurol 2019;78(3):257–67. https://doi.org/10.1093/jnen/nly130.

[104] Quan VL, Zhang B, Zhang Y, Mohan LS, Shi K, Wagner A, Kruse L, Taxter T, Beaubier N, White K, Zou L, Gerami P. Integrating next-generation sequencing with morphology improves prognostic and biologic classification of Spitz neoplasms. J Invest Dermatol 2020;140:1599–608. https://doi.org/10.1016/j.jid.2019.12.031.

[105] Ciesielski O, Biesiekierska M, Panthu B, Vialichka V, Pirola L, Balcerczyk A. The epigenetic profile of tumor endothelial cells. Effects of combined therapy with antiangiogenic and epigenetic drugs on cancer progression. Int J Mol Sci 2020;21(7):2606. https://doi.org/10.3390/ijms21072606.

[106] Marks DL, Olson RL, Fernandez-Zapico ME. Epigenetic control of the tumor microenvironment. Epigenomics 2016;8(12):1671–87. https://doi.org/10.2217/epi-2016-0110.

[107] Pirola L, Ciesielski O, Balcerczyk A. The methylation status of the epigenome: its emerging role in the regulation of tumor angiogenesis and tumor growth, and potential for drug targeting. Cancers (Basel) 2018;10(8):268. https://doi.org/10.3390/cancers10080268.

[108] Prokhortchouk E, Defossez PA. The cell biology of DNA methylation in mammals. Biochim Biophys Acta 2008;1783(11):2167–73. https://doi.org/10.1016/j.bbamcr.2008.07.015.

[109] Ribatti D, Tamma R. Epigenetic control of tumor angiogenesis. Microcirculation 2020;27(3). https://doi.org/10.1111/micc.12602, e12602.

[110] Jin B, Li Y, Robertson KD. DNA methylation: superior or subordinate in the epigenetic hierarchy? Genes Cancer 2011;2(6):607–17. https://doi.org/10.1177/1947601910393957.

[111] de Ruijter AJ, van Gennip AH, Caron HN, Kemp S, van Kuilenburg AB. Histone deacetylases (HDACs): characterization of the classical HDAC family. Biochem J 2003;370(Pt 3):737–49. https://doi.org/10.1042/BJ20021321.

[112] Ziche M, Morbidelli L, Choudhuri R, Zhang HT, Donnini S, Granger HJ, Bicknell R. Nitric oxide synthase lies downstream from vascular endothelial growth factor-induced but not basic fibroblast growth factor induced angiogenesis. J Clin Invest 1997;99:2625–34. https://doi.org/10.1172/JCI119451.

[113] Ziche M, Morbidelli L. Nitric oxide and angiogenesis. J Neuro-Oncol 2000;50:139–48. https://doi.org/10.1023/A:1006431309841.

[114] Fish JE, Matouk CC, Rachlis A, Lin S, Tai SC, D'Abreo C, Marsden PA. The expression of endothelial nitric-oxide synthase is controlled by a cell-specific histone code. J Biol Chem 2005;80(26):24824–38. https://doi.org/10.1074/jbc.M502115200.

[115] Hellebrekers DM, Melotte V, Viré E, Langenkamp E, Molema G, Fuks F, Herman JG, Van Criekinge W, Griffioen AW, van Engeland M. Identification of epigenetically silenced genes in tumor endothelial cells. Cancer Res 2007;67(9):4138–48. https://doi.org/10.1158/0008-5472.CAN-06-3032.

[116] Maiti A, Qi Q, Peng X, Yan L, Takabe K, Hait NC. Class I histone deacetylase inhibitor suppresses vasculogenic mimicry by enhancing the expression of tumor suppressor and anti-angiogenesis genes in aggressive human TNBC cells. Int J Oncol 2019;55(1):116–30. https://doi.org/10.3892/ijo.2019.4796.

[117] Aravindhan S, Ejam SS, Lafta MH, Markov A, Yumashev AV, Ahmadi M. Mesenchymal stem cells and cancer therapy: insights into targeting the tumor vasculature. Cancer Cell Int 2021;21(1):158. https://doi.org/10.1186/s12935-021-01836-9.

[118] van Beijnum JR, Giovannetti E, Poel D, Nowak-Sliwinska P, Griffioen AW. miRNAs: micro-managers of anti-cancer combination therapies. Angiogenesis 2017;20(2):269–85. https://doi.org/10.1007/s10456-017-9545-x.

[119] Adelaiye-Ogala R, Budka J, Damayanti NP, Arrington J, Ferris M, Hsu CC, Chintala S, Orillion A, Miles KM, Shen L, Elbanna M, Ciamporcero E, Arisa S, Pettazzoni P, Draetta GF, Seshadri M, Hancock B, Radovich M, Kota J, Buck M, Keilhack H, McCarthy BP, Persohn SA, Territo PR, Zang Y, Irudayaraj J, Tao WA, Hollenhorst P, Pili R. EZH2 modifies sunitinib resistance in renal cell carcinoma by Kinome reprogramming. Cancer Res 2017;77(23):6651–66. https://doi.org/10.1158/0008-5472.CAN-17-0899.

[120] Zupkovitz G, Tischler J, Posch M, Sadzak I, Ramsauer K, Egger G, Grausenburger R, Schweifer N, Chiocca S, Decker T, Seiser C. Negative and positive regulation of gene expression by mouse histone deacetylase 1. Mol Cell Biol 2006;26:7913–28.

[121] De Souza C, Chatterji BP. HDAC inhibitors as novel anti-cancer therapeutics. Recent Pat Anticancer Drug Discov 2015;10(2):145–62. https://doi.org/10.2174/1574892810666150317144511.

[122] Qian DZ, Wang X, Kachhap SK, Kato Y, Wei Y, Zhang L, Atadja P, Pili R. The histone deacetylase inhibitor NVP-LAQ824 inhibits angiogenesis and has a greater antitumor effect in combination with the vascular endothelial growth factor receptor tyrosine kinase inhibitor PTK787/ZK222584. Cancer Res 2004;64:6626–34. https://doi.org/10.1158/0008-5472.

[123] Berndsen RH, Abdul UK, Weiss A, Zoetemelk M, Te Winkel MT, Dyson PJ, Griffioen AW, Nowak-Sliwinska P. Epigenetic approach for angiostatic therapy: promising combinations for cancer treatment. Angiogenesis 2017;20(2):245–67. https://doi.org/10.1007/s10456-017-9551-z.

[124] Ayubi Joshagani MH, Dianat-Moghadam H, Seidi K, Jahanban-Esfahlan A, Zare P, Jahanban-Esfahlan R. Cell-free protein synthesis: the transition from batch reactions to minimal cells and microfluidic devices. Biotechnol Bioeng 2019;117:1204–29.

[125] Sleeboom JJF, Eslami Amirabadi H, Nair P, Sahlgren CM, den Toonder JMJ. Metastasis in context: modeling the tumor microenvironment with cancer-on-a-chip approaches. Dis Model Mech 2018;11(3). https://doi.org/10.1242/dmm.033100, dmm033100.

[126] Allgayer H, Leupold JH, Patil N. Defining the "Metastasome": perspectives from the genome and molecular landscape in colorectal cancer for metastasis evolution and clinical consequences. Semin Cancer Biol 2020;60:1–13. https://doi.org/10.1016/j.semcancer.2019.07.018.

[127] Tong D, Tian Y, Zhou T, Ye Q, Li J, Ding K, Li J. Improving prediction performance of colon cancer prognosis based on the integration of clinical and multi-omics data. BMC Med Inform Decis Mak 2020;20:22.

[128] Herrmann M, Probst P, Hornung R, Jurinovic V, Boulesteix AL. Large-scale benchmark study of survival prediction methods using multi-omics data. Brief Bioinform 2021;22, bbaa167.

[129] Patil H, Saxena SG, Barrow CJ, Kanwar JR, Kapat A, Kanwar RK. Chasing the personalized medicine dream through biomarker validation in colorectal cancer. Drug Discov Today 2017;22:111–9.

Index

Note: Page numbers followed by *f* indicate figures, *t* indicate tables, and *b* indicate boxes.

B

C

FSC
www.fsc.org
MIX
Papier | Fördert gute Waldnutzung
FSC® C083411

Kolibri 360 GmbH
Ferdinand-Jühlke-Straße 7
99095 Erfurt, Deutschland
produktsicherheit@kolibri360.de